Handbook of Developmental Biology

Handbook of Developmental Biology

Editor: Leonardo Reeves

STATES
ACADEMIC PRESS
www.statesacademicpress.com

States Academic Press,
109 South 5th Street,
Brooklyn, NY 11249, USA

Visit us on the World Wide Web at:
www.statesacademicpress.com

ISBN: 978-1-63989-258-7 (Hardback)

Cataloging-in-Publication Data

Handbook of developmental biology / edited by Leonardo Reeves.
p. cm.
Includes bibliographical references and index.
ISBN 978-1-63989-258-7
1. Developmental biology. 2. Growth. 3. Biology. 4. Ontogeny. I. Reeves, Leonardo.
QH491 .H36 2022
571.8--dc23

Table of Contents

Preface

This book was inspired by the evolution of our times; to answer the curiosity of inquisitive minds. Many developments have occurred across the globe in the recent past which has transformed the progress in the field.

The study of processes that enables plants and animals to grow and develop is known as developmental biology. Developmental biology includes the biology of asexual reproduction, regeneration and metamorphosis. It also studies the growth of stem cells in adult organisms. Some of the major processes of embryonic development studied in this discipline are morphogenesis, tissue growth, cell differentiation and regional specification. The developmental processes of organisms studied in this discipline are mainly divided into two categories: cell differentiation and regeneration. Cell differentiation is the development of different functional cell types. Regeneration refers to the ability to regrow a missing part. This book strives to provide a fair idea of this discipline and to help develop a better understanding of the latest advances within the field. It also unfolds the innovative aspects of developmental biology, which will be crucial for the progress of the field in the near future. Scientists and students actively engaged in this field will find this book full of crucial and unexplored concepts.

This book was developed from a mere concept to drafts to chapters and finally compiled together as a complete text to benefit the readers across all nations. To ensure the quality of the content we instilled two significant steps in our procedure. The first was to appoint an editorial team that would verify the data and statistics provided in the book and also select the most appropriate and valuable contributions from the plentiful contributions we received from authors worldwide. The next step was to appoint an expert of the topic as the Editor-in-Chief, who would head the project and finally make the necessary amendments and modifications to make the text reader-friendly. I was then commissioned to examine all the material to present the topics in the most comprehensible and productive format.

I would like to take this opportunity to thank all the contributing authors who were supportive enough to contribute their time and knowledge to this project. I also wish to convey my regards to my family who have been extremely supportive during the entire project.

Editor

1

Beyond the Mammalian Heart: Fish and Amphibians as a Model for Cardiac Repair and Regeneration

Kyle Jewhurst and Kelly A. McLaughlin *

Academic Editors: Robin Muise-Helmericks and Andy Wessels

Department of Biology, Tufts University, Medford, MA 02155, USA; Kyle.Jewhurst@tufts.edu
* Correspondence: Kelly.McLaughlin@tufts.edu

Abstract: The epidemic of heart disease, the leading cause of death worldwide, is made worse by the fact that the adult mammalian heart is especially poor at repair. Damage to the mammal heart—such as that caused by myocardial infarction—leads to scarring, resulting in cardiac dysfunction and heart failure. In contrast, the hearts of fish and urodele amphibians are capable of complete regeneration of cardiac tissue from multiple types of damage, with full restoration of functionality. In the last decades, research has revealed a wealth of information on how these animals are able to perform this remarkable feat, and non-mammalian models of heart repair have become a burgeoning new source of data on the morphological, cellular, and molecular processes necessary to heal cardiac damage. In this review we present the major findings from recent research on the underlying mechanisms of fish and amphibian heart regeneration. We also discuss the tools and techniques that have been developed to answer these important questions.

Keywords: heart; regeneration; apoptosis; necrosis; myocardium; epicardium; fibrosis; migration; proliferation; dedifferentiation; transdifferentiation; retinoic acid; TGF-β; FGF

---

1. Introduction

The human heart is an organ of both vital importance and startling fragility. Heart disease is the leading cause of death worldwide, responsible for one in every four deaths [1]. In the United States this figure is even higher, with one-third of all deaths attributable to heart disease. In addition, 8 million Americans have suffered a myocardial infarction at some point in their lives, and one third of adult Americans are afflicted with hypertension, which can have detrimental effects on heart physiology and function [2]. In short, heart disease is an epidemic unlike any other. Thus, a greater understanding of the causes of heart disease, and treatment for its most damaging effects, is of the greatest necessity.

The major difficulty in developing treatments for cardiac tissue damage is that the adult mammalian heart is particularly poor at repairing post-injury. Infarction or physical damage to mammal hearts often results in permanent scarring, leading to cardiac remodeling and heart dysfunction, and ultimately contributing to progressive heart failure [3]. Research into mammalian heart repair has generally followed two strategies—stem cell/cardiac progenitor-based therapies, and endogenous pathways of repair [4]. Once thought to be distinct processes, these two approaches may in fact be more related than previously believed. Some studies indicate that stem cell-based strategies work by acting indirectly on existing heart cells and stimulating endogenous repair processes, rather than any direct contribution to restoring tissue by the stem cells themselves [4,5]. An understanding of the inherent repair capabilities of the heart, therefore, seems to be the most comprehensive approach towards developing novel methods of treatment.

For decades, the mammalian heart was believed to be a post-mitotic organ, and therefore incapable of healing itself after damage. However, more recent research has shown that low levels of mitosis occur in mature cardiomyocytes, and new myocytes are routinely generated from pre-existing ones as part of the heart's natural maintenance of homeostasis [6]. Increased proliferation of myocytes adjacent to myocardial infarcts has also been seen in human patients, though this process is not sufficient to fully repair damage [7]. Cardiac regeneration—that is, restoration of muscle with limited to no scarring—has even been observed in neonatal mice under a wide variety of damage models [8–12], though the extent of regeneration these animals are capable of has been disputed [13]. Research into the inherent reparative and regenerative capacity of mammalian hearts thus shows promise for developing strategies to combat and repair heart damage in humans. However, since mammal embryos are difficult to manipulate and closely observe during repair, and adult mammal hearts are incapable of regenerating, the potential of mammalian models remains limited for understanding how heart tissue can be regenerated.

In recent years, therefore, the scientific community has begun to take a greater interest in non-mammalian hearts as a model for heart repair. In particular, the so-called "lower vertebrates", amphibians and fish, have proven themselves to be highly tractable model systems in this field of study. The fact that these animals can regenerate cardiac tissue has been known for some time; forty years ago, Oberpriller and Oberpriller discovered that damage to the hearts of adult newts caused cardiomyocytes near the wound site to proliferate, a sign of regenerative capacity [14]. That same year, Becker and colleagues showed that the salamander heart could fully restore itself after injury, both functionally and histologically [15]. In the following decades, further research was conducted on heart regeneration in urodele amphibians, most commonly the newt *Notophthalmus viridescens* [16–18], but also the axolotl *Amblystoma mexicanum* [19,20] and the salamander *Triturus viridescens* [15]. Many of the gross morphological and cellular events of cardiac regeneration were first characterized in amphibian systems, and they remain important as models for studying regenerative processes.

It was not until 2002 that a species outside of amphibians was found to be capable of cardiac regeneration in adult organisms—the zebrafish, *Danio rerio*. Poss and colleagues demonstrated that zebrafish hearts are capable of fully regenerating missing cardiac tissue, with no sign of scarring and full restoration of functionality, within 60 days after injury [21]. Since that discovery, *D. rerio* has become, by far, the most utilized and well-characterized model organism for studying heart regeneration. With a wealth of molecular tools and genomic data available for the species, profiling gene expression, creating transgenic organisms, and mutagenesis are made significantly easier than in less well-characterized model systems [21]. The morphological and cellular processes during zebrafish heart regeneration show a great deal of conservation with those same processes in amphibians, but the tools available to study zebrafish have made it possible to delve even further into the underlying molecular pathways. While zebrafish have been the primary model for cardiac regeneration, other species of teleost fish are capable of it as well, allowing for opportunities to study the evolutionary conservation of regenerative processes. Fish species in which heart regeneration has been studied include *Danio aequippinatus*, a close relative to the zebrafish [22], the goldfish *Carassius auratus*, a more distant relative of the *Danio* genus [23], and *Polypterus senegalus*, a member of the most basally branching group of ray-finned fishes [24]. However, the regenerative response is not present in all teleosts; at least one species, the medaka *Oryzias latipes*, scars instead of regenerating after cardiac trauma [25].

In the past decade, therefore, the field of cardiac regeneration has expanded rapidly. A wealth of research has been conducted in both amphibian and fish model systems, across numerous species, into how these animals are capable of restoring damaged or missing heart tissue. In this review, we discuss the major findings on heart regeneration in non-mammalian species, as well as the tools that have been developed to study this process.

2. Techniques for Studying Heart Regeneration and Repair

2.1. Mechanical Manipulation

As our knowledge of the processes of cardiac regeneration has advanced, so too have the tools we use to study it. The first developed, and by far most common, method of damaging hearts to study their repair is injury via mechanical manipulation. This broad category of techniques includes amputation of the ventricular apex [21], resection of the ventricular myocardium (away from the apex) [26], resection of the atrium [17], and mechanical squeezing or crushing [27].

Ventricular apical amputation is one of the most widely-used methods of injuring hearts, likely due to its highly visible phenotype and relatively easy access to the ventricular apex. The body of the animal and the pericardium are opened up to expose the heart. A pair of scissors is used to cut off the apex; in fish, approximately 20%–30% of the ventricle is removed [21,28], while in amphibians 10%–15% is standard [14,20]. In fish, the wound clots in less than a minute, and the animal is left to heal the incision in its abdomen [21]. In amphibians, sutures are required in the abdominal wall, and the animal is left at cold temperatures overnight to recover [14,20]. In all species, the survival rate of this procedure is 80%–90%, but amputating an excessive amount of the ventricle can greatly reduce this number [21]. Other variants of mechanical damage follow a similar procedure, with the major differences being where on the heart the incision takes place, or with what tools the damage is induced. These alternate methods may be used to target a different region of the heart, such as atrial resection [17], or to improve reproducibility and survivability, such as (non-apical) ventricular resection [26]. Because surgery through the abdominal wall may have its own effects on the repair response, sham-operated controls—consisting of an incision to access the heart, but no damage to the heart itself—are always necessary with this technique.

Although mechanical damage does not replicate any human heart disease in particular, it may be most applicable to the study of infarcts, where large regions of myocardial tissue undergo necrosis. However, the similarities are limited. In an infarct in the human heart, cellular debris remains at the wound site post-injury, and it must be removed before repair can begin; in a resected heart, no tissue is left behind, leaving one less step to be required before repair can take place [29]. Differences like these make it difficult to directly apply results from resection experiments into a human heart disease context. To this end, other damage models that better recapitulate certain aspects of heart disease have also been developed.

2.2. Cryoinjury

The cryoinjury model was developed to better mimic natural conditions of heart disease, specifically those found after myocardial infarction (MI). Unlike resection, which involves wholesale removal of tissue from the heart, cryoinjury induces tissue damage *in situ*, much like what would occur from ischemia during MI. Following cryoinjury, 25% of the ventricle undergoes necrosis or apoptosis, on a similar scale to MI-induced ischemia, after which fibrosis occurs much as it does in the mammalian heart. However, unlike in mammalian MI, regeneration follows in fish and amphibians, with the fibrotic tissue eventually replaced by functional myocardium [29]. Originally developed for mammalian studies, the cryoinjury model has been used for decades in studying repair of infarct-like wounding in mice and rats [30–32]. Since its adaptation for fish, cryoinjury has been extensively used in zebrafish models to study cardiac regeneration [29,33–37].

Cryoinjury results from application of extreme cold to living tissue. The change in temperature disrupts proteins and forms ice crystals that damage the plasma membrane, resulting in apoptosis and necrosis. Cryoinjury is commonly performed with a copper or stainless steel filament (<1 mm diameter) cooled by liquid nitrogen, known as the cryoprobe [29,34]. An alternate protocol uses a small (2 mm diameter) cone of dry ice instead [35]. After the animal is anesthetized, an incision is made in the body wall and pericardium to provide access to the heart. The cryoprobe is applied to the ventricle for a few seconds—the exact time varying by protocol—after which it is removed and the animal is left to heal naturally [29,34,35]. Mortality is greatly reduced compared to ventricular resection, with 5% or less subject death reported [35,36]. As with resection, sham-operated controls are necessary to account for the effects of opening the abdominal wall.

An interesting variation on this method—cautery, application of extreme heat—has been used in two heart regeneration studies, one in the giant danio (*D. aequippinatus*) and one in the goldfish (*Carassius auratus*). The procedure is performed in much the same way as cryoinjury, only using a heated filament; it has the same advantages as cryoinjury as well, leaving behind a region of necrotic tissue akin to that found in an infarct [22,23].

2.3. Inducible Transgenics

One of the major advantages of working with fish and amphibians over mammals is the relative ease of producing transgenic organisms. Because fish and amphibian embryos develop externally, techniques such as microinjection or electroporation allow for simpler introduction of transgenes than in mammals, which usually require transfection. To this end, a number of transgenic models of heart damage have been developed in the lower vertebrates. The inducible nature of these models, usually by exposure to or injection of a drug, prevents the transgene from interfering with normal heart development and allows for the creation of a disease phenotype within a highly specific time period.

The first such tool to be developed for use in a heart regeneration study was the Nitroreductase (NTR)/Metronidazole(Mtz) system, a genetically expressed, inducible, targetable method of cell ablation. *Escherichia coli* nitroreductase is expressed in the target organism, usually under a tissue- or cell type-specific promoter. Upon introduction of the prodrug Metronidazole [1-(2-hydroxyethyl)-2-methyl-5-nitroimidazole], a substrate of NTR, the Mtz is reduced to a potent DNA cross-linking agent. Death of the host cell via apoptosis follows within hours to days. The toxic form of Mtz does not cross cell boundaries, making this method highly cell-specific [38,39].

The NTR/Mtz system has so far been used for heart studies only in larval zebrafish. The NTR gene has been expressed under the *cardiac myosin light chain 2* (*cmlc2*) and *ventricular myosin heavy chain* (*vmhc*) promoters, allowing for either heart-wide expression or expression limited to specific regions of the heart. NTR can be conjugated to fluorescent proteins, including CFP and mCherry, allowing for visual identification of transgenic organisms and tissues. The zebrafish heart regenerates after Mtz-induced cell ablation, and since Mtz can be washed out, the regenerated cardiomyocytes will not themselves undergo further apoptosis [38,40].

Another tool that has been used for inducible, targeted cell ablation in zebrafish is the Z-CAT (zebrafish cardiac ablation transgenes) system, which uses two transgenes to restrict cell death to a target population. The first construct contains a fusion of Cre recombinase and estrogen receptor (CreER) to create a Cre that is inducible by 4-hydroxytamoxifen (4-HT). In order to restrict expression to cardiomyocytes, it is placed under a *cmlc2* promoter. The second transgene is the *bactin2:loxp-mCherry-STOP-loxp-DTA* construct; the *β-actin2* promoter further restricts activity to the myocardium, where, upon 4-HT induction, diphtheria toxin A-chain (DTA) is produced. This approach induces apoptosis in the target cells within 2 days of 4-HT exposure [41]. A total of ~60% of cardiomyocytes are ablated throughout the course of this process, resulting in severe heart failure phenotypes. Unlike NTR/Mtz, this method is not reversible; however, regeneration still follows within 1–3 weeks after 4-HT exposure [41–43].

A major advantage of genetic approaches is that they do not require any invasive surgery or manipulation, reducing the likelihood of off-target effects. The choice of promoter in the transgenic animal restricts expression of the transgene—and therefore cell death—to a specific cell type, which is both advantageous and disadvantageous. On the one hand, the ablation target is highly specific, allowing a researcher to study cardiomyocyte regeneration in isolation from the other layers of the heart. However, natural diseases like MI may result in the death of multiple cell types within a region, something these methods fail to replicate [29]. Regeneration from genetic ablation also takes place within a greatly accelerated timeframe compared to cryoinjury or resection, occurring within two weeks in the Z-CAT model [37]. The differences between repair of a contiguous region of missing (or necrotic) tissue, and repair of scattered ablation of cells, may therefore limit how applicable these results are to repair of infarcts. However, necrosis and apoptosis akin to the effects of genetic ablation do play a role in several other types of heart disease and heart failure, which these techniques can help to study. Ischemia, oxidative stress, alkalosis, and other conditions not caused by physical trauma to the heart can lead to an increase in both apoptosis and necrosis throughout the myocardium [44]. Furthermore, cell death is a causative factor in heart failure and associated conditions, and can cause mortality in as few as 0.08% of cardiomyocytes. Understanding how organisms regenerate from dispersed cell death can therefore be a highly useful tool in treating many forms of heart failure [45,46]. Overall, the choice of genetic or mechanical methods of ablation depends greatly on what sort of injury and repair process the researcher wishes to investigate.

2.4. Environmental Conditions

A number of cardiac diseases in humans are caused by either the presence of damaging agents in blood, such as reactive oxygen species, or deprivation of essential resources, such as oxygen. Oxidative stress, hypoxia [47], acidosis [48], and osmotic stress [49] are among the conditions with an adverse effect on cardiac function that result from a change in blood composition.

A major advantage of aquatic model organisms is the ability to replicate these conditions by simply changing the components of the solution that the animal is living in, altering pH, oxygen levels, or presence of cardiotoxins. For fish or amphibian larval stages, diffusion is usually sufficient to expose the heart to the solute of interest [50]. However, adult amphibian stages, especially of those species living on land, may require removal of the heart to perfuse it with solution [51]. In this manner, the gas content, pH, and presence of any solutes in the lumen of the heart can be precisely controlled. The disadvantage of this technique is that the heart can only be looked at in isolation, rather than as part of a functioning whole organism.

Much like genetic ablation models, such treatments cause dispersed cell damage or death, and as such are not appropriate for mimicking infarction or other acute forms of trauma. Because most environmental conditions do not induce as drastic a damage phenotype as mechanical methods like amputation or cryoinjury, these techniques are more commonly used for the study of non-regenerative heart repair. A series of studies on the ability of the frog *Rana ridibunda* to protect against heart damage has examined the effects of many diverse environmental factors, including: hyperosmosis, via increased NaCl and KCl concentration [52]; oxidative stress, via hydrogen peroxide [53]; hypoxia, either via replacement of oxygen in solution with nitrogen [51], or by addition of cobalt (II) chloride [54]; acidosis and alkalosis via buffers of the respective pH [55]; and mechanical overload, via increased perfusion pressure [56]. Environmental conditions have also been used in the study of heart repair in fish. By exposing zebrafish to hypoxic conditions, and then returning them to oxygenated water, it is possible to replicate the oxidative stress that occurs following ischemia/reperfusion in human hearts [50].

3. Heart Regeneration in Fish and Amphibians

While very different in gross morphology, the hearts of fish and amphibians have much in common with each other and with mammalian hearts. The vertebrate heart consists of three main layers—the endocardium, an endothelial layer lining the inner lumen of the heart; the myocardium, the muscle layer; and the epicardium, a protective sheet of epithelial cells covering the outside of the myocardium. The myocardium is itself composed of two layers: the compact myocardium that forms the walls of the heart, and the trabecular myocardium, which forms the contractile trabeculae on the interior of the heart chambers [21,24]. Additionally, the general pattern of heart development is well conserved among the vertebrates: a symmetrical pair of *nkx2.5*-positive mesodermal regions, known collectively as the heart field, help to establish the cardiac precursor cell population. During early development, these fields migrate to the midline of the organism and converge to form a linear heart tube. This tube then loops into an S-shape, as the cells along its axis differentiate into atrial and ventricular cells. Extracardiac populations of cells meanwhile migrate to the heart to form the endocardium and epicardium. In fish, no further chamber septation occurs; in amphibians, the atrium separates in two. Both taxa have only a single ventricle, unlike mammals [57].

Befitting such similarities in development and morphology, heart regeneration follows a highly stereotyped behavior, independent of species or most methods of damage. Within the first hours to days after injury, clotting begins to seal the wound, followed by an inflammatory response that leads to deposition of a collagen and fibrin extracellular matrix (ECM) by epicardium-derived fibroblasts. Next, cardiomyocytes outside the injury zone partially dedifferentiate to reenter the cell cycle, proliferate, and migrate to the wound site. The fibrotic material filling the wound is replaced over the following days to weeks with functional cardiomyocytes (summarized in Table 1). Depending on the exact damage model studied, additional steps may be required. For example, cryoinjury, which replicates the effects of an infarct, results in necrotic tissue that needs to be cleared before fibrosis and regeneration can continue [35]. Conversely, certain methods of damage may not require some of these steps during repair. Z-CAT genetic ablation, for instance, results in minimal collagen deposition [41].

Table 1. Timeline of main events in cardiac regeneration by organism and damage model. Times listed are the earliest point at which the process was observed (d = days post injury). A dash (—) indicates that no data exists for this model.

	Inflam-mation	Epicardial Activation	Fibrosis		Cardiomyocytes			Recovery (Function)	Recovery (Morph.)
			Fibrin	Collagen	Dediff.	Prolif.	Migration		
***D. rerio* (adult)**									
Amputation	<1 d [58]	1 d [59]	2 d [21]	14 d [21]	7 d [60]	7 d [21]	9 d [21]	30 d [61]	60 d [21]
Cryoinjury	1 d [35]	3 d [35]	4 d [36]	7 d [36]	4 d [62]	3 d [35]	14 d [37]	30 d [36]	60 d [36]
Ablation (Z-CAT)	3 d [41]	7 d [41]	—	14 d [41]	7 d [41]	7 d [41]	—	14 d [41]	30 d [41]
***D. rerio* (larva)**									
Ablation (NTR)	—	1 d [40]	—	—	1 d [40]	2 d [40]	2 d [40]	4 d [38]	4 d [38]
D. aequippinatus									
Cautery	1 d [22]	—	—	7 d [22]	—	7 d [22]	14 d [22]	—	45 d [22]
C. auratus									
Cautery	3 d [23]	—	7 d [23]	14 d [23]	14 d [23]	3 d [23]	—	—	45 d [23]
N. viridescens									
Amputation	5 d [17]	—	7 d [26]	7 d [26]	—	7 d [26]	30 d [14]	23 d [26]	60 d [26]
Crushing	—	—	—	1 d [63]	1 d [27]	—	—	14 d [27]	84d [63]

The major processes involved in heart regeneration—epicardial activation, fibrosis, and cardiomyocyte processes, including dedifferentiation, proliferation, and migration—are described in greater detail below. Most of the specific results were reported in zebrafish, although amphibian experiments have shown that many of these processes are conserved across vertebrate taxa. Where the regeneration process differs in other organisms, it is specifically noted.

3.1. Epicardial Activation

A key component of heart regeneration in numerous species and models of damage is the role of the epicardium. The epicardium and the tissue types derived from it play an important part in cardiac development, where epicardium-derived cells (EPDCs) differentiate into numerous cell types, including coronary vascular smooth muscle and fibroblasts [64]. During heart regeneration, epicardial processes include reactivation of developmental genes in epicardial cells and the transdifferentiation of epicardiocytes into other cell types necessary for regeneration to proceed, including fibroblasts. Epicardial activation begins with re-expression of *wt1*, a marker of stem cells and developing epicardium in particular, as well as other epicardial developmental genes like *raldh2* (necessary for retinoic acid synthesis) and *tbx18*, all within the first 1–2 days after injury. *Raldh2* is most active during early regeneration; it is first expressed in the atrial and ventricular epicardium days after the injury, is localized to the site of the injury by one week, and diminishes in expression by two weeks after injury. Meanwhile, *tbx18* expression persists much longer, at least one month post-injury [24,35,59,65,66].

Regeneration of the epicardium itself occurs relatively early in the regeneration process, occurring before restoration of the myocardium is visible [22]. Activated epicardial cells begin to proliferate approximately three days after injury, and by two weeks, the epicardium has enclosed the injured area, providing protection and support for regeneration of the myocardium [59,67].

Approximately one week after injury, epicardial cells upregulate expression of *snail2* and *twist1b*, markers of the epithelial-to-mesenchymal transition (EMT), in preparation for transdifferentiation [68]. Beginning at 2 weeks post-injury, activated epicardial cells then infiltrate the myocardium, where they downregulate *wt1* expression and differentiate into a variety of EPDC cell types [65]. Some migrate into the regenerating coronary vasculature to become vascular pericytes and organize neovascularization, a process mediated by platelet-derived growth factor (PDGF) and fibroblast growth factor (FGF) signaling [59,65,68,69]. Others express fibronectin and transdifferentiate to fibroblasts, which infiltrate the injury site to first deposit collagen, and later promote migration of cardiomyocytes into the wound to replace fibrotic tissue [42]. One important cell type not found in this list is cardiomyocytes; epicardial cells that undergo EMT do not subsequently express myocardial differentiation markers like myosin heavy chain (MHC) or *cmlc2*. Similarly, they do not express the endothelial marker *fli1*, indicating that endocardium is also not among the potential fates of EPDCs [65,69].

3.2. Cardiomyocyte Dedifferentiation and Proliferation

In order for injured or infarcted myocardium to regenerate, new cardiomyocytes must be produced; where, then, do they come from? Early studies in zebrafish heart regeneration indicated that the cardiomyocytes involved in regenerating missing myocardium were formed through differentiation of an undifferentiated progenitor cell population [59]. However, subsequent experiments have shown that all regenerated cardiomyocytes originate from existing cardiomyocytes [60,61]. Since adult cardiomyocytes are largely quiescent, regeneration requires them to reenter the cell cycle, which is controlled in part by the mitotic checkpoint kinases Mps1 and Plk1. Inhibition of either of these kinases prevents myocardium from replacing fibrotic tissue and leads to permanent scarring [21,60]. Also necessary for proliferation is the gap junction protein Cx43. Mps1 and Cx43, along with numerous other cell-cycle regulators, structural proteins, and cardiomyopathy-associated genes are regulated by the miRNA miR-133, downregulation of which is essential to create a permissive environment for cardiomyocyte proliferation to proceed [70].

Reentry of cardiomyocytes into the cell cycle also requires dedifferentiation and re-expression of several cardiac development genes [59,61]. The question of which genes are reactivated, and exactly to what extent dedifferentiation occurs, is unclear. Some groups have reported expression of the early developmental genes *nkx2.5*, *hand2*, *tbx5*, *tbx20*, and *mef2* in regenerating hearts [40,59], while others have observed no expression of these genes [28,60]. *Gata4*, an early cardiac development gene and one of the principle markers of this dedifferentiation process, has in contrast repeatedly been observed in both fish and amphibians [26,61,71]. *Gata4* expression is re-induced in cells of the outer compact myocardium, predominantly near the injury site, and these *gata4*$^+$ cells are what proliferate and fill the wound site [61]. Also upregulated in the regenerating myocardium is the homeobox gene *msx*, a gene that regulates regeneration across species and organs, including newt limbs and zebrafish fins. However, *msx* is not upregulated in heart development, indicating that the regeneration process is more complex than just a recapitulation of cardiac development [28].

Morphologically, dedifferentiation is visible in both fish and amphibian cardiomyocytes in the form of disorganized sarcomeres, loss of intercellular cohesion, a rounding of cell shape, and downregulation of terminal differentiation cardiac proteins like myosin and troponin [27,40,41,59,60,72]. Fewer experiments have looked at gene expression in amphibians during heart regeneration, but one of those that has showed an increase in *nkx2.5*, *hand2*, *gata4*, *gata5*, and *islet1* [26], the latter of which has not been observed in fish heart regeneration [40].

This combination of proliferation and dedifferentiation is indicative of epimorphic regeneration, in which an undifferentiated blastema forms at the site of injury, and undergoes differentiation to replace the missing or infarcted tissue. In zebrafish, this is manifested as an accumulation of highly proliferative blastemal cells expressing embryonic markers in the immediate vicinity of the infarct zone, beginning at 4 days post-injury. In addition to this immediate local response, there is also an increase in proliferation throughout the rest of the uninjured heart that persists to later stages of regeneration. The distribution of mitotic cells shifts away from the infarcted region as regeneration progresses, possibly indicating that later phases of regeneration are associated with heart-wide systemic repair [58]. This late, ventricular-wide increase in proliferation has also been observed in the newt heart during regeneration [67].

One particularly interesting source of cardiomyocytes during ventricular repair, observed in both amphibians and fish, is the atrium. Atrial myocardial proliferation in response to ventricular damage was first observed in the newt, *Notophthalmus viridescens*, where in fact proliferation was more prevalent in the atrium than in the ventricle [16]. Another amphibian species, the axolotl *Amblystoma mexicanum*, showed a comparable increase in atrial mitosis during regeneration of the ventricle [20].

In zebrafish, an atrial response is visible even before an increase in proliferation occurs. *Raldh2* is upregulated in the atrial endocardium within an hour after ventricular injury, hours before it is upregulated in the ventricle [24]. At one day post-injury, activation of the uninjured atrial epicardium begins simultaneously with that of the injured ventricle, with *raldh2* expression followed by *tbx18* upregulation. Expression of these markers, however, soon restricts itself to the area immediately adjacent to the ventricular wound. Meanwhile, fibroblast growth factor receptor (*fgfr*), associated with cardiomyocyte migration into the wound, is activated in contiguous patches throughout the atrial epicardium [59]. These findings suggest that activation of all three layers of the atrium is involved in cardiac regeneration.

The role of atrial cardiomyocytes in ventricular regeneration has been confirmed in zebrafish using a cell ablation model. When ventricular cardiomyocytes are ablated, atrial cardiomyocytes dedifferentiate into cardiac progenitor cells. These cells show disorganized structure, decreased expression of terminal differentiation markers like MHC, and re-expression of developmental genes including *gata4*, *hand2*, *nkx2.5*, *tbx5a*, *tbx20*, and *mef2*. The progenitor cells then reacquire a ventricular identity as they migrate into the ventricle to become mature ventricular cardiomyocytes. The migration process is dependent on Notch signaling, and upregulation of both Notch pathway components and *raldh2* occurs in the atrial endocardium, indicating a possible role for retinoic acid (RA) as well [40].

In addition to its role in epicardial activation, RA is necessary for cardiomyocyte proliferation. *Raldh2* is activated by the inflammatory response in the atrial and ventricular endocardium within hours after injury, and within a day becomes localized to the site of the injury. These *raldh2*-positive cells are also enriched for expression of cardiac developmental genes *hand2* and *gata5*. Along with *raldh2*-positive epicardial cells, the endocardium makes up one of the two major sources of RA during heart regeneration. Inhibition of RA signaling from endocardial and epicardial cells greatly reduces injury-induced cardiomyocyte proliferation. However, overexpression of RA has no effect on proliferation levels, indicating that RA signaling is necessary, but not sufficient for proper cardiomyocyte proliferation [24]. The endocardium is also responsible for producing Interleukin 11a (Il11a), a ligand of the Jak1/Stat3 signaling system. Jak1/Stat3, expressed in the myocardium, and its downstream effectors are likely involved in cell survival during the early, wound-healing portion of regeneration, and are required for later cardiomyocyte proliferation [73].

A number of other paracrine and juxtacrine signaling pathways, involving all three layers of the heart, are also necessary for cardiomyocyte proliferation. These include PDGF [74], Hedgehog, Insulin-like growth factor (IGF), TGF-β, and Notch [43]. Sonic hedgehog (Shh) is activated in the injury-adjacent epicardium, with downstream markers of Shh activity observed in adjacent cardiomyocytes. IGF is expressed in the endocardium, with its receptor present in the injured myocardium [43]. TGF-β is present in the myocardium, epicardium, and fibroblasts [37]. The Notch receptor, meanwhile, is upregulated in both the endocardium and epicardium, but remains absent in the intervening myocardium [71]. The expression pattern of Notch ligands is less well-studied; only *deltaC*, expressed in the endocardium alongside the *notch1b* isoform, has been examined. Notch receptor is expressed in the heart from 1 day to 1 week after injury, after which the gene is downregulated [28]. Notch signaling is necessary for proliferation of cardiomyocytes; when it is inhibited during the later period of Notch expression (6–7 days post-injury), the wound remains fibrotic and regeneration of the myocardium never takes place. Interestingly, hyperactivation of Notch signaling also prevents cardiomyocyte proliferation, indicating that a very specific level of signaling is necessary for regeneration to proceed properly [71].

Several *in vitro* experiments on cultured newt cardiomyocytes have further elaborated on the factors necessary for myocyte cell cycle reentry and proliferation. The major restriction point in beginning this process is the retinoblastoma protein (Rb), which in wild-type tissues prevents cells from entering S phase. Phosphorylation of Rb via cyclin-dependent kinase (Cdk) 4 and 6 deactivates this S-phase restriction and is necessary for newt cardiomyocytes to enter the cell cycle [75,76]. However, entry into S phase does not ensure subsequent proliferation. While most cardiomyocytes are able to enter S phase in culture, only one-third of those subsequently complete one or more rounds of cell division, with the majority arrested just before mitosis or before cytokinesis [76]. This separation between cell cycle reentry and proliferation is borne out by the fact that while PDGF and IGF promote cardiomyocyte proliferation, they cannot by themselves induce terminally differentiated myocytes to reenter the cell cycle [75]. Multiple ECM components have also been demonstrated to induce proliferation in cardiomyocytes in culture, including laminin, fibronectin [76], and tenascin C [67].

Proliferation, however, is not enough by itself; after cardiomyocytes proliferate, they must then be induced to migrate into the wound region. Inextricably bound to this process of cardiomyocyte migration is fibrosis, a crucial step in cardiac regeneration that bridges the gap between early wound healing and later regeneration of myocardium.

3.3. Fibrosis and Cardiomyocyte Migration

Among the first events required in cardiac regeneration, as part of a general wound-healing response, are clotting, activation of the inflammatory response, and fibrosis. In resection and cryoinjury models, these events occur primarily within the first week after injury. Clotting begins within seconds of injury, as the wound is sealed by erythrocytes [21]. Within the first day, activated thrombocytes enter the wound, and over the following weeks express PDGF to aid in neovascularization, and other factors necessary for regeneration [33,68]. Beginning at 3 h post-injury, and continuing over the next several days, the inflammatory response peaks as leukocytes, including macrophages and myeloperoxidase-positive neutrophils, begin to infiltrate the wound site [22,35,41,62,74]. Inhibition of the inflammatory response leads to a decrease in recruitment of phagocytes, impairment of angiogenesis, and decrease in cardiomyocyte proliferation, and ultimately prevents regeneration from completing [62].

At the same time as inflammation, apoptosis and/or necrosis begins within the injury area (or injury-adjacent area in resection experiments), as the cellular debris in the infarct is cleared away [33,35,36,63]. Simultaneous with inflammation and initiation of apoptosis, the erythrocyte clot is replaced by a temporary fibrin matrix, which retains structure in the wound until approximately one week post-injury, when the temporary matrix is lysed to be replaced by a primarily collagen matrix and initiate the phase of regeneration dominated by fibrosis [21,36].

Fibrosis, or scarring, is a process of wound repair that involves replacement of missing or necrotic tissue with connective tissue, primarily fibrin and collagen. In adult mammalian hearts, this is the only process by which lesions or infarcts can be repaired, but permanent scarring leads to changes in heart contractility, conductivity, and morphology that can greatly reduce the function of the heart [33]. In organisms that can regenerate their hearts, this fibrotic tissue is soon replaced by functional myocardium and no permanent scar persists. While scarring and regeneration are commonly seen as opposing outcomes, an initial period of fibrosis is necessary for proper regeneration in both fish and amphibians, providing a framework in the form of ECM for cardiomyocytes to assemble into working muscle [37,63,67]. During the period of fibrosis, a number of wound healing genes, such as *vegf* to promote vascularization and *granulin A* to promote cell growth, are upregulated [74]. In order for regeneration to be completed, newly proliferated cardiomyocytes need to be able to migrate into the injured region, and simultaneously, the fibrotic tissue needs to be removed. Beginning at one week, therefore, remodeling proteins such as matrix metalloproteinases (MMPs) are activated, and resolution of fibrosis begins [74].

The TGF-β signaling pathway is an essential component of both establishing and resolving fibrosis. During regeneration of the zebrafish heart, within the first two weeks after injury, TGF-β is expressed in the myocardium, epicardium, and fibroblasts of the injured region and infarct boundary, but is not found in the endocardium. [37,41] Multiple isoforms of the TGF-β receptor are also present, and the infarct zone and adjacent tissue show markers of TGF-β activity in the form of phosphorylated Smad3 [37].

TGF-β is required in both early and late phases of heart regeneration. Within the first two weeks, it is predominantly necessary for inducing fibroblasts to deposit fibronectin and collagen into the injury site, forming the ECM that provides a framework for later muscle regeneration. Inhibition of TGF-β signaling after two weeks does not affect the collagen matrix, but it does prevent cardiomyocytes from infiltrating and replacing the fibrotic ECM. TGF-β regulates expression of the glycoprotein tenascin C at the infarct boundary, which in turn is necessary for remodeling of the ECM and migration of cardiomyocytes into the wound region. [37] Fibronectin may also play a role in this migration process, as it appears to be required for signaling integrin receptor-expressing cardiomyocytes, and inhibition of it prevents regeneration [42]. This role for the ECM in guiding regeneration is conserved in amphibians, where fibronectin, tenascin C, and hyaluronic acid are upregulated in the wound region throughout regeneration, and in the case of the latter two up to 70 days post-injury [67]. In fish, the migrating cardiomyocytes are primarily directed by the chemokine Cxcl12-Cxcr4 system [66], and build new myocardium in the injured area from the outside of the wound (apical edge of the myocardium) inward [33,59]. New cardiomyocytes begin to become electrically coupled around 14 days post-injury, and are fully coupled by 30 days [36,61].

FGF signaling is also necessary for proper regeneration and resolution of scarring. FGF is secreted by myocardial cells adjacent to the injury site. This process activates FGF receptor-expressing epicardial cells, and is necessary for the collagen scar to be replaced by myocardium; while the mechanism for this signaling process has not been characterized, the similarities in effect to TGF-β inhibition suggest that FGF's role in epicardial activation may follow a similar pathway [59]. Interestingly, if FGF signaling is inhibited to prevent regeneration and induce scarring, the scarring process can be partially reversed by later restoration of FGF. While fibrotic tissue is never fully resorbed, myocardium is restored over the scar. This may indicate that scarring is a non-permanent outcome, and may provide an avenue for repairing existing infarct damage in patients [61].

4. Non-Regenerative Heart Repair

In addition to the prodigious regenerative capabilities possessed by certain species, many fish and amphibians are capable of repairing lower levels of heart damage that may not require a full regenerative response. The organs of ectothermic animals, including the heart, are more subject to changing environmental conditions than those of endothermic animals. As a result, they have developed adaptations that allow them to survive a wider range of temperatures, gas concentrations, and pH values. Some of these adaptations are protective, active before damaging agents are able to adversely affect cardiac function; these include robust antioxidant systems, ion pumps to prevent extreme pH values, and ability to maintain cardiac function during hypoxia [77]. However, those particular adaptations lie outside the scope of this review; for more information, see Driedzic *et al.* [77] Instead, we will focus on mechanisms associated with post-damage compensation and repair.

Since urodele amphibians and certain fish species are so adept at regenerating, most studies in these organisms have focused on full regeneration. As a result, they have not been used as often to study repair in a non-regenerative context, although some research has been conducted on the ability of zebrafish to repair after reperfusion injury [50]. Instead, lower vertebrate species that are not capable of regeneration have been used as the primary model for this type of repair, no group of animals more so than the anuran amphibians. While full cardiac regeneration has not been observed in anurans to date, they have proven to be a useful model for studying amphibian tolerance to low levels of damage, particularly oxidative and mechanical stress. A primary regulator of the stress response in amphibian, as well as mammalian, hearts is the mitogen-activated protein kinase (MAPK) family. MAPKs are activated by a variety of environmental and cytokine stimuli, and in turn phosphorylate important cardiac transcription factors (e.g., Mef2) and general cell regulation factors (e.g., c-Jun). These ultimately regulate a wide range of cell behaviors from apoptosis to cell survival and cardiomyocyte hypertrophy [78].

In the amphibian heart (mainly studied in the adult marsh frog, *Rana ridibunda*), MAPK activity is induced by stress stimuli including hyperosmosis, anoxia, and reperfusion [51]. One subset of MAPKs in particular, the p38-MAPK family, has been shown to be activated in frog hearts by a number of common disease-associated stimuli, including hypoxia [54], osmotic stress [52], reactive oxygen species [53,79], acidosis [55], and mechanical overload [56]. This p38-MAPK activity in turn confers protective effects on cardiomyocytes via phosphorylation of MAPK-activated protein kinase 2 (MAPKAPK2) and Heat shock protein 27 (Hsp27), both associated with cell survival [54,55,79]. One potentially important downstream target of this pathway is atrial naturetic peptide (ANP), a hormone responsible for regulating ion and fluid balance across the cell membrane, and commonly associated with stress compensation. ANP levels have been shown to increase in the frog heart after hypoxic or osmotic stress [52,54]. Additionally, changes in extracellular pH induce expression of the chaperone protein Hsp70, although this occurs independently of p38-MAPK activity [55]. Other classes of MAPKs, specifically the c-Jun N-terminal kinases (JNKs) and extracellularly responsive kinases (ERKs), have also been shown to be activated by stressors in the frog heart [51,53,56]. Overall, the MAPK-mediated stress response appears to be a crucial component of protecting and repairing low-level, environmental damage to the anuran heart.

What happens, however, if an anuran heart is damaged to an extent that would induce regeneration in a urodele or fish heart? Years before the regenerative capacity of newts was discovered, Rumyantsev observed that the adult common frog (*R. temporaria*) heart could partially repair damage induced by crushing the ventricle [80,81]. Within the first week after damage, phagocytes enter the wound region to remove necrotic myocardium and transcription in neighboring myocytes increases. In the second and third week, myocytes partially dedifferentiate and re-enter the cell cycle. Myofibers then extend into the wounded region, connecting the muscle on either side of the lesion. Unlike in regeneration, however, myocardium is never fully restored across the lesion, and scar tissue remains [81].

The question of why anurans and urodeles differ in regenerative capacity is still not well understood. Observation of mitotic kinetics in frog heart repair has shown that myocytes proliferate twice as slowly as fibrotic tissue [81]; however, post-damage myocyte proliferation is comparable between non-regenerating frogs and regenerating urodeles, indicating that this is likely not the main driving force [20,81]. Embryonic anurans are capable of regenerating many structures that do not regenerate in adult anurans, but do regenerate in adult urodeles, including limbs [82], the lens of the eye [83], and the brain [84]. It is likely that by understanding the differences in regenerative capacity between these three systems—embryonic anurans, adult anurans, and adult urodeles—we can better determine how to induce cardiac regeneration in non-regenerative organisms.

5. Conclusions

While much of the research on heart damage and repair has understandably been conducted in mammalian models to date, fish and amphibians provide a unique opportunity for studying a system that can completely regenerate missing or damaged cardiac tissue. Regeneration in mammalian hearts is limited-to-nonexistent, and scarring—with a resulting loss of heart function—is the most common result of cardiac damage. What little evidence of cardiac renewal exists occurs mostly in neonatal or embryonic mammals, though adult mammal hearts are not entirely devoid of restorative processes. Myocyte proliferation has been observed in response to human heart damage [7], *raldh2* activation occurs in the endocardium of infarcted mouse hearts [24], and epicardial activation and cardiomyocyte dedifferentiation, hallmarks of the regenerative response in zebrafish, occur after resection of neonatal mouse hearts [11]. In contrast, throughout their lifetimes, some species of fish and urodele amphibians can restore all three layers of the heart with no permanent scarring and full restoration of functionality.

In addition to the improved regenerative response of fish and amphibians, there exists a plethora of physical and molecular tools for studying heart regeneration in these model systems. Mechanical manipulations and other physical techniques like cryoinjury are significantly easier in fish and amphibians than mammals, particularly at embryonic stages. The wealth of genomic data for zebrafish also makes transgenic methods of damage, which can be used to minimize off-target effects and unnecessary surgery, an extremely viable solution. And while fish are the more ubiquitous model, amphibians should not be forgotten; the amphibian heart is more similar to the mammalian heart, molecularly and morphologically, than that of fish. Differences in regenerative capacity between urodele and anuran amphibians may also prove useful in understanding why some organisms can regenerate tissue while others, like mammals, use scarring as their main method of repair.

With a greater understanding of heart regeneration in amphibians and fish, and the effect that has had on understanding mammalian heart repair, potential new therapeutic strategies for mammals are being devised. One major obstacle is the relative lack of proliferation in mature mammalian cardiomyocytes, and so inducing them to reenter the cell cycle and proliferate has been a large focus of research. Targets that have shown success on this front, many of which have been observed to play a role in lower vertebrate heart regeneration and repair, include FGF, p38-MAPK [85], periostin [86], neuregulin [87], and cyclins and CDKs [88]. The level of conservation between these processes in lower vertebrates and mammals has been demonstrated by the ability of newt regeneration extract to induce dedifferentiation of mammalian myocytes [89]. MicroRNAs, which in addition to being involved in many cardiac remodeling processes have proven to be necessary for cardiomyocyte proliferation in zebrafish heart regeneration [70], have also been a promising target for inducing cardiac repair through regulation of endogenous gene expression [90]. The other major focus in inducing mammalian heart repair, besides myocyte proliferation, has been the role of EPDCs. While EPDCs do not directly lead to new myocardium, they promote a permissive environment through secretion of ECM by fibroblasts and promotion of angiogenesis. Improving the pro-regenerative capacity of mammalian EPDCs, while limiting fibrosis from the same, may prove to be as important a step towards inducing regeneration in the mammal heart as making new cardiomyocytes [91,92].

The regenerative abilities of fish and amphibians are just beginning to be understood. They have already proven to be a valuable model system for researching cardiac damage and repair, and their importance should only increase going into the future. By understanding how regeneration is possible in fish and amphibians—and why mammalian hearts fail to regenerate, despite the presence of conserved repair responses and pathways—we can hopefully discover novel ways to stimulate repair in mammalian hearts that can serve as therapeutics in human heart disease.

Acknowledgments: Research in the McLaughlin lab is supported by NSF grant IOS-0843355 and AHA award 14IRG18570000. The authors confirm that the funders had no influence over the study design, content of the article, or selection of this journal.

Author Contributions: K.J. and K.M. both contributed to the research and writing of this manuscript. All authors read and approved the final manuscript.

Conflicts of Interest: The authors declare no conflict of interest.

References

1. Lozano, R.; Naghavi, M.; Foreman, K.; Lim, S.; Shibuya, K.; Aboyans, V.; Abraham, J.; Adair, T.; Aggarwal, R.; Ahn, S.Y. Global and regional mortality from 235 causes of death for 20 age groups in 1990 and 2010: A systematic analysis for the Global Burden of Disease Study 2010. *Lancet* **2013**, *380*, 2095–2128. [CrossRef]
2. Mozaffarian, D.; Benjamin, E.J.; Go, A.S.; Arnett, D.K.; Blaha, M.J.; Cushman, M.; de Ferranti, S.; Despres, J.P.; Fullerton, H.J.; Howard, V.J.; *et al.* Heart disease and stroke statistics—2015 update: A report from the American Heart Association. *Circulation* **2015**, *131*, e29–e322. [CrossRef] [PubMed]
3. Azevedo, C.F.; Cheng, S.; Lima, J.A. Cardiac imaging to identify patients at risk for developing heart failure after myocardial infarction. *Curr. Heart Fail. Rep.* **2005**, *2*, 183–188. [CrossRef] [PubMed]
4. Garbern, J.C.; Lee, R.T. Cardiac stem cell therapy and the promise of heart regeneration. *Cell Stem Cell* **2013**, *12*, 689–698. [CrossRef] [PubMed]
5. Engel, F.B. Cardiomyocyte proliferation: A platform for mammalian cardiac repair. *Cell Cycle* **2005**, *4*, 1360–1363. [PubMed]
6. Senyo, S.E.; Steinhauser, M.L.; Pizzimenti, C.L.; Yang, V.K.; Cai, L.; Wang, M.; Wu, T.; Guerquin-Kern, J.; Lechene, C.P.; Lee, R.T. Mammalian heart renewal by pre-existing cardiomyocytes. *Nature* **2013**, *493*, 433–436. [CrossRef] [PubMed]
7. Beltrami, A.P.; Urbanek, K.; Kajstura, J.; Yan, S.; Finato, N.; Bussani, R.; Nadal-Ginard, B.; Silvestri, F.; Leri, A.; Beltrami, C.A. Evidence that human cardiac myocytes divide after myocardial infarction. *N. Engl. J. Med.* **2001**, *344*, 1750–1757. [CrossRef] [PubMed]
8. Xin, M.; Kim, Y.; Sutherland, L.B.; Murakami, M.; Qi, X.; McAnally, J.; Porrello, E.R.; Mahmoud, A.I.; Tan, W.; Shelton, J.M.; *et al.* Hippo pathway effector Yap promotes cardiac regeneration. *Proc. Natl. Acad. Sci. USA* **2013**, *110*, 13839–13844. [CrossRef] [PubMed]
9. Mahmoud, A.I.; Kocabas, F.; Muralidhar, S.A.; Kimura, W.; Koura, A.S.; Thet, S.; Porrello, E.R.; Sadek, H.A. Meis1 regulates postnatal cardiomyocyte cell cycle arrest. *Nature* **2013**, *497*, 249–253. [CrossRef] [PubMed]
10. Porrello, E.R.; Mahmoud, A.I.; Simpson, E.; Johnson, B.A.; Grinsfelder, D.; Canseco, D.; Mammen, P.P.; Rothermel, B.A.; Olson, E.N.; Sadek, H.A. Regulation of neonatal and adult mammalian heart regeneration by the miR-15 family. *Proc. Natl. Acad. Sci. USA* **2013**, *110*, 187–192. [CrossRef] [PubMed]
11. Porrello, E.R.; Mahmoud, A.I.; Simpson, E.; Hill, J.A.; Richardson, J.A.; Olson, E.N.; Sadek, H.A. Transient regenerative potential of the neonatal mouse heart. *Science* **2011**, *331*, 1078–1080. [CrossRef] [PubMed]
12. Haubner, B.J.; Adamowicz-Brice, M.; Khadayate, S.; Tiefenthaler, V.; Metzler, B.; Aitman, T.; Penninger, J.M. Complete cardiac regeneration in a mouse model of myocardial infarction. *Aging* **2012**, *4*, 966–977. [PubMed]
13. Andersen, D.C.; Ganesalingam, S.; Jensen, C.H.; Sheikh, S.P. Do neonatal mouse hearts regenerate following heart apex resection? *Stem Cell Rep.* **2014**, *2*, 406–413. [CrossRef] [PubMed]
14. Oberpriller, J.O.; Oberpriller, J.C. Response of the adult newt ventricle to injury. *J. Exp. Zool.* **1974**, *187*, 249–259. [CrossRef] [PubMed]
15. Becker, R.O.; Chapin, S.; Sherry, R. Regeneration of the ventricular myocardium in amphibians. *Nature* **1974**, *248*, 145–147. [CrossRef] [PubMed]
16. McDonnell, T.J.; Oberpriller, J.O. The atrial proliferative response following partial ventricular amputation in the heart of the adult newt. A light and electron microscopic autoradiographic study. *Tissue Cell* **1983**, *15*, 351–363. [CrossRef]
17. McDonnell, T.J.; Oberpriller, J.O. The response of the atrium to direct mechanical wounding in the adult heart of the newt, Notophthalmus viridescens. *Cell Tissue Res.* **1984**, *235*, 583–592. [CrossRef] [PubMed]
18. Oberpriller, J.O.; Oberpriller, J.C.; Matz, D.G.; Soonpaa, M.H. Stimulation of proliferative events in the adult amphibian cardiac myocyte. *Ann. N. Y. Acad. Sci.* **1995**, *752*, 30–46. [CrossRef] [PubMed]

19. Cano-Martínez, A.; Vargas-González, A.; Guarner-Lans, V.; Prado-Zayago, E.; Leon-Oleda, M.; Nieto-Lima, B. Functional and structural regeneration in the axolotl heart (*Ambystoma mexicanum*) after partial ventricular amputation. *Archivos de Cardiología de México* **2010**, *80*, 79–86. [PubMed]
20. Flink, I.L. Cell cycle reentry of ventricular and atrial cardiomyocytes and cells within the epicardium following amputation of the ventricular apex in the axolotl, *Amblystoma mexicanum*: Confocal microscopic immunofluorescent image analysis of bromodeoxyuridine-labeled nuclei. *Anat. Embryol.* **2002**, *205*, 235–244. [PubMed]
21. Poss, K.D.; Wilson, L.G.; Keating, M.T. Heart regeneration in zebrafish. *Science* **2002**, *298*, 2188–2190. [CrossRef] [PubMed]
22. Lafontant, P.J.; Burns, A.R.; Grivas, J.A.; Lesch, M.A.; Lala, T.D.; Reuter, S.P.; Field, L.J.; Frounfelter, T.D. The giant danio (*D. aequipinnatus*) as a model of cardiac remodeling and regeneration. *Anat. Rec.* **2012**, *295*, 234–248. [CrossRef] [PubMed]
23. Grivas, J.; Haag, M.; Johnson, A.; Manalo, T.; Roell, J.; Das, T.L.; Brown, E.; Burns, A.R.; Lafontant, P.J. Cardiac repair and regenerative potential in the goldfish (*Carassius auratus*) heart. *Comp. Biochem. Physiol. C Toxicol. Pharmacol.* **2014**, *163*, 14–23. [CrossRef] [PubMed]
24. Kikuchi, K.; Holdway, J.E.; Major, R.J.; Blum, N.; Dahn, R.D.; Begemann, G.; Poss, K.D. Retinoic acid production by endocardium and epicardium is an injury response essential for zebrafish heart regeneration. *Dev. Cell* **2011**, *20*, 397–404. [CrossRef] [PubMed]
25. Ito, K.; Morioka, M.; Kimura, S.; Tasaki, M.; Inohaya, K.; Kudo, A. Differential reparative phenotypes between zebrafish and medaka after cardiac injury. *Dev. Dyn.* **2014**, *243*, 1106–1115. [CrossRef] [PubMed]
26. Witman, N.; Murtuza, B.; Davis, B.; Arner, A.; Morrison, J.I. Recapitulation of developmental cardiogenesis governs the morphological and functional regeneration of adult newt hearts following injury. *Dev. Biol.* **2011**, *354*, 67–76. [CrossRef] [PubMed]
27. Laube, F.; Heister, M.; Scholz, C.; Borchardt, T.; Braun, T. Re-programming of newt cardiomyocytes is induced by tissue regeneration. *J. Cell Sci.* **2006**, *119*, 4719–4729. [CrossRef] [PubMed]
28. Raya, Á.; Koth, C.M.; Büscher, D.; Kawakami, Y.; Itoh, T.; Raya, R.M.; Sternik, G.; Tsai, H.; Rodríguez-Esteban, C.; Izpisúa-Belmonte, J.C. Activation of Notch signaling pathway precedes heart regeneration in zebrafish. *Proc. Natl. Acad. Sci. USA* **2003**, *100*, 11889–11895.
29. González-Rosa, J.M.; Mercader, N. Cryoinjury as a myocardial infarction model for the study of cardiac regeneration in the zebrafish. *Nat. Protoc.* **2012**, *7*, 782–788. [CrossRef] [PubMed]
30. Murry, C.E.; Wiseman, R.W.; Schwartz, S.M.; Hauschka, S.D. Skeletal myoblast transplantation for repair of myocardial necrosis. *J. Clin. Investig.* **1996**, *98*, 2512–2523. [CrossRef] [PubMed]
31. Ciulla, M.M.; Paliotti, R.; Ferrero, S.; Braidotti, P.; Esposito, A.; Gianelli, U.; Busca, G.; Cioffi, U.; Bulfamante, G.; Magrini, F. Left ventricular remodeling after experimental myocardial cryoinjury in rats. *J. Surg. Res.* **2004**, *116*, 91–97. [CrossRef] [PubMed]
32. Van den Bos, E.J.; Mees, B.M.; de Waard, M.C.; de Crom, R.; Duncker, D.J. A novel model of cryoinjury-induced myocardial infarction in the mouse: A comparison with coronary artery ligation. *Am. J. Physiol. Heart Circ. Physiol.* **2005**, *289*, H1291–H1300. [CrossRef] [PubMed]
33. González-Rosa, J.M.; Martín, V.; Peralta, M.; Torres, M.; Mercader, N. Extensive scar formation and regression during heart regeneration after cryoinjury in zebrafish. *Development* **2011**, *138*, 1663–1674. [CrossRef] [PubMed]
34. Chablais, F.; Jazwinska, A. Induction of myocardial infarction in adult zebrafish using cryoinjury. *J. Vis. Exp.* **2012**, *62*. [CrossRef] [PubMed]
35. Schnabel, K.; Wu, C.; Kurth, T.; Weidinger, G. Regeneration of cryoinjury induced necrotic heart lesions in zebrafish is associated with epicardial activation and cardiomyocyte proliferation. *PLoS ONE* **2011**, *6*, e18503. [CrossRef] [PubMed]
36. Chablais, F.; Veit, J.; Rainer, G.; Jaźwińska, A. The zebrafish heart regenerates after cryoinjury-induced myocardial infarction. *BMC Dev. Biol.* **2011**, *11*. [CrossRef] [PubMed]
37. Chablais, F.; Jaźwińska, A. The regenerative capacity of the zebrafish heart is dependent on TGFβ signaling. *Development* **2012**, *139*, 1921–1930. [CrossRef] [PubMed]
38. Curado, S.; Anderson, R.M.; Jungblut, B.; Mumm, J.; Schroeter, E.; Stainier, D.Y. Conditional targeted cell ablation in zebrafish: A new tool for regeneration studies. *Dev. Dyn.* **2007**, *236*, 1025–1035. [CrossRef] [PubMed]

39. Curado, S.; Stainier, D.Y.; Anderson, R.M. Nitroreductase-mediated cell/tissue ablation in zebrafish: A spatially and temporally controlled ablation method with applications in developmental and regeneration studies. *Nat. Protoc.* **2008**, *3*, 948–954. [CrossRef] [PubMed]
40. Zhang, R.; Han, P.; Yang, H.; Ouyang, K.; Lee, D.; Lin, Y.; Ocorr, K.; Kang, G.; Chen, J.; Stainier, D.Y. *In vivo* cardiac reprogramming contributes to zebrafish heart regeneration. *Nature* **2013**, *498*, 497–501. [CrossRef] [PubMed]
41. Wang, J.; Panáková, D.; Kikuchi, K.; Holdway, J.E.; Gemberling, M.; Burris, J.S.; Singh, S.P.; Dickson, A.L.; Lin, Y.; Sabeh, M.K. The regenerative capacity of zebrafish reverses cardiac failure caused by genetic cardiomyocyte depletion. *Development* **2011**, *138*, 3421–3430. [CrossRef] [PubMed]
42. Wang, J.; Karra, R.; Dickson, A.L.; Poss, K.D. Fibronectin is deposited by injury-activated epicardial cells and is necessary for zebrafish heart regeneration. *Dev. Biol.* **2013**, *382*, 427–435. [CrossRef] [PubMed]
43. Choi, W.; Gemberling, M.; Wang, J.; Holdway, J.E.; Shen, M.; Karlstrom, R.O.; Poss, K.D. *In vivo* monitoring of cardiomyocyte proliferation to identify chemical modifiers of heart regeneration. *Development* **2013**, *140*, 660–666. [CrossRef] [PubMed]
44. Konstantinidis, K.; Whelan, R.S.; Kitsis, R.N. Mechanisms of cell death in heart disease. *Arterioscler. Thromb. Vasc. Biol.* **2012**, *32*, 1552–1562. [CrossRef] [PubMed]
45. Wencker, D.; Chandra, M.; Nguyen, K.; Miao, W.; Garantziotis, S.; Factor, S.M.; Shirani, J.; Armstrong, R.C.; Kitsis, R.N. A mechanistic role for cardiac myocyte apoptosis in heart failure. *J. Clin. Investig.* **2003**, *111*, 1497–1504. [CrossRef] [PubMed]
46. Hayakawa, Y.; Chandra, M.; Miao, W.; Shirani, J.; Brown, J.H.; Dorn, G.W.; Armstrong, R.C.; Kitsis, R.N. Inhibition of cardiac myocyte apoptosis improves cardiac function and abolishes mortality in the peripartum cardiomyopathy of Gαq transgenic mice. *Circulation* **2003**, *108*, 3036–3041. [CrossRef] [PubMed]
47. Giordano, F.J. Oxygen, oxidative stress, hypoxia, and heart failure. *J. Clin. Investig.* **2005**, *115*, 500–508. [CrossRef] [PubMed]
48. Jeffrey, F.M.; Malloy, C.R.; Radda, G.K. Influence of intracellular acidosis on contractile function in the working rat heart. *Am. J. Physiol.* **1987**, *253*, H1499–H1505. [PubMed]
49. Falck, G.; Schjøtt, J.; Bruvold, M.; Krane, J.; Skarra, S.; Jynge, P. Hyperosmotic perfusion of the beating rat heart and the role of the Na/K/2Cl^- co-transporter and the Na/H exchanger. *Basic Res. Cardiol.* **2000**, *95*, 19–27. [CrossRef] [PubMed]
50. Parente, V.; Balasso, S.; Pompilio, G.; Verduci, L.; Colombo, G.I.; Milano, G.; Guerrini, U.; Squadroni, L.; Cotelli, F.; Pozzoli, O. Hypoxia/reoxygenation cardiac injury and regeneration in zebrafish adult heart. *PLoS ONE* **2013**, *8*, e53748. [CrossRef] [PubMed]
51. Aggeli, I.S.; Gaitanaki, C.; Lazou, A.; Beis, I. Activation of multiple MAPK pathways (ERKs, JNKs, p38-MAPK) by diverse stimuli in the amphibian heart. *Mol. Cell. Biochem.* **2001**, *221*, 63–69. [CrossRef] [PubMed]
52. Aggeli, I.K.; Gaitanaki, C.; Lazou, A.; Beis, I. Hyperosmotic and thermal stresses activate p38-MAPK in the perfused amphibian heart. *J. Exp. Biol.* **2002**, *205*, 443–454. [PubMed]
53. Gaitanaki, C.; Konstantina, S.; Chrysa, S.; Beis, I. Oxidative stress stimulates multiple MAPK signalling pathways and phosphorylation of the small HSP27 in the perfused amphibian heart. *J. Exp. Biol.* **2003**, *206*, 2759–2769. [CrossRef] [PubMed]
54. Gaitanaki, C.; Kalpachidou, T.; Aggeli, I.K.; Papazafiri, P.; Beis, I. $CoCl_2$ induces protective events via the p38-MAPK signalling pathway and ANP in the perfused amphibian heart. *J. Exp. Biol.* **2007**, *210*, 2267–2277. [CrossRef] [PubMed]
55. Stathopoulou, K.; Gaitanaki, C.; Beis, I. Extracellular pH changes activate the p38-MAPK signalling pathway in the amphibian heart. *J. Exp. Biol.* **2006**, *209*, 1344–1354. [CrossRef] [PubMed]
56. Aggeli, I.K.; Gaitanaki, C.; Lazou, A.; Beis, I. Stimulation of multiple MAPK pathways by mechanical overload in the perfused amphibian heart. *Am. J. Physiol. Regul. Integr. Comp. Physiol.* **2001**, *281*, R1689–R1698. [PubMed]
57. Bakkers, J. Zebrafish as a model to study cardiac development and human cardiac disease. *Cardiovasc. Res.* **2011**, *91*, 279–288. [CrossRef] [PubMed]
58. Sallin, P.; de Preux Charles, A.; Duruz, V.; Pfefferli, C.; Jaźwińska, A. A dual epimorphic and compensatory mode of heart regeneration in zebrafish. *Dev. Biol.* **2015**, *399*, 27–40. [CrossRef] [PubMed]

59. Lepilina, A.; Coon, A.N.; Kikuchi, K.; Holdway, J.E.; Roberts, R.W.; Burns, C.G.; Poss, K.D. A dynamic epicardial injury response supports progenitor cell activity during zebrafish heart regeneration. *Cell* **2006**, *127*, 607–619. [CrossRef] [PubMed]
60. Jopling, C.; Sleep, E.; Raya, M.; Martí, M.; Raya, A.; Belmonte, J.C.I. Zebrafish heart regeneration occurs by cardiomyocyte dedifferentiation and proliferation. *Nature* **2010**, *464*, 606–609. [CrossRef] [PubMed]
61. Kikuchi, K.; Holdway, J.E.; Werdich, A.A.; Anderson, R.M.; Fang, Y.; Egnaczyk, G.F.; Evans, T.; MacRae, C.A.; Stainier, D.Y.; Poss, K.D. Primary contribution to zebrafish heart regeneration by *gata4*$^+$ cardiomyocytes. *Nature* **2010**, *464*, 601–605. [CrossRef] [PubMed]
62. Huang, W.; Yang, C.; Chen, I.; Liu, Y.L.; Chang, S.; Chuang, Y. Treatment of glucocorticoids inhibited early immune responses and impaired cardiac repair in adult zebrafish. *PLoS ONE* **2013**, *8*, e66613. [CrossRef] [PubMed]
63. Piatkowski, T.; Mühlfeld, C.; Borchardt, T.; Braun, T. Reconstitution of the myocardium in regenerating newt hearts is preceded by transient deposition of extracellular matrix components. *Stem Cells Dev.* **2013**, *22*, 1921–1931. [CrossRef] [PubMed]
64. Gittenberger-de Groot, A.C.; Winter, E.M.; Poelmann, R.E. Epicardium-derived cells (EPDCs) in development, cardiac disease and repair of ischemia. *J. Cell Mol. Med.* **2010**, *14*, 1056–1060. [CrossRef] [PubMed]
65. González-Rosa, J.M.; Peralta, M.; Mercader, N. Pan-epicardial lineage tracing reveals that epicardium derived cells give rise to myofibroblasts and perivascular cells during zebrafish heart regeneration. *Dev. Biol.* **2012**, *370*, 173–186. [CrossRef] [PubMed]
66. Itou, J.; Oishi, I.; Kawakami, H.; Glass, T.J.; Richter, J.; Johnson, A.; Lund, T.C.; Kawakami, Y. Migration of cardiomyocytes is essential for heart regeneration in zebrafish. *Development* **2012**, *139*, 4133–4142. [CrossRef] [PubMed]
67. Mercer, S.E.; Odelberg, S.J.; Simon, H. A dynamic spatiotemporal extracellular matrix facilitates epicardial-mediated vertebrate heart regeneration. *Dev. Biol.* **2013**, *382*, 457–469. [CrossRef] [PubMed]
68. Kim, J.; Wu, Q.; Zhang, Y.; Wiens, K.M.; Huang, Y.; Rubin, N.; Shimada, H.; Handin, R.I.; Chao, M.Y.; Tuan, T. PDGF signaling is required for epicardial function and blood vessel formation in regenerating zebrafish hearts. *Proc. Natl. Acad. Sci. USA* **2010**, *107*, 17206–17210. [CrossRef] [PubMed]
69. Kikuchi, K.; Gupta, V.; Wang, J.; Holdway, J.E.; Wills, A.A.; Fang, Y.; Poss, K.D. *tcf21*$^+$ epicardial cells adopt non-myocardial fates during zebrafish heart development and regeneration. *Development* **2011**, *138*, 2895–2902. [CrossRef] [PubMed]
70. Yin, V.P.; Lepilina, A.; Smith, A.; Poss, K.D. Regulation of zebrafish heart regeneration by miR-133. *Dev. Biol.* **2012**, *365*, 319–327. [CrossRef] [PubMed]
71. Zhao, L.; Borikova, A.L.; Ben-Yair, R.; Guner-Ataman, B.; MacRae, C.A.; Lee, R.T.; Burns, C.G.; Burns, C.E. Notch signaling regulates cardiomyocyte proliferation during zebrafish heart regeneration. *Proc. Natl. Acad. Sci. USA* **2014**, *111*, 1403–1408. [CrossRef] [PubMed]
72. Tate, J.M.; Oberpriller, J.O. Primary cell culture and morphological characterization of ventricular myocytes from the adult newt, *Notophthalmus viridescens*. *Anat. Rec.* **1989**, *224*, 29–42. [CrossRef] [PubMed]
73. Fang, Y.; Gupta, V.; Karra, R.; Holdway, J.E.; Kikuchi, K.; Poss, K.D. Translational profiling of cardiomyocytes identifies an early Jak1/Stat3 injury response required for zebrafish heart regeneration. *Proc. Natl. Acad. Sci. USA* **2013**, *110*, 13416–13421. [CrossRef] [PubMed]
74. Lien, C.; Schebesta, M.; Makino, S.; Weber, G.J.; Keating, M.T. Gene expression analysis of zebrafish heart regeneration. *PLoS Biol.* **2006**, *4*, e260. [CrossRef] [PubMed]
75. Tanaka, E.M.; Gann, A.A.; Gates, P.B.; Brockes, J.P. Newt myotubes reenter the cell cycle by phosphorylation of the retinoblastoma protein. *J. Cell Biol.* **1997**, *136*, 155–165. [CrossRef] [PubMed]
76. Bettencourt-Dias, M.; Mittnacht, S.; Brockes, J.P. Heterogeneous proliferative potential in regenerative adult newt cardiomyocytes. *J. Cell Sci.* **2003**, *116*, 4001–4009. [CrossRef] [PubMed]
77. Driedzic, W.R.; Gesser, H. Energy metabolism and contractility in ectothermic vertebrate hearts: Hypoxia, acidosis, and low temperature. *Physiol. Rev.* **1994**, *74*, 221–258. [PubMed]
78. Bogoyevitch, M.A. Signalling via stress-activated mitogen-activated protein kinases in the cardiovascular system. *Cardiovasc. Res.* **2000**, *45*, 826–842. [CrossRef]

79. Gaitanaki, C.; Papatriantafyllou, M.; Stathopoulou, K.; Beis, I. Effects of various oxidants and antioxidants on the p38-MAPK signalling pathway in the perfused amphibian heart. *Mol. Cell. Biochem.* **2006**, *291*, 107–117. [CrossRef] [PubMed]
80. Rumyantsev, P. Post-injury DNA synthesis, mitosis and ultrastructural reorganization of adult frog cardiac myocytes. *Z. Zellforsch. Mikrosk. Anat.* **1973**, *139*, 431–450. [CrossRef] [PubMed]
81. Rumyantsev, P.P. Interrelations of the proliferation and differentiation processes during cardiac myogenesis and regeneration. *Int. Rev. Cytol.* **1977**, *51*, 187–273.
82. Gardiner, D.M.; Endo, T.; Bryant, S.V. The molecular basis of amphibian limb regeneration: Integrating the old with the new. *Semin. Cell Dev. Biol.* **2002**, *13*, 345–352. [CrossRef] [PubMed]
83. Rio-Tsonis, K.D.; Tsonis, P.A. Eye regeneration at the molecular age. *Dev. Dyn.* **2003**, *226*, 211–224. [CrossRef] [PubMed]
84. Endo, T.; Yoshino, J.; Kado, K.; Tochinai, S. Brain regeneration in anuran amphibians. *Dev. Growth Differ.* **2007**, *49*, 121–129. [CrossRef] [PubMed]
85. Engel, F.B.; Hsieh, P.C.; Lee, R.T.; Keating, M.T. FGF1/p38 MAP kinase inhibitor therapy induces cardiomyocyte mitosis, reduces scarring, and rescues function after myocardial infarction. *Proc. Natl. Acad. Sci. USA* **2006**, *103*, 15546–15551. [CrossRef] [PubMed]
86. Kühn, B.; del Monte, F.; Hajjar, R.J.; Chang, Y.; Lebeche, D.; Arab, S.; Keating, M.T. Periostin induces proliferation of differentiated cardiomyocytes and promotes cardiac repair. *Nat. Med.* **2007**, *13*, 962–969. [CrossRef] [PubMed]
87. Bersell, K.; Arab, S.; Haring, B.; Kühn, B. Neuregulin1/ErbB4 signaling induces cardiomyocyte proliferation and repair of heart injury. *Cell* **2009**, *138*, 257–270. [CrossRef] [PubMed]
88. Hassink, R.J.; Pasumarthi, K.B.; Nakajima, H.; Rubart, M.; Soonpaa, M.H.; de la Riviere, A.B.; Doevendans, P.A.; Field, L.J. Cardiomyocyte cell cycle activation improves cardiac function after myocardial infarction. *Cardiovasc. Res.* **2008**, *78*, 18–25. [CrossRef] [PubMed]
89. McGann, C.J.; Odelberg, S.J.; Keating, M.T. Mammalian myotube dedifferentiation induced by newt regeneration extract. *Proc. Natl. Acad. Sci. USA* **2001**, *98*, 13699–13704. [CrossRef] [PubMed]
90. Van Rooij, E.; Marshall, W.S.; Olson, E.N. Toward microRNA-based therapeutics for heart disease: The sense in antisense. *Circ. Res.* **2008**, *103*, 919–928. [CrossRef] [PubMed]
91. Martin-Puig, S.; Fuster, V.; Torres, M. Heart repair: From natural mechanisms of cardiomyocyte production to the design of new cardiac therapies. *Ann. N. Y. Acad. Sci.* **2012**, *1254*, 71–81. [CrossRef] [PubMed]
92. Ausoni, S.; Sartore, S. From fish to amphibians to mammals: In search of novel strategies to optimize cardiac regeneration. *J. Cell Biol.* **2009**, *184*, 357–364. [CrossRef] [PubMed]

2

Hedgehog Signaling in Endochondral Ossification

Shinsuke Ohba

Department of Bioengineering, Graduate School of Engineering, The University of Tokyo, 7-3-1 Hongo, Bunkyo-ku, Tokyo 113-0033, Japan; ohba@bioeng.t.u-tokyo.ac.jp

Academic Editors: Henk Roelink and Simon J. Conway

Abstract: Hedgehog (Hh) signaling plays crucial roles in the patterning and morphogenesis of various organs within the bodies of vertebrates and insects. Endochondral ossification is one of the notable developmental events in which Hh signaling acts as a master regulator. Among three Hh proteins in mammals, Indian hedgehog (Ihh) is known to work as a major Hh input that induces biological impact of Hh signaling on the endochondral ossification. Ihh is expressed in prehypertrophic and hypertrophic chondrocytes of developing endochondral bones. Genetic studies so far have demonstrated that the Ihh-mediated activation of Hh signaling synchronizes chondrogenesis and osteogenesis during endochondral ossification by regulating the following processes: (1) chondrocyte differentiation; (2) chondrocyte proliferation; and (3) specification of bone-forming osteoblasts. Ihh not only forms a negative feedback loop with parathyroid hormone-related protein (PTHrP) to maintain the growth plate length, but also directly promotes chondrocyte propagation. Ihh input is required for the specification of progenitors into osteoblast precursors. The combinatorial approaches of genome-wide analyses and mouse genetics will facilitate understanding of the regulatory mechanisms underlying the roles of Hh signaling in endochondral ossification, providing genome-level evidence of the potential of Hh signaling for the treatment of skeletal disorders.

Keywords: Indian hedgehog (Ihh); endochondral ossification; chondrocyte; osteoblast

1. Introduction

Hedgehog (Hh) signaling plays crucial roles in the patterning and morphogenesis of various organs within the bodies of vertebrates and insects [1]. In the conventional model of Hh signal transduction, the seven-pass transmembrane protein Smoothened (Smo) has an intrinsic intracellular signaling activity, which is repressed by the 12-pass transmembrane Hh receptor Patched (Ptch). The binding of Hh ligands to Ptch initiates Hh-Ptch signal transduction by relieving the repressive effect of Ptch on Smo. In vertebrates, three zinc finger transcription factors Gli1, Gli2, and Gli3 mediate Hh-responsive transcription [1]. Gli1 acts as a strong transcriptional activator. Gli2 and Gli3 are processed into transcriptional repressor forms upon Hh input. Gli2 has been suggested to function primarily as a transcriptional activator, and Gli3 as a transcriptional repressor, although a few studies have shown the opposite (for review see [1]).

Substantial progress has been made in understanding how Hh signaling engages skeletal development and interacts with other signaling factors, mainly based on mouse genetic studies manipulating the Hh signaling molecules mentioned above. This review aims to summarize the roles of Hh and its mode of action in skeletal development, with a particular focus on endochondral ossification.

2. Indian Hedgehog (Ihh) Synchronizes Chondrogenesis and Osteogenesis during Endochondral Ossification

Mammalian skeletons are formed by two distinct processes: intramembranous ossification and endochondral ossification. In intramembranous ossification, mesenchymal cells directly differentiate into the bone-forming cells called osteoblasts. This process forms the frontal bone, parietal bone, maxilla, mandibular bone, part of the clavicle, and so on. In contrast, axial skeletons and the long bones in the limbs develop through endochondral ossification, a process in which cartilage molds are initially formed and replaced by bones. During the endochondral ossification process, cartilage formation (chondrogenesis) and bone formation (osteogenesis) are orchestrated in a spatially and temporally controlled manner.

Endochondral ossification starts with the condensation of immature mesenchymal cells at a region where future bone develops. Cells in the center of the condensation give rise to chondrocytes expressing cartilage matrixes, including type II collagen, whereas cells at the periphery of the condensation form the perichondrium. Cartilage tissues grow through the proliferation and maturation of chondrocytes. Chondrocyte maturation is represented by its post-mitotic hypertrophy and the deposition of calcified matrices. Hypertrophic chondrocytes undergo apoptosis, followed by the absorption of residual calcified matrix by chondroclasts that come into the skeleton upon vascular invasion.

Cells at a certain region in the perichondrium (see later) are specified into an osteoblast lineage to form the bone collar, a predecessor of cortical bones. Osteoblasts are generated through several precursors. Immature mesenchymal cells initially differentiate into osteoblast precursors expressing runt-related transcription factor 2 (Runx2); the Runx2-positive precursors then express Sp7 (osterix) as well as Runx2, further differentiating into mature osteoblasts. Runx2 and Sp7 are key determinants for the osteoblast lineage. Mature osteoblasts contribute to bone formation by secreting bone matrixes. Bone gla protein (Bglap, osteocalcin) is a bona-fide marker for mature osteoblasts.

In the fetal cartilage of endochondral skeletons, referred to as the growth plate hereafter, chondrocytes are sequentially layered and aligned according to their differentiation phases: the proliferating phase, prehypertrophic phase, and hypertrophic phase (Figure 1). Bone collar formation in the perichondrium synchronizes with chondrocyte hypertrophy; it occurs at the perichondrial region adjacent to hypertrophic chondrocytes (Figure 1). Thus, the process of chondrocyte maturation and osteoblast differentiation can be observed in their chronological order from the epiphysis toward the diaphysis (Figure 1).

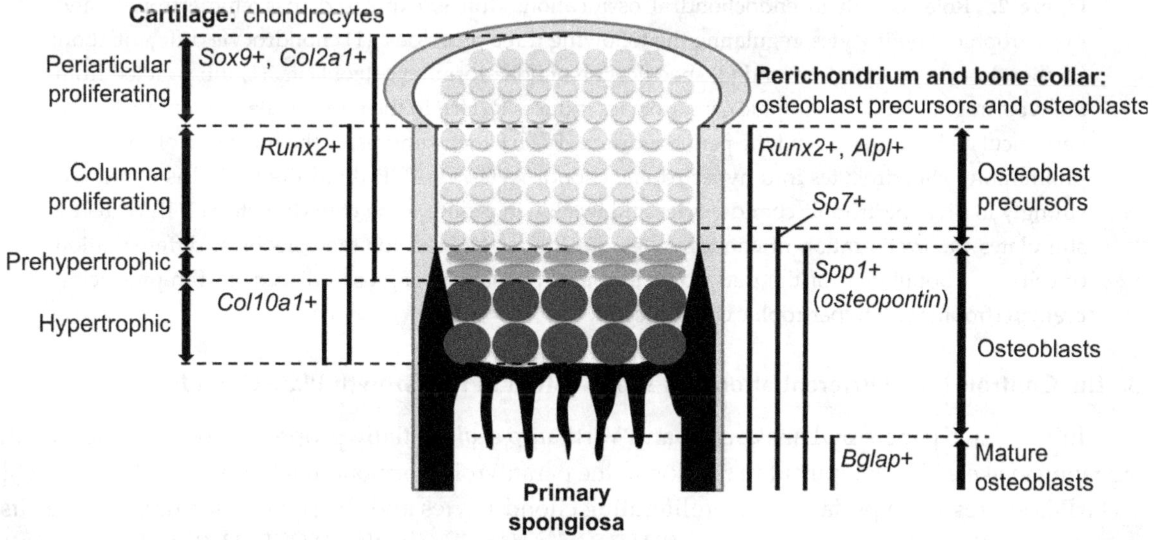

Figure 1. Structure of the growth plate in mouse embryos. An epiphysis of a fetal mouse long bone is depicted. The expression domains of marker genes and the locations of different populations are indicated (left, cartilage; right, perichondrium-bone collar).

Among the three Hh proteins in mammals, Ihh is known to work as a major Hh input that induces biological impact of Hh signaling on the endochondral ossification. In developing endochondral bones, prehypertrophic and hypertrophic chondrocytes in the growth plate express the Ihh gene. Target genes of Hh signaling, Ptch1 and Gli1, are expressed in proliferating chondrocytes, perichondrial cells, and primary spongiosa. Cells closer to the Ihh-producing cells express the target genes at higher levels. Genetic studies so far have demonstrated that the Ihh-mediated activation of Hh signaling synchronizes chondrogenesis and osteogenesis during endochondral ossification by regulating the following processes (Figure 2): (1) chondrocyte differentiation; (2) chondrocyte proliferation; and (3) specification of bone-forming osteoblasts; the first action is suggested to be primarily exerted in an indirect manner via parathyroid hormone-related protein (PTHrP) (see below), whereas the second and third ones are exerted in a direct manner.

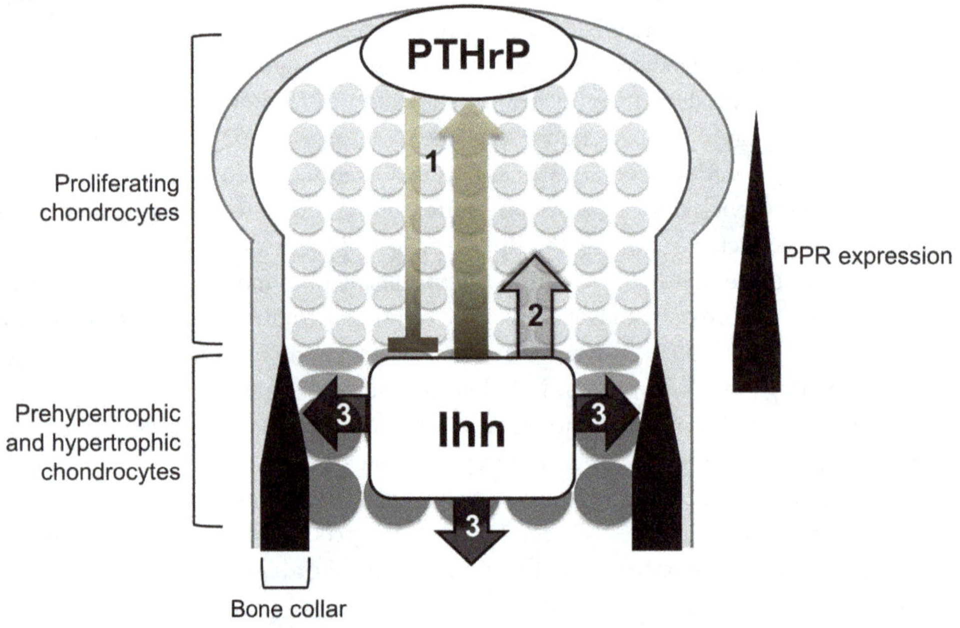

Figure 2. Roles of Ihh in endochondral ossification. Ihh is expressed in prehypertrophic and hypertrophic chondrocytes, regulating the following three processes: (1) chondrocyte differentiation; (2) chondrocyte proliferation; and (3) specification of bone-forming osteoblasts. (1) Ihh secreted from prehypertrophic and hypertrophic chondrocytes induces PTHrP in the periarticular regions including periarticular chondrocytes and the perichondrium. PTHrP then suppresses the transition of columnar proliferating chondrocytes into hypertrophic chondrocytes via PPR (bold line). PPR is expressed strongly in prehypertrophic chondrocytes and weakly in proliferating chondrocytes; (2) Ihh directly stimulates the proliferation of chondrocytes; (3) Ihh is required for the osteoblast differentiation of cells in a population of the perichondrium and primary spongiosa adjacent to Ihh-producing prehypertrophic and hypertrophic chondrocytes.

3. Ihh Controls both Differentiation and Proliferation of the Growth Plate Chondrocytes

In 1996, two papers by Lanske *et al.* and Vortkamp *et al.* initially proposed possible roles of Hh signaling in skeletal development in relation to the parathyroid hormone-related protein (PTHrP) [2,3]. PTHrP is expressed in periarticular proliferating chondrocytes and the perichondrium, whereas its receptor, parathyroid hormone 1 receptor (PTH1R; also described as the PTH/PTHrP receptor, or PPR), is expressed strongly in prehypertrophic chondrocytes and weakly in proliferating chondrocytes [4,5]. A series of loss-of-function studies in mice with the deletion of PTHrP, PPR, or both PTH and PTHrP [2,6–8] and gain-of-function studies in mice with the chondrocyte-specific overexpression of PTHrP or the constitutively active PPR [9–11] indicated that PTHrP kept chondrocytes proliferating by suppressing their hypertrophy via PPR.

In the paper by Vortkamp *et al.*, Ihh overexpression in chick limbs decreased hypertrophic chondrocytes while increasing PTHrP expression in the periarticular perichondrium [3]. Although premature hypertrophy of $PTHrP^{-/-}$ limbs was rescued by the addition of PTHrP, Shh had no effect on the $PTHrP^{-/-}$ limbs. By utilizing mouse limb explant cultures, Lanske *et al.* also showed that although PTHrP treatment and Shh treatment both induced the elongation of the growth plate with the suppression of chondrocyte hypertrophy in wild-type limbs, neither of them had effects on $PPR^{-/-}$ limbs [2]. Together with the findings on PTHrP function mentioned in the last paragraph, the data presented by Vortkamp *et al.* and Lanske *et al.* suggested that the transition from proliferating chondrocytes to hypertrophic ones was regulated by both PTHrP-PPR signaling and Ihh in a common feedback loop, where the PTHrP-PPR signaling mediated the effect of Ihh on the transition.

Skeletal phenotypes of $Ihh^{-/-}$ mice were then reported in 1999 [12]. $Ihh^{-/-}$ mice showed shortening of the proliferating chondrocyte layer and acceleration of chondrocyte hypertrophy, as seen in the loss of function of PTHrP signaling in mice. More importantly, these phenotypes accompanied a loss of PTHrP expression in the periarticular regions. In contrast, when Hh signaling was activated specifically in chondrocytes by the loss of *Ptch1* in mice, chondrocyte hypertrophy was delayed and PTHrP expression was upregulated in the periarticular regions [13]. These phenotypes partly supported the hypothesis on the interaction between Ihh and PTHrP.

The interaction was genetically determined by Karp *et al.* through the analysis of compound mutant mouse lines [14]. $Ihh^{-/-}$; $PTHrP^{-/-}$ mutant limbs showed abnormalities similar to $Ihh^{-/-}$ mutant limbs; both mutants exhibited premature chondrocyte hypertrophy and the absence of the growth plate, trabecular bone, and bone collar. Karp and colleagues next activated PTHrP signaling in $Ihh^{-/-}$ chondrocytes in order to clarify which functions of Ihh were mediated by PTHrP. The compound mutants inhibited premature chondrocyte hypertrophy, but did not increase the number of mitotically active chondrocytes compared to $Ihh^{-/-}$ mutants. These results suggest that: (1) Ihh is required for both the differentiation and the proliferation of growth plate chondrocytes; (2) PTHrP mediates part of the function of Ihh to regulate the maintenance of a pool of proliferating chondrocytes; and (3) Ihh can promote the proliferation of these cells in a PTHrP-independent manner.

Growth plate phenotypes of chimeric mice carrying $PPR^{-/-}$ cells and WT cells further provided a clue about the interaction between Ihh and PTHrP [15] (Table 1). Embryonic stem (ES) cell lines homozygous for the *PPR*-null mutation ($PPR^{-/-}$ cells) were injected into WT blastocysts to produce the chimera [15]. $PPR^{-/-}$ chondrocytes ectopically became hypertrophic in columnar proliferating regions and expressed Ihh. The ectopic hypertrophy of $PPR^{-/-}$ cells was accompanied by the upregulation of PTHrP expression in periarticular WT cells and an elongation of the growth plate. Chimeric mice carrying $PPR^{-/-}$; $Ihh^{-/-}$ cells and WT cells revealed the role of Ihh in the interaction [16]. Although $PPR^{-/-}$; $Ihh^{-/-}$ chondrocytes still exhibited ectopic hypertrophy in the manner of the $PPR^{-/-}$ chondrocytes, the upregulation of PTHrP expression and the elongation of the growth plate were no longer observed in the chimeras.

Table 1. Comparison of phenotypes between $PPR^{-/-}$/WT chimeric and $PPR^{-/-}$; $Ihh^{-/-}$/WT chimeric.mice.

	$PPR^{-/-}$/WT Chimeras	**$PPR^{-/-}$; $Ihh^{-/-}$/WT Chimeras**
Ectopic hypertrophy of mutant cells	+	+
Upregulation of PTHrP in periarticular WT cells	+	−
Elongation of the growth plate	+	−
Ectopic bone collar formation in the perichondrum	+	−

Thus, the genetic studies introduced so far support the idea that Ihh constitutes a negative feedback loop with PTHrP to maintain the growth plate length (Figure 2): Ihh secreted from prehypertrophic chondrocytes induces the expression of PTHrP in the periarticular regions including periarticular chondrocytes and the perichondrium, possibly in a concentration-dependent manner as morphogens typically do. PTHrP suppresses chondrocyte hypertrophy by acting on prehypertrophic chondrocytes that strongly express PPR. The suppression of hypertrophy keeps chondrocytes proliferating, and the distance between PTHrP-producing cells and Ihh-producing cells increases.

To correct for this change in the distance, a change in the expression of PTHrP is induced. Thus, the feedback loop maintains a pool of proliferating chondrocytes, and thereby maintains a certain length of the growth plate in order to maximize skeletal growth. In this context, direct Ihh input is likely required for PTHrP expression in periarticular proliferating chondrocytes, since the local removal of *Smo* caused little or no expression of PTHrP within the periarticular domain that showed no activation of Hh signaling [17]. There are also evidences for the induction of the PTHrP expression by Ihh in the perichondrium. As noted earlier, the study by Vortkamp *et al.* demonstrated that Ihh overexpression increased PTHrP expression in the periarticular perichondrium [3], suggesting that Ihh acts on the perichondrium to express PTHrP, *i.e.*, the initiation of the feedback loop. In addition, removal of the perichondrium from chicken embryonic tibiotarsi resulted in the expansion of the hypertrophic chondrocyte layer; the addition of parathyroid hormone, which binds to PPR as PTHrP does, cancelled the expansion [18]. These finding suggest PPR-dependent roles of the perichondrium in the regulation of cartilage differentiation. The study by Long *et al.*, together with the one by Vortkamp *et al.*, may further support the importance of the perichondrium as a source of PTHrP in the negative feedback loop.

In addition to its indirect action on chondrocyte differentiation via PTHrP, Ihh has been shown to have direct actions on growth plate chondrocytes. Decreased proliferation of chondrocytes was observed in $Ihh^{-/-}$ mice and chondrocyte-specific $Smo^{-/-}$ mice [12,19], indicating that Ihh input directly promotes chondrocyte propagation in the growth plate as suggested by Karp *et al.* (see earlier and Figure 2) [14]. In addition, two studies have suggested that Ihh can also directly regulate the differentiation of growth plate chondrocytes at multiple steps (periarticular proliferating to columnar proliferating; columnar proliferating to hypertrophic) in a PTHrP-independent manner. Kobayashi *et al.* proposed that Ihh promoted the differentiation of periarticular proliferating chondrocytes into columnar proliferating chondrocytes independently of PTHrP [20,21], while Mak *et al.* showed that chondrocyte-specific removal of Smo delayed chondrocyte hypertrophy without PTHrP in mice [22].

Which Gli factors mediate the biological effects of Ihh on the growth plate chondrocytes? Gli3 is likely to primarily act upon Ihh input in this context. Abnormalities in the proliferation and maturation of chondrocytes in $Ihh^{-/-}$ mice were rescued by removal of *Gli3* on an $Ihh^{-/-}$ background, accompanying the recovery of PTHrP expression in periarticular regions [23,24]. In contrast, no obvious defects in the growth plate were observed in either $Gli1^{-/-}$, $Gli2^{-/-}$ or $Gli1^{-/-}$; $Gli2^{-/-}$ mouse embryos [8,25–27]. Thus, the suppression of Gli3 repressor activity by Ihh plays more dominant roles in the maintenance of the growth plate than the activation of Gli activators by Ihh.

Several studies shed light on postnatal roles of Hh signaling in cartilage and the growth plate. Postnatal deletion of *Ihh* from *Col2a1*-expressing cells caused ectopic chondrocyte hypertrophy, decreased proliferation of chondrocytes, and the disorganized growth plate in mice [28]. When *Ihh* was deleted from *Prrx1*-expressing skeletal progenitors, the growth plate and secondary ossification center were absent at postnatal stages in mice [29]. These data indicate the postnatal requirement of Ihh during endochondral bone development. Maeda *et al.* further showed that Ihh functioned in both PTHrP-dependent and -independent manners in this context [30]. The ectopic chondrocyte hypertrophy led by the deletion of *Ihh* from *Col2a1*-expressing cells was temporally rescued by the forced expression of constitutively active *PPR* in the *Ihh*-deleted population using a *Col2a1* promoter fragment, whereas decreased proliferation of chondrocytes was not corrected in the compound mutant. These data suggest that: (1) Ihh has positive effects on chondrocyte proliferation at the postnatal stage, but the effects are unlikely mediated by PPR; and (2) Ihh and PTHrP may functionally interact at postnatal stages, as they do at embryonic stages. However, given the temporal rescue of growth plate phenotypes in the compound mutant, contributions of the interaction to chondrocyte hypertrophy, if any, are likely to be less at postnatal stages than at embryonic stages; other mechanisms or signaling molecules may also mediate the process. Hirai *et al.* showed that chondrocyte-specific ablation of *PPR* in postnatal mice led to the acceleration of hypertrophy, followed by premature closure of the growth plate, in association with increased chondrocyte apoptosis [31]. They suggest the possibility that the above mechanism may also underlie the growth plate closure caused by the postnatal deletion of *Ihh* from *Col2a1*-expressing cells, because Ihh induces PTHrP.

4. Ihh is Required for Osteoblast Differentiation

Ihh is a master regulator of osteoblast specification during endochondral ossification. In endochondral ossification, osteoblasts first appear in a region of the perichondrium adjacent to pre-hypertrophic and hypertrophic chondrocytes. The following mouse genetic studies have indicated that direct Ihh input from these chondrocytes to progenitors is required for osteoblast specification in the perichondrium and the primary spongiosa. Both bone collars and the expression of Runx2 and Bglap were absent in the perichondrium of $Ihh^{-/-}$ mice [12,23]. Involvement of Ihh and Ihh-producing cells, *i.e.*, pre- and hypertrophic chondrocytes, in bone collar formation in the perichondrium was further supported by phenotypes of $PPR^{-/-}$/WT chimeric mice and $PPR^{-/-}$; $Ihh^{-/-}$/WT chimeric mice [15,16] (Table 1). $PPR^{-/-}$; $Ihh^{-/-}$/WT chimeric mice were generated by the injection of $PPR^{-/-}$; $Ihh^{-/-}$ ES cells into WT blastocysts [16]. In $PPR^{-/-}$/WT chimeras, perichondrial cells adjacent to ectopically hypertrophic $PPR^{-/-}$ cells underwent ectopic calcification. The ectopic calcification was cancelled by the removal of *Ihh* in the ectopically hypertrophic $PPR^{-/-}$ cells, although ectopic hypertrophy was still observed in the $PPR^{-/-}$; $Ihh^{-/-}$ cells with robust expressions of bone morphogenetic proteins (BMP) 2 and 6. Thus, Ihh produced by pre-hypertrophic and hypertrophic chondrocytes is necessary for the differentiation of perichondrial cells into osteoblasts.

When Hh signaling activity was removed by the *Smo* deletion in perichondrial cells, Runx2 expression and bone collar formation were completely absent in the perichondrium [32]. In $Smo^{-/-}$/WT chimeric mice, $Smo^{-/-}$ perichondrial cells in the region where the bone collar is formed under physiological conditions expressed chondrocyte marker genes including type II collagen (Col2a1) and type X collagen (Col10a1), and did not undergo osteoblast differentiation; $Smo^{-/-}$ cells in the primary spongiosa did not contribute to bone formation [32]. Chondrocyte-specific overexpression of *Ihh* or the ablation of *Ptch1* from perichondrial cells, which caused supra-physiological activation of Hh signaling in each population, resulted in the accelerated bone collar formation [13,32].

Closer analyses of $Ihh^{-/-}$ mice by Colnot *et al.* provide clues regarding the mechanism by which Hh signaling regulates the osteoblast differentiation of perichondrial cells at the cellular level [33]. They found that perichondrial cells are thin and disorganized in $Ihh^{-/-}$ mice; the abnormality became evident within 24 h after the physiological onset of Ihh expression at E12.5. In E14 $Ihh^{-/-}$ embryos, *Col1a1* expression was diffuse and not restricted to cells adjacent to *Col2a1*-expressing chondrogenic condensation in the perichondrial region. Given that Runx2 appears at E12.5 in a physiological context, the organization of perichondrial cells may be associated with the cell fate specification of progenitors into Runx2-positive osteoblast precursors.

Thus, Ihh produced by pre- and hypertrophic chondrocytes is essential for osteoblast development in the perichondrium and the primary spongiosa during endochondral ossification (Figure 2); more specifically, Hh signaling is required for the specification of progenitors into Runx2-positive osteoblast precursors. Differentiation of the Runx2-positive osteoblast precursors into Sp7-positive ones is unlikely to require Hh signaling, since the ablation of *Smo* from *Sp7*-expressing cells does not affect osteoblast differentiation [34]. Given that perichondrial cells that cannot receive Hh input acquire chondrocytic phenotypes [32], the signaling may act as a molecular switch for the cell fate specification of the osteo-chondroprogenitor population.

We and others have shown that Hh signaling regulates not only skeletal development, but also postnatal bone homeostasis. Disruption of Ihh in postnatal chondrocytes led to premature vascularization, a delay in the formation of the ossification center, and less trabecular bone [28], suggesting that chondrocyte-derived Ihh was involved in the maintenance of growth plates and trabecular bones in the postnatal stage. $Ptch1^{+/-}$ mice had high bone mass with an acceleration of both bone formation and bone resorption; a similar trend to that in the mouse phenotypes was observed in patients with basal cell carcinoma syndrome (BCCS) caused by an inactivating mutation of one of the *Ptch1* alleles [35]. Although high bone turnover phenotypes were also observed in mice in which *Ptch1* was disrupted in *Bglap*-expressing mature osteoblasts, they showed low bone mass unlike *Ptch1*$^{+/-}$ mutants [36]; it is possible that the indirect effects of Hh signaling on osteoclasts are dominant over its direct effects on osteoblasts, when Hh signaling was activated specifically in mature osteoblasts. In addition, heterozygous deletion of *Gli1* led to decreased bone mass with an uncoupling of bone formation and resorption in mice, suggesting that Gli activators participate in postnatal bone homeostasis upon Hh input [37].

5. Hierarchy between Hh Signaling and Other Osteogenic Factors in Osteoblast Differentiation

Genetic studies may suggest a further conclusion about the hierarchy between Hh signaling and other osteogenic signaling factors such as Wnt and bone morphogenetic proteins (BMPs) during osteoblast differentiation. The differentiation was largely arrested at the Runx2-positive stage, when canonical Wnt signaling activity was removed in the limb bud [38], the mesenchymal condensation [39,40], or the perichondrium [34,40]; the removal of canonical Wnt signaling activity in *Sp7*-expressing osteoblast precursors resulted in a lack of *Bglap*-expressing mature osteoblasts [34]. These data indicate that canonical Wnt signaling is required for the transition of Runx2-positive osteoblast precursors into Sp7-positive ones and for the transition of Sp7-positive precursors into mature osteoblasts, both of which occur later than the phase requiring Hh signaling. Indeed, canonical Wnt signaling was impaired in *Ihh*$^{-/-}$ mice [39], and the activation of Hh signaling in the perichondrium did not rescue the reduction of bone formation in mice defective for canonical Wnt signaling activity [40].

By taking advantage of *ex vivo* organ cultures as well as mouse genetics, we analyzed the hierarchy between Hh and BMP signaling in skeletal development [27]. The activation of BMP signaling accelerated both osteogenesis and chondrogenesis in the perichondrium, but only after Hh-dependent lineage specification into osteoblasts or chondrocytes took place [27]. In the physiological context, osteogenic function of BMP is likely to require Hh-Gli activator-mediated specification of progenitors into osteoblasts in the perichondrium; BMP can act as a chondrogenic factor on the perichondrial cells only when they do not receive Hh input [27].

Thus, Hh signaling acts epistatically on both canonical Wnt signaling and BMP signaling in the osteoblast differentiation cascade, and these pathways sequentially regulate the cascades. Genetic studies so far collectively support the following model for osteoblast differentiation and maturation: Skeletal progenitors are specified into the Runx2-positive osteoblast precursor upon Hh input [12,32]. Canonical Wnt signaling then sequentially induces both the transition of the Runx2-positive osteoblast precursor into the Sp7-positive one [38–40] and the generation of Bglap (bone gla protein)-positive, bone-forming osteoblasts from the Sp7-positive precursor [34]. BMP signaling is likely to accelerate a series of these processes after the Hh-mediated specification into an osteoblast lineage [27]. Mutant mice where *Bmp2* and *Bmp4* were deleted in *Prrx1*-expressing skeletal progenitors showed comparable expression of Runx2 to WT mice [41], but lost Sp7 expression at 3 weeks of age, suggesting that BMP signaling may be involved in the transition of the Runx2-positive osteoblast precursor into the Sp7-positive one, which occurs later than the phase requiring Hh signaling. Given that decapentaplegic (dpp), a homolog of BMP has been shown to act downstream of hedgehog in Drosophila (reviewed in [1]), one may think that BMP signaling mediates the action of Ihh in mammalian osteoblast development. BMP2 and BMP6 are expressed in hypertrophic chondrocytes as Ihh is in developing endochondral bones [16]. When the ectopic calcification in the perichondrium was cancelled by the removal of *Ihh* in the ectopically hypertrophic *PPR*$^{-/-}$ cells as mentioned earlier (Table 1), BMP2 and BMP6 were still expressed in the ectopically hypertrophic *PPR*$^{-/-}$ cells [16]. This result provides a genetic support for that BMP signaling is unlikely to act downstream of Hh signaling in mammalian osteoblast development.

6. Molecules Acting Downstream of Hh Input in Osteoblast Differentiation

Regarding molecules mediating the biological actions of Hh signaling on osteoblast differentiation, it seems reasonable to imagine that Gli and Runx2 are major mediators of these actions, given that Gli acts downstream of Hh signaling and that Runx2 expression is lost in the perichondrium of mice lacking Hh signaling.

A series of studies indicated that Gli1, Gli2, and Gli3 are involved in Hh-mediated osteoblast differentiation. Shimoyama *et al.* showed that Ihh promoted osteoblast differentiation in a Gli2-dependent manner; and Ihh treatment or Gli2 overexpression induced the expression of Runx2, which physically interacted with Gli2 [42]. In line with this, the skeletal phenotypes of *Ihh*$^{-/-}$ embryos were completely rescued in *Ihh*$^{-/-}$; *Gli3*$^{-/-}$; *C2-NGli2* embryos, in which *NGli2* (an N-terminally truncated, constitutively active form of Gli2) was exogenously expressed in *Col2a1*-expressing cells [43]. On the basis of these findings, Joeng and Long proposed that the Gli2 activator and the Gli3 repressor collectively mediated all major aspects of Ihh function in endochondral ossification.

We recently found that bone formation was impaired in *Gli1*$^{-/-}$ mice; *Gli1*$^{-/-}$ perichondrial cells expressed Col2a1 and Col10a1, but not Runx2 or Sp7. *Gli1*$^{-/-}$; *Gli2*$^{-/-}$ mice showed more severe skeletal phenotypes than either *Gli1*$^{-/-}$ or *Gli2*$^{-/-}$ mice, and osteoblast differentiation was impaired in *Gli1*$^{-/-}$; *Gli3*$^{-/-}$ perichondrial cells compared to that in *Gli3*$^{-/-}$ cells *in vitro* [26]. Indeed, Gli1 activated the transcription of early and middle marker genes for osteoblasts by directly binding to 5′ regulatory regions of the genes [26], and it interfered with the Sox9-mediated transactivation of chondrocyte marker genes by suppressing the DNA binding of Sox9 [27]. We also found that the Gli3 repressor suppressed the Runx2-mediated transcription of osteoblastic genes by antagonizing the DNA binding of Runx2 [35]. These data suggest that Gli1 functions collectively with Gli2 and Gli3 in osteoblast formation, particularly in the specification of progenitors into osteoblasts.

Regarding the involvement of Runx2 in the action of Hh, the recovery of Runx2 expression in the *Ihh*$^{-/-}$ perichondrium did not cancel the abnormality of osteoblast differentiation in *Ihh*$^{-/-}$ mice [44], suggesting that Hh signaling used effectors other than *Runx2* to promote osteoblast differentiation. Gli transcription factors are possible candidates for these effectors.

Recently, Regard *et al.* found that an abnormal activation of Hh signaling underlay heterotopic ossification, which is caused by the inactivating mutation of *Gnas* [45]. GNAS encodes Gαs, a transducer of the G protein-coupled receptor (GPCR); Gαs was suggested to act downstream of Smo and upstream of Gli, and the Gαs-cAMP-PKA axis suppressed Hh signaling activities by regulating Gli activation and processing in this context. Taken together, these facts suggested that Gαs negatively modulated the Hh signaling activity [45].

7. Conclusions

As summarized in this review, over the last two decades mouse genetic approaches have unraveled the functions of Hh signaling and their underlying molecular mechanisms in endochondral ossification. Ihh regulates chondrocyte differentiation by constituting a negative feedback loop with PTHrP, which is a central system in the proper growth of endochondral skeletons; Ihh plays a prerequisite role in osteoblast formation. However, the gene regulatory networks underlying these actions are still poorly understood. In particular, what genes are regulated by Gli factors and translate Hh input into the biological actions on skeleton? What transcription factors engage the Gli-mediated network during skeletal development? A combinatorial approach using genome-wide analyses and mouse genetics will facilitate understanding of these regulatory mechanisms, providing genome-level evidence of the potential of Hh signaling for the treatment of skeletal disorders.

Acknowledgments: The author thanks Ung-il Chung and the members of the Chung-Ohba Laboratory for their helpful discussions, and N. Nagumo for administrative assistance. This work is supported by the Grant-in-Aid for Young Scientists (A) (26713054) from Japan Society for the Promotion of Science.

Author Contributions: S.O. conceived, designed, and wrote the paper.

Conflicts of Interest: The author has no conflict of interest.

References

1. Ingham, P.W.; McMahon, A.P. Hedgehog signaling in animal development: Paradigms and principles. *Genes Dev.* **2001**, *15*, 3059–3087. [CrossRef] [PubMed]
2. Lanske, B.; Karaplis, A.C.; Lee, K.; Luz, A.; Vortkamp, A.; Pirro, A.; Karperien, M.; Defize, L.H.; Ho, C.; Mulligan, R.C.; *et al.* PTH/PTHrP receptor in early development and Indian hedgehog-regulated bone growth. *Science* **1996**, *273*, 663–666. [CrossRef] [PubMed]
3. Vortkamp, A.; Lee, K.; Lanske, B.; Segre, G.V.; Kronenberg, H.M.; Tabin, C.J. Regulation of rate of cartilage differentiation by Indian hedgehog and PTH-related protein. *Science* **1996**, *273*, 613–622. [CrossRef] [PubMed]
4. Lee, K.; Deeds, J.D.; Segre, G.V. Expression of parathyroid hormone-related peptide and its receptor messenger ribonucleic acids during fetal development of rats. *Endocrinology* **1995**, *136*, 453–463. [PubMed]
5. Lee, K.; Lanske, B.; Karaplis, A.C.; Deeds, J.D.; Kohno, H.; Nissenson, R.A.; Kronenberg, H.M.; Segre, G.V. Parathyroid hormone-related peptide delays terminal differentiation of chondrocytes during endochondral bone development. *Endocrinology* **1996**, *137*, 5109–5118. [PubMed]
6. Amizuka, N.; Warshawsky, H.; Henderson, J.E.; Goltzman, D.; Karaplis, A.C. Parathyroid hormone-related peptide-depleted mice show abnormal epiphyseal cartilage development and altered endochondral bone formation. *J. Cell Biol.* **1994**, *126*, 1611–1623. [CrossRef] [PubMed]
7. Karaplis, A.C.; Luz, A.; Glowacki, J.; Bronson, R.T.; Tybulewicz, V.L.; Kronenberg, H.M.; Mulligan, R.C. Lethal skeletal dysplasia from targeted disruption of the parathyroid hormone-related peptide gene. *Genes Dev.* **1994**, *8*, 277–289. [CrossRef] [PubMed]
8. Miao, D.; Liu, H.; Plut, P.; Niu, M.; Huo, R.; Goltzman, D.; Henderson, J.E. Impaired endochondral bone development and osteopenia in Gli2-deficient mice. *Exp. Cell. Res.* **2004**, *294*, 210–222. [CrossRef] [PubMed]
9. Weir, E.C.; Philbrick, W.M.; Amling, M.; Neff, L.A.; Baron, R.; Broadus, A.E. Targeted overexpression of parathyroid hormone-related peptide in chondrocytes causes chondrodysplasia and delayed endochondral bone formation. *Proc. Natl. Acad. Sci. USA* **1996**, *93*, 10240–10245. [CrossRef] [PubMed]
10. Schipani, E.; Lanske, B.; Hunzelman, J.; Luz, A.; Kovacs, C.S.; Lee, K.; Pirro, A.; Kronenberg, H.M.; Juppner, H. Targeted expression of constitutively active receptors for parathyroid hormone and parathyroid hormone-related peptide delays endochondral bone formation and rescues mice that lack parathyroid hormone-related peptide. *Proc. Natl. Acad. Sci. USA* **1997**, *94*, 13689–13694. [CrossRef] [PubMed]
11. Philbrick, W.M.; Dreyer, B.E.; Nakchbandi, I.A.; Karaplis, A.C. Parathyroid hormone-related protein is required for tooth eruption. *Proc. Natl. Acad. Sci. USA* **1998**, *95*, 11846–11851. [CrossRef] [PubMed]
12. St-Jacques, B.; Hammerschmidt, M.; McMahon, A.P. Indian hedgehog signaling regulates proliferation and differentiation of chondrocytes and is essential for bone formation. *Genes Dev.* **1999**, *13*, 2072–2086. [CrossRef] [PubMed]
13. Mak, K.K.; Chen, M.H.; Day, T.F.; Chuang, P.T.; Yang, Y. Wnt/beta-catenin signaling interacts differentially with Ihh signaling in controlling endochondral bone and synovial joint formation. *Development* **2006**, *133*, 3695–3707. [CrossRef] [PubMed]
14. Karp, S.J.; Schipani, E.; St-Jacques, B.; Hunzelman, J.; Kronenberg, H.; McMahon, A.P. Indian hedgehog coordinates endochondral bone growth and morphogenesis via parathyroid hormone related-protein-dependent and -independent pathways. *Development* **2000**, *127*, 543–548. [PubMed]
15. Chung, U.I.; Lanske, B.; Lee, K.; Li, E.; Kronenberg, H. The parathyroid hormone/parathyroid hormone-related peptide receptor coordinates endochondral bone development by directly controlling chondrocyte differentiation. *Proc. Natl. Acad. Sci. USA* **1998**, *95*, 13030–13035. [CrossRef] [PubMed]
16. Chung, U.I.; Schipani, E.; McMahon, A.P.; Kronenberg, H.M. Indian hedgehog couples chondrogenesis to osteogenesis in endochondral bone development. *J. Clin. Invest.* **2001**, *107*, 295–304. [CrossRef] [PubMed]
17. Hilton, M.J.; Tu, X.; Long, F. Tamoxifen-inducible gene deletion reveals a distinct cell type associated with trabecular bone, and direct regulation of PTHrP expression and chondrocyte morphology by Ihh in growth region cartilage. *Dev. Biol.* **2007**, *308*, 93–105. [CrossRef] [PubMed]
18. Long, F.; Linsenmayer, T.F. Regulation of growth region cartilage proliferation and differentiation by perichondrium. *Development* **1998**, *125*, 1067–1073. [PubMed]

19. Long, F.; Zhang, X.M.; Karp, S.; Yang, Y.; McMahon, A.P. Genetic manipulation of hedgehog signaling in the endochondral skeleton reveals a direct role in the regulation of chondrocyte proliferation. *Development* **2001**, *128*, 5099–5108. [PubMed]
20. Kobayashi, T.; Chung, U.I.; Schipani, E.; Starbuck, M.; Karsenty, G.; Katagiri, T.; Goad, D.L.; Lanske, B.; Kronenberg, H.M. PTHrP and Indian hedgehog control differentiation of growth plate chondrocytes at multiple steps. *Development* **2002**, *129*, 2977–2986. [PubMed]
21. Kobayashi, T.; Soegiarto, D.W.; Yang, Y.; Lanske, B.; Schipani, E.; McMahon, A.P.; Kronenberg, H.M. Indian hedgehog stimulates periarticular chondrocyte differentiation to regulate growth plate length independently of PTHrP. *J. Clin. Invest.* **2005**, *115*, 1734–1742. [CrossRef] [PubMed]
22. Mak, K.K.; Kronenberg, H.M.; Chuang, P.T.; Mackem, S.; Yang, Y. Indian hedgehog signals independently of PTHrP to promote chondrocyte hypertrophy. *Development* **2008**, *135*, 1947–1956. [CrossRef] [PubMed]
23. Hilton, M.J.; Tu, X.; Cook, J.; Hu, H.; Long, F. Ihh controls cartilage development by antagonizing Gli3, but requires additional effectors to regulate osteoblast and vascular development. *Development* **2005**, *132*, 4339–4351. [CrossRef] [PubMed]
24. Koziel, L.; Wuelling, M.; Schneider, S.; Vortkamp, A. Gli3 acts as a repressor downstream of Ihh in regulating two distinct steps of chondrocyte differentiation. *Development* **2005**, *132*, 5249–5260. [CrossRef] [PubMed]
25. Park, H.L.; Bai, C.; Platt, K.A.; Matise, M.P.; Beeghly, A.; Hui, C.C.; Nakashima, M.; Joyner, A.L. Mouse Gli1 mutants are viable but have defects in shh signaling in combination with a Gli2 mutation. *Development* **2000**, *127*, 1593–1605. [PubMed]
26. Hojo, H.; Ohba, S.; Yano, F.; Saito, T.; Ikeda, T.; Nakajima, K.; Komiyama, Y.; Nakagata, N.; Suzuki, K.; Takato, T.; *et al.* Gli1 protein participates in hedgehog-mediated specification of osteoblast lineage during endochondral ossification. *J. Biol. Chem.* **2012**, *287*, 17860–17869. [CrossRef] [PubMed]
27. Hojo, H.; Ohba, S.; Taniguchi, K.; Shirai, M.; Yano, F.; Saito, T.; Ikeda, T.; Nakajima, K.; Komiyama, Y.; Nakagata, N.; *et al.* Hedgehog-Gli activators direct osteo-chondrogenic function of bone morphogenetic protein toward osteogenesis in the perichondrium. *J. Biol. Chem.* **2013**, *288*, 9924–9932. [CrossRef] [PubMed]
28. Maeda, Y.; Nakamura, E.; Nguyen, M.T.; Suva, L.J.; Swain, F.L.; Razzaque, M.S.; Mackem, S.; Lanske, B. Indian hedgehog produced by postnatal chondrocytes is essential for maintaining a growth plate and trabecular bone. *Proc. Natl. Acad. Sci. USA* **2007**, *104*, 6382–6387. [CrossRef] [PubMed]
29. Amano, K.; Densmore, M.J.; Lanske, B. Conditional deletion of Indian hedgehog in limb mesenchyme results in complete loss of growth plate formation but allows mature osteoblast differentiation. *J. Bone Miner. Res.* **2015**, *30*, 2262–2272. [CrossRef] [PubMed]
30. Maeda, Y.; Schipani, E.; Densmore, M.J.; Lanske, B. Partial rescue of postnatal growth plate abnormalities in Ihh mutants by expression of a constitutively active PTH/PTHrP receptor. *Bone* **2010**, *46*, 472–478. [CrossRef] [PubMed]
31. Hirai, T.; Chagin, A.S.; Kobayashi, T.; Mackem, S.; Kronenberg, H.M. Parathyroid hormone/parathyroid hormone-related protein receptor signaling is required for maintenance of the growth plate in postnatal life. *Proc. Natl. Acad. Sci. USA* **2011**, *108*, 191–196. [CrossRef] [PubMed]
32. Long, F.; Chung, U.I.; Ohba, S.; McMahon, J.; Kronenberg, H.M.; McMahon, A.P. Ihh signaling is directly required for the osteoblast lineage in the endochondral skeleton. *Development* **2004**, *131*, 1309–1318. [CrossRef] [PubMed]
33. Colnot, C.; de la Fuente, L.; Huang, S.; Hu, D.; Lu, C.Y.; St-Jacques, B.; Helms, J.A. Indian hedgehog synchronizes skeletal angiogenesis and perichondrial maturation with cartilage development. *Development* **2005**, *132*, 1057–1067. [CrossRef] [PubMed]
34. Rodda, S.J.; McMahon, A.P. Distinct roles for hedgehog and canonical Wnt signaling in specification, differentiation and maintenance of osteoblast progenitors. *Development* **2006**, *133*, 3231–3244. [CrossRef] [PubMed]
35. Ohba, S.; Kawaguchi, H.; Kugimiya, F.; Ogasawara, T.; Kawamura, N.; Saito, T.; Ikeda, T.; Fujii, K.; Miyajima, T.; Kuramochi, A.; *et al.* Patched1 haploinsufficiency increases adult bone mass and modulates Gli3 repressor activity. *Dev. Cell* **2008**, *14*, 689–699. [CrossRef] [PubMed]
36. Mak, K.K.; Bi, Y.; Wan, C.; Chuang, P.T.; Clemens, T.; Young, M.; Yang, Y. Hedgehog signaling in mature osteoblasts regulates bone formation and resorption by controlling PTHrP and RANKL expression. *Dev. Cell* **2008**, *14*, 674–688. [CrossRef] [PubMed]

37. Kitaura, Y.; Hojo, H.; Komiyama, Y.; Takato, T.; Chung, U.I.; Ohba, S. Gli1 haploinsufficiency leads to decreased bone mass with an uncoupling of bone metabolism in adult mice. *PLoS ONE* **2014**, *9*, e109597. [CrossRef]
38. Hill, T.P.; Spater, D.; Taketo, M.M.; Birchmeier, W.; Hartmann, C. Canonical Wnt/beta-catenin signaling prevents osteoblasts from differentiating into chondrocytes. *Dev. Cell* **2005**, *8*, 727–738. [CrossRef] [PubMed]
39. Hu, H.; Hilton, M.J.; Tu, X.; Yu, K.; Ornitz, D.M.; Long, F. Sequential roles of hedgehog and Wnt signaling in osteoblast development. *Development* **2005**, *132*, 49–60. [CrossRef] [PubMed]
40. Day, T.F.; Guo, X.; Garrett-Beal, L.; Yang, Y. Wnt/beta-catenin signaling in mesenchymal progenitors controls osteoblast and chondrocyte differentiation during vertebrate skeletogenesis. *Dev. Cell* **2005**, *8*, 739–750. [CrossRef] [PubMed]
41. Bandyopadhyay, A.; Tsuji, K.; Cox, K.; Harfe, B.D.; Rosen, V.; Tabin, C.J. Genetic analysis of the roles of BMP2, BMP4, and BMP7 in limb patterning and skeletogenesis. *PLoS Genet.* **2006**, *2*, e216. [CrossRef] [PubMed]
42. Shimoyama, A.; Wada, M.; Ikeda, F.; Hata, K.; Matsubara, T.; Nifuji, A.; Noda, M.; Amano, K.; Yamaguchi, A.; Nishimura, R.; *et al.* Ihh/Gli2 signaling promotes osteoblast differentiation by regulating *Runx2* expression and function. *Mol. Biol. Cell* **2007**, *18*, 2411–2418. [CrossRef] [PubMed]
43. Joeng, K.S.; Long, F.X. The Gli2 transcriptional activator is a crucial effector for Ihh signaling in osteoblast development and cartilage vascularization. *Development* **2009**, *136*, 4177–4185. [CrossRef] [PubMed]
44. Tu, X.; Joeng, K.S.; Long, F. Indian hedgehog requires additional effectors besides *Runx2* to induce osteoblast differentiation. *Dev. Biol.* **2012**, *362*, 76–82. [CrossRef] [PubMed]
45. Regard, J.B.; Malhotra, D.; Gvozdenovic-Jeremic, J.; Josey, M.; Chen, M.; Weinstein, L.S.; Lu, J.; Shore, E.M.; Kaplan, F.S.; Yang, Y. Activation of hedgehog signaling by loss of gnas causes heterotopic ossification. *Nat. Med.* **2013**, *19*, 1505–1512. [CrossRef] [PubMed]

3

Control of Hedgehog Signalling by the Cilia-Regulated Proteasome

Christoph Gerhardt, Antonia Wiegering, Tristan Leu and Ulrich Rüther *

Institute for Animal Developmental and Molecular Biology, Heinrich-Heine University Düsseldorf, 40225 Düsseldorf, Germany; christoph.gerhardt@hhu.de (C.G.); antonia.wiegering@hhu.de (A.W.); tristan.leu@hhu.de (T.L.)

* Correspondence: ruether@hhu.de

Academic Editors: Henk Roelink and Simon J. Conway

Abstract: The Hedgehog signalling pathway is evolutionarily highly conserved and essential for embryonic development of invertebrates and vertebrates. Consequently, impaired Hedgehog signalling results in very severe human diseases, ranging from holoprosencephaly to Pallister-Hall syndrome. Due to this great importance for human health, the focus of numerous research groups is placed on the investigation of the detailed mechanisms underlying Hedgehog signalling. Today, it is known that tiny cell protrusions, known as primary cilia, are necessary to mediate Hedgehog signalling in vertebrates. Although the Hedgehog pathway is one of the best studied signalling pathways, many questions remain. One of these questions is: How do primary cilia control Hedgehog signalling in vertebrates? Recently, it was shown that primary cilia regulate a special kind of proteasome which is essential for proper Hedgehog signalling. This review article will cover this novel cilia-proteasome association in embryonic Hedgehog signalling and discuss the possibilities provided by future investigations on this topic.

Keywords: ciliary proteasome; proteolytic processing; GLI2; GLI3; cilia; signalling; RPGRIP1L; mouse; *Drosophila*; SHH

---

1. Introduction

In the year 1980, Christiane Nüsslein-Volhard and Eric Frank Wieschaus reported the identification of 15 loci whose mutations affect the development of the fruit fly *Drosophila melanogaster* larvae [1]. In the course of these investigations, the *hedgehog* (*hh*) gene was also discovered. Soon, it was evident that this gene is not only essential for the development of invertebrates, like *Drosophila*, but also for the development of vertebrates [2–5]. The product of this gene serves as a ligand for an evolutionarily highly-conserved signalling pathway which was termed hedgehog (HH) signalling pathway. Important components of the vertebrate HH signal transduction are the glioblastoma (GLI) proteins—GLI1, GLI2, and GLI3. While GLI1 is a target gene of GLI2-A and GLI3-A, and exclusively acts as a constitutive transcriptional activator, itself [6,7], GLI2 and GLI3 can function as activators or repressors [8], revealing the complex nature of regulating GLI processing. The conversion of the full-length GLI2 and GLI3 proteins into their truncated repressor forms is called proteolytic processing. This processing event is realised by the ubiquitin-proteasome system (UPS).

The UPS is the major protein degradation system in eukaryotes, important for up to 80%–90% of all eukaryotic proteins, revealing its importance [9]. It consists of, in simplified terms, ubiquitin-activating enzymes (E1), ubiquitin-conjugation enzymes (E2), ubiquitin ligases (E3), and the proteasome. Proteins designated to get degraded become phosphorylated and subsequently ubiquitinated. The polyubiquitin conjugation is realised by the cooperative action of E1, E2, and E3. The proteasome is the catalytic component of the ubiquitin-proteasome system consisting of two regulatory 19S subunits and a catalytic 20S subunit. Ubiquitinated proteins are recognised by the 19S regulatory subunits and subsequently degraded by the 20S subunit which contains multiple peptidases [10]. However, not every proteasomal action results in protein degradation. A few proteins get proteolytically processed from their full-length form to a shorter repressor form.

As the UPS is not only important for protein processing, but also for protein degradation, it can influence proteins in different manners. Interestingly, this is also true for HH signalling. For example, GLI2 and GLI3 are proteolytically processed [11], while GLI1 is not processed, but degraded, by the UPS [12–14]. Thus, the GLI proteins serve as a very interesting model to analyse the molecular mechanisms underlying the differences between protein degradation and protein processing. While the subcellular localisation where GLI1 gets phosphorylated is unknown, the phosphorylation process of GLI2 and GLI3 is thought to take place at the basal body (BB) of cilia [15]. In the processing event of GLI2 and GLI3, ubiquitination takes place immediately after phosphorylation. However, it is still unknown if the polyubiquitination of GLI2 and GLI3 occurs at the BB. Additionally, the ubiquitin E3 ligase JADE-1, which is known to be involved in the ubiquitination of β-catenin [16], has been shown to be localised at the BB [17]. Furthermore, the key ubiquitin E3 ligase, APC, was found to be located at the BB [18] supporting the hypothesis that both, phosphorylation and polyubiquitination processes take place at the BB.

Apart of differences in GLI proteins of being degraded or processed, also differences in efficiency of processing are noticeable. For example, GLI2 is not as efficiently processed as GLI3 [19]. Previous studies, therefore, tried to examine the differences between these proteins in order to understand the reason leading to this variation in processing efficiency. Pan et al. hypothesised that differential GLI2 and GLI3 processing is controlled by the processing determinant domain (PDD) [20] (Figure 1). Accordingly, they could show that replacement of GLI2-PDD by GLI3-PDD led to a more efficient processing of GLI2 [21]. Moreover, Schrader et al. examined the difference between GLI1 and GLI3. They found a three-part signal mechanism which is dependent on a zinc-finger domain, the linker sequence, and the degron, which is essential for GLI3 processing [22] (Figure 1). The first processing signal is the zinc-finger domain of GLI3. However, this domain is not the crucial difference to distinguish proteins from being either processed or degraded, because GLI1 also includes a similar zinc-finger domain. The second processing signal is the linker sequence which is located between the zinc-finger domain and the lysines of the degron sequence. It is proposed that the proteasome binds to the linker area which is assumed to be a proteasome initiation region. Finally the degron is the third processing signal and the starting point of proteasomal processing. The linker sequence in combination with the degron is missing in GLI1 and, hence, seems to be the critical difference leading to processing of GLI3, and not of GLI1 [22]. Consequently, the sequence of the respective protein determines whether it becomes degraded or processed, while the proteasome is not the factor which distinguishes degradation from processing.

In vertebrates, HH signalling is mediated by cytoplasmic protrusions known as primary cilia. Primary cilia have a length of 1–15 μm depending on the cell type and basically consist of three different compartments: the BB, the axoneme and the transition zone (TZ). The BB is a modified centrosome from which the ciliary microtubule-based framework (axoneme) grows out. The axoneme, which is comprised of nine doublet microtubules arranged in a circle, gives stability to the cilium and allows protein transport inside the cilium. The region in between the BB and the axoneme is called TZ. The TZ acts as a selective barrier controlling ciliary import and export of proteins. Ciliary assembly and maintenance relies on the bidirectional microtubule-based intraflagellar transport (IFT) that allows the permanent delivery of axonemal precursors to the assembly site at the ciliary tip [23]. On the basis of their structure, primary cilia function as sensory antennae at the surface of cells by concentrating multiple receptors. These receptors then, in turn, mediate the transduction of many signalling cascades including the HH signalling cascade [24,25].

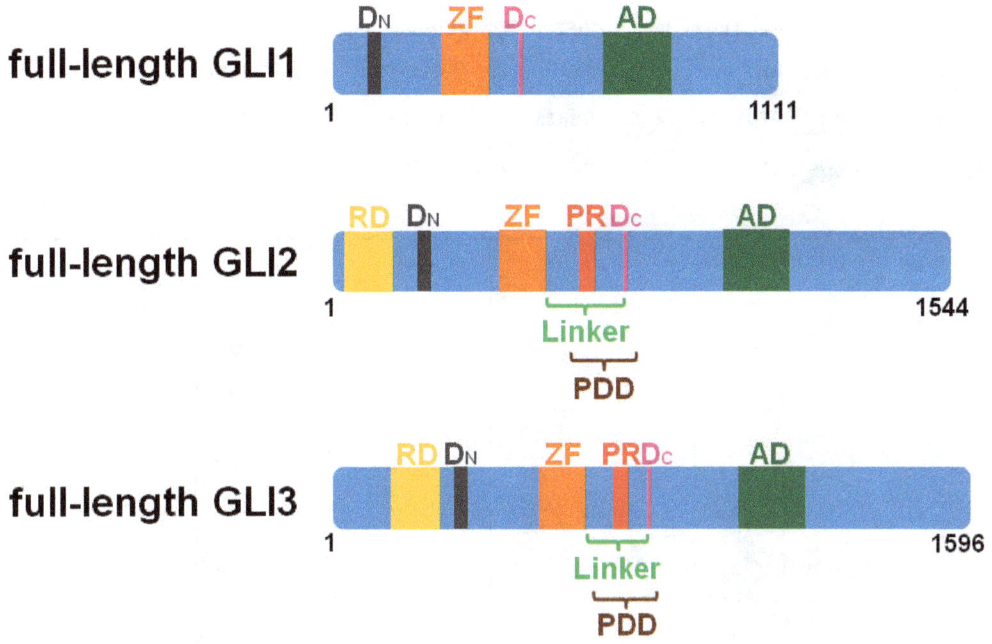

Figure 1. Schematic illustration of domains in murine GLI1, GLI2, and GLI3. GLI1, GLI2, and GLI3 bind to DNA via their zinc-finger domains (ZF). While all three proteins contain an activator domain (AD), only GLI2 and GLI3 display a repressor domain (RD), a processing determinant domain (PDD) and a processing region (PR). The N-terminal degron (D_N) plays an important role in the degradation of all GLI proteins. The C-terminal degron (D_C) acts as a degradation signal in GLI1 and as a processing signal in GLI2 and GLI3.

In summary, a simplified model of HH signal transduction at the primary cilium is conducted by different proteins (also see Figure 2): The twelve-pass transmembrane receptor PATCHED1 (PTC1) is located in the ciliary membrane of vertebrates. After being bound by its ligand HH, the HH/PTC1 complex leaves the cilium. Subsequently, the seven-pass transmembrane protein SMOOTHENED (SMO) accumulates in the ciliary membrane and converts full-length GLI2 and GLI3 (GLI2-FL and GLI3-FL) into transcriptional activators (GLI2-A and GLI3-A) most likely by modifying them [26,27]. Afterwards, GLI2-A and GLI3-A induce the expression of HH target genes (e.g., GLI1). Without HH, PTC1 remains within the ciliary membrane and SMO is not allowed to enter the cilium. In this case, the full-length GLI2 and GLI3 proteins are proteolytically processed into transcriptional repressors (GLI2-R and GLI3-R) [11], demonstrating that GLI2 and GLI3 processing depends on cilia [28]. In detail, the proteolytic processing event of GLI2 and GLI3 initiates with the phosphorylation of both proteins by protein kinase A (PKA) at the BB [15]. In addition to PKA, casein kinase 1 (CK1) and glycogen synthase kinase 3 (GSK3) are involved in the phosphorylation of GLI2 and GLI3 [19,29]. Although GSK3 was detected at the BB [30], there is no evidence yet whether CK1 and GSK3 phosphorylate GLI2 and GLI3 at the base of cilia. An additional player in regulating GLI processing and activity is the kinesin family member 7 (KIF7) protein which localises along the entire cilium [31–35]. However, the mechanism how KIF7 affects GLI processing is only poorly understood.

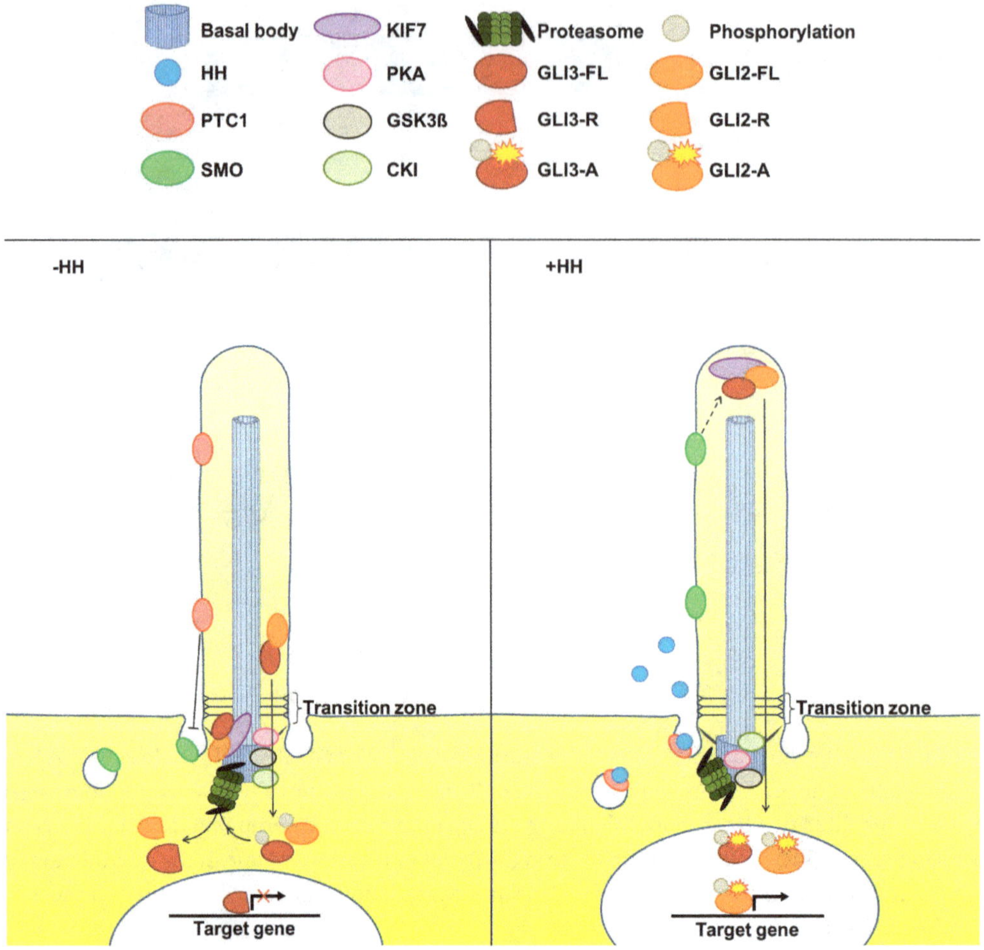

Figure 2. Simplified scheme of the cilia-mediated HH signal transduction pathway in vertebrates. Without HH, PTC1 is located in the ciliary membrane and prevents the ciliary entry of SMO. As a consequence, the full-length proteins of GLI2 and GLI3, which are bound to the ciliary tip organizer KIF7, are phosphorylated by PKA, CK1, and GSK3, and finally proteolytically processed by the cilia-regulated proteasome. KIF7 is essential for this processing event but the mechanism by which KIF7 controls GLI processing remains elusive [36]. The products of GLI2 and GLI3 processing are their repressor forms (GLI2-R and GLI3-R) which enter the nucleus and block HH target gene expression. The repressor form of GLI3 is, thereby, predominant. In the presence of HH, the HH ligand binds to PTC1 and, in turn, the HH/PTC1 complex leaves the cilium allowing ciliary entry of SMO. By a poorly understood mechanism, SMO causes the conversion of the full-length GLI2 and GLI3 proteins (GLI2-FL and GLI3-FL) into GLI2 and GLI3 activator forms (GLI2-A and GLI3-A), which induce HH target gene expression.

2. The Cilia-Regulated Proteasome Is Essential for GLI Processing

In principle, the mechanism of GLI3 processing is well understood, but the subcellular localisation of GLI3 processing is still unknown. As described above, the vertebrate HH signalling machinery tightly depends on primary cilia and all key components of the HH signalling pathway (including GLI2 and GLI3) display a ciliary localisation. In vertebrate cells, proteasomes exist almost ubiquitously within the cytoplasm and the nucleus [37]. Recently, it was shown that proteasomal components localise at the primary cilium. All analysed components of the 19S proteasomal subunit were detected at the BB, while Psma5, a component of the 20S proteasomal subunit, was located along the entire cilium [38]. These localisation studies and the finding that PKA acts at the BB of cilia [15] suggest that the processing of GLI2 and GLI3 takes place at the BB. As the existence of a centrosome-associated proteasome was already shown before [39,40], the question arises whether the cilium is important for the function of the BB-associated proteasome or whether the centrosome-associated proteasome and the BB-associated proteasome are one and the same. In 2003, it was suggested that proteasomes might exert different functions depending on their subcellular localisations and that these differences might be governed by their association and interaction with specific regulatory proteins [41]. Indeed, in silico studies using a systematic network-based approach to work out the "cilia/centrosome complex interactome (CCCI)" showed that the greatest community of the CCCI consists of proteasomal components [42] indicating that the association of ciliary proteins and the proteasome might be of particular significance. Accordingly, the ciliary proteins Bardet-Biedl syndrome (BBS) 1, BBS2, BBS4, BBS6, BBS7, BBS8, inversin (INVS; also known as NPHP2), IQ motif-containing protein B1 (IQCB1; also known as NPHP5), oral-facial-digital syndrome 1 (OFD1), and retinitis pigmentosa gtpase regulator interacting protein 1-like (RPGRIP1L; also known as FTM, NPHP8, or MKS5) interact directly with different proteasomal components [38,43–45] (Figure 3). In case of BBS4, Gerdes et al. demonstrated for the first time that a ciliary protein is involved in the proteasomal regulation [46]. It is now known that the loss of BBS4, BBS7, OFD1, or RPGRIP1L leads to a reduced proteasomal activity thereby impairing intercellular signalling pathways [38,43,46,47]. Interestingly, the absence of the proteins BBS4, BBS7, OFD1, or RPGRIP1L result in a severe ciliopathy phenotype in mouse embryos, being associated with multiple defects and increased lethality. For instance, *Bbs4*-negative embryos exhibit defects in brain, eye, kidney, and liver development and become obese [48–52], while *Bbs7*$^{-/-}$ embryos suffer from brain, eye, and sperm defects, as well as from getting obese [53]. *Ofd1*-deficient embryos show defects in brain, limb, lung, heart, and kidney development [54,55], and *Rpgrip1l*-negative murine embryos display brain, eye, limb, lung, heart, kidney, and liver defects [56–58]. Analysing effects on lethality, *Bbs7*$^{-/-}$ mice are viable [53], while *Bbs4*$^{-/-}$ mice display a higher perinatal lethality [49] and *Rpgrip1l*$^{-/-}$, as well as *Ofd1*$^{-/-}$, mice are not viable at all [54–56].

The proteasome is essential for the development and function of numerous organs and structures of the human body [59–69]. Thus, reduced activity of the cilia-regulated proteasome is a possible cause of ciliopathies. Further evidence highlighting the importance of cilia-regulated proteasomes arises from rescue experiments in vivo. The injection of proteasomal component mRNA or treatment with the proteasome activators mevalonolactone or sulforaphane restored defective convergent extension and somatic definition in zebrafish embryos treated with *bbs4* or *ofd1* morpholinos [43]. However, it would be presumptuous to assume that ciliopathies are exclusively caused by the reduction of proteasomal activity. For example, ciliary length defects can occur independently of the reduced proteasomal activity in *Rpgrip1l*-negative murine embryos [38]. Potentially, the decreased proteasomal activity enhances the severity of a ciliopathy. To test this hypothesis, it would be necessary to quantify proteasomal activity in mouse embryos suffering from milder ciliopathies as for example *Nphp1*-, *Nphp4*-, or *Nek8*-mutant embryos which exclusively display eye or kidney defects [70–72]. Moreover, future studies are indispensable to answer the question whether the data about the cilia-regulated proteasome of different model organisms is translatable to humans.

Remarkably, ciliary proteins seem to use different mechanisms with which they regulate proteasomal activity. In the absence of BBS4, BBS7 and OFD1, the reduced proteasomal activity is based on a decreased amount of different proteasomal components revealing that these proteins control the composition of the proteasome [43]. On the contrary, RPGRIP1L deficiency leads to an accumulation of 19S and 20S proteasomal subunit components at the ciliary base [38] indicating that RPGRIP1L regulates proteasomal activity differently. Since RPGRIP1L directs proteasomal activity by the interaction with Psmd2, a component of the 19S proteasomal subunit, it was hypothesised that RPGRIP1L changes the conformation of Psmd2 in order to control proteasomal activity [38]. While analysing the relationship between RPGRIP1L and the proteasome, a novel kind of proteasome was found. Loss of RPGRIP1L provoked a decreased proteasomal activity exclusively at the base of cilia [38]. Consequently, the proteasome located at the base of primary cilia seems to be differently regulated from all proteasomes that are present at other subcellular localisations. Importantly, this kind of proteasome which was termed "ciliary proteasome" is essential for GLI3 processing [38].

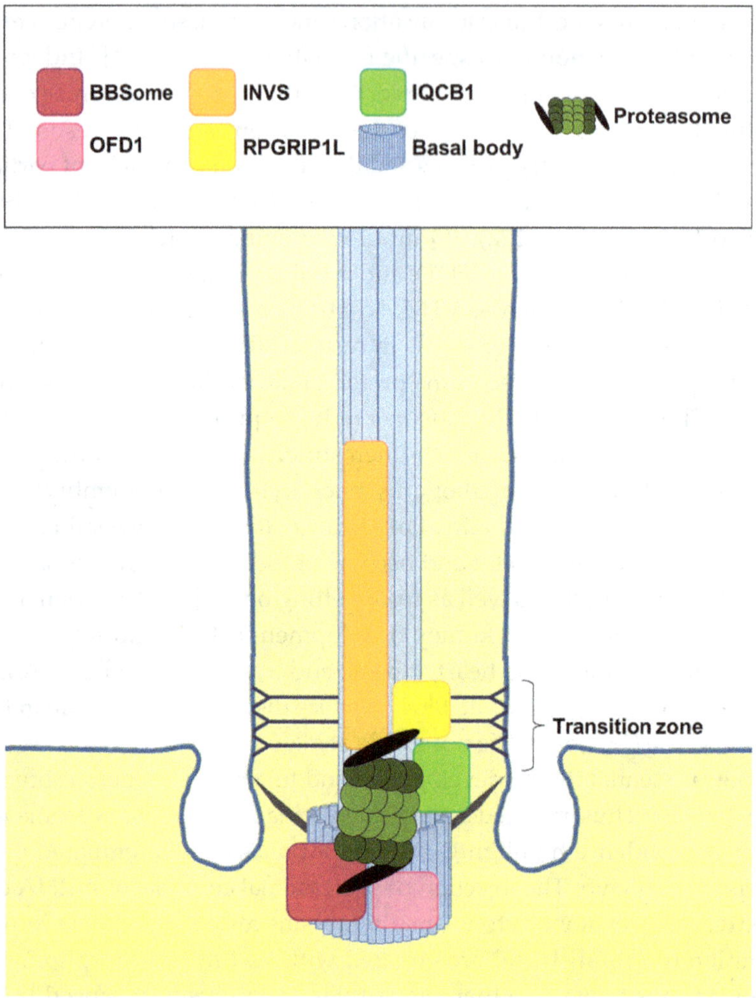

Figure 3. Interactions between ciliary proteins and the cilia-regulated proteasome. INVS is located in the Inversin compartment and in the transition zone and interacts with components of the 19S proteasomal subunit (black lid of the proteasome). The transition zone protein RPGRIP1L interacts with components of the 19S proteasomal subunit. IQCB1 is present at the transition zone and the basal body and interacts with components of the 20S proteasomal subunit. OFD1 localises to the basal body and interacts with components of the 19S proteasomal subunit. Components of the BBSome are located at the basal body and interact with components of the 19S and 20S proteasomal subunit.

3. GLI2-R and GLI3-R Are Essential for Proper Embryonic Development in Vertebrates

The relationship between the cilia-regulated proteasome and HH signalling is essential for homeostasis and embryonic development in vertebrates. For example, an alteration of proteasomal activity is thought to be involved in the development of cancer [73–93]. In this context, the cilia-regulated proteasome might have an essential role, a topic which was extensively reviewed elsewhere [45]. Here, we will focus on the meaning of HH signalling in vertebrate embryogenesis. Regarding HH signalling, the products of the cilia-regulated proteasome are GLI2-R and GLI3-R. Although it was reported that GLI2 is not processed as efficiently as GLI3 [19], both proteins play decisive roles during vertebrate development in their truncated forms. GLI2-R takes part in the development of the brain and the limbs in mice as well as the neural tube in zebrafish [94–96]. Together with GLI3-R, GLI2-R controls anterior limb patterning and digit number [96]. However, the scope of GLI3-R in vertebrate development is much broader than that of GLI2-R. In addition to limb development [97], GLI3-R was shown to be involved in the development of the vertebrate lung, kidneys, ureter, mammary gland, ears, and the CNS. In mice, GLI3-R controls proper lung organogenesis [98], the nephron number [99,100], the functional development of the ureter [101], the induction of mammogenesis [102], and the dorsoventral patterning of the inner ear [103]. As another review article has previously discussed, the cilia-regulated proteasome seems to be involved in most if not all cilia-mediated signalling pathways [45]. Consequently, all defects caused by a decreased activity of the cilia-regulated proteasome are most likely a consequence of numerous disturbed cilia-mediated signalling pathways. However, the outstanding importance of the association between the cilia-regulated proteasome and HH signalling for vertebrate development is highlighted by investigations using the $Gli3^{\Delta 699}$ mouse mutant which is considered to express a constitutively active GLI3-R protein. The $Gli3^{\Delta 699}$ allele was generated by an insertion of a selection marker cassette into the *Gli3* locus [104]. This mutant allele terminates just C-terminally of the zinc-finger domain (amino acid position 699). $Gli3^{\Delta 699/+}$ mice are viable and fertile with exhibiting polydactyly of the forelimbs at very low frequency. However, $Gli3^{\Delta 699/\Delta 699}$ mice died within 12–18 h after birth. Mouse embryos homozygous for $Gli3^{\Delta 699}$ suffer from skeletal and limb defects, imperforate anus, visceral abnormalities, loss of the adrenal gland, and kidney defects. Remarkably, the introduction of $GLI3^{\Delta 699}$ restores several aspects of brain development, like telencephalic patterning, olfactory bulb morphogenesis, and the agenesis of the corpus callosum in *Rpgrip1l*-negative mouse embryos [105,106].

Until now, the existence of the cilia-regulated proteasome was reported in the model organisms mouse and zebrafish [38,43], suggesting that this kind of proteasome is present in all vertebrates. However, to consider the impairment of the cilia-controlled proteasome as a candidate factor for human diseases (e.g., ciliopathies or cancer), it is inevitable to perform future experiments demonstrating the existence of a cilia-regulated proteasome in humans. To analyse the link between the cilia-governed proteasome and HH signalling, these future studies should also address the question whether human GLI2 and GLI3 are processed by this kind of proteasome.

4. Is the Role of the Cilia-Regulated Proteasome Evolutionarily Conserved?

The cilia-regulated proteasome is essential for HH signalling and for embryonic development in mice [38,43]. Until 2014, the general view was that HH signalling was extremely important for embryonic development of the invertebrate *Drosophila*, but operating in a cilia-independent manner [107]. It was thought that HH signalling in *Drosophila* mainly occurs over the entire plasma membrane of its cells. Thus, PTC localises at any place of the plasma membrane and inhibits the presence of SMO in the membrane. The binding of HH to PTC allows SMO to enter the plasma membrane after membrane exit of the HH/PTC complex. As a consequence, cubitus interruptus (CI), the *Drosophila* homolog of the GLI proteins, is converted into CI-A in the cytoplasm. In the absence of HH, CI is processed to CI-R by proteasomes in the cytoplasm [108–113]. To our knowledge, a specific subcytoplasmic localisation of these proteasomes has never been reported. After the conversion of CI, CI-A or CI-R translocates into the nucleus to regulate HH target gene expression [113–115]. In contrast to vertebrates, *Drosophila* seems to form cilia exclusively on sensory neurons and spermatozoa [116,117]. In 2014, Kuzhandaivel et al. reported the transduction of HH signals by the immotile cilia of olfactory sensory neurons (OSNs) in *Drosophila* [118]. PTC, SMO, and COS2 were demonstrated to localise along the whole cilium of OSNs and it was shown that the ciliary presence of SMO is important for proper HH signalling. Although the general view is that HH signalling cannot function without CI in *Drosophila* [119], CI was not detectable in OSN cilia [118]. However, ciliary localisation of SMO induces the expression of HH target genes in OSNs. Since CI gets proteolytically processed by the proteasome [108–113], the question arises whether the proteasome which processes CI is regulated by ciliary proteins in *Drosophila* OSNs. Until now, it is even unknown whether proteasomes are present at the base of OSN cilia. A detection of proteasomal components at the base of OSN cilia would make the existence of an evolutionarily conserved cilia-regulated proteasome likely. If future investigations demonstrated that proteasomal components are absent from the ciliary base of OSNs, it would be interesting to analyse where the processing of CI takes place in OSNs. Furthermore, if proteasomal components are missing at the base of OSN cilia, the fact that CI was not detected in OSN cilia [118] raises the question whether HH signal transduction in OSN cilia is able to function without CI. Since this finding would be rather surprising, and highly interesting, it should be a topic of future studies. However, previous studies could show an association between HH signalling and cilia in planarians [120], but the localisation of HH pathway components at cilia or even the mediation of HH signalling by cilia was never shown for other invertebrates than *Drosophila*. The fact that RPGRIP1L, the ciliary protein being known to exclusively control the cilia-localised proteasome in mice [38], is missing in *Drosophila* [56] gives two possibilities: the first possibility is that the cilia-regulated proteasome does not exist in *Drosophila* and, consequently, RPGRIP1L is not needed in *Drosophila*. The second possibility is that other proteins than RPGRIP1L control the activity of the cilia-governed proteasome in *Drosophila*. If future studies in vertebrates lead to the identification of new proteins that control the cilia-regulated proteasome, the presence of these proteins should be analysed in *Drosophila*. Furthermore, it would be of great interest to understand how these ciliary proteins control the ciliary proteasome. These findings might reveal a conserved regulatory mechanism in evolution, but if *Drosophila* lacks the cilia-controlled proteasome, another hypothesis is conceivable: Since RPGRIP1L is conserved from cnidarians to humans, but is absent in arthropods and nematodes [56], *Drosophila* could be an exception regarding the presence of the cilia-regulated proteasome. Remarkably, PSMA5, a component of the 20S proteasomal subunit, was detected along the whole cilium in mice [38]. Within flagella of the green alga *Chlamydomonas reinhardtii*, the components of the ubiquitin conjugation system, but no proteasomal components, have been found [121] suggesting that the cilia-localised proteasome might have developed later in evolution.

5. Is the Cilia-Regulated Proteasome Involved in the Regulation of PTC and/or SMO Action?

Even at the level of PTC and SMO, proteasomes exert essential functions in HH signalling. In vitro studies in vertebrate systems using primary cells, as well as cell lines from different vertebrate species, demonstrated that, in the presence of HH, PTC1 was ubiquitinated at K1413 in its C-terminal domain by the E3 ligases ITCH and WWP2 and finally degraded by the proteasome [122]. In the absence of HH, SMO is not allowed to enter the cell membrane and to localise at the cell surface. Several examinations in *Drosophila* revealed that SMO gets ubiquitinated, internalised, and degraded by lysosome- and proteasome-dependent mechanisms in lack of HH [123–126]. Interestingly, SMO is able to form a complex with an E3 ligase called SMURF which directly controls proteasomal degradation of HH-unbound PTC [127].

Consequently, positive, as well as negative, regulation of HH signalling depends on proteasomal intervention. Due to the close relation of vertebrate HH signalling to cilia, the question arises whether the cilia-regulated proteasome or another kind of proteasome realises the degradation of PTC and/or SMO. Considering that loss of RPGRIP1L reduces exclusively the activity of the BB-localised proteasome, SMO would be expected to accumulate at cilia in the absence of RPGRIP1L. However, the localisation of SMO is unaltered upon RPGRIP1L deficiency [38] suggesting that the cilia-regulated proteasome might not be responsible for the degradation of SMO. To definitely evaluate the contribution of the cilia-regulated proteasome, it is imperative that further investigations take place.

6. Conclusions

The HH signalling transduction cascade is involved in the development of almost every vertebrate organ and structure [128]. Consequently, impaired HH signalling results in severe human diseases [24,129–131]. In this context, the precise coordination between positive and negative regulation of HH signalling is of eminent importance. Since it is yet unknown whether the cilia-regulated proteasome is involved in the regulation of PTC and SMO, this review focuses on the negative regulation mechanisms via proteolytic processing of the GLI2 and GLI3 transcription factors by the cilia-regulated proteasome. Two major questions have to be addressed on this topic:

(1) Is the proteolytic processing of the GLI2 and GLI3 proteins by the cilia-regulated proteasome evolutionarily conserved?

The HH transduction pathway is evolutionarily highly conserved. In vertebrates, HH signalling is mainly transduced by primary cilia. Within this cilia-mediated HH signalling, the cilia-regulated proteasome plays a decisive role. Previously, it was shown that cilia of sensory neurons are able to mediate HH signalling [118]. Thus, *Drosophila* is a well-suited model organism to examine if the cilia-regulated proteasome also exists in invertebrates and, if this is the case, to analyse whether the proteolytic processing of CI is facilitated by this kind of proteasome. Since several experiments are easier to perform in *Drosophila* than in mice, the presence of the cilia-controlled proteasome in *Drosophila* could give the opportunity to gain more insight into the mechanisms underlying the regulation of this proteasome by ciliary proteins.

(2) Does the cilia-regulated proteasome play a role in the control of HH signalling during human embryonic development?

As early as 1995, Denys Wheatley proposed primary cilia to be necessary for human homeostasis [132]. Today, the outstanding importance of primary cilia for human development is indisputable, but it has to be shown whether the cilia-regulated proteasome participates in human embryonic development. If this is true, it should be analysed whether this participation is involved in the control of HH signalling (GLI3 processing) and to what extent the cilia-governed proteasome affects embryonic development in humans. The precondition for these studies is the existence of the cilia-controlled proteasome in humans. Previously, interactions between proteasomal components and ciliary proteins have been shown in murine as well as in human cells [38,43,44] indicating that the cilia-regulated proteasome might exist in humans. If future studies reveal that the cilia-regulated proteasome is essential for human embryonic development, novel therapies against severe human ciliopathies are possible. In this context, proteasome activators were successfully used for treatment of ciliopathic zebrafishs and cells isolated from ciliopathic mice [38,43]. It is plausible that a prenatal application of proteasome activators to pregnant women could be a curative treatment of ciliopathies in the future.

Acknowledgments: The authors thank Lauren Biwer and Neña Krüger for critical reading of the manuscript. This work was supported by the Deutsche Forschungsgemeinschaft (Sonderforschungsbereiche 590 and 612) to U.R.

Conflicts of Interest: The authors declare that they have no competing interests.

References

1. Nusslein-Volhard, C.; Wieschaus, E. Mutations affecting segment number and polarity in Drosophila. *Nature* **1980**, *287*, 795–801. [CrossRef] [PubMed]
2. Riddle, R.; Johnson, R.; Laufer, E.; Tabin, C. Sonic hedgehog mediates the polarizing activity of the ZPA. *Cell* **1993**, *75*, 1401–1416. [CrossRef]
3. Echelard, Y.; Epstein, D.; St-Jacques, B.; Shen, L.; Mohler, J.; McMahon, J.; McMahon, A. Sonic hedgehog, a member of a family of putative signaling molecules, is implicated in the regulation of CNS polarity. *Cell* **1993**, *75*, 1417–1430. [CrossRef]
4. Krauss, S.; Concordet, J.; Ingham, P. A functionally conserved homolog of the Drosophila segment polarity gene hh is expressed in tissues with polarizing activity in zebrafish embryos. *Cell* **1993**, *75*, 1431–1444. [CrossRef]
5. Ingham, P.; McMahon, A. Hedgehog signaling in animal development: Paradigms and principles. *Genes Dev.* **2001**, *15*, 3059–3087. [CrossRef] [PubMed]

6. Hynes, M.; Stone, D.; Dowd, M.; Pitts-Meek, S.; Goddard, A.; Gurney, A.; Rosenthal, A. Control of cell pattern in the neural tube by the zinc finger transcription factor and oncogene Gli-1. *Neuron* **1997**, *19*, 15–26. [CrossRef]
7. Ruiz i Altaba, A. The works of GLI and the power of hedgehog. *Nat. Cell Biol.* **1999**, *1*, E147–E148. [CrossRef] [PubMed]
8. Sasaki, H.; Nishizaki, Y.; Hui, C.; Nakafuku, M.; Kondoh, H. Regulation of Gli2 and Gli3 activities by an amino-terminal repression domain: Implication of Gli2 and Gli3 as primary mediators of Shh signaling. *Development* **1999**, *126*, 3915–3924.
9. Lilienbaum, A. Relationship between the proteasomal system and autophagy. *Int. J. Biochem. Mol. Biol.* **2013**, *4*, 1–26.
10. Coux, O.; Tanaka, K.; Goldberg, A. Structure and functions of the 20S and 26S proteasomes. *Annu. Rev. Biochem.* **1996**, *65*, 801–847. [CrossRef] [PubMed]
11. Wang, B.; Fallon, J.; Beachy, P. Hedgehog-regulated processing of Gli3 produces an anterior/posterior repressor gradient in the developing vertebrate limb. *Cell* **2000**, *100*, 423–434. [CrossRef]
12. Dai, P.; Akimaru, H.; Tanaka, Y.; Maekawa, T.; Nakafuku, M.; Ishii, S. Sonic Hedgehog-induced activation of the Gli1 promoter is mediated by GLI3. *J. Biol. Chem.* **1999**, *274*, 8143–8152. [CrossRef] [PubMed]
13. Kaesler, S.; Lüscher, B.; Rüther, U. Transcriptional activity of GLI1 is negatively regulated by protein kinase A. *Biol. Chem.* **2000**, *381*, 545–551. [CrossRef] [PubMed]
14. Park, H.; Bai, C.; Platt, K.; Matise, M.; Beeghly, A.; Hui, C.; Nakashima, M.; Joyner, A. Mouse Gli1 mutants are viable but have defects in SHH signaling in combination with a Gli2 mutation. *Development* **2000**, *127*, 1593–1605. [PubMed]
15. Tuson, M.; He, M.; Anderson, K. Protein kinase A acts at the basal body of the primary cilium to prevent Gli2 activation and ventralization of the mouse neural tube. *Development* **2011**, *138*, 4921–4930. [CrossRef] [PubMed]
16. Chitalia, V.; Foy, R.; Bachschmid, M.; Zeng, L.; Panchenko, M.; Zhou, M.; Bharti, A.; Seldin, D.; Lecker, S.; Dominguez, I.; et al. Jade-1 inhibits Wnt signalling by ubiquitylating beta-catenin and mediates Wnt pathway inhibition by pVHL. *Nat. Cell Biol.* **2008**, *10*, 1208–1216. [CrossRef] [PubMed]
17. Borgal, L.; Habbig, S.; Hatzold, J.; Liebau, M.; Dafinger, C.; Sacarea, I.; Hammerschmidt, M.; Benzing, T.; Schermer, B. The Ciliary Protein Nephrocystin-4 Translocates the Canonical Wnt-Regulator Jade-1 to the Nucleus to Negatively Regulate Beta-Catenin Signaling. *J. Biol. Chem.* **2012**, *287*, 25370–25380. [CrossRef] [PubMed]
18. Wang, W.; Wu, T.; Kirschner, M. The master cell cycle regulator APC-Cdc20 regulates ciliary length and disassembly of the primary cilium. *eLife* **2014**, *3*, e03083. [CrossRef] [PubMed]
19. Pan, Y.; Bai, C.; Joyner, A.; Wang, B. Sonic hedgehog signaling regulates Gli2 transcriptional activity by suppressing its processing and degradation. *Mol. Cell. Biol.* **2006**, *26*, 3365–3377. [CrossRef] [PubMed]
20. Pan, Y.; Wang, B. A novel protein-processing domain in Gli2 and Gli3 differentially blocks complete protein degradation by the proteasome. *J. Biol. Chem.* **2007**, *282*, 10846–10852. [CrossRef] [PubMed]
21. Li, J.; Wang, C.; Pan, Y.; Bai, Z.; Wang, B. Increased proteolytic processing of full-length Gli2 transcription factor reduces the hedgehog pathway activity in vivo. *Dev. Dyn.* **2011**, *240*, 766–774. [CrossRef] [PubMed]
22. Schrader, E.; Harstad, K.; Holmgren, R.; Matouschek, A. A three-part signal governs differential processing of Gli1 and Gli3 proteins by the proteasome. *J. Biol. Chem.* **2011**, *286*, 39051–39058. [CrossRef] [PubMed]
23. Rosenbaum, J.; Witman, G. Intraflagellar transport. *Nat. Rev. Mol. Cell Biol.* **2002**, *3*, 813–825. [CrossRef]
24. Goetz, S.; Anderson, K. The primary cilium: A signalling centre during vertebrate development. *Nat. Rev. Genet.* **2010**, *11*, 331–344. [CrossRef] [PubMed]
25. Eggenschwiler, J.; Anderson, K. Cilia and developmental signaling. *Annu. Rev. Cell Dev. Biol.* **2007**, *23*, 345–373. [CrossRef] [PubMed]
26. Chen, M.; Wilson, C.; Li, Y.; Law, K.; Lu, C.; Gacayan, R.; Zhang, X.; Hui, C.; Chuang, P. Cilium-independent regulation of Gli protein function by Sufu in Hedgehog signaling is evolutionarily conserved. *Genes Dev.* **2009**, *23*, 1910–1928. [CrossRef]
27. Humke, E.; Dorn, K.; Milenkovic, L.; Scott, M.; Rohatgi, R. The output of Hedgehog signaling is controlled by the dynamic association between Suppressor of Fused and the Gli proteins. *Genes Dev.* **2010**, *24*, 670–682. [CrossRef] [PubMed]

28. Haycraft, C.; Banizs, B.; Aydin-Son, Y.; Zhang, Q.; Michaud, E.; Yoder, B. Gli2 and Gli3 localize to cilia and require the intraflagellar transport protein polaris for processing and function. *PLoS Genet.* **2005**, *1*, e53. [CrossRef]
29. Wang, B.; Li, Y. Evidence for the direct involvement of {beta}TrCP in Gli3 protein processing. *Proc. Natl. Acad. Sci. USA* **2006**, *103*, 33–38. [CrossRef]
30. Zhang, B.; Zhang, T.; Wang, G.; Wang, G.; Chi, W.; Jiang, Q.; Zhang, C. GSK3β-Dzip1-Rab8 cascade regulates ciliogenesis after mitosis. *PLoS Biol.* **2015**, *13*, e1002129. [CrossRef] [PubMed]
31. He, M.; Subramanian, R.; Bangs, F.; Omelchenko, T.; Liem, K.J.; Kapoor, T.; Anderson, K. The kinesin-4 protein Kif7 regulates mammalian Hedgehog signalling by organizing the cilium tip compartment. *Nat. Cell Biol.* **2014**, *16*, 663–672. [CrossRef] [PubMed]
32. Cheung, H.; Zhang, X.; Ribeiro, A.; Mo, R.; Makino, S.; Puviindran, V.; Law, K.; Briscoe, J.; Hui, C. The kinesin protein Kif7 is a critical regulator of Gli transcription factors in mammalian hedgehog signaling. *Sci. Signal.* **2009**, *2*, ra29. [CrossRef] [PubMed]
33. Endoh-Yamagami, S.; Evangelista, M.; Wilson, D.; Wen, X.; Theunissen, J.; Phamluong, K.; Davis, M.; Scales, S.; Solloway, M.; de Sauvage, F.; et al. The mammalian Cos2 homolog Kif7 plays an essential role in modulating Hh signal transduction during development. *Curr. Biol.* **2009**, *19*, 1320–1326. [CrossRef] [PubMed]
34. Liem, K.J.; He, M.; Ocbina, P.; Anderson, K. Mouse Kif7/Costal2 is a cilia-associated protein that regulates Sonic hedgehog signaling. *Proc. Natl. Acad. Sci. USA* **2009**, *106*, 13377–13382. [CrossRef] [PubMed]
35. Emechebe, U.; Kumar, P.P.; Rozenberg, J.; Moore, B.; Firment, A.; Mirshahi, T.; Moon, A. T-box3 is a ciliary protein and regulates stability of the Gli3 transcription factor to control digit number. *eLife* **2016**, *5*, e07897. [CrossRef]
36. Pedersen, L.; Akhmanova, A. Kif7 keeps cilia tips in shape. *Nat. Cell Biol.* **2014**, *16*, 623–625. [CrossRef] [PubMed]
37. Brooks, P.; Fuertes, G.; Murray, R.; Bose, S.; Knecht, E.; Rechsteiner, M.; Hendil, K.; Tanaka, K.; Dyson, J.; Rivett, J. Subcellular localization of proteasomes and their regulatory complexes in mammalian cells. *Biochem. J.* **2000**, *346*, 155–161. [CrossRef] [PubMed]
38. Gerhardt, C.; Lier, J.; Burmühl, S.; Struchtrup, A.; Deutschmann, K.; Vetter, M.; Leu, T.; Reeg, S.; Grune, T.; Rüther, U. The transition zone protein Rpgrip1l regulates proteasomal activity at the primary cilium. *J. Cell Biol.* **2015**, *210*, 115–133. [CrossRef]
39. Fabunmi, R.; Wigley, W.; Thomas, P.; DeMartino, G. Activity and regulation of the centrosome-associated proteasome. *J. Biol. Chem.* **2000**, *275*, 409–413. [CrossRef] [PubMed]
40. Wigley, W.; Fabunmi, R.; Lee, M.; Marino, C.; Muallem, S.; DeMartino, G.; Thomas, P. Dynamic association of proteasomal machinery with the centrosome. *J. Cell Biol.* **1999**, *145*, 481–490. [CrossRef] [PubMed]
41. Wójcik, C.; DeMartino, G. Intracellular localization of proteasomes. *Int. J. Biochem. Cell Biol.* **2003**, *35*, 579–589. [CrossRef]
42. Amato, R.; Morleo, M.; Giaquinto, L.; di Bernardo, D.; Franco, B. A network-based approach to dissect the cilia/centrosome complex interactome. *BMC Genom.* **2014**, *15*, 658. [CrossRef]
43. Liu, Y.; Tsai, I.; Morleo, M.; Oh, E.; Leitch, C.; Massa, F.; Lee, B.; Parker, D.; Finley, D.; Zaghloul, N.; et al. Ciliopathy proteins regulate paracrine signaling by modulating proteasomal degradation of mediators. *J. Clin. Investig.* **2014**, *124*, 2059–2070. [CrossRef] [PubMed]
44. Sang, L.; Miller, J.; Corbit, K.; Giles, R.; Brauer, M.; Otto, E.; Baye, L.; Wen, X.; Scales, S.; Kwong, M.; et al. Mapping the NPHP-JBTS-MKS protein network reveals ciliopathy disease genes and pathways. *Cell* **2011**, *145*, 513–528. [CrossRef] [PubMed]
45. Gerhardt, C.; Leu, T.; Lier, J.; Rüther, U. The cilia-regulated proteasome and its role in the development of ciliopathies and cancer. *Cilia* **2016**, *5*, 14. [CrossRef]
46. Gerdes, J.; Liu, Y.; Zaghloul, N.; Leitch, C.; Lawson, S.; Kato, M.; Beachy, P.; Beales, P.; DeMartino, G.; Fisher, S.; et al. Disruption of the basal body compromises proteasomal function and perturbs intracellular Wnt response. *Nat. Genet.* **2007**, *39*, 1350–1360. [CrossRef] [PubMed]
47. Gascue, C.; Tan, P.; Cardenas-Rodriguez, M.; Libisch, G.; Fernandez-Calero, T.; Liu, Y.P.; Astrada, S.; Robello, C.; Naya, H.; Katsanis, N.; et al. Direct role of Bardet-Biedl syndrome proteins in transcriptional regulation. *J. Cell Sci.* **2012**, *125*, 362–375. [CrossRef]

48. Chamling, X.; Seo, S.; Bugge, K.; Searby, C.; Guo, D.; Drack, A.; Rahmouni, K.; Sheffield, V. Ectopic expression of human BBS4 can rescue Bardet-Biedl syndrome phenotypes in Bbs4 null mice. *PLoS ONE* **2013**, *8*, e59101. [CrossRef] [PubMed]
49. Eichers, E.; Abd-El-Barr, M.; Paylor, R.; Lewis, R.; Bi, W.; Lin, X.; Meehan, T.; Stockton, D.; Wu, S.; Lindsay, E.; et al. Phenotypic characterization of Bbs4 null mice reveals age-dependent penetrance and variable expressivity. *Hum. Genet.* **2006**, *120*, 211–226. [CrossRef] [PubMed]
50. Mykytyn, K.; Mullins, R.; Andrews, M.; Chiang, A.; Swiderski, R.; Yang, B.; Braun, T.; Casavant, T.; Stone, E.; Sheffield, V. Bardet-Biedl syndrome type 4 (BBS4)-null mice implicate Bbs4 in flagella formation but not global cilia assembly. *Proc. Natl. Acad. Sci. USA* **2004**, *101*, 8664–8669. [CrossRef] [PubMed]
51. Ross, A.; May-Simera, H.; Eichers, E.; Kai, M.; Hill, J.; Jagger, D.; Leitch, C.; Chapple, J.; Munro, P.; Fisher, S.; et al. Disruption of Bardet-Biedl syndrome ciliary proteins perturbs planar cell polarity in vertebrates. *Nat. Genet.* **2005**, *37*, 1135–1140. [CrossRef] [PubMed]
52. Rahmouni, K.; Fath, M.; Seo, S.; Thedens, D.; Berry, C.; Weiss, R.; Nishimura, D.; Sheffield, V. Leptin resistance contributes to obesity and hypertension in mouse models of Bardet-Biedl syndrome. *J. Clin. Investig.* **2008**, *118*, 1458–1467. [CrossRef] [PubMed]
53. Zhang, Q.; Nishimura, D.; Vogel, T.; Shao, J.; Swiderski, R.; Yin, T.; Searby, C.; Carter, C.; Kim, G.; Bugge, K.; et al. BBS7 is required for BBSome formation and its absence in mice results in Bardet-Biedl syndrome phenotypes and selective abnormalities in membrane protein trafficking. *J. Cell Sci.* **2013**, *126*, 2372–2380. [CrossRef] [PubMed]
54. Bimonte, S.; de Angelis, A.; Quagliata, L.; Giusti, F.; Tammaro, R.; Dallai, R.; Ascenzi, M.; Diez-Roux, G.; Franco, B. Ofd1 is required in limb bud patterning and endochondral bone development. *Dev. Biol.* **2011**, *349*, 179–191. [CrossRef]
55. Ferrante, M.; Zullo, A.; Barra, A.; Bimonte, S.; Messaddeq, N.; Studer, M.; Dollé, P.; Franco, B. Oral-facial-digital type I protein is required for primary cilia formation and left-right axis specification. *Nat. Genet.* **2006**, *38*, 112–117. [CrossRef] [PubMed]
56. Vierkotten, J.; Dildrop, R.; Peters, T.; Wang, B.; Rüther, U. Ftm is a novel basal body protein of cilia involved in Shh signalling. *Development* **2007**, *134*, 2569–2577. [CrossRef] [PubMed]
57. Gerhardt, C.; Lier, J.; Kuschel, S.; Rüther, U. The ciliary protein Ftm is required for ventricular wall and septal development. *PLoS ONE* **2013**, *8*, e57545. [CrossRef] [PubMed]
58. Delous, M.; Baala, L.; Salomon, R.; Laclef, C.; Vierkotten, J.; Tory, K.; Golzio, C.; Lacoste, T.; Besse, L.; Ozilou, C.; et al. The ciliary gene RPGRIP1L is mutated in cerebello-oculo-renal syndrome (Joubert syndrome type B) and Meckel syndrome. *Nat. Genet.* **2007**, *39*, 875–881. [CrossRef] [PubMed]
59. Rubinsztein, D. The roles of intracellular protein-degradation pathways in neurodegeneration. *Nature* **2006**, *443*, 780–786. [CrossRef] [PubMed]
60. Breusing, N.; Arndt, J.; Voss, P.; Bresgen, N.; Wiswedel, I.; Gardemann, A.; Siems, W.; Grune, T. Inverse correlation of protein oxidation and proteasome activity in liver and lung. *Mech. Ageing Dev.* **2009**, *130*, 748–753. [CrossRef]
61. Wang, X.; Robbins, J. Proteasomal and lysosomal protein degradation and heart disease. *J. Mol. Cell Cardiol.* **2014**, *71*, 16–24. [CrossRef] [PubMed]
62. Bedford, L.; Hay, D.; Devoy, A.; Paine, S.; Powe, D.; Seth, R.; Gray, T.; Topham, I.; Fone, K.; Rezvani, N.; et al. Depletion of 26S proteasomes in mouse brain neurons causes neurodegeneration and Lewy-like inclusions resembling human pale bodies. *J. Neurosci.* **2008**, *28*, 8189–8198. [CrossRef] [PubMed]
63. Kitajima, Y.; Tashiro, Y.; Suzuki, N.; Warita, H.; Kato, M.; Tateyama, M.; Ando, R.; Izumi, R.; Yamazaki, M.; Abe, M.; et al. Proteasome dysfunction induces muscle growth defects and protein aggregation. *J. Cell Sci.* **2014**, *127*, 5204–5217. [CrossRef] [PubMed]
64. Han, M.; Hu, Z.; Chen, C.; Chen, Y.; Gucek, M.; Li, Z.; Markey, S. Dysbindin-associated proteome in the p2 synaptosome fraction of mouse brain. *J. Proteome Res.* **2014**, *13*, 4567–4580. [CrossRef] [PubMed]
65. Sasaki, K.; Hamazaki, J.; Koike, M.; Hirano, Y.; Komatsu, M.; Uchiyama, Y.; Tanaka, K.; Murata, S. PAC1 gene knockout reveals an essential role of chaperone-mediated 20S proteasome biogenesis and latent 20S proteasomes in cellular homeostasis. *Mol. Cell. Biol.* **2010**, *30*, 3864–3874. [CrossRef] [PubMed]
66. Mtango, N.; Latham, K. Ubiquitin proteasome pathway gene expression varies in rhesus monkey oocytes and embryos of different developmental potential. *Physiol. Genom.* **2007**, *31*, 1–14. [CrossRef] [PubMed]

67. Shin, S.; Shimizu, N.; Tokoro, M.; Nishikawa, S.; Hatanaka, Y.; Anzai, M.; Hamazaki, J.; Kishigami, S.; Saeki, K.; Hosoi, Y.; et al. Mouse zygote-specific proteasome assembly chaperone important for maternal-to-zygotic transition. *Biol. Open* **2013**, *2*, 170–182. [CrossRef] [PubMed]
68. Al-Shami, A.; Jhaver, K.; Vogel, P.; Wilkins, C.; Humphries, J.; Davis, J.; Xu, N.; Potter, D.; Gerhardt, B.; Mullinax, R.; et al. Regulators of the proteasome pathway, Uch37 and Rpn13, play distinct roles in mouse development. *PLoS ONE* **2010**, *5*, e13654. [CrossRef]
69. Kim, K.; Adelstein, R.; Kawamoto, S. Isoform-specific proteasomal degradation of Rbfox3 during chicken embryonic development. *Biochem. Biophys. Res. Commun.* **2014**, *450*, 1662–1667. [CrossRef] [PubMed]
70. Won, J.; Marín de Evsikova, C.; Smith, R.; Hicks, W.; Edwards, M.; Longo-Guess, C.; Li, T.; Naggert, J.; Nishina, P. NPHP4 is necessary for normal photoreceptor ribbon synapse maintenance and outer segment formation, and for sperm development. *Hum. Mol. Genet.* **2011**, *20*, 482–496. [CrossRef] [PubMed]
71. Louie, C.; Caridi, G.; Lopes, V.; Brancati, F.; Kispert, A.; Lancaster, M.; Schlossman, A.; Otto, E.; Leitges, M.; Gröne, H.; et al. AHI1 is required for photoreceptor outer segment development and is a modifier for retinal degeneration in nephronophthisis. *Nat. Genet.* **2010**, *42*, 175–180. [CrossRef] [PubMed]
72. Atala, A.; Freeman, M.; Mandell, J.; Beier, D. Juvenile cystic kidneys (jck): A new mouse mutation which causes polycystic kidneys. *Kidney Int.* **1993**, *43*, 1081–1085. [CrossRef]
73. Kamat, A.; Karashima, T.; Davis, D.; Lashinger, L.; Bar-Eli, M.; Millikan, R.; Shen, Y.; Dinney, C.; McConkey, D. The proteasome inhibitor bortezomib synergizes with gemcitabine to block the growth of human 253JB-V bladder tumors in vivo. *Mol. Cancer Ther.* **2004**, *3*, 279–290. [PubMed]
74. Papageorgiou, A.; Kamat, A.; Benedict, W.; Dinney, C.; McConkey, D. Combination therapy with IFN-alpha plus bortezomib induces apoptosis and inhibits angiogenesis in human bladder cancer cells. *Mol. Cancer Ther.* **2006**, *5*, 3032–3041. [CrossRef]
75. Konac, E.; Varol, N.; Kilicciogl u, I.; Bilen, C. Synergistic effects of cisplatin and proteasome inhibitor bortezomib on human bladder cancer cells. *Oncol. Lett.* **2015**, *10*, 560–564. [CrossRef] [PubMed]
76. Agyin, J.; Santhamma, B.; Nair, H.; Roy, S.; Tekmal, R. BU-32: A novel proteasome inhibitor for breast cancer. *Breast Cancer Res. Treat.* **2009**, *11*, R74. [CrossRef] [PubMed]
77. Bonfili, L.; Cuccioloni, M.; Cecarini, V.; Mozzicafreddo, M.; Palermo, F.; Cocci, P.; Angeletti, M.; Eleuteri, A. Ghrelin induces apoptosis in colon adenocarcinoma cells via proteasome inhibition and autophagy induction. *Apoptosis* **2013**, *18*, 1188–1200. [CrossRef] [PubMed]
78. Mackay, H.; Hedley, D.; Major, P.; Townsley, C.; Mackenzie, M.; Vincent, M.; Degendorfer, P.; Tsao, M.; Nicklee, T.; Birle, D.; et al. A phase II trial with pharmacodynamic endpoints of the proteasome inhibitor bortezomib in patients with metastatic colorectal cancer. *Clin. Cancer Res.* **2005**, *11*, 5526–5533. [CrossRef]
79. Ao, L.; Reichel, D.; Hu, D.; Jeong, H.; Kim, K.; Bae, Y.; Lee, W. Polymer Micelle Formulations of Proteasome Inhibitor Carfilzomib for Improved Metabolic Stability and Anti-Cancer Efficacy in Human Multiple Myeloma and Lung Cancer Cell Lines. *J. Pharmacol. Exp. Ther.* **2015**, *355*, 168–173. [CrossRef] [PubMed]
80. Zhu, W.; Liu, J.; Nie, J.; Sheng, W.; Cao, H.; Shen, W.; Dong, A.; Zhou, J.; Jiao, Y.; Zhang, S.; et al. MG132 enhances the radiosensitivity of lung cancer cells in vitro and in vivo. *Oncol. Rep.* **2015**, *34*, 2083–2089. [CrossRef] [PubMed]
81. Bazzaro, M.; Lee, M.; Zoso, A.; Stirling, W.; Santillan, A.; Shih, I.; Roden, R. Ubiquitin-proteasome system stress sensitizes ovarian cancer to proteasome inhibitor-induced apoptosis. *Cancer Res.* **2006**, *66*, 3754–3763. [CrossRef]
82. Mlynarczuk-Bialy, I.; Doeppner, T.; Golab, J.; Nowis, D.; Wilczynski, G.; Parobczak, K.; Wigand, M.; Hajdamowicz, M.; Biały, L.; Aniolek, O.; et al. Biodistribution and Efficacy Studies of the Proteasome Inhibitor BSc2118 in a Mouse Melanoma Model. *Transl. Oncol.* **2014**, *7*, 570–579. [CrossRef]
83. Bold, R.; Virudachalam, S.; McConkey, D. Chemosensitization of pancreatic cancer by inhibition of the 26S proteasome. *J. Surg. Res.* **2001**, *100*, 11–17. [CrossRef] [PubMed]
84. Befani, C.; Vlachostergios, P.; Hatzidaki, E.; Patrikidou, A.; Bonanou, S.; Simos, G.; Papandreou, C.; Liakos, P. Bortezomib represses HIF-1α protein expression and nuclear accumulation by inhibiting both PI3K/Akt/TOR and MAPK pathways in prostate cancer cells. *J. Mol. Med. (Berl.)* **2012**, *90*, 45–54. [CrossRef] [PubMed]
85. Davis, N.; Taber, D.; Ansari, R.; Ryan, C.; George, C.; Vokes, E.; Vogelzang, N.; Stadler, W. Phase II trial of PS-341 in patients with renal cell cancer: A University of Chicago phase II consortium study. *J. Clin. Oncol.* **2004**, *22*, 115–119. [CrossRef] [PubMed]

86. Kondagunta, G.; Drucker, B.; Schwartz, L.; Bacik, J.; Marion, S.; Russo, P.; Mazumdar, M.; Motzer, R. Phase II trial of bortezomib for patients with advanced renal cell carcinoma. *J. Clin. Oncol.* **2004**, *22*, 3720–3725. [CrossRef] [PubMed]
87. Vaziri, S.; Grabowski, D.; Hill, J.; Rybicki, L.; Burk, R.; Bukowski, R.; Ganapathi, M.; Ganapathi, R. Inhibition of proteasome activity by bortezomib in renal cancer cells is p53 dependent and VHL independent. *Anticancer Res.* **2009**, *29*, 2961–2969. [PubMed]
88. Vlashi, E.; Kim, K.; Lagadec, C.; Donna, L.; McDonald, J.; Eghbali, M.; Sayre, J.; Stefani, E.; McBride, W.; Pajonk, F. In vivo imaging, tracking, and targeting of cancer stem cells. *J. Natl. Cancer Inst.* **2009**, *101*, 350–359. [CrossRef] [PubMed]
89. Vlashi, E.; Lagadec, C.; Chan, M.; Frohnen, P.; McDonald, A.; Pajonk, F. Targeted elimination of breast cancer cells with low proteasome activity is sufficient for tumor regression. *Breast Cancer Res. Treat.* **2013**, *141*, 197–203. [CrossRef] [PubMed]
90. Li, Y.; Zhang, T.; Korkaya, H.; Liu, S.; Lee, H.; Newman, B.; Yu, Y.; Clouthier, S.; Schwartz, S.; Wicha, M.; et al. Sulforaphane, a dietary component of broccoli/broccoli sprouts, inhibits breast cancer stem cells. *Clin. Cancer Res.* **2010**, *16*, 2580–2590. [CrossRef] [PubMed]
91. Li, Y.; Zhang, T. Targeting cancer stem cells with sulforaphane, a dietary component from broccoli and broccoli sprouts. *Future Oncol.* **2013**, *9*, 1097–1103. [CrossRef]
92. Lagadec, C.; Vlashi, E.; Bhuta, S.; Lai, C.; Mischel, P.; Werner, M.; Henke, M.; Pajonk, F. Tumor cells with low proteasome subunit expression predict overall survival in head and neck cancer patients. *BMC Cancer* **2014**, *14*, 152. [CrossRef] [PubMed]
93. Li, S.; Fu, J.; Watkins, D.; Srivastava, R.; Shankar, S. Sulforaphane regulates self-renewal of pancreatic cancer stem cells through the modulation of Sonic hedgehog-GLI pathway. *Mol. Cell. Biochem.* **2013**, *373*, 217–227. [CrossRef] [PubMed]
94. Komada, M.; Saitsu, H.; Shiota, K.; Ishibashi, M. Expression of Fgf15 is regulated by both activator and repressor forms of Gli2 in vitro. *Biochem. Biophys. Res. Commun.* **2008**, *369*, 350–356. [CrossRef] [PubMed]
95. Ke, Z.; Kondrichin, I.; Gong, Z.; Korzh, V. Combined activity of the two Gli2 genes of zebrafish play a major role in Hedgehog signaling during zebrafish neurodevelopment. *Mol. Cell. Neurosci.* **2008**, *37*, 388–401. [CrossRef] [PubMed]
96. Bowers, M.; Eng, L.; Lao, Z.; Turnbull, R.; Bao, X.; Riedel, E.; Mackem, S.; Joyner, A. Limb anterior-posterior polarity integrates activator and repressor functions of GLI2 as well as GLI3. *Dev. Biol.* **2012**, *370*, 110–124. [CrossRef]
97. Wang, C.; Rüther, U.; Wang, B. The Shh-independent activator function of the full-length Gli3 protein and its role in vertebrate limb digit patterning. *Dev. Biol.* **2007**, *305*, 460–469. [CrossRef] [PubMed]
98. Li, Y.; Zhang, H.; Choi, S.; Litingtung, Y.; Chiang, C. Sonic hedgehog signaling regulates Gli3 processing, mesenchymal proliferation, and differentiation during mouse lung organogenesis. *Dev. Biol.* **2004**, *270*, 214–231. [CrossRef] [PubMed]
99. Chi, L.; Galtseva, A.; Chen, L.; Mo, R.; Hui, C.; Rosenblum, N. Kif3a controls murine nephron number via GLI3 repressor, cell survival, and gene expression in a lineage-specific manner. *PLoS ONE* **2013**, *8*, e65448. [CrossRef] [PubMed]
100. Cain, J.; Islam, E.; Haxho, F.; Chen, L.; Bridgewater, D.; Nieuwenhuis, E.; Hui, C.; Rosenblum, N. GLI3 repressor controls nephron number via regulation of Wnt11 and Ret in ureteric tip cells. *PLoS ONE* **2009**, *4*, e7313. [CrossRef] [PubMed]
101. Cain, J.; Islam, E.; Haxho, F.; Blake, J.; Rosenblum, N. GLI3 repressor controls functional development of the mouse ureter. *J. Clin. Investig.* **2011**, *121*, 1199–1206. [CrossRef] [PubMed]
102. Lee, M.; Sun, L.; Veltmaat, J. Hedgehog and Gli signaling in embryonic mammary gland development. *J. Mammary Gland Biol. Neoplasia* **2013**, *18*, 133–138. [CrossRef]
103. Bok, J.; Dolson, D.; Hill, P.; Rüther, U.; Epstein, D.; Wu, D. Opposing gradients of Gli repressor and activators mediate Shh signaling along the dorsoventral axis of the inner ear. *Development* **2007**, *134*, 1713–1722. [CrossRef] [PubMed]
104. Böse, J.; Grotewold, L.; Rüther, U. Pallister-Hall syndrome phenotype in mice mutant for Gli3. *Hum. Mol. Genet.* **2002**, *11*, 1129–1135. [CrossRef] [PubMed]

105. Besse, L.; Neti, M.; Anselme, I.; Gerhardt, C.; Rüther, U.; Laclef, C.; Schneider-Maunoury, S. Primary cilia control telencephalic patterning and morphogenesis via Gli3 proteolytic processing. *Development* **2011**, *138*, 2079–2088. [CrossRef] [PubMed]
106. Laclef, C.; Anselme, I.; Besse, L.; Catala, M.; Palmyre, A.; Baas, D.; Paschaki, M.; Pedraza, M.; Métin, C.; Durand, B.; et al. The role of primary cilia in corpus callosum formation is mediated by production of the Gli3 repressor. *Hum. Mol. Genet.* **2015**, *24*, 4997–5014. [CrossRef] [PubMed]
107. Basto, R.; Lau, J.; Vinogradova, T.; Gardiol, A.; Woods, C.; Khodjakov, A.; Raff, J. Flies without centrioles. *Cell* **2006**, *125*, 1375–1386. [CrossRef] [PubMed]
108. Zhang, W.; Zhao, Y.; Tong, C.; Wang, G.; Wang, B.; Jia, J.; Jiang, J. Hedgehog-regulated Costal2-kinase complexes control phosphorylation and proteolytic processing of Cubitus interruptus. *Dev. Cell* **2005**, *8*, 267–278. [CrossRef] [PubMed]
109. Price, M.; Kalderon, D. Proteolysis of the Hedgehog signaling effector Cubitus interruptus requires phosphorylation by Glycogen Synthase Kinase 3 and Casein Kinase 1. *Cell* **2002**, *108*, 823–835. [CrossRef]
110. Chen, Y.; Gallaher, N.; Goodman, R.; Smolik, S. Protein kinase A directly regulates the activity and proteolysis of cubitus interruptus. *Proc. Natl. Acad. Sci. USA* **1998**, *95*, 2349–2354. [CrossRef] [PubMed]
111. Price, M.; Kalderon, D. Proteolysis of cubitus interruptus in Drosophila requires phosphorylation by protein kinase A. *Development* **1999**, *126*, 4331–4339. [PubMed]
112. Wang, G.; Wang, B.; Jiang, J. Protein kinase A antagonizes Hedgehog signaling by regulating both the activator and repressor forms of Cubitus interruptus. *Genes Dev.* **1999**, *13*, 2828–2837. [CrossRef] [PubMed]
113. Aza-Blanc, P.; Ramírez-Weber, F.; Laget, M.; Schwartz, C.; Kornberg, T. Proteolysis that is inhibited by hedgehog targets Cubitus interruptus protein to the nucleus and converts it to a repressor. *Cell* **1997**, *89*, 1043–1053. [CrossRef]
114. Wang, Q.; Holmgren, R. The subcellular localization and activity of Drosophila cubitus interruptus are regulated at multiple levels. *Development* **1999**, *126*, 5097–5106. [PubMed]
115. Wang, G.; Jiang, J. Multiple Cos2/Ci interactions regulate Ci subcellular localization through microtubule dependent and independent mechanisms. *Dev. Biol.* **2004**, *268*, 493–505. [CrossRef] [PubMed]
116. Keil, T. Functional morphology of insect mechanoreceptors. *Microsc. Res. Tech.* **1997**, *39*, 506–531. [CrossRef]
117. Fuller, M. Spermatogenesis. In *Development of Drosophila*; Bate, M., Martinez Arias, A., Eds.; Cold Spring Harbor Laboratory Press: New York, NY, USA, 1993; Volume 2, pp. 71–147.
118. Kuzhandaivel, A.; Schultz, S.; Alkhori, L.; Alenius, M. Cilia-mediated hedgehog signaling in Drosophila. *Cell Rep.* **2014**, *7*, 672–680. [CrossRef] [PubMed]
119. Méthot, N.; Basler, K. An absolute requirement for Cubitus interruptus in Hedgehog signaling. *Development* **2001**, *128*, 733–742. [PubMed]
120. Rink, J.; Gurley, K.; Elliott, S.; Sánchez Alvarado, A. Planarian Hh signaling regulates regeneration polarity and links Hh pathway evolution to cilia. *Science* **2009**, *326*, 1406–1410. [CrossRef] [PubMed]
121. Huang, K.; Diener, D.; Rosenbaum, J. The ubiquitin conjugation system is involved in the disassembly of cilia and flagella. *J. Cell Biol.* **2009**, *186*, 601–613. [CrossRef] [PubMed]
122. Chen, X.; Chinchilla, P.; Fombonne, J.; Ho, L.; Guix, C.; Keen, J.; Mehlen, P.; Riobo, N. Patched-1 proapoptotic activity is downregulated by modification of K1413 by the E3 ubiquitin-protein ligase Itchy homolog. *Mol. Cell. Biol.* **2014**, *34*, 3855–3866. [CrossRef] [PubMed]
123. Li, S.; Chen, Y.; Shi, Q.; Yue, T.; Wang, B.; Jiang, J. Hedgehog-regulated ubiquitination controls smoothened trafficking and cell surface expression in Drosophila. *PLoS Biol.* **2012**, *10*, e1001239. [CrossRef] [PubMed]
124. Xia, R.; Jia, H.; Fan, J.; Liu, Y.; Jia, J. USP8 promotes smoothened signaling by preventing its ubiquitination and changing its subcellular localization. *PLoS Biol.* **2012**, *10*, e1001238. [CrossRef] [PubMed]
125. Fan, J.; Jiang, K.; Liu, Y.; Jia, J. Hrs promotes ubiquitination and mediates endosomal trafficking of smoothened in Drosophila hedgehog signaling. *PLoS ONE* **2013**, *8*, e79021. [CrossRef] [PubMed]
126. Yang, X.; Mao, F.; Lv, X.; Zhang, Z.; Fu, L.; Lu, Y.; Wu, W.; Zhou, Z.; Zhang, L.; Zhao, Y. Drosophila Vps36 regulates Smo trafficking in Hedgehog signaling. *J. Cell Sci.* **2013**, *126*, 4230–4238. [CrossRef] [PubMed]
127. Huang, S.; Zhang, Z.; Zhang, C.; Lv, X.; Zheng, X.; Chen, Z.; Sun, L.; Wang, H.; Zhu, Y.; Zhang, J.; et al. Activation of Smurf E3 ligase promoted by smoothened regulates hedgehog signaling through targeting patched turnover. *PLoS Biol.* **2013**, *11*, e1001721. [CrossRef]
128. Lee, R.; Zhao, Z.; Ingham, P. Hedgehog signalling. *Development* **2016**, *143*, 367–372. [CrossRef] [PubMed]

129. Briscoe, J.; Thérond, P. The mechanisms of Hedgehog signalling and its roles in development and disease. *Nat. Rev. Mol. Cell Biol.* **2013**, *14*, 416–429. [CrossRef] [PubMed]
130. Waters, A.; Beales, P. Ciliopathies: An expanding disease spectrum. *Pediatr. Nephrol.* **2011**, *26*, 1039–1056. [CrossRef] [PubMed]
131. Villavicencio, E.; Walterhouse, D.; Iannaccone, P. The sonic hedgehog-patched-gli pathway in human development and disease. *Am. J. Hum. Genet.* **2000**, *67*, 1047–1054. [CrossRef]
132. Wheatley, D. Primary cilia in normal and pathological tissues. *Pathobiology* **1995**, *63*, 222–238. [CrossRef] [PubMed]

4

Canonical Sonic Hedgehog Signaling in Early Lung Development

Hugo Fernandes-Silva [1,2], Jorge Correia-Pinto [1,2,3] and Rute Silva Moura [1,2,4,*]

1 Life and Health Sciences Research Institute (ICVS), School of Medicine, University of Minho, 4710-057 Braga, Portugal; hugosilva@med.uminho.pt (H.F.-S.); jcp@med.uminho.pt (J.C.-P.)
2 ICVS/3B's-PT Government Associate Laboratory, 4710-057 Braga/Guimarães, Portugal
3 Department of Pediatric Surgery, Hospital de Braga, 4710-243 Braga, Portugal
4 Biology Department, School of Sciences, University of Minho, 4710-057 Braga, Portugal
* Correspondence: rutemoura@med.uminho.pt

Academic Editors: Henk Roelink and Simon J. Conway

Abstract: The canonical hedgehog (HH) signaling pathway is of major importance during embryonic development. HH is a key regulatory morphogen of numerous cellular processes, namely, cell growth and survival, differentiation, migration, and tissue polarity. Overall, it is able to trigger tissue-specific responses that, ultimately, contribute to the formation of a fully functional organism. Of all three HH proteins, Sonic Hedgehog (SHH) plays an essential role during lung development. In fact, abnormal levels of this secreted protein lead to severe foregut defects and lung hypoplasia. Canonical SHH signal transduction relies on the presence of transmembrane receptors, such as Patched1 and Smoothened, accessory proteins, as Hedgehog-interacting protein 1, and intracellular effector proteins, like GLI transcription factors. Altogether, this complex signaling machinery contributes to conveying SHH response. Pulmonary morphogenesis is deeply dependent on SHH and on its molecular interactions with other signaling pathways. In this review, the role of SHH in early stages of lung development, specifically in lung specification, primary bud formation, and branching morphogenesis is thoroughly reviewed.

Keywords: Sonic Hedgehog (SHH); lung development; endoderm specification; branching morphogenesis; signaling; transcriptional regulation

---

1. Introduction

The development of a fully functional organism is a result of a highly regulated process, orchestrated by several signaling pathways and countless interactions that ultimately compose an intricate molecular network. These signaling pathways sharply control pivotal cellular processes, such as cell survival, proliferation, differentiation, migration, and apoptosis, among others and, thus, contribute to the formation of organs and systems.

Hedgehog (HH) signaling participates and regulates key cellular processes, such as growth, self-renewal, cell survival, differentiation, migration, and tissue polarity, during embryonic development [1]. Furthermore, it is also involved in vertebrate body patterning and morphogenesis of several organs, such as the lung [2]. Due to its role in different steps of development, HH signaling activity must be rigorously controlled or, otherwise, abnormalities may appear. In fact, aberrant activation of HH pathway is associated with different types of developmental disorders and cancers in brain, gastrointestinal, lung, breast, and prostate, to name a few; actually, there is an active field of research that aims to find HH pathway inhibitors and study their potential impact on distinct illnesses [3,4]. In addition, HH is also linked with stem and progenitor cells in organ repair and homeostatic mechanisms [5]. Therefore, HH signaling is of major importance from early embryo development to adult tissue maintenance, requiring a very tight spatial and temporal regulation to ensure its correct function.

Lung development relies on continuous interactions between the epithelial and mesenchymal compartment, coordinated by numerous transcription and growth factors that, overall, contribute to the proper formation of the respiratory system [6]. This developmental process requires the precise activation of several signaling cascades, namely, FGF (fibroblast growth factor), TGFβ-BMP (transforming growth factor β-bone morphogenetic protein), WNT (Wingless-related integration site), HIPPO, RA (retinoic acid), NOTCH, and SHH (Sonic Hedgehog) (reviewed by [7,8]). In this review, we aim to provide a description of the current knowledge and relevance of the canonical Sonic Hedgehog signaling pathway during early stages of lung development.

2. Canonical Sonic Hedgehog Signaling Pathway: Overview

HH gene was discovered in 1980 by Nusslein-Volhard and Wieschaus during a screening for mutations that altered fruit fly (*Drosophila melanogaster*) body plan [9]. In their studies, the authors showed the importance of this gene in *Drosophila* dorsal-ventral patterning and segmentation process. In vertebrates, there are three homologs of *Drosophila* segment polarity gene: desert hedgehog (*dhh*), indian hedgehog (*ihh*) and sonic hedgehog (*shh*) [10]. *shh* is the most widely expressed HH gene and it is implicated in the developmental mechanisms underlying the formation of many organs, including the lung. On the other hand, *dhh* and *ihh* have more specific roles: *ihh* is associated with pancreas and bone development [11,12] whereas *dhh* is associated with male fertility, particularly spermatogenesis and testis organogenesis, and with peripheral nerve formation [13–15].

In mammals, SHH functional protein results from a precursor that undergoes a series of post-translational modifications. The process starts with an autoproteolytic step [16,17], that gives rise to a 25 kDa C-terminal peptide and a 19 kDa N-terminal product responsible for the signaling activity. Then, the N-terminal is modified by covalent addition of a hydrophobic cholesterol molecule [18,19] and by palmitoylation. These alterations contribute to SHH solubility and, consequently, to its long-range signaling capacity, as well as to the ability to form tissue gradients [20]. HH lipid modifications and their properties are extensively reviewed in [21]. After post-translation modifications, the functional protein is transported outside the cell with the assistance of a transporter membrane protein named Dispatched1 (DISP1) [22].

SHH may act in both autocrine and paracrine way, however, classically it is generally associated with paracrine signaling. In the canonical signaling pathway, SHH glycoprotein reaches the target cell and it is handled by the 12-transmembrane protein Patched1 (PTCH1). Normally, PTCH1 is responsible for the inhibition of the G-protein-coupled 7-transmembrane spanning protein Smoothened (SMO) [23]. When the signaling pathway is "ON", SHH binds to PTCH1 and releases SMO inhibition that can move to the primary cilia, an essential cellular structure for the transduction of HH signal in vertebrates that acts as a SHH sensor [24,25]. Due to this event, zinc finger glioma-associated transcription factors (GLI) (*cubitus interruptus* family, in *Drosophila*) are transformed into their activator forms and dissociate from suppressor of fused (SUFU), a protein encoded by a tumor suppressor gene [26]. The activator form of GLIs, the final effectors of the pathway, are able to move to the nucleus and positively influence the transcription of specific target genes [27]. In the absence of SHH ligand, "OFF" state, PTCH1 inhibits SMO; GLIs are phosphorylated through PKA/CK1/GSK3 (cAMP-dependent protein kinase/casein kinase 1/glycogen synthase kinase 3) complex and undergo a proteolytic cleavage that allows the detachment from SUFU. As a consequence, GLIs are converted in their repressor form and subsequently transfer to the nucleus to repress the expression of particular target genes [28,29]. In Figure 1, the major events underlying SHH signaling are illustrated. Sonic Hedgehog machinery and transduction of signal are extensively reviewed in [30].

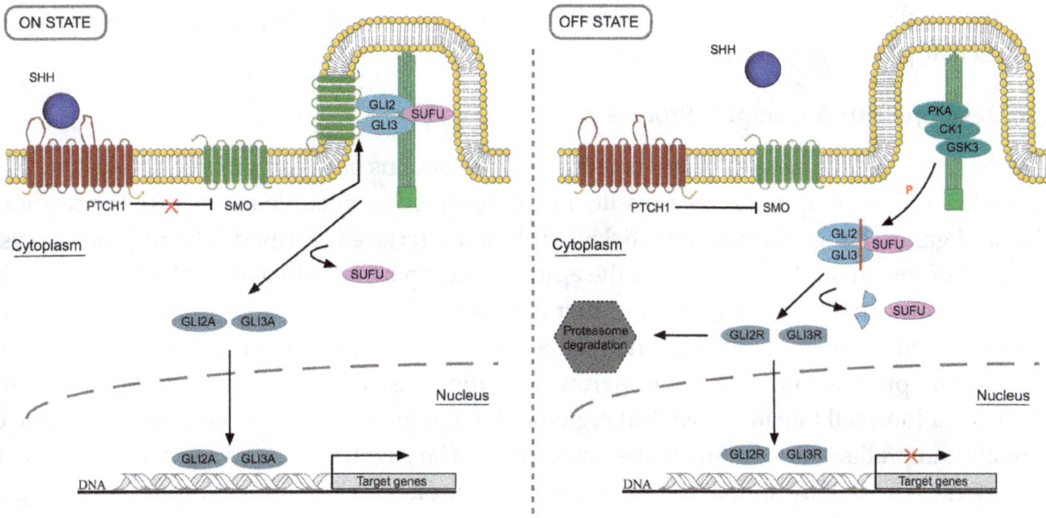

Figure 1. Sonic hedgehog signaling pathway. (1) ON state: SHH protein binds to PTCH1 receptor, at the cell surface level. This event abolishes SMO inhibition that can move to the primary cilia to induce GLI activation through SUFU detachment. Then, GLI2A and GLI3A translocate to the nucleus in order to promote the transcription of target genes; GLI2A acts as the main activator. (2) OFF state: In the absence of SHH ligand, PTCH1 inhibits SMO, and GLIs are phosphorylated by PKA/CK1/GSK3 complex. GLI2R and GLI3R are formed and can follow two possible destinations: proteasome degradation or translocation into the nucleus to repress the transcription of targets genes. In this case, GLI3R is the major repressor.

In vertebrates, there are three types of GLI transcription factors, GLI1, GLI2 and GLI3 that may be posttranslationally converted into their activator or repressor forms. In the presence of SHH ligand, GLI activator form is produced; among the three types, GLI2A is considered to be the primary pathway activator. On the other hand, in the absence of SHH ligand, GLI is truncated and converted in the repressor form. In this context, GLI3R is the main repressor and can translocate to the nucleus to suppress the expression of specific genes [31]. Basically, SHH signaling pathway modulates the balance between the activator and repressor forms of GLI, mainly GLI2 and GLI3, and, as result, GLI family members are able to regulate the expression of SHH target genes.

In addition to the abovementioned components of HH signaling pathway, there are other players such as PTCH2, cell surface protein growth arrest specific (GAS), cell adhesion CAM regulated by oncogenes (CDO) and brother of CDO (BOC) that function as co-receptors, and positively influence the pathway activity [32].

Regarding SHH signaling targets, they include pathway components as, for instance, *gli1* which acts only as a transcriptional activator through a positive feedback loop and does not contribute for HH signaling transduction [33]; furthermore, *ptch1* and *hip1* (that codes for the hedgehog-interacting protein, HIP1), are also downstream targets of SHH signaling but, in this case, they act by negative feedback [34,35]. In the particular case of membrane-bound protein HIP1, it recruits extracellular SHH and, hence, prevents its binding to the transmembrane receptor PTCH1 thus limiting SHH diffusion and signaling. This negative feedback loop explains why very high levels of SHH ligand lead to pathway repression instead of its activation [36]. Additionally, SHH signaling induces the expression of different sets of target genes, in a tissue-specific manner, as for example: secreted signaling proteins like BPM4 [37], cell cycle genes like N-Myc [38], and transcription factors such as *foxa2* (*forkhead box A2*, also known as *hnf-3β*, *hepatocyte nuclear factor-3β*) [39].

The HH canonical signaling pathway is, by far, the most studied. Nonetheless, there is increasing evidence that points to the existence of a non-canonical HH signaling. In this case, HH cellular response does not rely on the activation of GLI transcription factors, or HH signaling components interact with other molecular pathways (reviewed by [40,41]).

3. Lung Development: A Complex Process

The formation of the lung depends on complex interactions between different molecular factors during embryonic development. As a result, a well-designed organ able to perform the respiratory function and gas exchange through an efficient air-blood interface is formed. The respiratory system is composed of two main distinct tissues: the epithelial compartment that arises from the endoderm layer, and the mesenchymal compartment that has its origin in the mesoderm layer. During lung development, epithelium-mesenchyme interactions are constant and the intricate molecular network underlying this process depends on numerous signaling cascades. This process begins with the specification of lung cell fate in the ventral region of the foregut endoderm. Subsequently, these cells differentiate and proliferate, resulting in the formation of a laryngotracheal groove. From this structure, trachea emerges proximally and, in the distal region, ventrolateral buds start to sprout, and will later form the primary lung bronchi. Concurrently, the separation of two tubular structures leaning against each other occurs giving rise to the trachea and the esophagus. Finally, the mature lung structure results from successive branching of lung epithelium, by repetitive splitting and growth into the adjacent splanchnic mesenchyme, to form the airway tree [42]. The full maturation of the respiratory tract extends through post-natal life with the development of alveoli.

Lung development can be divided into five different stages, organized according to major events throughout gestation and that are common to both mice and humans: embryonic (embryonic day (E)9.5–12.5/mouse; 0–7 weeks/human), pseudoglandular (E12.5–16.5; 7–17 weeks), canalicular (E16.5–17.5; 17–27 weeks), saccular (E17.5 to Postnatal (P) 5; 28–36 weeks) and alveolar (P5–P30; 36 weeks to two years) [43,44]. The maturation of the human lung extends through postnatal life until the beginning of adulthood [45]. The first two developmental stages (early stages) are associated with the establishment of the conducting airways, and the last three (late stages) are linked with vascular development, alveolar development and with the reduction of mesenchymal tissue, that is crucial for the formation of the thin air-blood interface indispensable to gas exchange. The following sections will be devoted to exploring the importance of the Sonic Hedgehog signaling pathway during early stages of lung development.

3.1. Lung Endoderm Specification

The lung, thyroid, liver, and pancreas arise from the anterior-ventral region of the primitive foregut. This epithelial tubular structure, endoderm-derived, will differentially express tissue-specific genes along its anterior-posterior axis, which will determine the region where the future organs will emerge. Around E9.0 in the mouse, and day 28 in humans, it is possible to identify the expression of the homeodomain tissue-specific gene *nkx2.1* in the endoderm cells of the anterior foregut ventral region. This transcription factor, at this stage, specifies the foregut domain corresponding to the lung and thyroid, and that is why it is also known as thyroid transcription factor 1 (*ttf1*) or as thyroid-specific enhancer-binding protein (*t/ebp*) [46,47]. Additionally, at this gestational age, endoderm expresses *foxa2* that is required for primitive foregut tube closure [48], and *gata6* (a member of the zinc-finger family of transcription factors). The exact mechanism underlying the induction of lung cell endoderm specification, in both time and space, is still unknown. Even so, it is clear that numerous molecular players transduce signals between neighboring tissues and contribute to the initiation of this process. The respiratory lineage is determined by endodermal cells expressing *nkx2.1* [49]. Both *foxa2* and *gata6* cooperate with *nkx2.1* and contribute to the differentiation of primitive foregut endoderm into respiratory epithelial cell lineages [50,51]. Canonical WNT2/2b ligands, present in the adjacent lateral plate mesoderm, induce the expression of *nkx2.1* in the foregut endoderm [52]. Moreover, BMPs contribute to restricting its expression to the site of lung bud initiation [53]. In addition, FGFs are also thought to be implicated in foregut specification and seem to work in a concentration-dependent manner [54].

During lung specification, HH components, such as *shh*, are expressed in the ventral foregut endoderm and tracheal diverticulum [55]. On the other hand, *gli1*, *gli2*, and *gli3* are expressed in the splanchnic lateral plate mesoderm [56,57]. This pattern of expression is a clear illustration of SHH paracrine signaling, and it also indicates that epithelial-mesenchymal interactions occur as early as foregut specification. Recently, a complex molecular network that involves RA-SHH-BMP-WNT has been identified as indispensable for lung specification [58]. Basically, SHH endoderm expression is induced by RA produced in the neighboring mesoderm. Then, SHH signals back to the mesoderm in order to activate GLI2/3 transcription factors that, consequently, stimulate the expression of WNT2/2b ligands and BMP4. RA acts upstream SHH which, in its turn, acts upstream canonical WNT2/2b and BMPs that then induce *nkx2.1* expression and, consequently, lung specification programs. This signaling cascade is conserved in mice, *Xenopus*, and humans [58].

Knockout mice studies confirmed the importance of SHH during foregut development and lung specification. *shh*-null mutants display a series of characteristics comparable to foregut defects observed in humans, namely, esophageal atresia/stenosis, tracheoesophageal fistula, and tracheal and lung anomalies; these malformations are so severe that these animals perish at birth. In these mutants, the tracheoesophageal septum fails to develop, causing a juxtaposition of the two tubes, which means that SHH is crucial for proper development of the esophagus, trachea, and lung. Likewise, this mouse mutant exhibits lower levels of *ptch* and *gli* in the surrounding mesoderm [55]. In fact, $gli2^{-/-}$ and $gli3^{-/-}$ mutants exhibit different types of foregut abnormalities that overlap with the defects observed in $shh^{-/-}$ mutants [57,59]. These findings suggest that SHH response is conveyed, at least in part, by GLI transcription factors. Curiously, $gli2^{-/-}$ $gli3^{-/-}$ double mutants display an even more severe phenotype since they do not form trachea, lung, and esophagus, which suggests the involvement of other hedgehog family molecules, or that Gli2 and Gli3 have additional hedgehog-independent functions [59]. On the other hand, *gli1*-null mutants are apparently normal and viable [60]. In Figure 2, the main interactions of SHH signaling underlying endoderm specification are represented.

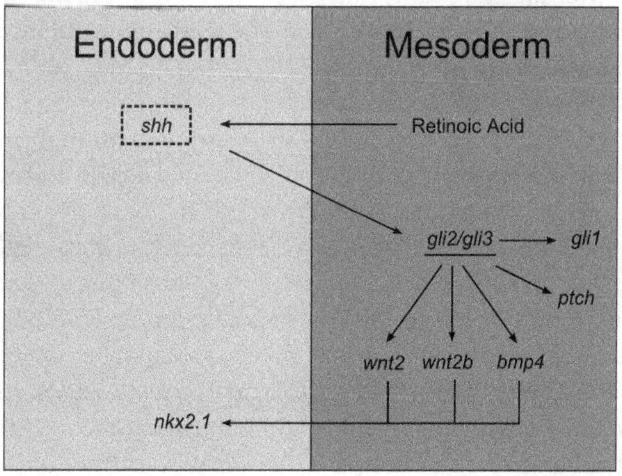

Figure 2. Schematic representation of SHH main interactions during lung endoderm specification.

3.2. Branching Morphogenesis

3.2.1. Primary Bud Formation

After the respiratory lineage has been determined, respiratory progenitors keep on proliferating and begin to differentiate, at different rates, to give rise to the unique cells that compose the lung. As a result, the laryngotracheal groove appears with the subsequent formation of the tracheoesophageal septum that will separate the esophagus from the trachea. Concomitantly, this new tube will grow distally and, eventually, primary lung buds will emerge. In mice, the sprouting of the ventrolateral buds is visible around E9.5, as a result of intricate epithelial-mesenchymal interactions and the activity of crucial transcriptions factors.

SHH signaling pathway contributes for primary bud outgrowth. Undeniably, *shh* is present in the lung from E10.5 and until E16.5 (late pseudoglandular stage). *shh* is highly expressed in the distal tips of the lung bud epithelium and absent from the most proximal regions [61]. Regarding its cellular surface receptor, *ptch1*, it is detected at high levels in the mesenchyme surrounding the distal epithelium, in E11.5 mouse lung [62]. Moreover, *gli1-3* are also expressed in the mesoderm. The importance of SHH signaling in primary bud morphogenesis can be disclosed through the analysis of *shh* and *gli* mouse mutants. $shh^{-/-}$ mouse mutant displays, among other phenotypes, single-lobe hypoplastic lungs, however, proximodistal differentiation of airway epithelium is preserved [63]. Furthermore, *gli2*, *gli3*, and *hip1* knockouts also exhibit lobation defects [36,56,59]. Likewise, in $shh^{-/-}$ embryos, development of lung buds is delayed about half a day, only appearing at E10.5. These abnormalities might be a consequence of primary bud malformation which, subsequently, affects lung branching morphogenesis [59].

3.2.2. Branching Morphogenesis

After primary bud formation, bronchi elongate caudally and the epithelium undergoes a series of successive divisions that will give rise to the airways of the respiratory tree in a process known as branching morphogenesis. Simultaneously, the vascular network is formed and mesenchymal cells differentiate to give rise to the cartilage tissue. Crosstalk between the two cellular compartments, a key feature of branching morphogenesis, relies on diverse signaling pathways to deliver the appropriate response.

During branching morphogenesis, SHH signal transduction seems to be cilia-dependent. The analysis of *talpid*[3] chicken mutants revealed smaller lungs, with fewer branches and surrounded by fibrotic mesenchymal tissue as early as E6 [64]. The absence of TALPID3 (a centrosomal protein) causes a loss both of motile and non-motile primary cilia [65,66] leading to the disruption of SHH signaling [67,68]. The abnormal lung morphology observed in these mutants is a consequence of a defective SHH signaling, resulting in defects in both the epithelium and mesenchyme, although the distal structures remain unaltered [64].

The FGF signaling pathway is a major regulator of branching morphogenesis. The importance of FGF signaling for the initiation of bud branching was determined through analysis of the *Drosophila* tracheal system [69] and it seems to be conserved in both mammalian [70] and avian lung [71]. There are several FGF ligands present in the embryonic lung, but FGF10 stands out as a positive growth factor that promotes epithelial cell proliferation and expansion of the respiratory system [72,73]. In fact, $fgf10^{-/-}$ mutants fail to develop lungs due to branching morphogenesis disruption and impairment of primary bud formation [74]. Furthermore, FGF10 cognate receptor, FGFR2b, is also critical for lung morphogenesis since mice deficient in this particular gene display defective lungs [75]. FGF signaling is a classic example of the epithelial-mesenchymal interactions underlying pulmonary branching. *fgf10* is expressed in the lung mesenchyme surrounding the distal-most epithelium; *fgfr2b* is mainly found in the epithelial compartment. The spatial localization of the *fgf10* is normally associated with its role to induce endoderm proliferation and bud outgrowth, by acting as a chemotactic factor, thus modulating the induction and the direction of the airway branching [70,71,76]. Likewise, FGF9 is essential for proper lung development as a key regulator of mesenchymal proliferation. Disruption of *fgf9* signaling yields pulmonary hypoplasia and decreased airway branching due to a clear reduction of lung mesenchymal tissue and, consequently, decreased mesenchymal *fgf10* expression at branching regions of the lung [77]. *fgf9* is expressed in the epithelial compartment from E10.5–12.5 and in the prospective visceral pleura (mesothelium) from E10.5 onwards, and it signals to the mesenchymal compartment thereby contributing to the epithelium-mesenchyme interaction that controls lung branching [77].

The molecular mechanisms underlying lung branching must be finely regulated or otherwise, abnormalities will occur. In fact, *fgf10* expression levels are modulated by SHH signaling. *shh* is highly expressed in the distal epithelium of the growing buds, adjacent to the normal *fgf10* expression in the bud mesenchyme, whereas *hip1, ptch1, smo,* and *gli* are expressed in the mesenchymal compartment [36,45,57,62,63,78]. Despite the high levels of ligand in the distal bud tip, SHH signaling is not active due to the presence of HIP1, a SHH cell surface receptor, that binds to SHH and prevents signaling activation. As a consequence, *fgf10* expression is allowed in the growing mesenchymal tip. In the interbud/non-branching regions, SHH signaling is active and represses *fgf10* expression. Overall, this mechanism limits FGF10 signaling to active branching zones and controls lung bud outgrowth [70,79]. Actually, *hip1*$^{-/-}$ mouse mutants develop primary buds, nonetheless, the initial secondary branching fails to occur with an impairment in the lobulation pattern; these mice exhibit smaller lungs with a stunted airway tree [36]. In the absence of HIP1 (a SHH negative regulator), SHH signaling is enhanced and *fgf10* expression is almost completely repressed and, consequently, the initiation and growth of new secondary branches do not occur. Conversely, *fgfr2b* expression is not HIP1-dependent [36]. Additionally, overexpression of SHH in lung endoderm impairs *fgf10* expression [70]. Curiously, *hip1*$^{-/-}$; *ptch1*$^{+/-}$ double mutants display a more severe phenotype (smaller lungs and thickened mesenchyme) than *hip1*$^{-/-}$ [36]. Indeed, *ptch1* overexpression in *hip1*$^{-/-}$ lungs partially rescued the lung phenotype (slight increase in lung size and lobation improvement). These findings suggest an overlapping role for these two receptors during lung branching [36]. On the other hand, *shh*$^{-/-}$ mutants display an increase in *fgf10* levels and, importantly, a widespread expression pattern which suggests that SHH is not necessary for *fgf10* expression but it is crucial to spatially restrict *fgf10* expression domain to the distal mesenchyme [63]. Furthermore, it has been shown that FGF10–FGFR2b signaling positively regulates *shh* epithelial expression [80]. The SHH-FGF10 feedback loop is highly complex since FGF10 and SHH regulate each other's expression on the transcript level. These interactions seem to be mediated by ETV transcription factors (PEA3 group ETS domain transcription factor) that modulate *shh* expression. The FGF-ETV-SHH regulatory axis appears to be involved in controlling branching periodicity [81]. As a conclusion, SHH-FGF regulatory axis confines *fgf10* expression to active branching zones. It is worth mentioning that the expression pattern of SHH signaling members is quite conserved in the avian lung except for the high expression levels in distal tips. As a matter of fact, *shh* and its signaling machinery is completely absent from the distal tips. Indeed, in this region, SHH signaling is not active and *fgf10* is expressed in the mesenchyme, as it occurs in the mammalian lung [78]. Moreover, in vitro and in vivo studies have demonstrated that epithelial FGF9 induces/maintains *shh* expression contributing to the expansion and proliferation of subepithelial mesenchyme at early stages of branching [82].

In addition, WNT and BMP signaling are crucial for proper lung branching and are also downstream targets of SHH signaling. BMP4, a TGF-β superfamily member, is expressed in the distal bud epithelium and in the bordering mesenchyme; it inhibits epithelial cell proliferation and induces cell death in the mesenchyme [83]. *shh* mouse mutants, at E11.5, exhibit lower expression levels of *bmp4* in the mesenchymal compartment while in the epithelium remain unaltered [52]. Regarding canonical WNT signaling, there are many WNT ligands involved in branching morphogenesis. For instance, *wnt2* and *wnt7b* are expressed in the mesenchymal and epithelial compartment, respectively [63,84]. In *shh*$^{-/-}$ mutants, *wnt7b* expression levels are similar to wild-type lungs, while *wnt2* expression is downregulated [63]. Taken together, this evidence implies that SHH signaling primarily acts in pulmonary mesenchyme by modifying the expression levels of key signaling molecules. Non-canonical WNT signaling is also important for branching morphogenesis, namely *wnt5a* [85,86]; in fact, it has been demonstrated that lung-specific overexpression of *wnt5a*, in both mouse (E12-13) and chick (E10), causes a decrease in *shh* expression levels [86,87].

In the branching lung, SHH regulates the expression of transcription factors such as *foxf1* (that belong to the forkhead family of transcription factors) or T-box family of transcription factors (*tbx*). *foxf1*-null mice are lethal at very early stages of embryonic development; *foxf1* heterozygotes display a phenotype similar to *shh*-null mutants, namely respiratory failure due to lung hypoplasia and impairment in lung maturation, and foregut malformations [88]. In fact, *foxf1* expression is absent from lung tissue in $shh^{-/-}$ mutants. *foxf1* mRNA is present in the subepithelial mesenchyme regions of the airways, with higher expression levels in the mesenchyme surrounding the distal epithelium of the buds. This spatial distribution, complementary to *shh* expression pattern, favors a putative interaction between the two factors; actually, exogenous SHH is able to activate *foxf1* transcription in the developing lung [88]. Interestingly, in $shh^{-/-}$; $gli3^{-/-}$ lungs *foxf1* expression levels are upregulated. Considering that the absence of SHH signaling causes an increase in Gli3R levels, it is likely that Gli3R contributes to the repression of *foxf1* in $shh^{-/-}$ lung mesenchyme and to growth inhibition [89]. These findings suggest that *foxf1* promotes proliferation and that it is probably one of the effectors of the SHH mitogen effect on the mesenchymal compartment.

TBXs are also downstream targets of SHH signaling. *tbx* are expressed in the developing lung as early as E9.5: *tbx1* is present in the lung epithelium while *tbx2–5* are present in the mesenchymal compartment [90,91]. In particular, TBX4 seems to be crucial for lung branching initiation through the interaction with FGF10 [92]. Moreover, *tbx2* regulates lung growth by repressing cyclin-dependent kinase inhibitor genes (*cdkn1a,b*) that repress cell cycle. In fact, *tbx2*-deficient mice exhibit hypoplastic lungs and diminished branching morphogenesis [91]. *tbx2,3* have redundant functions controlling proliferation in the pulmonary mesenchyme and act upstream of WNT signaling but downstream of SHH signaling [93]. In fact, $shh^{-/-}$ and $smo^{-/-}$ lungs display reduced levels of *tbx2-3* corroborating this interaction [89,93]. Furthermore, in *smo*-deficient lungs, *axin2* expression levels (WNT target gene) are strongly reduced; nonetheless, *tbx2* mesenchymal re-expression reverts this scenario [93]. *tbx2,3* convey SHH signals during lung branching morphogenesis. SHH-*tbx2/3* molecular network contributes to the proliferative expansion of lung mesenchyme and branching morphogenesis by repressing WNT signaling antagonists, and by regulating the expression of cell-cycle regulating genes such as *cdkn* [93].

On the other hand, *shh* is a target of *foxa1/2* transcription factors. Deletion of both *foxa1* and *foxa2* leads to an impairment on branching morphogenesis, as early as E12.5, which may be partially due a decrease/absence of *shh* [94]. The contrary is not true since *foxa1/2* are not downstream targets of SHH signaling [94]. Additionally, transcriptional regulation can be modulated by small, noncoding RNAs (microRNAs) that affect gene expression post-transcriptionally. Recently it has been demonstrated that miRNA-326 is not only a downstream target but also a negative modulator of SHH signaling by direct modification of *gli2* and *smo* [95]. In Figure 3, the principal interactions of SHH signaling in branching morphogenesis are indicated.

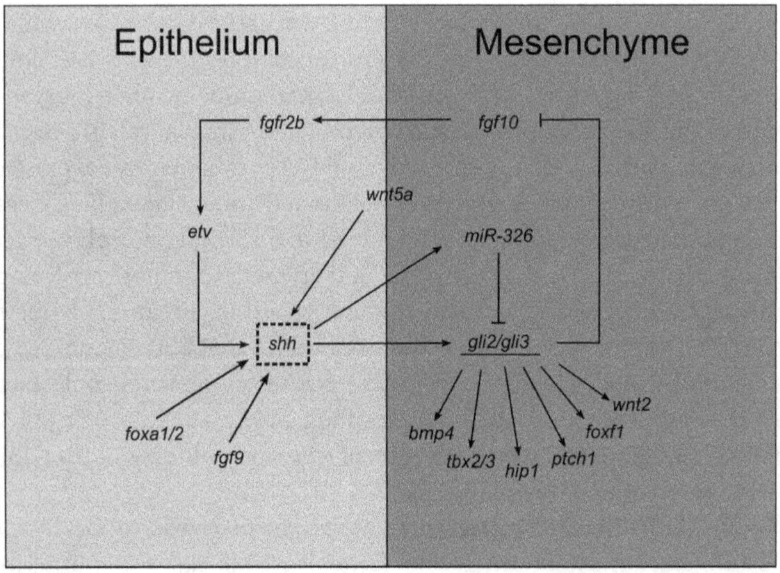

Figure 3. Schematic representation of SHH main interactions during lung endoderm specification.

Additionally, in Table 1, we have summarized SHH upstream and downstream genes that contribute to early lung development.

Table 1. Summary of SHH signaling upstream and downstream targets that contribute to early lung development.

Developmental Stage	Gene/Signaling Pathway	Upstream	Downstream	Molecular Interaction	Reference
Lung Specification	*bmp4*		×	Direct	[58]
	gli1		×	Direct	[55]
	gli2		×	Direct	[58,59]
	gli3		×	Direct	[55,58,59]
	nkx2.1 (*ttf1* or *t/ebp*)		×	Indirect	[58]
	ptch		×	Direct	[58]
	RA pathway	×		Direct	[58]
	wnt2		×	Direct	[58]
	wnt2b		×	Direct	[58]
Branching Morphogenesis	*bmp4*		×	Direct	[55,63,83]
	cdkn1a		×	Indirect	[93]
	cdkn1b		×	Indirect	[93]
	etv	×		Direct	[81]
	fgf9	×		Direct	[82]
	fgf10	×	×	Direct	[63,70,79–81]
	foxa1/2	×		Direct	[94]
	foxf1		×	Direct	[88,89]
	gli2		×	Direct	[59]
	gli3		×	Direct	[59]
	hip1		×	Direct	[36]
	miR-326	×	×	Direct/indirect	[95]
	ptch1		×	Direct	[36]
	tbx2		×	Direct	[89,93]
	tbx3		×	Direct	[89,93]
	wnt2		×	Direct	[63]
	wnt5a	×		Direct	[85,86]

4. Final Remarks

SHH is a key morphogen that regulates pulmonary morphogenesis by triggering tissue-specific responses that contribute to the formation of a complex respiratory structure. In this review, we have highlighted the major roles of the canonical SHH signaling pathway in early stages of lung development. Regardless of the stage, SHH contributes to the epithelial-mesenchymal interactions that underlie lung development. In this particular case, SHH produced by the epithelial compartment acts on the pulmonary mesenchyme by regulating the expression of key signaling molecules, such as FGF10 and BMP4, just to name a few. The SHH molecular network must be sharply controlled in both time and space or, otherwise, severe congenital defects will occur. Additionally, canonical SHH signaling is also involved in lung alveolarization and maturation, and it has been implicated in adult lung diseases, such as pulmonary fibrosis, asthma, and chronic obstructive pulmonary disease (revised in [96]). Moreover, signaling pathways that regulate developmental processes and organ homeostasis usually play critical roles in tumorigenesis. Abnormal hyperactivation of the Hedgehog signaling pathway has been described in different subtypes of lung cancer. For instance, in small-cell lung cancer, it has been reported an increase in SHH ligands or an overexpression of GLI proteins [97]. On the other hand, in lung cancer cell line A549 and in some lung tumors, the loss of HIP1 (SHH pathway antagonist) has been documented [98]. In all these cases, aberrant SHH signaling activation upregulates cancer cell proliferation, maintains cancer stem cells and, eventually in some cases, enhance their metastatic potential (reviewed in [99,100]).

Lung development is orchestrated by many signaling cascades perfectly coordinated to give rise to a fully functional organ. Perhaps, there are still unknown molecular interactions that need to be uncovered, and it is likely that SHH may be a player in these hypothetical networks.

Acknowledgments: This work has been funded by FEDER funds, through the Competitiveness Factors Operational Program (COMPETE), and by National funds, through the Foundation for Science and Technology (FCT), under the scope of the project POCI-01-0145-FEDER-007038; and by the project NORTE-01-0145-FEDER-000013, supported by the Northern Portugal Regional Operational Program (NORTE 2020), under the Portugal 2020 Partnership Agreement, through the European Regional Development Fund (FEDER). The funders had no role in decision to publish or preparation of the manuscript.

Author Contributions: H.F.-S. wrote the manuscript and prepared the figures. J.C.-P. contributed to the discussion. R.S.M. conceived, participated in writing and revision of the paper. All authors commented on the manuscript.

Conflicts of Interest: The authors declare no conflict of interest.

References

1. Choudhry, Z.; Rikani, A.A.; Choudhry, A.M.; Tariq, S.; Zakaria, F.; Asghar, M.W.; Sarfraz, M.K.; Haider, K.; Shafiq, A.A.; Mobassarah, N.J. Sonic hedgehog signalling pathway: A complex network. *Ann. Neurosci.* **2014**, *21*, 28–31. [CrossRef] [PubMed]
2. Ingham, P.W.; McMahon, A.P. Hedgehog signaling in animal development: Paradigms and principles. *Genes Dev.* **2001**, *15*, 3059–3087. [CrossRef] [PubMed]
3. Gupta, S.; Takebe, N.; LoRusso, P. Targeting the Hedgehog pathway in cancer. *Ther. Adv. Med. Oncol.* **2010**, *2*, 237–250. [CrossRef] [PubMed]
4. Rimkus, T.K.; Carpenter, R.L.; Qasem, S.; Chan, M.; Lo, H.W. Targeting the Sonic Hedgehog Signaling Pathway: Review of Smoothened and GLI Inhibitors. *Cancers (Basel)* **2016**, *8*, 22. [CrossRef] [PubMed]
5. Petrova, R.; Joyner, A.L. Roles for Hedgehog signaling in adult organ homeostasis and repair. *Development* **2014**, *141*, 3445–3457. [CrossRef] [PubMed]
6. Warburton, D.; El-Hashash, A.; Carraro, G.; Tiozzo, C.; Sala, F.; Rogers, O.; De Langhe, S.; Kemp, P.J.; Riccardi, D.; Torday, J.; et al. Lung organogenesis. *Curr. Top. Dev. Biol.* **2010**, *90*, 73–158. [CrossRef] [PubMed]
7. Ornitz, D.M.; Yin, Y. Signaling networks regulating development of the lower respiratory tract. *Cold Spring Harb. Perspect. Biol.* **2012**, *4*, a008318. [CrossRef] [PubMed]
8. Herriges, M.; Morrisey, E.E. Lung development: Orchestrating the generation and regeneration of a complex organ. *Development* **2014**, *141*, 502–513. [CrossRef] [PubMed]
9. Nusslein-Volhard, C.; Wieschaus, E. Mutations affecting segment number and polarity in Drosophila. *Nature* **1980**, *287*, 795–801. [CrossRef] [PubMed]
10. Echelard, Y.; Epstein, D.J.; St-Jacques, B.; Shen, L.; Mohler, J.; McMahon, J.A.; McMahon, A.P. Sonic hedgehog, a member of a family of putative signaling molecules, is implicated in the regulation of CNS polarity. *Cell* **1993**, *75*, 1417–1430. [CrossRef]
11. Karp, S.J.; Schipani, E.; St-Jacques, B.; Hunzelman, J.; Kronenberg, H.; McMahon, A.P. Indian hedgehog coordinates endochondral bone growth and morphogenesis via parathyroid hormone related-protein-dependent and -independent pathways. *Development* **2000**, *127*, 543–548. [PubMed]
12. Kawahira, H.; Ma, N.H.; Tzanakakis, E.S.; McMahon, A.P.; Chuang, P.T.; Hebrok, M. Combined activities of hedgehog signaling inhibitors regulate pancreas development. *Development* **2003**, *130*, 4871–4879. [CrossRef] [PubMed]
13. Bitgood, M.J.; Shen, L.; McMahon, A.P. Sertoli cell signaling by Desert hedgehog regulates the male germline. *Curr. Biol.* **1996**, *6*, 298–304. [CrossRef]
14. Parmantier, E.; Lynn, B.; Lawson, D.; Turmaine, M.; Namini, S.S.; Chakrabarti, L.; McMahon, A.P.; Jessen, K.R.; Mirsky, R. Schwann cell-derived Desert hedgehog controls the development of peripheral nerve sheaths. *Neuron* **1999**, *23*, 713–724. [CrossRef]
15. Yao, H.H.; Whoriskey, W.; Capel, B. Desert Hedgehog/Patched 1 signaling specifies fetal Leydig cell fate in testis organogenesis. *Genes Dev.* **2002**, *16*, 1433–1440. [CrossRef] [PubMed]
16. Lee, J.J.; Ekker, S.C.; von Kessler, D.P.; Porter, J.A.; Sun, B.I.; Beachy, P.A. Autoproteolysis in hedgehog protein biogenesis. *Science* **1994**, *266*, 1528–1537. [CrossRef] [PubMed]
17. Porter, J.A.; von Kessler, D.P.; Ekker, S.C.; Young, K.E.; Lee, J.J.; Moses, K.; Beachy, P.A. The product of hedgehog autoproteolytic cleavage active in local and long-range signalling. *Nature* **1995**, *374*, 363–366. [CrossRef] [PubMed]

18. Taylor, F.R.; Wen, D.; Garber, E.A.; Carmillo, A.N.; Baker, D.P.; Arduini, R.M.; Williams, K.P.; Weinreb, P.H.; Rayhorn, P.; Hronowski, X.; et al. Enhanced potency of human Sonic hedgehog by hydrophobic modification. *Biochemistry* **2001**, *40*, 4359–4371. [CrossRef] [PubMed]
19. Tukachinsky, H.; Kuzmickas, R.P.; Jao, C.Y.; Liu, J.; Salic, A. Dispatched and scube mediate the efficient secretion of the cholesterol-modified hedgehog ligand. *Cell Rep.* **2012**, *2*, 308–320. [CrossRef] [PubMed]
20. Chen, M.H.; Li, Y.J.; Kawakami, T.; Xu, S.M.; Chuang, P.T. Palmitoylation is required for the production of a soluble multimeric Hedgehog protein complex and long-range signaling in vertebrates. *Genes Dev.* **2004**, *18*, 641–659. [CrossRef] [PubMed]
21. Mann, R.K.; Beachy, P.A. Novel lipid modifications of secreted protein signals. *Annu. Rev. Biochem.* **2004**, *73*, 891–923. [CrossRef] [PubMed]
22. Tian, H.; Jeong, J.; Harfe, B.D.; Tabin, C.J.; McMahon, A.P. Mouse Disp1 is required in sonic hedgehog-expressing cells for paracrine activity of the cholesterol-modified ligand. *Development* **2005**, *132*, 133–142. [CrossRef] [PubMed]
23. Taipale, J.; Cooper, M.K.; Maiti, T.; Beachy, P.A. Patched acts catalytically to suppress the activity of Smoothened. *Nature* **2002**, *418*, 892–897. [CrossRef] [PubMed]
24. Rohatgi, R.; Milenkovic, L.; Scott, M.P. Patched1 regulates hedgehog signaling at the primary cilium. *Science* **2007**, *317*, 372–376. [CrossRef] [PubMed]
25. Kovacs, J.J.; Whalen, E.J.; Liu, R.; Xiao, K.; Kim, J.; Chen, M.; Wang, J.; Chen, W.; Lefkowitz, R.J. Beta-arrestin-mediated localization of smoothened to the primary cilium. *Science* **2008**, *320*, 1777–1781. [CrossRef] [PubMed]
26. Humke, E.W.; Dorn, K.V.; Milenkovic, L.; Scott, M.P.; Rohatgi, R. The output of Hedgehog signaling is controlled by the dynamic association between Suppressor of Fused and the Gli proteins. *Genes Dev.* **2010**, *24*, 670–682. [CrossRef] [PubMed]
27. Alexandre, C.; Jacinto, A.; Ingham, P.W. Transcriptional activation of hedgehog target genes in Drosophila is mediated directly by the cubitus interruptus protein, a member of the GLI family of zinc finger DNA-binding proteins. *Genes Dev.* **1996**, *10*, 2003–2013. [CrossRef] [PubMed]
28. Chen, Y.; Gallaher, N.; Goodman, R.H.; Smolik, S.M. Protein kinase A directly regulates the activity and proteolysis of cubitus interruptus. *Proc. Natl. Acad. Sci. USA* **1998**, *95*, 2349–2354. [CrossRef] [PubMed]
29. Price, M.A.; Kalderon, D. Proteolysis of the Hedgehog signaling effector Cubitus interruptus requires phosphorylation by Glycogen Synthase Kinase 3 and Casein Kinase 1. *Cell* **2002**, *108*, 823–835. [CrossRef]
30. Briscoe, J.; Thérond, P.P. The mechanisms of Hedgehog signalling and its roles in development and disease. *Nat. Rev. Mol. Cell Biol.* **2013**, *14*, 416–429. [CrossRef] [PubMed]
31. Sasaki, H.; Nishizaki, Y.; Hui, C.; Nakafuku, M.; Kondoh, H. Regulation of Gli2 and Gli3 activities by an amino-terminal repression domain: Implication of Gli2 and Gli3 as primary mediators of Shh signaling. *Development* **1999**, *126*, 3915–3924. [PubMed]
32. Allen, B.L.; Tenzen, T.; McMahon, A.P. The Hedgehog-binding proteins Gas1 and Cdo cooperate to positively regulate Shh signaling during mouse development. *Genes Dev.* **2007**, *21*, 1244–1257. [CrossRef] [PubMed]
33. Bai, C.B.; Auerbach, W.; Lee, J.S.; Stephen, D.; Joyner, A.L. Gli2, but not Gli1, is required for initial Shh signaling and ectopic activation of the Shh pathway. *Development* **2002**, *129*, 4753–4761. [PubMed]
34. Goodrich, L.V.; Johnson, R.L.; Milenkovic, L.; McMahon, J.A.; Scott, M.P. Conservation of the hedgehog/patched signaling pathway from flies to mice: Induction of a mouse patched gene by Hedgehog. *Genes Dev.* **1996**, *10*, 301–312. [CrossRef] [PubMed]
35. Chuang, P.T.; McMahon, A.P. Vertebrate Hedgehog signalling modulated by induction of a Hedgehog-binding protein. *Nature* **1999**, *397*, 617–621. [CrossRef] [PubMed]
36. Chuang, P.T.; Kawcak, T.; McMahon, A.P. Feedback control of mammalian Hedgehog signaling by the Hedgehog-binding protein, Hip1, modulates Fgf signaling during branching morphogenesis of the lung. *Genes Dev.* **2003**, *17*, 342–347. [CrossRef] [PubMed]
37. Astorga, J.; Carlsson, P. Hedgehog induction of murine vasculogenesis is mediated by Foxf1 and Bmp4. *Development* **2007**, *134*, 3753–3761. [CrossRef] [PubMed]
38. Oliver, T.G.; Grasfeder, L.L.; Carroll, A.L.; Kaiser, C.; Gillingham, C.L.; Lin, S.M.; Wickramasinghe, R.; Scott, M.P.; Wechsler-Reya, R.J. Transcriptional profiling of the Sonic hedgehog response: A critical role for N-myc in proliferation of neuronal precursors. *Proc. Natl. Acad. Sci. USA* **2003**, *100*, 7331–7336. [CrossRef] [PubMed]

39. Sasaki, H.; Hui, C.; Nakafuku, M.; Kondoh, H. A binding site for Gli proteins is essential for HNF-3beta floor plate enhancer activity in transgenics and can respond to Shh in vitro. *Development* **1997**, *124*, 1313–1322. [PubMed]
40. Jenkins, D. Hedgehog signalling: Emerging evidence for non-canonical pathways. *Cell. Signal.* **2009**, *21*, 1023–1034. [CrossRef] [PubMed]
41. Brennan, D.; Chen, X.; Cheng, L.; Mahoney, M.; Riobo, N.A. Noncanonical Hedgehog signaling. *Vitam. Horm.* **2012**, *88*, 55–72. [CrossRef]
42. Cardoso, W.V.; Lü, J. Regulation of early lung morphogenesis: Questions, facts and controversies. *Development* **2006**, *133*, 1611–1624. [CrossRef] [PubMed]
43. Kimura, J.; Deutsch, G.H. Key mechanisms of early lung development. *Pediatr. Dev. Pathol.* **2007**, *10*, 335–347. [CrossRef] [PubMed]
44. Joshi, S.; Kotecha, S. Lung growth and development. *Early Hum. Dev.* **2007**, *83*, 789–794. [CrossRef] [PubMed]
45. Zhang, M.; Wang, H.; Teng, H.; Shi, J.; Zhang, Y. Expression of SHH signaling pathway components in the developing human lung. *Histochem. Cell Biol.* **2010**, *134*, 327–335. [CrossRef] [PubMed]
46. Lazzaro, D.; Price, M.; de Felice, M.; Di Lauro, R. The transcription factor TTF-1 is expressed at the onset of thyroid and lung morphogenesis and in restricted regions of the foetal brain. *Development* **1991**, *113*, 1093–1104. [PubMed]
47. Kimura, S.; Hara, Y.; Pineau, T.; Fernandez-Salguero, P.; Fox, C.H.; Ward, J.M.; Gonzalez, F.J. The T/ebp null mouse: Thyroid-specific enhancer-binding protein is essential for the organogenesis of the thyroid, lung, ventral forebrain, and pituitary. *Genes Dev.* **1996**, *10*, 60–69. [CrossRef] [PubMed]
48. Ang, S. L.; Rossant, J. HNF-3b is essential for node and notochord formation in mouse development. *Cell* **1994**, *78*, 561–574. [CrossRef]
49. Minoo, P.; Su, G.; Drum, H.; Bringas, P.; Kimura, S. Defects in tracheoesophageal and lung morphogenesis in Nkx2.1(−/−) mouse embryos. *Dev. Biol.* **1999**, *209*, 60–71. [CrossRef] [PubMed]
50. Ikeda, K.; Shaw-White, J.R.; Wert, S.E.; Whitsett, J.A. Hepatocyte nuclear factor 3 activates transcription of thyroid transcription factor 1 in respiratory epithelial cells. *Mol. Cell. Biol.* **1996**, *16*, 3626–3636. [CrossRef] [PubMed]
51. Shaw-White, J.R.; Bruno, M.D.; Whitsett, J.A. GATA-6 activates transcription of thyroid transcription factor-1. *J. Biol. Chem.* **1999**, *274*, 2658–2664. [CrossRef] [PubMed]
52. Goss, A.M.; Tian, Y.; Tsukiyama, T.; Cohen, E.D.; Zhou, D.; Lu, M.M.; Yamaguchi, T.P.; Morrisey, E.E. Wnt2/2b and beta-catenin signaling are necessary and sufficient to specify lung progenitors in the foregut. *Dev. Cell* **2009**, *17*, 290–298. [CrossRef] [PubMed]
53. Domyan, E.T.; Ferretti, E.; Throckmorton, K.; Mishina, Y.; Nicolis, S.K.; Sun, X. Signaling through BMP receptors promotes respiratory identity in the foregut via repression of Sox2. *Development* **2011**, *138*, 971–981. [CrossRef] [PubMed]
54. Serls, A.E.; Doherty, S.; Parvatiyar, P.; Wells, J.M.; Deutsch, G.H. Different thresholds of fibroblast growth factors pattern the ventral foregut into liver and lung. *Development* **2005**, *132*, 35–47. [CrossRef] [PubMed]
55. Litingtung, Y.; Lei, L.; Westphal, H.; Chiang, C. Sonic hedgehog is essential to foregut development. *Nat. Genet.* **1998**, *20*, 58–61. [CrossRef] [PubMed]
56. Hui, C.C.; Slusarski, D.; Platt, K.A.; Holmgren, R.; Joyner, A.L. Expression of three mouse homologs of the Drosophila segment polarity gene cubitus interruptus, Gli, Gli-2, and Gli-3, in ectoderm- and mesoderm-derived tissues suggests multiple roles during postimplantation development. *Dev. Biol.* **1994**, *162*, 402–413. [CrossRef] [PubMed]
57. Grindley, J.C.; Bellusci, S.; Perkins, D.; Hogan, B.L. Evidence for the involvement of the Gli gene family in embryonic mouse lung development. *Dev. Biol.* **1997**, *188*, 337–348. [CrossRef] [PubMed]
58. Rankin, S.A.; Han, L.; McCracken, K.W.; Kenny, A.P.; Anglin, C.T.; Grigg, E.A.; Crawford, C.M.; Wells, J.M.; Shannon, J.M.; Zorn, A.M. A Retinoic Acid-Hedgehog Cascade Coordinates Mesoderm-Inducing Signals and Endoderm Competence during Lung Specification. *Cell Rep.* **2016**, *16*, 66–78. [CrossRef] [PubMed]
59. Motoyama, J.; Liu, J.; Mo, R.; Ding, Q.; Post, M.; Hui, C.C. Essential function of Gli2 and Gli3 in the formation of lung, trachea and oesophagus. *Nat. Genet.* **1998**, *20*, 54–57. [CrossRef]
60. Park, H.L.; Bai, C.; Platt, K.A.; Matise, M.P.; Beeghly, A.; Hui, C.C.; Nakashima, M.; Joyner, A.L. Mouse Gli1 mutants are viable but have defects in SHH signaling in combination with a Gli2 mutation. *Development* **2000**, *127*, 1593–1605. [PubMed]

61. Miller, L.A.; Wert, S.E.; Whitsett, J.A. Immunolocalization of sonic hedgehog (Shh) in developing mouse lung. *J. Histochem. Cytochem.* **2001**, *49*, 1593–1604. [CrossRef] [PubMed]
62. Bellusci, S.; Furuta, Y.; Rush, M.G.; Henderson, R.; Winnier, G.; Hogan, B.L. Involvement of Sonic hedgehog (Shh) in mouse embryonic lung growth and morphogenesis. *Development* **1997**, *124*, 53–63. [PubMed]
63. Pepicelli, C.V.; Lewis, P.M.; McMahon, A.P. Sonic hedgehog regulates branching morphogenesis in the mammalian lung. *Curr. Biol.* **1998**, *8*, 1083–1086. [CrossRef]
64. Davey, M.G.; McTeir, L.; Barrie, A.M.; Freem, L.J.; Stephen, L.A. Loss of cilia causes embryonic lung hypoplasia, liver fibrosis, and cholestasis in the talpid3 ciliopathy mutant. *Organogenesis* **2014**, *10*, 177–185. [CrossRef] [PubMed]
65. Yin, Y.; Bangs, F.; Paton, I.R.; Prescott, A.; James, J.; Davey, M.G.; Whitley, P.; Genikhovich, G.; Technau, U.; Burt, D.W.; et al. The Talpid3 gene (KIAA0586) encodes a centrosomal protein that is essential for primary cilia formation. *Development* **2009**, *136*, 655–664. [CrossRef] [PubMed]
66. Stephen, L.A.; Davis, G.M.; McTeir, K.E.; James, J.; McTeir, L.; Kierans, M.; Bain, A.; Davey, M.G. Failure of centrosome migration causes a loss of motile cilia in talpid(3) mutants. *Dev. Dyn.* **2013**, *242*, 923–931. [CrossRef] [PubMed]
67. Davey, M.G.; Paton, I.R.; Yin, Y.; Schmidt, M.; Bangs, F.K.; Morrice, D.R.; Smith, T.G.; Buxton, P.; Stamataki, D.; Tanaka, M.; et al. The chicken talpid3 gene encodes a novel protein essential for Hedgehog signaling. *Genes Dev.* **2006**, *20*, 1365–1377. [CrossRef] [PubMed]
68. Bangs, F.; Antonio, N.; Thongnuek, P.; Welten, M.; Davey, M.G.; Briscoe, J.; Tickle, C. Generation of mice with functional inactivation of talpid3, a gene first identified in chicken. *Development* **2011**, *138*, 3261–3272. [CrossRef] [PubMed]
69. Sutherland, D.; Samakovlis, C.; Krasnow, M.A. branchless encodes a Drosophila FGF homolog that controls tracheal cell migration and the pattern of branching. *Cell* **1996**, *87*, 1091–1101. [CrossRef]
70. Bellusci, S.; Grindley, J.; Emoto, H.; Itoh, N.; Hogan, B.L. Fibroblast growth factor 10 (FGF10) and branching morphogenesis in the embryonic mouse lung. *Development* **1997**, *124*, 4867–4878. [PubMed]
71. Moura, R.S.; Coutinho-Borges, J.P.; Pacheco, A.P.; Damota, P.O.; Correia-Pinto, J. FGF signaling pathway in the developing chick lung: Expression and inhibition studies. *PLoS ONE* **2011**, *6*, e17660. [CrossRef] [PubMed]
72. Nogawa, H.; Ito, T. Branching morphogenesis of embryonic mouse lung epithelium in mesenchyme-free culture. *Development* **1995**, *121*, 1015–1022. [PubMed]
73. Metzger, R.J.; Krasnow, M.A. Genetic control of branching morphogenesis. *Science* **1999**, *284*, 1635–1639. [CrossRef] [PubMed]
74. Sekine, K.; Ohuchi, H.; Fujiwara, M.; Yamasaki, M.; Yoshizawa, T.; Sato, T.; Yagishita, N.; Matsui, D.; Koga, Y.; Itoh, N.; et al. Fgf10 is essential for limb and lung formation. *Nat. Genet.* **1999**, *21*, 138–141. [CrossRef] [PubMed]
75. De Moerlooze, L.; Spencer-Dene, B.; Revest, J.M.; Hajihosseini, M.; Rosewell, I.; Dickson, C. An important role for the IIIb isoform of fibroblast growth factor receptor 2 (FGFR2) in mesenchymal-epithelial signalling during mouse organogenesis. *Development* **2000**, *127*, 483–492. [PubMed]
76. Park, W.Y.; Miranda, B.; Lebeche, D.; Hashimoto, G.; Cardoso, W.V. FGF-10 is a chemotactic factor for distal epithelial buds during lung development. *Dev. Biol.* **1998**, *201*, 125–134. [CrossRef] [PubMed]
77. Colvin, J.S.; White, A.C.; Pratt, S.J.; Ornitz, D.M. Lung hypoplasia and neonatal death in Fgf9-null mice identify this gene as an essential regulator of lung mesenchyme. *Development* **2001**, *128*, 2095–2106. [CrossRef] [PubMed]
78. Moura, R.S.; Silva-Gonçalves, C.; Vaz-Cunha, P.; Correia-Pinto, J. Expression analysis of Shh signaling members in early stages of chick lung development. *Histochem. Cell Biol.* **2016**, *146*, 457–466. [CrossRef] [PubMed]
79. Lebeche, D.; Malpel, S.; Cardoso, W.V. Fibroblast growth factor interactions in the developing lung. *Mech. Dev.* **1999**, *86*, 125–136. [CrossRef]
80. Abler, L.L.; Mansour, S.L.; Sun, X. Conditional gene inactivation reveals roles for Fgf10 and Fgfr2 in establishing a normal pattern of epithelial branching in the mouse lung. *Dev. Dyn.* **2009**, *238*, 1999–2013. [CrossRef] [PubMed]
81. Herriges, J.C.; Verheyden, J.M.; Zhang, Z.; Sui, P.; Zhang, Y.; Anderson, M.J.; Swing, D.A.; Zhang, Y.; Lewandoski, M.; Sun, X. FGF-Regulated ETV Transcription Factors Control FGF-SHH Feedback Loop in Lung Branching. *Dev. Cell* **2015**, *35*, 322–332. [CrossRef] [PubMed]

82. White, A.C.; Xu, J.; Yin, Y.; Smith, C.; Schmid, G.; Ornitz, D.M. FGF9 and SHH signaling coordinate lung growth and development through regulation of distinct mesenchymal domains. *Development* **2006**, *133*, 1507–1517. [CrossRef] [PubMed]
83. Bellusci, S.; Henderson, R.; Winnier, G.; Oikawa, T.; Hogan, B.L. Evidence from normal expression and targeted misexpression that bone morphogenetic protein (Bmp-4) plays a role in mouse embryonic lung morphogenesis. *Development* **1996**, *122*, 1693–1702. [PubMed]
84. Moura, R.S.; Carvalho-Correia, E.; daMota, P.; Correia-Pinto, J. Canonical Wnt signaling activity in early stages of chick lung development. *PLoS ONE* **2014**, *9*, e112388. [CrossRef] [PubMed]
85. Li, C.; Xiao, J.; Hormi, K.; Borok, Z.; Minoo, P. Wnt5a participates in distal lung morphogenesis. *Dev. Biol.* **2002**, *248*, 68–81. [CrossRef] [PubMed]
86. Li, C.; Hu, L.; Xiao, J.; Chen, H.; Li, J.T.; Bellusci, S.; Delanghe, S.; Minoo, P. Wnt5a regulates Shh and Fgf10 signaling during lung development. *Dev. Biol.* **2005**, *287*, 86–97. [CrossRef] [PubMed]
87. Loscertales, M.; Mikels, A.J.; Hu, J.K.; Donahoe, P.K.; Roberts, D.J. Chick pulmonary Wnt5a directs airway and vascular tubulogenesis. *Development* **2008**, *135*, 1365–1376. [CrossRef] [PubMed]
88. Mahlapuu, M.; Enerbäck, S.; Carlsson, P. Haploinsufficiency of the forkhead gene Foxf1, a target for sonic hedgehog signaling, causes lung and foregut malformations. *Development* **2001**, *128*, 2397–2406. [PubMed]
89. Li, Y.; Zhang, H.; Choi, S.C.; Litingtung, Y.; Chiang, C. Sonic hedgehog signaling regulates Gli3 processing, mesenchymal proliferation, and differentiation during mouse lung organogenesis. *Dev. Biol.* **2004**, *270*, 214–231. [CrossRef] [PubMed]
90. Chapman, D.L.; Garvey, N.; Hancock, S.; Alexiou, M.; Agulnik, S.I.; Gibson-Brown, J.J.; Cebra-Thomas, J.; Bollag, R.J.; Silver, L.M.; Papaioannou, V.E. Expression of the T-box family genes, Tbx1-Tbx5, during early mouse development. *Dev. Dyn.* **1996**, *206*, 379–390. [CrossRef]
91. Lüdtke, T.H.; Farin, H.F.; Rudat, C.; Schuster-Gossler, K.; Petry, M.; Barnett, P.; Christoffels, V.M.; Kispert, A. Tbx2 controls lung growth by direct repression of the cell cycle inhibitor genes Cdkn1a and Cdkn1b. *PLoS Genet.* **2013**, *9*, e1003189. [CrossRef]
92. Sakiyama, J.; Yamagishi, A.; Kuroiwa, A. Tbx4-Fgf10 system controls lung bud formation during chicken embryonic development. *Development* **2003**, *130*, 1225–1234. [CrossRef] [PubMed]
93. Lüdtke, T.H.; Rudat, C.; Wojahn, I.; Weiss, A.C.; Kleppa, M.J.; Kurz, J.; Farin, H.F.; Moon, A.; Christoffels, V.M.; Kispert, A. Tbx2 and Tbx3 Act Downstream of Shh to Maintain Canonical Wnt Signaling during Branching Morphogenesis of the Murine Lung. *Dev. Cell* **2016**, *39*, 239–253. [CrossRef] [PubMed]
94. Wan, H.; Dingle, S.; Xu, Y.; Besnard, V.; Kaestner, K.H.; Ang, S.L.; Wert, S.; Stahlman, M.T.; Whitsett, J.A. Compensatory roles of Foxa1 and Foxa2 during lung morphogenesis. *J. Biol. Chem.* **2005**, *280*, 13809–13816. [CrossRef] [PubMed]
95. Jiang, Z.; Cushing, L.; Ai, X.; Lü, J. miR-326 is downstream of Sonic hedgehog signaling and regulates the expression of Gli2 and smoothened. *Am. J. Respir. Cell Mol. Biol.* **2014**, *51*, 273–283. [CrossRef] [PubMed]
96. Kugler, M.C.; Joyner, A.L.; Loomis, C.A.; Munger, J.S. Sonic hedgehog signaling in the lung. From development to disease. *Am. J. Respir. Cell Mol. Biol.* **2015**, *52*, 1–13. [CrossRef] [PubMed]
97. Watkins, D.N.; Berman, D.M.; Burkholder, S.G.; Wang, B.; Beachy, P.A.; Baylin, S.B. Hedgehog signalling within airway epithelial progenitors and in small-cell lung cancer. *Nature* **2003**, *422*, 313–317. [CrossRef] [PubMed]
98. Olsen, C.L.; Hsu, P.P.; Glienke, J.; Rubanyi, G.M.; Brooks, A.R. Hedgehog-interacting protein is highly expressed in endothelial cells but down-regulated during angiogenesis and in several human tumors. *BMC Cancer* **2004**, *4*, 43. [CrossRef] [PubMed]
99. Velcheti, V.; Govindan, R. Hedgehog signaling pathway and lung cancer. *J. Thorac. Oncol.* **2007**, *2*, 7–10. [CrossRef] [PubMed]
100. Abe, Y.; Tanaka, N. The Hedgehog Signaling Networks in Lung Cancer: The Mechanisms and Roles in Tumor Progression and Implications for Cancer Therapy. *Biomed. Res. Int.* **2016**, *2016*, 7969286. [CrossRef] [PubMed]

5

Contributions of Noncanonical Smoothened Signaling During Embryonic Development

Tanushree Pandit and Stacey K. Ogden *

Department of Cell and Molecular Biology, St. Jude Children's Research Hospital, Memphis, TN 38105, USA; tanushree.pandit@stjude.org
* Correspondence: stacey.ogden@stjude.org

Abstract: The Sonic Hedgehog (Shh) signaling pathway is active during embryonic development in metazoans, and provides instructional cues necessary for proper tissue patterning. The pathway signal transducing component, Smoothened (Smo), is a G protein-coupled receptor (GPCR) that has been demonstrated to signal through at least two effector routes. The first is a G protein–independent canonical route that signals to Gli transcriptional effectors to establish transcriptional programs specifying cell fate during early embryonic development. The second, commonly referred to as the noncanonical Smo signal, induces rapid, transcription-independent responses that are essential for establishing and maintaining distinct cell behaviors during development. Herein, we discuss contributions of this noncanonical route during embryonic development. We also highlight important open questions regarding noncanonical Smo signal route selection during development, and consider implications of noncanonical signal corruption in disease.

Keywords: Sonic Hedgehog; smoothened; signal transduction; development; morphogenesis; noncanonical signaling

1. Introduction

During embryogenesis, the Sonic Hedgehog (Shh) signal transduction network instructs cell fate decisions to drive tissue patterning. Corruption of Shh pathway regulation leads to developmental disorders including holoprosencephaly (HPE), Gorlin's Syndrome, Greig cephalopolysyndactyly, and Pallister-Hall syndrome, and in some cases results in early embryonic lethality [1]. As such, significant effort has been dedicated to understanding how the Shh pathway is activated and how its signal is propagated in ligand-receiving cells.

Vertebrates encode three Hh family ligands: Shh, Desert Hh, and Indian Hh, with Shh being expressed most broadly [2]. All Hh family ligands induce signaling in target cells by binding to Patched (Ptch) membrane receptors. Ptch receptors are twelve-pass transmembrane proteins that share homology with the bacterial Resistance, Nodulation, and Division Transporter superfamily [3,4]. In the absence of Shh, Ptch maintains the signal transducing G protein-coupled receptor (GPCR) Smoothened (Smo) in an inactive state. The mechanism by which Ptch represses Smo signaling is not yet known. However, based upon its relationship to select transporter proteins, the conventional model posits that Ptch controls availability of small molecule modulators of Smo activity [5]. Ptch contains a sterol sensing domain that is required for Smo repression [6], suggesting it may target molecules with sterol rings to control Smo activity. Consistent with this hypothesis, cholesterol and various oxysterols function as direct Smo agonists [7–9]. Sterols bind through a ligand binding pocket in the extracellular amino-terminal cysteine-rich domain (CRD) of Smo [10–13]. A second ligand binding pocket is present in the seven-transmembrane core of Smo, and has been demonstrated to bind natural and synthetic small molecules, endocannabinoids, and the free fatty acid arachidonic acid [14–18].

The precise mechanisms by which activation signals are transduced downstream of Smo have yet to be clearly established. However, it is generally accepted that Smo controls at least two effector routes. The first is the extensively studied canonical signal, which controls the Gli transcriptional effector function. Canonical signaling requires Smo translocation into a sensory organelle called the primary cilium, where it controls proteolytic processing of Gli transcription factors [19,20]. In the absence of a Smo signal, Gli2 and Gli3 are phosphorylated by cAMP-dependent protein kinase (PKA), which tags them for processing into truncated transcriptional repressors [21–23]. Ciliary Smo signaling halts Gli processing, allowing for stabilization of Gli2 and Gli3 as full-length transcriptional activators responsible for induction of Shh target genes. One such target is *Gli1*, which functions as a feed-forward activator to sustain or amplify target gene expression [22]. The majority of what is known about how Shh signaling impacts embryonic development stems from study of this effector route. Research focused on ciliary Smo signaling to Gli has been extensively reviewed [2,24–27] and so will not be discussed here. We will instead focus on Gli-independent Smo signaling, and discuss the comparatively limited body of work centered on contributions of this noncanonical signaling effector route during embryonic development.

2. Noncanonical Smoothened Signaling

The specific signaling events downstream of Smo that control noncanonical effectors are not yet clear, but in most cases, are thought to involve GPCR function of Smo [5,26]. Classical GPCRs initiate downstream signaling by activating heterotrimeric G proteins. These protein complexes consist of a guanine nucleotide binding Gα subunit bound to a G$\beta\gamma$ heterodimer. Activation of the heterotrimeric G protein complex occurs in response to ligand-induced conformational shifts of its partner GPCR, which stimulates exchange of GDP for GTP on the Gα subunit [28]. This triggers release of the heterotrimeric G protein complex from the GPCR, and separation of Gα from G$\beta\gamma$. Gα-GTP and free G$\beta\gamma$ subunits then signal to affect second messengers such as cAMP, cGMP, calcium, inositol triphosphate, diacylglycerol, and select gasses to instruct an appropriate cellular response [29–31].

Based upon the stereotypical GPCR topology of Smo, and the prominent role of PKA in regulating Gli repressor formation, it was initially predicted that Smo would signal to Gli through a Gαi heterotrimeric G protein intermediate [32,33]. Active Gαi proteins negatively regulate adenylyl cyclases to reduce intracellular cAMP and inhibit PKA activity, providing a direct route whereby Smo might block Gli repressor formation. Early studies using zebrafish embryos and *Xenopus* melanophores provided indirect support for this functionality [34,35]. A role for Gαi downstream of Smo was first suggested by studies in the zebrafish model system. Injection of mRNA encoding the potent Gαi inhibitor pertussis toxin (PTX) into zebrafish embryos led to patterning defects indicative of compromised Hh signaling. These defects included fusion of eyes, loss of ventral forebrain, and abnormal somite patterning [34]. Effects on somite patterning could be rescued by dominant negative PKA, but not by co-injection of Indian Hh RNA, suggesting PTX action downstream of ligand and upstream of PKA. In *Xenopus*, overexpression of Smo in pigment cells led to persistent pigment aggregation, a phenotype that can result from constitutive Gαi activity [35]. This effect was blocked by expression of dominant negative Gαi or by treatment with Gαi-inactivating PTX, supporting Gαi involvement in Smo-induced pigment aggregation.

Although the aforementioned studies provided circumstantial evidence implicating Gαi action in the Shh signaling cascade, results from in vivo studies aimed at directly testing G protein involvement in Gli regulation failed to reach a consensus. During neural tube development, Shh signaling specifies cell fate by inducing expression of specific transcription factors in discrete domains of the ventral neural tube [2]. In vivo studies aimed at determining whether Gαi modulation in the chick neural tube would alter Shh-induced fate determination suggested that neither inhibition nor constitutive activation of Gαi corrupted neural tube cell fate specification [36]. These results argued against a role for Gαi in regulation of Gli. However, subsequent studies in the chick neural tube, and in *Drosophila*, provided evidence that constitutively active Gαi could trigger phenotypes suggestive of ectopic Gli induction [37–39]. Moreover, in vitro interrogation of Smo-G protein coupling revealed that, in the absence of Ptch, Smo constitutively activated all Gαi/o family members including Gαz in a manner that was sensitive to the inverse Smo agonist cyclopamine [40]. Intriguingly, truncation of the intracellular carboxyl-terminal tail of Smo, which is necessary for entry into the primary cilium where it controls Gli processing, did not block Smo-activated Gαi signaling [41]. This finding indicated that signals to G proteins and Gli effectors could be separated, suggesting the existence of two independent Smo signaling arms: a canonical route controlling Gli and a noncanonical route involving Gαi, and controlling Shh-induced cellular responses that occur in a Gli-independent manner. Responses identified as being under the control of noncanonical Smo signaling including cytoskeletal modulation leading to cellular migration and axon guidance, axon fasciculation, neurotransmitter selection, cellular proliferation, and lipid metabolism are discussed in detail below.

3. Cytoskeletal Dynamics, Cellular Migration, and Axon Guidance

The Shh-regulated developmental processes most commonly linked to noncanonical Smo signaling are cellular migration and axon guidance, which are related cellular responses that can be regulated through analogous mechanisms [42]. In mouse fibroblasts, Smo signals through Gαi to activate Rac1 and RhoA GTPases, which promote cell migration by modulating cytoskeletal behavior [41]. This signal route is likely active during Shh-induced fibroblast chemotaxis, which has been demonstrated to occur in a Smo-dependent, Gli-independent manner. Although not specifically linked to the Smo-Gαi effector route, Shh-induced chemotaxis was found to be PTX-sensitive, and was maintained in cells expressing a Smo mutant that does not enter the primary cilium or signal to Gli [43]. Smo mutants that fail to enter the primary cilium maintain the ability to couple with Gαi [41], making it feasible that the chemotactic response occurs in a Gαi-dependent manner. Gαi involvement is further supported by the observation that metabolism of arachidonic acid through the 5-lipoxygenase pathway contributes to fibroblast chemotaxis [44]. We recently demonstrated that Smo signaling through Gαiβγ activates the lipid remodeling enzyme cPLA2α to produce arachidonic acid [18], providing a mechanism by which the free fatty acid would be made available for lipoxygenase activity.

Notably, the ability of Smo to affect cell migration through cytoskeletal behavior is not limited to fibroblasts. Smo has been demonstrated to signal though the Gαi-RhoA effector route to promote tubulogenesis of human endothelial cells [45], and to influence the cytoskeleton in osteoblast precursor cells [46]. In osteoblasts, over-activation of the Smo-Gαi signal occurs in response to genetic disruption of primary cilia function, and results in inappropriate induction of stress fibers [46]. These results indicate that ciliary dysfunction can lead to uncontrolled noncanonical Smo signaling, suggesting the balance of canonical and noncanonical signal output is influenced, at least in part, by Smo ciliary localization.

Neuronal cell types represent an additional cellular context in which noncanonical Smo signaling impacts cytoskeletal behavior. In motor neurons, Smo signaling through the 5-lipoxygenase pathway promotes formation of neurite projections, priming the neurons for establishment of proper connectivity at later developmental stages [47]. Shh signaling through Smo has also been shown to promote migration of oligodendrocyte precursors of the optic nerve in both mouse and chick systems [48]. As such, Smo signaling through G protein effectors may be active in multiple neuronal contexts to link Shh signaling with cytoskeletal behavior during development.

Another process involving cytoskeletal rearrangements that has been linked to noncanonical Smo signaling is commissural axon guidance (Figure 1). During spinal cord development, axons encounter attractive and repulsive guidance signals that must be properly integrated to specify an appropriate targeting route. For example, in the neural tube, dI1 dorsal interneurons must first extend ventrally toward the floor plate, cross the midline, and then turn rostrally to establish functional positioning (Figure 2) [49]. A number of guidance cues contribute to this process, including dorsal repulsive factors BMP and Draxin, and ventral attractive factors Netrin, Shh, and VEGF [49]. Shh was first identified as a floor plate-derived attractive cue based upon its ability to reorient axons of spinal cord explants [50].

Genetic and chemical inhibition studies revealed that Shh-mediated attraction was dependent upon Smo. However, the rapid response of axons to Shh suggested a transcription-independent mechanism, rendering Smo signaling through its canonical effector route unlikely. Consistent with this notion, Smo was demonstrated to function in a Gli-independent manner to guide axons [51]. In vitro axon guidance assays revealed that Shh, via Smo, induced phosphorylation and activation of Src family kinases (SFKs) to alter axon trajectories. Although the Shh co-receptor Boc was found to be a necessary upstream component for this functionality, the downstream noncanonical Smo effector route signaling to SFKs to alter the actin cytoskeleton was not identified [51,52]. Smo signaling though Gαi to activate SFKs is feasible, as Gαi family heterotrimeric G proteins have been documented to activate Src though their activated Gβγ subunits [53]. However, SFKs can also be directly activated by select GPCRs without the use of a heterotrimeric G protein intermediate [53], warranting further investigation into the precise mechanisms by which Smo signals to SFKs.

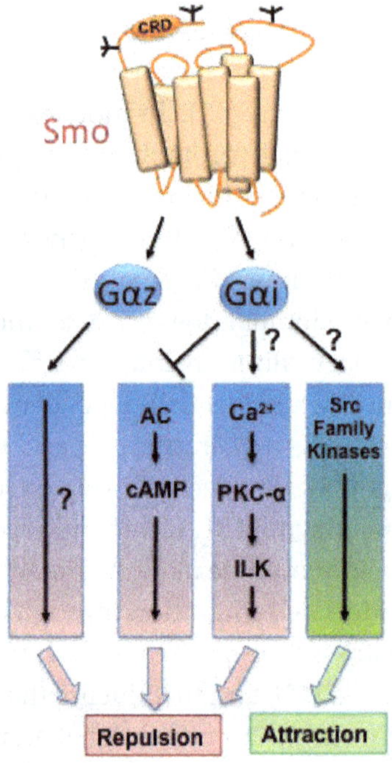

Figure 1. Noncanonical Smoothened signals contributing to axon guidance. The diagram summarizes noncanonical Smo signals that have been reported to influence axon guidance. The mechanism(s) by which Smo signals to modulate intracellular Ca^{2+} levels or control Src Family Kinase activation to influence neuronal axon guidance is not yet established, but we speculate signaling could occur through Gαi. Smo signals through Gαz to repel enteric neurons. Effectors acting downstream of Gαz in this context have not been defined. Repulsion is indicated in pink and attraction is indicated in green. CRD—cysteine rich domain ligand binding pocket. N-linked glycans in the Smo extracellular domain are indicated in black.

In addition to functioning as an attractant prior to midline crossing, floor plate-derived Shh also functions to repel commissural axons in a Smo-dependent manner post-crossing (Figure 2) [49]. In ovo RNAi experiments in the chick neural tube have suggested the repulsive role of Shh occurs independently of Smo signaling, and instead results from function of Hedgehog interacting protein (Hip) masking the Shh signal [54]. However, more recent work using rat spinal cords demonstrated that Shh signals in a Smo-dependent manner to activate axons for responsiveness to Semaphorins, which mediate repulsion upon midline crossing [55]. The shift to repulsion correlates with a Shh-induced decrease in axonal cAMP concentration, consistent with the established propensity of low cAMP and reduced PKA activity to favor axon repulsion [55,56]. Alterations of axon trajectories that were observed in rat spinal cords following loss of Shh or Smo could be recapitulated by forskolin-mediated cAMP production, which can functionally mimic the loss of Gαi activity [55]. As such, the Smo-Gαi effector route may contribute to commissural axon guidance both pre- and post-midline crossing. However, reduced PKA activity post-crossing has been demonstrated to involve 14-3-3 proteins binding PKA [57]. PKA-14-3-3 binding was found to stabilize PKA regulatory and catalytic subunit association, thereby locking PKA in an inactive state [57]. Function of 14-3-3 in axon repulsion was not determined to be under the control of Shh signaling, suggesting parallel PKA-regulatory pathways likely synergize with Shh to assure proper axon turning post-midline crossing.

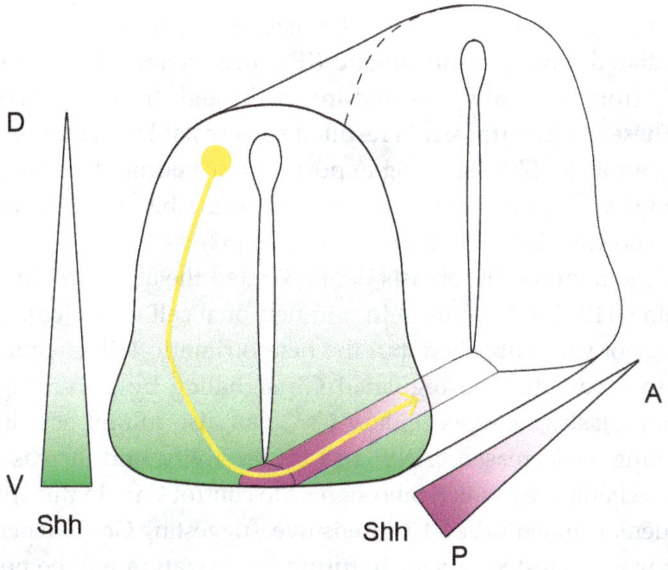

Figure 2. Shh-mediated commissural axon guidance cues. The diagram indicates the trajectory of a commissural axon (yellow) responding to Shh guidance cues in the neural tube. Dorsally (D)-localized dI1 neurons project their axons ventrally (V) toward chemo-attractive Shh (green) that is secreted from the floor plate (magenta). Upon crossing the floor plate, commissural axons change their response to Shh, resulting in axon repulsion. Shh-mediated repulsion prompts axons to target from posterior (P, Shh-high, magenta) to anterior (A, Shh-low, white) to facilitate proper terminal positioning.

Recent work suggests that, in addition to affecting axon guidance in the central nervous system, noncanonical Smo signaling is also utilized in the enteric nervous system to repel enteric axons of the gut. In this context, Smo signaling through the Gαi family member Gαz repels axons to prevent them from projecting into intestinal villi [58]. Gαz is the only Gαi family member that is resistant to PTX and is the family member most potently activated by Smo in in vitro assays [40]. Despite high Smo activity toward Gαz, enteric axon guidance is the only context in which Smo and Gαz have been functionally coupled. Further investigation is needed to ascertain whether Gαz has been overlooked in other contexts due to PTX insensitivity, or because Smo engages Gαz in a limited number of cell types. Intriguingly, Shh-induced enteric axon repulsion could not utilize Boc or Cdo co-receptors, and instead required the Gas1 co-receptor to be present at the enteric axonal terminals to induce the Gαz response [58]. These results may suggest that upstream receptors could contribute to G protein selectivity by Smo in different tissue contexts during development.

4. Axon Fasciculation

As axons travel toward their final targets, neighboring axons bundle as they grow, allowing them to track along paths determined by earlier pioneer axons [59]. Although an effect of Shh on fasciculation has not been reported in the neural tube, Shh-mediated cAMP reduction in chick retinal ganglion cell axons has been found to halt growth cone extension to allow for fasciculation of the optic commissure as it crosses the optic chiasm [60]. This response is dependent upon a Shh-induced Ca^{2+}

increase that triggers protein kinase Cα and integrin-linked kinase activation [61]. The documented ability of Smo to increase cellular Ca^{2+} concentration through Gαi [62] suggests these responses may occur in response to signaling through its noncanonical Gαi effector route.

5. Neurotransmitter Selection

Studies in *Xenopus* support that noncanonical Smo-Gαi signaling regulates post-mitotic neuronal differentiation through controlling neurotransmitter expression [62]. During spinal cord development, determination of the neurotransmitter phenotype is controlled by Ca^{2+} spikes in response to specific extracellular signals [62–64]. Shh treatment of *Xenopus* spinal cords has been observed to induce phospholipase C-mediated inositol triphosphate (IP3) production at the primary cilium, leading to the release of Ca^{2+} from internal stores and influx through transient receptor potential cation channel 1 (TRCP1). These calcium transients resulted in the establishment of a GABAergic neuronal phenotype, indicating a role for Shh signaling in post-mitotic neuron differentiation. Importantly, in this study, Shh-regulated Ca^{2+} modulation was found to be inhibited by PTX or constitutively active PKA, indicating that Smo-Gαi signaling is likely required [62].

Subsequent studies in mouse fibroblasts demonstrated the ability of Smo signaling to control ciliary Ca^{2+} by opening TRP Ca^{2+} channels in non-neuronal cell types [65]. The specific channels controlled by Smo are not yet established, but the heterotrimeric TRP channel PKD1-L1/PKD2-L1 has been proposed to modulate Smo-regulated Gli signaling by governing Shh effector ciliary trafficking [66]. In fibroblasts, increased ciliary Ca^{2+} can also inhibit activity of ciliary adenylyl cyclases 5 and 6, leading to decreased cAMP and PKA activity, and increased Gli transcriptional activity. The precise mechanism by which Smo signals to control Ca^{2+} in this cellular context was not determined, but was demonstrated to be PTX-insensitive, suggesting Gαi independence [65]. However, PTX-insensitive Gαz was not tested. As such, future interrogation will be necessary to determine whether Smo-regulated fibroblast ciliary Ca^{2+} modulation occurs through Smo-Gαz coupling or through a novel noncanonical Smo signal.

6. Cellular Proliferation

The ability of Shh to induce proliferative responses by activating mitogenic transcriptional programs is well established [67]. Recent studies demonstrate that, in addition to initiating a canonical signal to control cellular proliferation, Smo can signal through Gαi to induce a proliferative response [38,68]. Functional interrogation of Gαi in cerebellar granule neuronal precursors (CGNP) revealed that overexpression of active Gαi enhanced Shh-induced proliferation in this neuronal cell type. Knockdown of Gαi2 or Gαi3 in CGNPs reduced proliferation, supporting a requisite contribution of Gαi to the Shh/Smo induced proliferative response [38]. Similarly, in vivo expression of constitutively active SMOM2 in basal epithelial cells of the mammary gland was observed to induce proliferation of neighboring cells in a Gαi-dependent manner [68]. Chemical ablation of Gli revealed that the proliferative response in mammary epithelial cells was fully independent of Gli function, providing in vivo support for the ability of the noncanonical proliferative signal to occur in the absence of a Shh transcriptional response [68]. The Gαi-activated signal driving proliferation in adjacently localized mammary epithelial cells was not determined. Nevertheless, the ability of Smo-Gαi signaling to drive proliferation in diverse cell types suggests that the signal response may occur in multiple contexts during development to drive tissue morphogenesis.

7. Metabolism

An increasing body of work connects canonical and noncanonical Shh signaling with metabolism in both vertebrate and invertebrate systems [69]. Although early reports presented contradictory roles as to whether Shh signaling was a positive or negative regulator of adipogenesis, recent studies confirm that Shh signaling functions to inhibit adipose fate determination [69–72]. The inhibitory role of the Shh pathway in fat formation was first identified in *Drosophila* [70]. In this study, formation of the fly fat body was found to be inhibited by over-activation of Hh signaling, and promoted by inhibition of the pathway. In vitro studies in adipocytes revealed that these effects were conserved in vertebrate cells, with Shh signaling effectively diverting fate determination of pre-adipocytes towards an osteogenic program [70]. In a subsequent study, a genome-wide screen in *Drosophila* identified the Hh pathway as the top-scoring signaling cascade capable of modulating adipocyte fate determination [72]. This study revealed that Hh signaling modulated triglyceride levels in the fly fat body, and that these effects were conserved in vivo in mice. In both of these studies, effects occurred in a predominantly Gli-dependent manner. Moreover, a recent study revealed a specific Gli-code that is associated with altered lipid metabolism in the liver [73]. However, increasing cAMP levels in pre-adipocytes in vitro can rescue adipocyte differentiation in the presence of excessive Shh signaling [70,72], raising the possibility that a Smo-Gαi-cAMP signal axis may be capable of influencing adipocyte differentiation in coordination with canonical Smo signaling activity. The ability of the noncanonical arm to influence the canonical arm through Gαi activation is feasible; we recently reported that arachidonic acid produced in response to Smo signaling through Gαi$\beta\gamma$ directly associates with Smo to promote its ciliary entry and enhance signaling to Gli [18]. As such, activation of the noncanonical arm may function to amplify canonical signaling in select cell types during cell fate specification and tissue morphogenesis.

An aspect of metabolism that is more clearly linked to Shh-Smo-Gαi signaling is the rapid induction of Warburg-like metabolism in adipocytes [74]. This metabolic response is reminiscent of behavior observed in cancer cells that switch from oxidative phosphorylation to aerobic glycolysis in the presence of oxygen. The adipocyte metabolic switch was found to be Smo-dependent, PTX-sensitive, and rapid, ruling out a dependence on Gli transcriptional activity. Rather, the observed metabolic reprogramming occurred as a consequence of Smo signaling via Gαi to open plasma membrane Ca^{2+} channels, leading to phosphorylation-dependent activation of AMPK [74]. Importantly, this response could be induced by traditional Smo agonists, and by inverse agonists of the Gli effector route behaving as partial agonists for Gαi activation. These results revealed that canonical and noncanonical signal output can be functionally uncoupled, suggesting Smo can bias its signal output in response to differential small molecule binding.

8. Future Perspectives

The ability of Smo to signal through at least two distinct effector routes provides Shh signaling the opportunity to integrate with additional signaling pathways and impact wide-ranging biological responses during development. Although knowledge of these novel signaling connections is growing, a number of open questions regarding noncanonical Smo signaling have yet to be addressed. Key among them is identification of the molecular mechanisms controlling Smo effector route selection. As discussed above, studies in pre-adipocytes suggest that Smo effector route selection can be modulated through differential small molecule binding [74]. As such, Smo ligand selectivity may vary across cellular or temporal contexts to assure a proper balance of canonical and noncanonical effector activity during embryonic development.

An additional mechanism by which signal bias might be controlled is through post-translational modification of Smo. We recently identified that loss of Smo N-linked glycosylation can shift signaling away from Gαi and toward Gli [75]. We did not determine the mechanism by which de-glycosylation imparted Smo signal bias. However, we speculate that, as is likely the case with differential ligand binding, differential glycosylation shifts the active Smo conformation to favor one set of downstream effectors over another. Given that cellular glycosylation machinery varies across tissues [76], it is possible that the Smo N-glycosylation signature may exhibit tissue specificity. Future studies are needed to determine whether differential Smo N-glycosylation does indeed occur during development, and whether it correlates with overt signal bias.

It is well established in the literature that Smo sub-cellular localization controls its activity. In order to signal to Gli, Smo must traffic to the tip of the primary cilium [19]. Conversely, Smo is competent to signal to Gαi from outside the primary cilium and from the ciliary base [45,46,69]. As such, the speed of Smo ciliary entry and passage through the ciliary transition zone may dictate the duration or amplitude of noncanonical signal output. This hypothesis is supported by the aforementioned studies in osteoblast precursors demonstrating that genetic ablation of primary cilium function shifts Smo toward uncontrolled Gαi activation [46]. A physiological mechanism by which Smo ciliary entry is controlled may be through the modulation of ciliary membrane lipids. Consistent with this notion, the phosphatidylinositol 3-kinase PI3K-C2α has been demonstrated to be essential for Smo ciliary localization [77]. As such, PI3P may be targeted to influence Smo noncanonical/canonical signal transition. Future studies are needed to test this hypothesis, and to identify additional factors influencing Smo ciliary entry and impacting its noncanonical signaling activity.

Continued analysis of noncanonical Smo signaling is essential for a better understanding of the physiological role of Gli-independent Smo signaling during development. Going forward, the impact of noncanonical Smo signaling to disease will also need to be considered. Aberrant Shh signaling is causative in medulloblastoma and basal cell carcinoma, and is commonly activated in a number of additional tumor types [78]. Consequently, the Shh pathway is considered a strong candidate for targeted cancer therapy. Vismodegib, an FDA-approved Smo antagonist, has shown efficacy in the clinic. However, undesirable side effects and emergence of tumor resistance have occurred [79], necessitating improved methods for therapeutic intervention. Given the ability of Smo-Gli inhibitors to behave as partial agonists for Smo-Gαi, Gli-independent Smo effectors will need to be considered for future protocol iterations. As such, continued evaluation of physiological noncanonical Smo signaling in the context of development will be of utmost importance.

Acknowledgments: This work was supported by NIH NIGMS grant R35GM122546 to S.K.O. and by ALSAC of St. Jude Children's Research Hospital. We thank members of the Ogden lab for comments on the manuscript and S. Marada for assistance with figure preparation.

Author Contributions: S.K.O. and T.P. wrote the manuscript. T.P. made the figures.

Conflicts of Interest: The authors declare no conflict of interest.

References

1. Cohen, M.M., Jr. Hedgehog signaling update. *Am. J. Med. Genet. A* **2010**, *152A*, 1875–1914. [CrossRef] [PubMed]
2. Ingham, P.W.; McMahon, A.P. Hedgehog signaling in animal development: Paradigms and principles. *Genes Dev.* **2001**, *15*, 3059–3087. [CrossRef] [PubMed]
3. Marigo, V.; Davey, R.A.; Zuo, Y.; Cunningham, J.M.; Tabin, C.J. Biochemical evidence that patched is the Hedgehog receptor. *Nature* **1996**, *384*, 176–179. [CrossRef] [PubMed]
4. Motoyama, J.; Takabatake, T.; Takeshima, K.; Hui, C. Ptch2, a second mouse Patched gene is co-expressed with Sonic hedgehog. *Nat. Genet.* **1998**, *18*, 104–106. [CrossRef] [PubMed]
5. Arensdorf, A.M.; Marada, S.; Ogden, S.K. Smoothened Regulation: A Tale of Two Signals. *Trends Pharmacol. Sci.* **2016**, *37*, 62–72. [CrossRef] [PubMed]
6. Strutt, H.; Thomas, C.; Nakano, Y.; Stark, D.; Neave, B.; Taylor, A.M.; Ingham, P.W. Mutations in the sterol-sensing domain of Patched suggest a role for vesicular trafficking in Smoothened regulation. *Curr. Biol.* **2001**, *11*, 608–613. [CrossRef]
7. Huang, P.; Nedelcu, D.; Watanabe, M.; Jao, C.; Kim, Y.; Liu, J.; Salic, A. Cellular Cholesterol Directly Activates Smoothened in Hedgehog Signaling. *Cell* **2016**, *166*, 1176–1187. [CrossRef] [PubMed]
8. Luchetti, G.; Sircar, R.; Kong, J.H.; Nachtergaele, S.; Sagner, A.; Byrne, E.F.; Covey, D.F.; Siebold, C.; Rohatgi, R. Cholesterol activates the G-protein coupled receptor Smoothened to promote Hedgehog signaling. *eLife* **2016**, *5*. [CrossRef] [PubMed]
9. Nachtergaele, S.; Mydock, L.K.; Krishnan, K.; Rammohan, J.; Schlesinger, P.H.; Covey, D.F.; Rohatgi, R. Oxysterols are allosteric activators of the oncoprotein Smoothened. *Nat. Chem. Biol.* **2012**, *8*, 211–220. [CrossRef] [PubMed]

10. Byrne, E.F.; Sircar, R.; Miller, P.S.; Hedger, G.; Luchetti, G.; Nachtergaele, S.; Tully, M.D.; Mydock-McGrane, L.; Covey, D.F.; Rambo, R.P.; et al. Structural basis of Smoothened regulation by its extracellular domains. *Nature* **2016**, *535*, 517–522. [CrossRef] [PubMed]
11. Nachtergaele, S.; Whalen, D.M.; Mydock, L.K.; Zhao, Z.; Malinauskas, T.; Krishnan, K.; Ingham, P.W.; Covey, D.F.; Siebold, C.; Rohatgi, R. Structure and function of the Smoothened extracellular domain in vertebrate Hedgehog signaling. *eLife* **2013**, *2*, e01340. [CrossRef] [PubMed]
12. Nedelcu, D.; Liu, J.; Xu, Y.; Jao, C.; Salic, A. Oxysterol binding to the extracellular domain of Smoothened in Hedgehog signaling. *Nat. Chem. Biol.* **2013**, *9*, 557–564. [CrossRef] [PubMed]
13. Myers, B.R.; Sever, N.; Chong, Y.C.; Kim, J.; Belani, J.D.; Rychnovsky, S.; Bazan, J.F.; Beachy, P.A. Hedgehog pathway modulation by multiple lipid binding sites on the smoothened effector of signal response. *Dev. Cell* **2013**, *26*, 346–357. [CrossRef] [PubMed]
14. Khaliullina, H.; Bilgin, M.; Sampaio, J.L.; Shevchenko, A.; Eaton, S. Endocannabinoids are conserved inhibitors of the Hedgehog pathway. *Proc. Natl. Acad. Sci. USA* **2015**, *112*, 3415–3420. [CrossRef] [PubMed]
15. Taipale, J.; Chen, J.K.; Cooper, M.K.; Wang, B.; Mann, R.K.; Milenkovic, L.; Scott, M.P.; Beachy, P.A. Effects of oncogenic mutations in Smoothened and Patched can be reversed by cyclopamine. *Nature* **2000**, *406*, 1005–1009. [PubMed]
16. Chen, J.K.; Taipale, J.; Cooper, M.K.; Beachy, P.A. Inhibition of Hedgehog signaling by direct binding of cyclopamine to Smoothened. *Genes Dev.* **2002**, *16*, 2743–2748. [CrossRef] [PubMed]
17. Chen, J.K.; Taipale, J.; Young, K.E.; Maiti, T.; Beachy, P.A. Small molecule modulation of Smoothened activity. *Proc. Natl. Acad. Sci. USA* **2002**, *99*, 14071–14076. [CrossRef] [PubMed]
18. Arensdorf, A.M.; Dillard, M.E.; Menke, J.M.; Frank, M.W.; Rock, C.O.; Ogden, S.K. Sonic Hedgehog Activates Phospholipase A2 to Enhance Smoothened Ciliary Translocation. *Cell Rep.* **2017**, *19*, 2074–2087. [CrossRef] [PubMed]
19. May, S.R.; Ashique, A.M.; Karlen, M.; Wang, B.; Shen, Y.; Zarbalis, K.; Reiter, J.; Ericson, J.; Peterson, A.S. Loss of the retrograde motor for IFT disrupts localization of Smo to cilia and prevents the expression of both activator and repressor functions of Gli. *Dev. Biol.* **2005**, *287*, 378–389. [CrossRef] [PubMed]
20. Corbit, K.C.; Aanstad, P.; Singla, V.; Norman, A.R.; Stainier, D.Y.; Reiter, J.F. Vertebrate Smoothened functions at the primary cilium. *Nature* **2005**, *437*, 1018–1021. [CrossRef] [PubMed]
21. Aza-Blanc, P.; Ramirez-Weber, F.A.; Laget, M.P.; Schwartz, C.; Kornberg, T.B. Proteolysis that is inhibited by hedgehog targets Cubitus interruptus protein to the nucleus and converts it to a repressor. *Cell* **1997**, *89*, 1043–1053. [CrossRef]
22. Ruiz i Altaba, A. Gli proteins encode context-dependent positive and negative functions: Implications for development and disease. *Development* **1999**, *126*, 3205–3216. [PubMed]
23. Wang, B.; Fallon, J.F.; Beachy, P.A. Hedgehog-regulated processing of Gli3 produces an anterior/posterior repressor gradient in the developing vertebrate limb. *Cell* **2000**, *100*, 423–434. [CrossRef]
24. Jiang, J.; Hui, C.C. Hedgehog signaling in development and cancer. *Dev. Cell* **2008**, *15*, 801–812. [CrossRef] [PubMed]
25. Lee, R.T.; Zhao, Z.; Ingham, P.W. Hedgehog signalling. *Development* **2016**, *143*, 367–372. [CrossRef] [PubMed]
26. Robbins, D.J.; Fei, D.L.; Riobo, N.A. The Hedgehog signal transduction network. *Sci. Signal.* **2012**, *5*, re6. [CrossRef] [PubMed]
27. Mukhopadhyay, S.; Rohatgi, R. G-protein-coupled receptors, Hedgehog signaling and primary cilia. *Semin. Cell Dev. Biol.* **2014**. [CrossRef] [PubMed]
28. Lefkowitz, R.J. Historical review: A brief history and personal retrospective of seven-transmembrane receptors. *Trends Pharmacol. Sci.* **2004**, *25*, 413–422. [CrossRef] [PubMed]
29. Khan, S.M.; Sung, J.Y.; Hebert, T.E. Gbetagamma subunits-Different spaces, different faces. *Pharmacol. Res. Off. J. Ital. Pharmacol. Soc.* **2016**, *111*, 434–441.
30. Dupre, D.J.; Robitaille, M.; Rebois, R.V.; Hebert, T.E. The role of Gbetagamma subunits in the organization, assembly, and function of GPCR signaling complexes. *Annu. Rev. Pharmacol. Toxicol.* **2009**, *49*, 31–56. [CrossRef] [PubMed]
31. Wacker, D.; Stevens, R.C.; Roth, B.L. How Ligands Illuminate GPCR Molecular Pharmacology. *Cell* **2017**, *170*, 414–427. [CrossRef] [PubMed]
32. Alcedo, J.; Ayzenzon, M.; Von Ohlen, T.; Noll, M.; Hooper, J.E. The Drosophila smoothened gene encodes a seven-pass membrane protein, a putative receptor for the hedgehog signal. *Cell* **1996**, *86*, 221–232. [CrossRef]

33. Van den Heuvel, M.; Ingham, P.W. smoothened encodes a receptor-like serpentine protein required for hedgehog signalling. *Nature* **1996**, *382*, 547–551. [CrossRef] [PubMed]
34. Hammerschmidt, M.; McMahon, A.P. The effect of pertussis toxin on zebrafish development: A possible role for inhibitory G-proteins in hedgehog signaling. *Dev. Biol.* **1998**, *194*, 166–171. [CrossRef] [PubMed]
35. DeCamp, D.L.; Thompson, T.M.; de Sauvage, F.J.; Lerner, M.R. Smoothened activates Galphai-mediated signaling in frog melanophores. *J. Biol. Chem.* **2000**, *275*, 26322–26327. [CrossRef] [PubMed]
36. Low, W.C.; Wang, C.; Pan, Y.; Huang, X.Y.; Chen, J.K.; Wang, B. The decoupling of Smoothened from Galphai proteins has little effect on Gli3 protein processing and Hedgehog-regulated chick neural tube patterning. *Dev. Biol.* **2008**, *321*, 188–196. [CrossRef] [PubMed]
37. Ogden, S.K.; Fei, D.L.; Schilling, N.S.; Ahmed, Y.F.; Hwa, J.; Robbins, D.J. G protein Galphai functions immediately downstream of Smoothened in Hedgehog signalling. *Nature* **2008**, *456*, 967–970. [CrossRef] [PubMed]
38. Barzi, M.; Kostrz, D.; Menendez, A.; Pons, S. Sonic Hedgehog-induced proliferation requires specific Galpha inhibitory proteins. *J. Biol. Chem.* **2011**, *286*, 8067–8074. [CrossRef] [PubMed]
39. Vuolo, L.; Herrera, A.; Torroba, B.; Menendez, A.; Pons, S. Ciliary adenylyl cyclases control the Hedgehog pathway. *J. Cell Sci.* **2015**, *128*, 2928–2937. [CrossRef] [PubMed]
40. Riobo, N.A.; Saucy, B.; Dilizio, C.; Manning, D.R. Activation of heterotrimeric G proteins by Smoothened. *Proc. Natl. Acad. Sci. USA* **2006**, *103*, 12607–12612. [CrossRef] [PubMed]
41. Polizio, A.H.; Chinchilla, P.; Chen, X.; Kim, S.; Manning, D.R.; Riobo, N.A. Heterotrimeric Gi proteins link Hedgehog signaling to activation of Rho small GTPases to promote fibroblast migration. *J. Biol. Chem.* **2011**, *286*, 19589–19596. [CrossRef] [PubMed]
42. Meyer, G.; Feldman, E.L. Signaling mechanisms that regulate actin-based motility processes in the nervous system. *J. Neurochem.* **2002**, *83*, 490–503. [CrossRef] [PubMed]
43. Bijlsma, M.F.; Damhofer, H.; Roelink, H. Hedgehog-stimulated chemotaxis is mediated by smoothened located outside the primary cilium. *Sci. Signal.* **2012**, *5*, ra60. [CrossRef] [PubMed]
44. Bijlsma, M.F.; Borensztajn, K.S.; Roelink, H.; Peppelenbosch, M.P.; Spek, C.A. Sonic hedgehog induces transcription-independent cytoskeletal rearrangement and migration regulated by arachidonate metabolites. *Cell Signal.* **2007**, *19*, 2596–2604. [CrossRef] [PubMed]
45. Chinchilla, P.; Xiao, L.; Kazanietz, M.G.; Riobo, N.A. Hedgehog proteins activate pro-angiogenic responses in endothelial cells through non-canonical signaling pathways. *Cell Cycle* **2010**, *9*, 570–579. [CrossRef] [PubMed]
46. Yuan, X.; Cao, J.; He, X.; Serra, R.; Qu, J.; Cao, X.; Yang, S. Ciliary IFT80 balances canonical versus non-canonical hedgehog signalling for osteoblast differentiation. *Nat. Commun.* **2016**, *7*, 11024. [CrossRef] [PubMed]
47. Bijlsma, M.F.; Peppelenbosch, M.P.; Spek, C.A.; Roelink, H. Leukotriene synthesis is required for hedgehog-dependent neurite projection in neuralized embryoid bodies but not for motor neuron differentiation. *Stem Cells* **2008**, *26*, 1138–1145. [CrossRef] [PubMed]
48. Merchan, P.; Bribian, A.; Sanchez-Camacho, C.; Lezameta, M.; Bovolenta, P.; de Castro, F. Sonic hedgehog promotes the migration and proliferation of optic nerve oligodendrocyte precursors. *Mol. Cell. Neurosci.* **2007**, *36*, 355–368. [CrossRef] [PubMed]
49. De Ramon Francas, G.; Zuniga, N.R.; Stoeckli, E.T. The spinal cord shows the way—How axons navigate intermediate targets. *Dev. Biol.* **2016**. [CrossRef] [PubMed]
50. Charron, F.; Stein, E.; Jeong, J.; McMahon, A.P.; Tessier-Lavigne, M. The morphogen sonic hedgehog is an axonal chemoattractant that collaborates with netrin-1 in midline axon guidance. *Cell* **2003**, *113*, 11–23. [CrossRef]
51. Yam, P.T.; Langlois, S.D.; Morin, S.; Charron, F. Sonic hedgehog guides axons through a noncanonical, Src-family-kinase-dependent signaling pathway. *Neuron* **2009**, *62*, 349–362. [CrossRef] [PubMed]
52. Okada, A.; Charron, F.; Morin, S.; Shin, D.S.; Wong, K.; Fabre, P.J.; Tessier-Lavigne, M.; McConnell, S.K. Boc is a receptor for sonic hedgehog in the guidance of commissural axons. *Nature* **2006**, *444*, 369–373. [CrossRef] [PubMed]
53. Luttrell, D.K.; Luttrell, L.M. Not so strange bedfellows: G-protein-coupled receptors and Src family kinases. *Oncogene* **2004**, *23*, 7969–7978. [CrossRef] [PubMed]

54. Bourikas, D.; Pekarik, V.; Baeriswyl, T.; Grunditz, A.; Sadhu, R.; Nardo, M.; Stoeckli, E.T. Sonic hedgehog guides commissural axons along the longitudinal axis of the spinal cord. *Nat. Neurosci.* **2005**, *8*, 297–304. [CrossRef] [PubMed]
55. Parra, L.M.; Zou, Y. Sonic hedgehog induces response of commissural axons to Semaphorin repulsion during midline crossing. *Nat. Neurosci.* **2010**, *13*, 29–35. [CrossRef] [PubMed]
56. Song, S.K.; Choi, S.Y.; Kim, K.T. Opposing effects of protein kinase A and C on capacitative calcium entry into HL-60 promyelocytes. *Biochem. Pharmacol.* **1998**, *56*, 561–567. [CrossRef]
57. Yam, P.T.; Kent, C.B.; Morin, S.; Farmer, W.T.; Alchini, R.; Lepelletier, L.; Colman, D.R.; Tessier-Lavigne, M.; Fournier, A.E.; Charron, F. 14–3-3 proteins regulate a cell-intrinsic switch from sonic hedgehog-mediated commissural axon attraction to repulsion after midline crossing. *Neuron* **2012**, *76*, 735–749. [CrossRef] [PubMed]
58. Jin, S.; Martinelli, D.C.; Zheng, X.; Tessier-Lavigne, M.; Fan, C.M. Gas1 is a receptor for sonic hedgehog to repel enteric axons. *Proc. Natl. Acad. Sci. USA* **2015**, *112*, E73–E80. [CrossRef] [PubMed]
59. Learte, A.R.; Hidalgo, A. The role of glial cells in axon guidance, fasciculation and targeting. *Adv. Exp. Med. Biol.* **2007**, *621*, 156–166. [PubMed]
60. Trousse, F.; Marti, E.; Gruss, P.; Torres, M.; Bovolenta, P. Control of retinal ganglion cell axon growth: A new role for Sonic hedgehog. *Development* **2001**, *128*, 3927–3936. [PubMed]
61. Guo, D.; Standley, C.; Bellve, K.; Fogarty, K.; Bao, Z.Z. Protein kinase Calpha and integrin-linked kinase mediate the negative axon guidance effects of Sonic hedgehog. *Mol. Cell. Neurosci.* **2012**, *50*, 82–92. [CrossRef] [PubMed]
62. Belgacem, Y.H.; Borodinsky, L.N. Sonic hedgehog signaling is decoded by calcium spike activity in the developing spinal cord. *Proc. Natl. Acad. Sci. USA* **2011**, *108*, 4482–4487. [CrossRef] [PubMed]
63. Borodinsky, L.N.; Root, C.M.; Cronin, J.A.; Sann, S.B.; Gu, X.; Spitzer, N.C. Activity-dependent homeostatic specification of transmitter expression in embryonic neurons. *Nature* **2004**, *429*, 523–530. [CrossRef] [PubMed]
64. Marek, K.W.; Kurtz, L.M.; Spitzer, N.C. cJun integrates calcium activity and tlx3 expression to regulate neurotransmitter specification. *Nat. Neurosci.* **2010**, *13*, 944–950. [CrossRef] [PubMed]
65. Moore, B.S.; Stepanchick, A.N.; Tewson, P.H.; Hartle, C.M.; Zhang, J.; Quinn, A.M.; Hughes, T.E.; Mirshahi, T. Cilia have high cAMP levels that are inhibited by Sonic Hedgehog-regulated calcium dynamics. *Proc. Natl. Acad. Sci. USA* **2016**, *113*, 13069–13074. [CrossRef] [PubMed]
66. Delling, M.; DeCaen, P.G.; Doerner, J.F.; Febvay, S.; Clapham, D.E. Primary cilia are specialized calcium signalling organelles. *Nature* **2013**, *504*, 311–314. [CrossRef] [PubMed]
67. Ulloa, F.; Briscoe, J. Morphogens and the control of cell proliferation and patterning in the spinal cord. *Cell Cycle* **2007**, *6*, 2640–2649. [CrossRef] [PubMed]
68. Villanueva, H.; Visbal, A.P.; Obeid, N.F.; Ta, A.Q.; Faruki, A.A.; Wu, M.-F.; Hilsenbeck, S.G.; Shaw, C.A.; Yu, P.; Plummer, N.W.; et al. An essential role for G-alpha-i2 in Smoothened-stimulated epithelial cell proliferation in the mammary gland. *Sci. Signal.* **2015**, *8*. [CrossRef] [PubMed]
69. Teperino, R.; Aberger, F.; Esterbauer, H.; Riobo, N.; Pospisilik, J.A. Canonical and non-canonical Hedgehog signalling and the control of metabolism. *Semin. Cell Dev. Biol.* **2014**. [CrossRef] [PubMed]
70. Suh, J.M.; Gao, X.; McKay, J.; McKay, R.; Salo, Z.; Graff, J.M. Hedgehog signaling plays a conserved role in inhibiting fat formation. *Cell Metab.* **2006**, *3*, 25–34. [CrossRef] [PubMed]
71. Kopinke, D.; Roberson, E.C.; Reiter, J.F. Ciliary Hedgehog Signaling Restricts Injury-Induced Adipogenesis. *Cell* **2017**, *170*, 340–351. [CrossRef] [PubMed]
72. Pospisilik, J.A.; Schramek, D.; Schnidar, H.; Cronin, S.J.; Nehme, N.T.; Zhang, X.; Knauf, C.; Cani, P.D.; Aumayr, K.; Todoric, J.; et al. Drosophila genome-wide obesity screen reveals hedgehog as a determinant of brown versus white adipose cell fate. *Cell* **2010**, *140*, 148–160. [CrossRef] [PubMed]
73. Matz-Soja, M.; Rennert, C.; Schonefeld, K.; Aleithe, S.; Boettger, J.; Schmidt-Heck, W.; Weiss, T.S.; Hovhannisyan, A.; Zellmer, S.; Kloting, N.; et al. Hedgehog signaling is a potent regulator of liver lipid metabolism and reveals a GLI-code associated with steatosis. *eLife* **2016**, *5*. [CrossRef] [PubMed]
74. Teperino, R.; Amann, S.; Bayer, M.; McGee, S.L.; Loipetzberger, A.; Connor, T.; Jaeger, C.; Kammerer, B.; Winter, L.; Wiche, G.; et al. Hedgehog partial agonism drives Warburg-like metabolism in muscle and brown fat. *Cell* **2012**, *151*, 414–426. [CrossRef] [PubMed]

75. Marada, S.; Navarro, G.; Truong, A.; Stewart, D.P.; Arensdorf, A.M.; Nachtergaele, S.; Angelats, E.; Opferman, J.T.; Rohatgi, R.; McCormick, P.J.; et al. Functional Divergence in the Role of N-Linked Glycosylation in Smoothened Signaling. *PLoS Genet.* **2015**, *11*, e1005473. [CrossRef] [PubMed]
76. Boscher, C.; Dennis, J.W.; Nabi, I.R. Glycosylation, galectins and cellular signaling. *Curr. Opin. Cell Biol.* **2011**, *23*, 383–392. [CrossRef] [PubMed]
77. Franco, I.; Gulluni, F.; Campa, C.C.; Costa, C.; Margaria, J.P.; Ciraolo, E.; Martini, M.; Monteyne, D.; De Luca, E.; Germena, G.; et al. PI3K class II alpha controls spatially restricted endosomal PtdIns3P and Rab11 activation to promote primary cilium function. *Dev. Cell* **2014**, *28*, 647–658. [CrossRef] [PubMed]
78. Barakat, M.T.; Humke, E.W.; Scott, M.P. Learning from Jekyll to control Hyde: Hedgehog signaling in development and cancer. *Trends Mol. Med.* **2011**, *16*, 337–348. [CrossRef] [PubMed]
79. Yauch, R.L.; Dijkgraaf, G.J.; Alicke, B.; Januario, T.; Ahn, C.P.; Holcomb, T.; Pujara, K.; Stinson, J.; Callahan, C.A.; Tang, T.; et al. Smoothened mutation confers resistance to a Hedgehog pathway inhibitor in medulloblastoma. *Science* **2009**, *326*, 572–574. [CrossRef] [PubMed]

6

The Hedgehog Signaling Pathway Emerges as a Pathogenic Target

Margery G. Smelkinson

Biological Imaging Section, Research Technologies Branch, National Institute of Allergy and Infectious Diseases, National Institutes of Health, Bethesda, MD 20892, USA; msmelkinson@gmail.com

Abstract: The Hedgehog (Hh) signaling pathway plays an essential role in the growth, development, and homeostatis of many tissues in vertebrates and invertebrates. Much of what is known about Hh signaling is in the context of embryonic development and tumor formation. However, a growing body of evidence is emerging indicating that Hh signaling is also involved in postnatal processes such as tissue repair and adult immune responses. To that extent, Hh signaling has also been shown to be a target for some pathogens that presumably utilize the pathway to control the local infected environment. In this review, we discuss what is currently known regarding pathogenic interactions with Hh signaling and speculate on the reasons for this pathway being a target. We also hope to shed light on the possibility of using small molecule modulators of Hh signaling as effective therapies for a wider range of human diseases beyond their current use in a limited number of cancers.

Keywords: Hedgehog; Gli; influenza; Epstein Barr virus; Hepatitis B and C; HIV; H. pylori; fibrosis; immunity

---

1. Introduction: Basics of the Hedgehog Signaling Pathway and Its Evolutionary Conservation

The Hedgehog (Hh) family of ligands are secreted signaling molecules essential for embryogenesis and tissue homeostatis in the adult [1–3]. The Hh pathway was originally identified for its developmental role in patterning the *Drosophila* embryo, and Hh pathway components have subsequently been found to be remarkably conserved between invertebrates and vertebrates where the pathway also plays key roles in several types of cancers, some of which are being treated with drugs that inhibit signaling [3,4].

The Hh ligand acts as a morphogen to control cellular fates by signaling at narrowly defined durations and concentrations [1,2]. In the absence of Hh, the primary transcriptional effector(s)—Cubitus interruptus (Ci) in Drosophila and Gli-2 and Gli-3 in mammals—are tethered to microtubules in the fly or primary cilia in mammals through a transcription factor inhibitory complex (TFIC) (Figure 1A). This complex contains kinases responsible for phosphorylation and consequent partial degradation of Gli/Ci into transcriptional repressors that lack the C-terminal transactivation domain and nuclear export sequences [1–3]. As these shorter protein sequences retain their N-terminal nuclear localization signals, they readily enter the nucleus and inhibit expression of a subset of target genes.

Cells that express the 12-pass transmembrane protein, Patched (ptc), along with coreceptors can receive the Hh ligand and transduce the signal [1,2]. Hh binding to Ptc initiates a phosphorylation cascade of the G-protein coupled receptor-like, seven-pass transmembrane domain protein Smoothened (Smo) resulting in its surface accumulation and a conformational change that is important for triggering downstream effects (Figure 1B). Although there is some divergence between flies and mammals in steps that relay Smo activation to downstream components [2,5], the culmination in all species is a disruption of the TFIC. This leads to the activation of some Hh target genes simply by inhibiting repressor formation. However, further activation via Fused in the fly [6–8] or de-repression by Su(fu) in mammals [9–11] results in the conversion of Gli/Ci into a labile transcription factor that can activate the full range of targets.

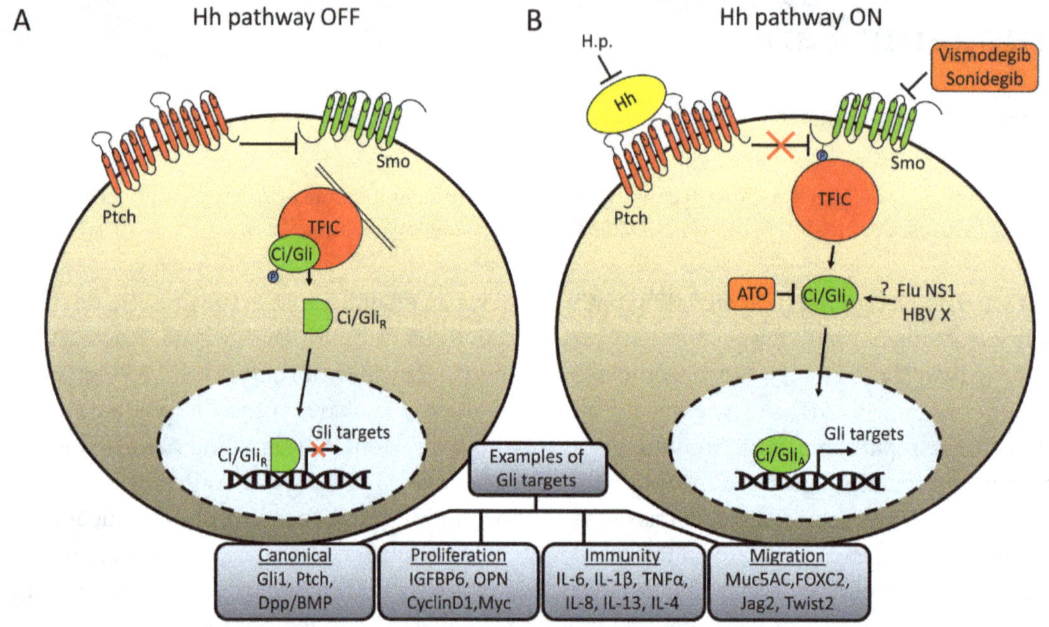

Figure 1. Schematic illustration of Hh signaling and how pathogens may modify pathway activity. (**A**) In the absence of Hh ligand, the receptor Ptch inhibits Smo activation. Ci/Gli is retained in the cytoplasm tethered to microtubules (in flies) or primary cilia (in mammals) through a complex of several proteins, simplistically termed here as "transcription factor inhibitory complex" (TFIC). This complex promotes Ci/Gli phosphorylation, which results in partial proteolysis to a repressor form that can readily enter the nucleus and repress expression of some Gli targets. (**B**) When Hh ligand binds to Ptch, inhibition upon Smo is relieved and the C-terminus of Smo is phosphorylated which promotes the release of Ci/Gli from the TFIC. The activated form of Gli/Ci can enter the nucleus and activate expression of Hh targets. Examples of canonical Gli target genes as well as those involved specifically in proliferation, immunity, and migration are given. Helicobacter Pylori (H.p.) has been proposed to act, in part, by repressing expression of the Hh ligand. Influenza NS1 and HBV X protein have been proposed to interact directly with Gli/Ci, but the precise mechanism by which they affect transcriptional activity has not been fully elucidated. The diagram also shows the pathway components which can be inhibited by FDA-approved small molecules: Vidmodegib and Sonidegib inhibit the activity of Smo, whereas arsenic trioxide (ATO) inhibits the activity of Gli1/2.

One of the key differences between flies and mammals is the redundancy in pathway components [2]. Whereas Drosophila has only one Hh ligand and one primary receptor, mammals have three ligands (Sonic Hedgehog-Shh, Indian Hedgehog-Ihh, and Desert Hedgehog-Dhh) and two primary receptors (Ptch1 and Ptch2). Similarly, Drosophila has only one transcriptional effector, Ci, which can act as both an activator and repressor, whereas mammals have three, Gli1, Gli2, and Gli3, where Gli2 is the primary activator, Gli3 is the primary repressor, and Gli1 is a target gene that acts as an activator in a positive feedback loop [12]. There are also differences in how the message is relayed from Smo to downstream components. Namely, mammalian cells require intracellular transport components housed in primary cilia [13]; this subcellular compartmentalization is not required in Drosophila [5]. Regardless of these differences, the method by which canonical Hh signals as a morphogen in all systems is understood to be by creating precise balances between repressor and activator forms of Gli/Ci, which activate target genes with varying numbers and affinities of transcription factor binding sites [14,15].

Non-canonical pathways have also been described more recently which are independent of Gli activity or independent of Hh activation, suggesting that the influence of the Hh pathway on cellular processes may be far more expansive than originally thought [16]. If this is the case, presumably regulating this pathway provides a broad target to pathogens, such as viruses, that have a limited number of genes and effector avenues for controlling the host environment.

2. Hh Signaling as a Target of Pathogens

While much is known about Hh signaling in embryogenesis [12,17], it has been shown, more recently, that signaling also occurs postnatally in several tissue types such as the skin, lung, gut, and within the haemopoatic and immune systems [18–25]. A clear role has been established for this postnatal signaling in maintaining tissue homeostasis, stem cell maintenance, and regulating haematopoiesis and lymphopoiesis [25–30]. However, more recently, it has also been shown that Hh signaling is activated in response to encounters with pathogens, wounding, and/or damaging agents. As the body of evidence supporting Hh signaling in these latter processes grows, perhaps unsurprisingly, so does the data demonstrating the Hh pathway as a target of several pathogens. Examples of Hh-interacting pathogens are discussed below.

Hepatitis B and C (HBV and HCV, respectively) are major causes of liver cirrhosis and hepatocellular carcinoma (HCC) worldwide [31]. Initial findings by Pereira et al. showed that livers from patients with chronic HBV and HCV infection displayed an increase in hepatocyte production of Hh ligands and an accumulation of Hh-responsive cells with higher levels of pathway activity correlating to more dire outcomes [32]. Complementary studies further confirmed that treatment of liver cells in vitro with the whole HBV replicon or with serum from HCV-infected patients increased expression of Hh targets in a Gli-dependent manner [33,34] and led to pro-fibrotic effects [33]. In the case of HBV, the viral protein causing this effect was revealed to be HBV X (Figure 1B) which increases Gli1 protein stability and promotes nuclear accumulation, an interaction shown likely to be direct through a series of in vitro binding assays [34–36]. While the precise motive by which these viral activities promote infections remains unclear, Choi et al. showed that increasing Hh signaling in liver cells promoted permissiveness for HCV replication, implying the presence of a positive feedback loop between pathway activation and viral production [37].

Hh target genes are also increased in Epstein–Barr virus (EBV)-derived nasopharyngeal carcinoma (NPC) tissue, NPC-derived cell lines, and in EBV infected epithelial cells, in vitro [38]. Further mechanistic studies revealed that the EBV-dependent increase in Gli1 expression correlated to a decrease in expression of Human Leukocyte Antigen, which is involved in presentation of viral antigens to cytotoxic T cells, which may limit the recognition of EBV by the immune system [39]. Hh activity is also upregulated in mice kidneys afflicted with HIV-associated nephropathy and in a human podocyte cell line infected with HIV [40]. This increase was associated with enhanced expression of proliferation and migration markers, loss of kidney filtration barrier function, and increased permeability, which could presumably augment viral dissemination and decrease host defenses, among other effects.

Helicobacter Pylori (H.p.) is a gram-negative bacterium found in the stomach and is the major cause of chronic atrophic gastritis and gastric cancer worldwide [41,42]. During the early stages of H.p. infection, Hh was shown to be upregulated in HCl-secreting parietal cells in an in vivo mouse infection model and in epithelial cell cultures grown in vitro from whole dissociated gastric glands [43–45]. This upregulation of signaling appeared to be due, in part, to the inflammation/repair response as this promoted macrophage recruitment to the infected area and was dependent upon NFkB induction. Prolonged exposure of H.p., however, resulted in a decrease of Hh expression in the gastric epithelium and an associated loss of parietal cells in the gastric gland in Mongolian gerbils, an established animal model for H. pylori infection, and in patients with H.p.-dependent chronic gastritis analyzed postmortem [46,47]. Eradication of H.p. infection could often restore Hh expression and frequently reversed tumor transformations [48–51]. This decrease of Hh signaling over the course of H.p. infection is partially in response to macrophage secretion of regulatory cytokines [43]. However, it has been also been proposed that H.p.-dependent upregulation of caudal-type homeobox 2 (CXD2), a protein that can bind directly to the promotor of Shh, is responsible for repressing expression, suggesting that this may be an evolved mechanism for regulating Hh pathway activity [52–54].

More recently, we reported that the Hh signaling pathway is also a direct target of the influenza A virus, linking pathway activation to the pathology of the virus [55]. This observation was originally made using *Drosophila melanogaster* as a model organism to screen for cell non-autonomous activities of disease genes and also an organism where Hh signaling has been shown recently to be linked to innate immunity [56–60].

In larval wing precursors (Figure 2A) expressing the influenza gene Nonstructural protein 1 (NS1), we detected a dramatic increase in expression of the Hh target gene, *decapentaplegic* (*dpp*) (Figure 2B), as well as the Dpp target, pMad, and a corresponding increase in distance between the adult wing veins L3 and L4, phenotypes indicative of increased Hh signaling [55,61,62]. Expression of BMP2, the mammalian homologue of *dpp*, and the Hh target, Ptch1, was also enhanced cell-autonomously in infected mouse lungs, indicating that the effect of NS1 on Hh signaling is conserved between species (Figure 2C,D) [55]. As NS1 is one of 14 proteins encoded by the influenza A virus whose main role is as a virulence factor to interact with host proteins to promote viral growth and maturation, it is perhaps expected that this protein may have evolved to exploit a conserved signaling pathway [63–65].

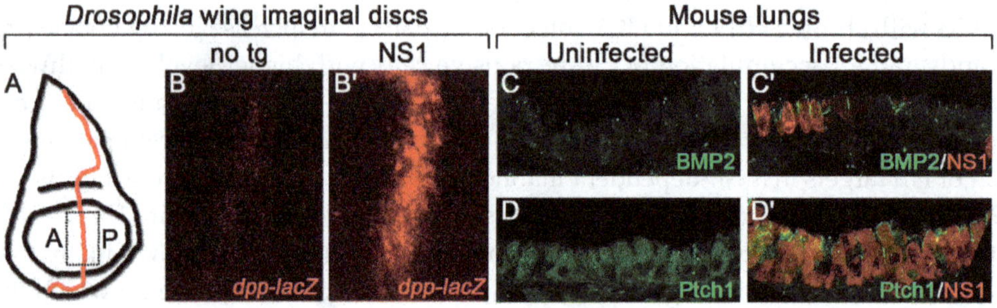

Figure 2. Influenza NS1 increases Hh target gene expression in *Drosophila* and mice. (**A**) Schematic diagram of a *Drosophila* wing imaginal disc with the anterior/posterior (A/P) border, the domain where Hh signaling is active, demarcated in red. The dashed box represents the area of the disc that was imaged in B. (**B**) Influenza NS1, expressed from a transgene in the wing imaginal disc, increases expression of the Hh target gene reporter, *dpp-lacZ*, at the A/P border compared with a disc with no transgene (no tg). (**C,D**) Infected mouse lungs show a cell-autonomous increase in expression of the Hh targets, BMP2 (**C**) and Ptch1 (**D**) compared with uninfected lungs. Target proteins are in green, NS1 is in red.

Using *Drosophila*, a forward genetic screen was performed to identify point mutations that might disrupt the Hh-dependent activity of NS1. A point mutation was recovered resulting in an alanine to valine substitution at position 122 that reduced this activity in flies and in transfected cells [55]. Interestingly, however, when the A122V mutation was incorporated into a mouse-adapted influenza A virus, it cell-autonomously enhanced expression of some Hh targets in the mouse lung and significantly hastened lethality. These results indicate that, in addition to its multiple intracellular functions, NS1 plays a vital role in activating—but at the same time, mitigating—the activity of a highly conserved signaling pathway during infection. Interestingly, no mutation at position 122 of NS1 has been identified previously in any influenza strain, which may reflect the critical role of NS1 in restraining signaling in order to protect the host. This muting of Hh signaling may be utilized to dampen deleterious effects that could be caused by Hh signaling (discussed below) to ensure optimal viral maturation prior to dissemination.

The precise mechanism by which Hepatitis viruses, HIV, EBV, and H.p. interact with Hh signaling has not been completely elucidated, however, all interactions and often the deleterious effects could be blocked by eliminating Gli [34,39,66] and/or by blocking Hh signaling with potent inhibitors [33,35,38,40].

Similarly, in flies, NS1-dependent pathway activation was blocked by over-expressing the receptor Ptc (which recruits Hh ligand and prevents it from dissipating to other cells), an RNA interference (RNAi) molecule directed to Ci, or other factors with inhibitory roles in signal transduction [55]. Likewise, constitutively activating signaling using a phosphomimetic form of Smoothened promoted activity. Thus, similar to other viruses, influenza NS1 requires pathway activation to upregulate Hh target genes.

The mechanism by which NS1 modulates Hh signaling was analyzed using the extensive array of genetic and molecular tools available when using *Drosophila* as a model organism to study disease genes. For example, a series of genetic epistasis experiments and quantitative imaging experiments using Förster resonance energy transfer-Fluorecence lifetime imaging (FRET-FLIM) assays revealed that NS1 interacts directly with the transcriptional mediator, Ci/Gli1, parroting the HBV X/Gli1 interaction [36,55]. Furthermore, this analysis showed that the A122V mutation significantly impeded this interaction. Importantly, pathway activity was not required simply to stabilize the full-length form of Ci and eliminate repressor formation, but additional pathway components with positive roles in signaling were also required for NS1 activity. Taken together, these data suggest that viral proteins likely do not activate the Hh pathway or expression of target genes themselves directly, but rather that Hh signaling is first activated by canonical signaling and then a viral interaction intervenes to alter the readout of the pathway activity. How Hh signaling is initially activated during infection remains an intriguing question.

3. Why Might Hh Signaling Be a Frequent Target for Pathogens?

To ensure efficient use of their genomes, pathogenic interactions are usually established with more centrally connected host proteins so that several processes may be targeted concomitantly. These virulence factors constantly evolve their interface residues, either to evade or to optimize their binding capabilities to host proteins [67]. Thus, some viruses may have evolved the ability to modulate Hh signaling due to its substantial involvement in several processes such as wound healing and the role it plays in immunity.

3.1. Hh Signaling Modulates Wound Repair and Tissue Fibrosis

As Hh signaling has significant roles in tissue homeostasis and remodeling, we speculated that Hh signaling may be activated as a part of a defense mechanism during infection to promote tissue repair (Figure 3). Indeed, a firm role for Hh signaling in wound repair has been established in several cells types such as the lung, skin, and pancreas where damaged areas correlate with Hh activation.

In the lung, acute activation of Shh signaling in epithelial cells was observed upon naphthalene injury, which was strikingly similar to the activation observed during prenatal development [68]. Shh is also upregulated in lung homogenates during hyperoxia-induced injury and recovery [69] and bleomycin administration [70], as well as around damaged airways in lungs chronically exposed to fluorescein isothiocyanate (FITC) [22,71].

In the skin, Asai et al. found that Ptch1 expression is upregulated in wounds in mice and that *in vitro* stimulation of cultured skin cells with Hh ligand promoted production of angiogenic signals, increased proliferation of fibroblasts, and increased migration, adhesion, and tube formation of endothelial progenitor cells to aid in wound closure [72]. Furthermore, topical gene therapy treatment on wounds using DNA encoding Shh accelerated wound recovery, while commensurate studies showed that wound closure, vascularization, and proliferation were inhibited at wound sites by intraperitoneally injected cylopamine, a potent Smoothened antagonist [21,72–74].

Similarly, Shh signaling in pancreatic fibroblasts is required for tissue repair during the onset of pancreatic cancer and pancreatitis due to a mutation in the proto-oncogene, K-ras, as a deletion of one allele of Gli1 or mice lacking expression of Shh in the pancreatic epithelium significantly impaired tissue remodeling and cytokine expression in this model [75]. This study also confirmed the presence of Ptch1 on the surface of several immune components including peripheral T cells, suggesting that Hh ligand secreted from the damaged areas can activate the local immune network to work simultaneously to repair the tissue. Interestingly, Gli1 also plays a key role in pancreatic tumorigenesis by enhancing expression and activation of Signal transducer and activator of transcription 3 (STAT3) by direct and Hh-dependent upregulation of IL-6 expression [76]. Therefore, a positive feedback loop exists whereby Hh signaling helps initiates tumor formation which, in turn, triggers Hh signaling to assist in tissue repair.

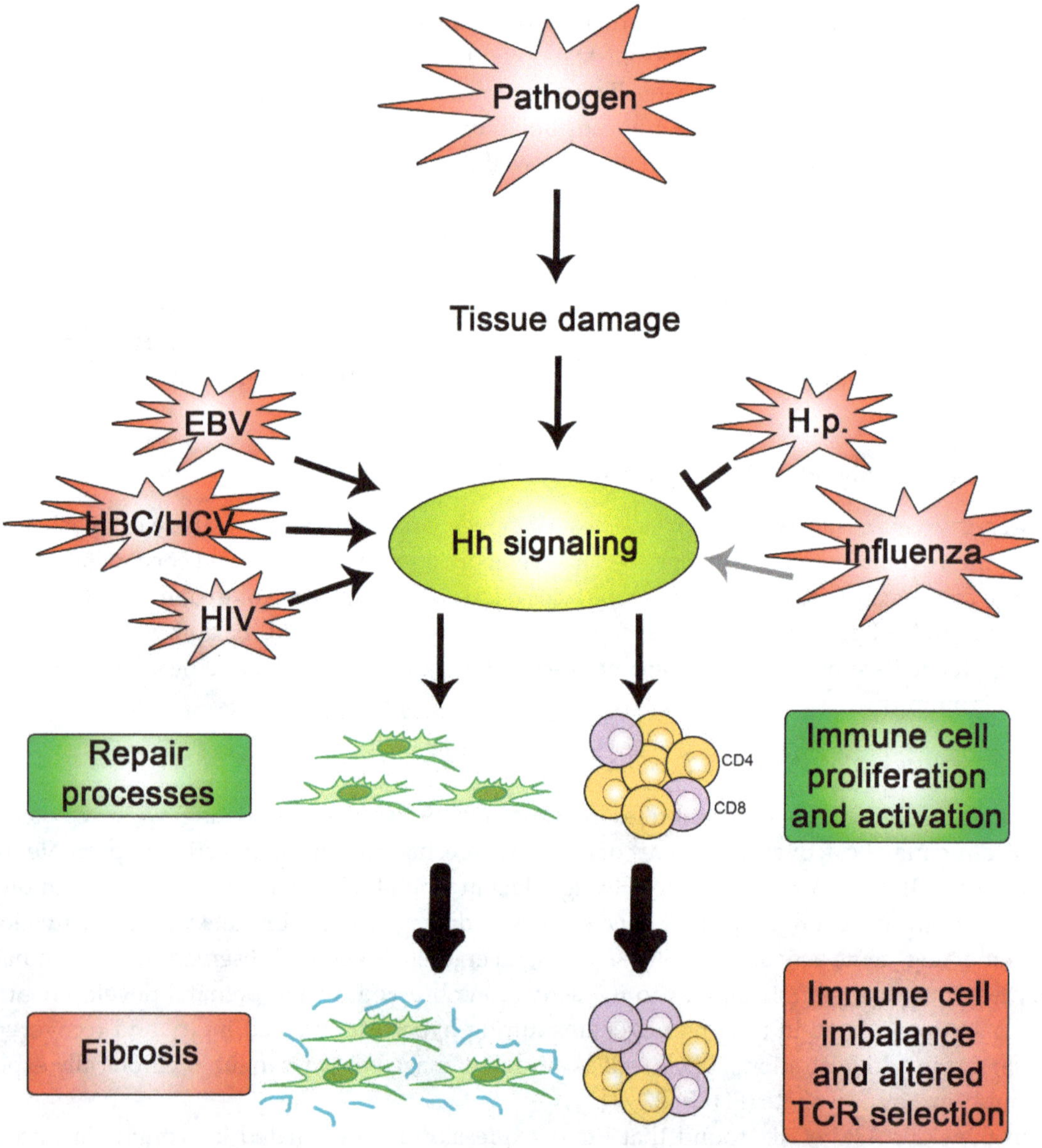

Figure 3. Hypothesis for what occurs during a pathogenic interaction with Hh signaling. Pathogens that cause damage to the host tissue promote activation of Hh signaling due to the key roles it plays in repair processes and immunity (green boxes). If not regulated properly, these cellular processes can cause fibrosis and an imbalanced immune response, respectively (red boxes). Pathogens, such as Influenza, EBV, HBC, HCV, HIV, and H.p., have been shown to directly modulate pathway activity once signaling is activated and may do so in order to exacerbate or restrain these detrimental outcomes.

Where repair processes fall short, however, tissue fibrosis can occur, marked by the formation of detrimental scar tissue. Perhaps not surprisingly, there is also a strong link between tissue fibrosis and Hh signaling in several tissues inflicted with chronic ailments, primarily driven by the Hh-dependent activation of epithelial to mesenchymal transition of cells and the resultant excessive extracellular matrix deposition [77]. Pulmonary fibrotic diseases such as idiopathic pulmonary fibrosis (IPF) [78,79], interstitial lung disease [80–82], and usual and nonspecific interstitial pneumonia [83] are all characterized by expression of Shh and Ptch1 in areas of fibrosis. Additionally, Gli1nlacZ/+ mice show an increased number of Hh-responsive cells in bleomycin-induced fibrosis where overexpression of Shh can further augment damage [84].

Hh ligand secreted by liver cells in response to damage causes cellular differentiation of fibroblasts and myofibroblasts which are critical for normal liver regeneration [37,85,86]. However, similarly to the lung, excessive signaling leads to fibrosis, which can cause cirrhosis and HCC if not controlled [37,87]. Syn et al. demonstrated that Hh upregulates expression of osteopontin (OPN), a cytokine involved in wound healing, in a Nonalcoholic Steathohepatitis-related liver fibrosis, where a reduction in either Hh signaling or OPN reduced fibrogenesis [88]. This was postulated to be through a direct interaction between Gli and Gli-binding sites in the OPN promotor [89]. Moreover, HBV and HCV infections are leading causes of liver fibrosis which is likely mediated through Hh-signaling enhancement (the role of Hh signaling in HBV and HCV pathogenesis is discussed above) [32,33,35].

Hh ligand and target genes, including those involved in extracellular matrix (ECM) production, have also been shown to be upregulated in a mouse model of renal fibrosis following injury where protection was provided by cyclopamine treatment or by deficiency in Gli1 [90,91]. Similarly, transgenic zebrafish over-expressing Hh ligands developed pancreatic fibrosis, and pathway components are found upregulated in patients suffering from pancreatitis [92,93]. Correspondingly, Hh-mediated expression of ECM genes were also upregulated in cultured renal cells and in culture-activated pancreatic stellate, both of which showed augmented proliferation and/or migration [91,94].

These data clearly establish the role of Hh signaling in tissue repair processes where unstrained signaling can cause harmful fibrotic outcomes. This aligns with a model in which pathogen-dependent tissue damage causes a Hh-mediated repair response (Figure 3). Some viruses may augment this already active signal to promote irreparable fibrotic damage to ensure viral spread. However, Hh activation at sub-maximal capacity, such as what appears to occur during influenza and in late-stage H.p. infection, may limit formation of fibrotic tissue which may also benefit the pathogen by increasing the time allotted for replication, maturation, and/or dissemination while, at the same time, also maintaining an ample pool of viable hosts available for reinfection.

3.2. Hh Signaling Involved in Immune Pathways and Immune Diseases

The Hh pathway also has postnatal roles in defining the immune response which is conserved from flies to humans, although the relevant signaling output and how it is applied as a defense mechanism differs between vertebrates and invertebrates. In a study conducted in *Drosophila* by Lee et al., Hedgehog signaling was found to be required as a first line of defense against harmful uracil-secreting pathogens in the gut by triggering the production of microbicidal reactive oxygen species (ROS) [60,95,96]. Flies expressing RNAis directed to several canonical Hh pathway components showed reduced ROS production upon bacterial infection and a consequent higher mortality rate. Exogenous expression of the Hh target gene, *Cadherin 99C (Cad99C)*, however, could rescue this phenotype, and subsequent experiments determined that Cad99C-dependent formation of signaling endosomes stimulated the enzyme, dual oxidase (DUOX), to produce ROS in response to uracil detection.

In mammals, Hh signaling takes a multipronged approach in regulating the immune system by controlling several T-cell features such as differentiation, proliferation, and activity [97,98]. Shh ligands, secreted from thymic epithelial cells to T-cell progenitors, influence cell lineage and proliferation [97]. Pro-thymocytes undergoing a series of differentiation stages ultimately mature into either helper or cytotoxic T cells defined by the expression of the cell-surface markers, CD4 or CD8, respectively [99]. During this process, thymocytes progress from double-negative (DN) to double-positive (DP) expressing markers, with several intermediate steps in between classified by CD25 and CD44 expression profiles, then on to mature single-positive cells that can then exit the thymus to the bloodstream [100]. Throughout this process, T-cell receptor maturation and selection is occurring via gene rearrangement such that a diverse array of receptors are produced that can recognize foreign peptides [101].

Analysis of thymocyte phenotypes in mouse mutants of Hh pathway components show that Hh signaling is important for regulating T-cell proliferation rates, determining the final differentiation step between CD4 vs. CD8, controlling various steps throughout the conversion process of double negative cells to single positive cell, as well as modulating the strength of T-cell receptor (TCR) signaling to influence TCR repertoire selection [27–29,102,103]. Additionally, Ihh expression directly in developing DP T-cells themselves is critical for negatively regulating proliferation and differentiation of thymocytes at earlier stages in the developmental process [104]. Ihh is also expressed directly in mature CD8 T-cells and, through an autocrine manner of signaling, is involved in regulating activity by controlling immunological synapse formation and mediating target cell lysis [98]. Thus, both positive [27,80,81,98,103,104] and negative [28,29,104] roles for Hh signaling have been described for these processes, suggesting that there may be a narrow range of signaling that is optimal for proper T-cell maturation andactivity and for generating proportional representation of subpopulations.

Direct induction of cytokine expression—secreted factors of immune cells that have an effect on other cells—is another capability of Hh signaling. Peripheral CD4 T-cells retain expression of Hh pathway components and can respond to Hh signaling following TCR-activation [80,81]. Hh signaling can induce clonal expansion of this cell population by enhancing expression of cytokines, such as IL-2, IL-10, and IFNγ, which promote entry into the S-G2 phase of the cell cycle. Additionally, human macrophages stimulated in vitro with recombinant Shh respond by upregulating expression of some cytokines and chemokines, such as IL-6, IL-8, MCP-1, and IL-10, whereas expression of others was reduced in a conditional Hh KO H.p. infection model [105,106]. Shh also induces expression of the pro-fibrotic cytokines, IL-13 and IL-4, in natural killer T-cells during liver fibrosis [107–109].

Consistent with Hh controlling immune cell proliferation and cytokine induction, some auto-immune disorders [110–113] and leukemias [114,115] have been associated with aberrant pathway activation. Allergic asthma is an example of an autoimmune condition where the pathophysiology of the disease is connected to an erroneous immune reaction to aeroallergen inhalation [116]. This disease has been directly linked to Shh production in a murine model where pathway activation results in an over-abundance of conversion of naïve T cells to T_{helper} 2 (Th2) cells [110]. This occurs through a Hh-dependent upregulation of several target genes that specify Th2 cells' fates such as the cytokines, IL-4 and IL-1rl1 [30,110]. Thus, while Th2 cells are important for protection against extracellular parasites and tissue remodeling upon damage, they are also involved in the pathogenesis of some allergic and inflammatory diseases [117].

Similarly, influenza infection strongly induced expression of cytokines CXCL-10 and IL6, the latter at least partially by a direct, cell-autonomous interaction between Hh signaling and NS1 [55]. Interestingly, IL6 (along with other Hh target genes discussed above) was present at comparatively higher levels in animals infected with the more pathogenic virus carrying the A122V point mutation in NS1, thus we speculate that the hastened lethality caused by the mutant virus may be due, in part, to a Hh-dependent over-production of cytokines (often called cytokine storms), which have been thought to be the cause of past influenza pandemics [118].

Collectively, these studies indicate that select populations of immune cells are primed to respond to the Hh signal, often resulting in proliferation and an upregulation of a subset of cytokines. Increasing Hh signaling, as occurs in certain tissues during HBV, HCV, EBV, and HIV infection, may help the virus evade the immune response by disrupting the balance of immune components available to extract the infection (Figure 3). In contrast, limiting full pathway activation, as occurs during influenza infection, may suppress the immune response to evade it and/or protect the host from detrimental outcomes, such as cytokine storms [75,76,97].

4. Therapeutic Strategies

Small molecule modulators of Hh signaling have been used in basic research for several years now to detect links between signaling and specific phenotypes of interest. Currently, CDC-0449 (Vismodegib) and LDE225 (Sonidegib), both Smoothened inhibitors, and arsenic trioxide (ATO), a Gli1/2 inhibitor, have been approved by the Food and Drug Administration (FDA) to treat basal cell carcinoma (BCC) and certain leukemias, respectively, whereas many others are still in clinical trials (Figure 1B) [3,119,120]. The idea that these approved and yet-to-be-approved molecules might be repurposed to have therapeutic value in humans beyond certain types of cancers is certainly worth exploring [24].

Treatment both in vivo and in vitro with Vismodegib or the Gli inhibitors, Gant-58 and Gant-61, has successfully reversed the detrimental phenotypes caused by several of the ailments mentioned in this review. For example, Vismodegib reduced liver fibrosis and tumor formation in a mouse model of fibrosis-associated HCC [121] and ameliorated early liver fibrosis in a rat model of common bile duct ligation [122]. Vismodegib also reduced the growth of HBV X-expressing tumor xenografts in nude mice and HCC formation in transgenic mice expressing the HBV X protein [35]. Similarly, Gli inhibitors reduced the pro-fibrotic effects and autophagy inhibition in HCV-exposed fibroblasts [33], reduced tumor-sphere formation in several EBV-infected cell lines [38], and decreased the proliferation of Human Papilloma Virus-derived cervical cancer cells [123]. Likewise, ATO was shown to inhibit tumor growth in several types of cancers in both in vitro and in vivo models [124–129].

Thus, it stands to reason that therapeutic uses of FDA-approved molecules that inhibit Hh signaling may be expanded for use as potent inhibitors to treat several pathogenic infections. In contrast to the currently available therapies, such as vaccines and antivirals, which target strain-specific and rapidly-mutating viral proteins, treatments that target highly-conserved host targets may ultimately provide superior and continual protection across a broader spectrum of strains.

Interestingly, HIV-infected cells and tissues appear to be responsive to treatment with both Hh antagonists and agonists—the former directed to infected cells, the latter directed to the local uninfected environment. In the former case, Gli inhibitors were shown to decrease HIV-dependent proliferation and migration of mouse kidney podocytes in vitro [40]. In the latter case, treatment of humanized mice with a Smoothened Agonist reduced leukocyte infiltration into the brain by fortifying the blood–brain barrier, thus limiting the viral niche [130,131]. These differences illustrate a circumstance in which the location of the infection (kidney vs brain) can be controlled through contrasting Hh-modulating mechanisms.

These HIV studies reinforce the importance of carefully selecting a relevant treatment since the cellular processes regulated by this signaling pathway are expansive and unforeseeably complex. This may require a combinatorial approach in which several drugs are targeted to different Hh-dependent processes, which may further strengthen or reduce signaling in a context-dependent fashion. This may be possible by using a combination of drugs that target pathway components with both positive and negative roles in signaling. Additionally, defining the precise mechanism in which viral factors interact with the pathway would, no doubt, restrict the search for useful drugs.

5. Summary

The Hh signaling pathway has emerged as a target of several pathogens in recent years, where cases of both pathogen-dependent increases and decreases in pathway activity have been observed. These changes in signaling correlate to an exacerbated detriment to the host and sometimes also as protection, as in the case of influenza. We speculate that the key postnatal roles that Hh signaling plays in wound repair and the immune response are critical features that would be appealing for pathogenic control. Potential new therapies involving Hh inhibitory and stimulatory compounds that disrupt or reverse these interactions could derive from these findings and suggest an important new avenue for further investigation.

Acknowledgments: I would like to thank Ethan Bier for his critical comments, guidance, and advice on this manuscript.

Conflicts of Interest: M.G.S. is an author on Patent# US9095579 B2 that relates to treating influenza with small molecule modulators of the Hedgehog signaling pathway.

References

1. Pak, E.; Segal, R.A. Hedgehog Signal Transduction: Key Players, Oncogenic Drivers, and Cancer Therapy. *Dev. Cell* **2016**, *38*, 333–344. [CrossRef] [PubMed]
2. Lee, R.T.; Zhao, Z.; Ingham, P.W. Hedgehog signalling. *Development* **2016**, *143*, 367–372. [CrossRef] [PubMed]
3. Wu, F.; Zhang, Y.; Sun, B.; McMahon, A.P.; Wang, Y. Hedgehog Signaling: From Basic Biology to Cancer Therapy. *Cell Chem. Biol.* **2017**, *24*, 252–280. [CrossRef] [PubMed]
4. Nusslein-Volhard, C.; Wieschaus, E. Mutations affecting segment number and polarity in Drosophila. *Nature* **1980**, *287*, 795–801. [CrossRef] [PubMed]
5. Huangfu, D.; Anderson, K.V. Signaling from Smo to Ci/Gli: Conservation and divergence of Hedgehog pathways from Drosophila to vertebrates. *Development* **2006**, *133*, 3–14. [CrossRef] [PubMed]
6. Ohlmeyer, J.T.; Kalderon, D. Hedgehog stimulates maturation of Cubitus interruptus into a labile transcriptional activator. *Nature* **1998**, *396*, 749–753. [CrossRef] [PubMed]
7. Zhou, Q.; Kalderon, D. Hedgehog activates fused through phosphorylation to elicit a full spectrum of pathway responses. *Dev. Cell* **2011**, *20*, 802–814. [CrossRef] [PubMed]
8. Alves, G.; Limbourg-Bouchon, B.; Tricoire, H.; Brissard-Zahraoui, J.; Lamour-Isnard, C.; Busson, D. Modulation of Hedgehog target gene expression by the Fused serine-threonine kinase in wing imaginal discs. *Mech. Dev.* **1998**, *78*, 17–31. [CrossRef]
9. Wang, C.; Pan, Y.; Wang, B. Suppressor of fused and Spop regulate the stability, processing and function of Gli2 and Gli3 full-length activators but not their repressors. *Development* **2010**, *137*, 2001–2009. [CrossRef] [PubMed]
10. Humke, E.W.; Dorn, K.V.; Milenkovic, L.; Scott, M.P.; Rohatgi, R. The output of Hedgehog signaling is controlled by the dynamic association between Suppressor of Fused and the Gli proteins. *Genes Dev.* **2010**, *24*, 670–682. [CrossRef] [PubMed]
11. Tukachinsky, H.; Lopez, L.V.; Salic, A. A mechanism for vertebrate Hedgehog signaling: Recruitment to cilia and dissociation of SuFu-Gli protein complexes. *J. Cell Biol.* **2010**, *191*, 415–428. [CrossRef] [PubMed]
12. Jiang, J.; Hui, C.C. Hedgehog signaling in development and cancer. *Dev. Cell* **2008**, *15*, 801–812. [CrossRef] [PubMed]
13. Huangfu, D.; Liu, A.; Rakeman, A.S.; Murcia, N.S.; Niswander, L.; Anderson, K.V. Hedgehog signalling in the mouse requires intraflagellar transport proteins. *Nature* **2003**, *426*, 83–87. [CrossRef] [PubMed]
14. Aberger, F.; Ruiz, I.A.A. Context-dependent signal integration by the GLI code: The oncogenic load, pathways, modifiers and implications for cancer therapy. *Semin. Cell Dev. Biol.* **2014**, *33*, 93–104. [CrossRef] [PubMed]
15. Winklmayr, M.; Schmid, C.; Laner-Plamberger, S.; Kaser, A.; Aberger, F.; Eichberger, T.; Frischauf, A.-M. Non-consensus GLI binding sites in Hedgehog target gene regulation. *BMC Mol. Biol.* **2010**, *11*, 2. [CrossRef] [PubMed]
16. Jenkins, D. Hedgehog signalling: Emerging evidence for non-canonical pathways. *Cell. Signal.* **2009**, *21*, 1023–1034. [CrossRef] [PubMed]
17. Ingham, P.W.; McMahon, A.P. Hedgehog signaling in animal development: Paradigms and principles. *Genes Dev.* **2001**, *15*, 3059–3087. [CrossRef] [PubMed]
18. Outram, S.V.; Varas, A.; Pepicelli, C.V.; Crompton, T. Hedgehog signaling regulates differentiation from double-negative to double-positive thymocyte. *Immunity* **2000**, *13*, 187–197. [CrossRef]
19. Dierks, C.; Grbic, J.; Zirlik, K.; Beigi, R.; Englund, N.P.; Guo, G.-R.; Veelken, H.; Engelhardt, M.; Mertelsmann, R.; Kelleher, J.F.; et al. Essential role of stromally induced hedgehog signaling in B-cell malignancies. *Nat. Med.* **2007**, *13*, 944–951. [CrossRef] [PubMed]
20. Sacedon, R.; Diez, B.; Nunez, V.; Hernandez-Lopez, C.; Gutierrez-Frias, C.; Cejalvo, T.; Outram, S.V.; Crompton, T.; Zapata, A.G.; Vicente, A.; et al. Sonic hedgehog is produced by follicular dendritic cells and protects germinal center B cells from apoptosis. *J. Immunol.* **2005**, *174*, 1456–1461. [CrossRef] [PubMed]

21. Le, H.; Kleinerman, R.; Lerman, O.Z.; Brown, D.; Galiano, R.; Gurtner, G.C.; Warren, S.M.; Levine, J.P.; Saadeh, P.B. Hedgehog signaling is essential for normal wound healing. *Wound Repair Regener.* **2008**, *16*, 768–773. [CrossRef] [PubMed]
22. Stewart, G.A.; Hoyne, G.F.; Ahmad, S.A.; Jarman, E.; Wallace, W.A.; Harrison, D.J.; Haslett, C.; Lamb, J.R.; Howie, S.E. Expression of the developmental Sonic hedgehog (Shh) signalling pathway is up-regulated in chronic lung fibrosis and the Shh receptor patched 1 is present in circulating T lymphocytes. *J. Pathol.* **2003**, *199*, 488–495. [CrossRef] [PubMed]
23. Nielsen, C.M.; Williams, J.; van den Brink, G.R.; Lauwers, G.Y.; Roberts, D.J. Hh pathway expression in human gut tissues and in inflammatory gut diseases. *Lab. Investig.* **2004**, *84*, 1631–1642. [CrossRef] [PubMed]
24. Bhardwaj, G.; Murdoch, B.; Wu, D.; Baker, D.P.; Williams, K.P.; Chadwick, K.; Ling, L.E.; Karanu, F.N.; Bhatia, M. Sonic hedgehog induces the proliferation of primitive human hematopoietic cells via BMP regulation. *Nat. Immunol.* **2001**, *2*, 172–180. [CrossRef] [PubMed]
25. Lau, C.I.; Outram, S.V.; Saldana, J.I.; Furmanski, A.L.; Dessens, J.T.; Crompton, T. Regulation of murine normal and stress-induced erythropoiesis by Desert Hedgehog. *Blood* **2012**, *119*, 4741–4751. [CrossRef] [PubMed]
26. Petrova, R.; Joyner, A.L. Roles for Hedgehog signaling in adult organ homeostasis and repair. *Development* **2014**, *141*, 3445–3457. [CrossRef] [PubMed]
27. El Andaloussi, A.; Graves, S.; Meng, F.; Mandal, M.; Mashayekhi, M.; Aifantis, I. Hedgehog signaling controls thymocyte progenitor homeostasis and differentiation in the thymus. *Nat. Immunol.* **2006**, *7*, 418–426. [CrossRef] [PubMed]
28. Rowbotham, N.J.; Furmanski, A.L.; Hager-Theodorides, A.L.; Ross, S.E.; Drakopoulou, E.; Koufaris, C.; Outram, S.V.; Crompton, T. Repression of hedgehog signal transduction in T-lineage cells increases TCR-induced activation and proliferation. *Cell Cycle* **2008**, *7*, 904–908. [CrossRef] [PubMed]
29. Rowbotham, N.J.; Hager-Theodorides, A.L.; Cebecauer, M.; Shah, D.K.; Drakopoulou, E.; Dyson, J.; Outram, S.V.; Crompton, T. Activation of the Hedgehog signaling pathway in T-lineage cells inhibits TCR repertoire selection in the thymus and peripheral T-cell activation. *Blood* **2007**, *109*, 3757–3766. [CrossRef] [PubMed]
30. Furmanski, A.L.; Saldana, J.I.; Ono, M.; Sahni, H.; Paschalidis, N.; D'Acquisto, F.; Crompton, T. Tissue-derived hedgehog proteins modulate Th differentiation and disease. *J. Immunol.* **2013**, *190*, 2641–2649. [CrossRef] [PubMed]
31. Caldwell, S.; Park, S.H. The epidemiology of hepatocellular cancer: From the perspectives of public health problem to tumor biology. *J. Gastroenterol.* **2009**, *44* (Suppl. 19), 96–101. [CrossRef] [PubMed]
32. De Almeida Pereira, T.; Witek, R.P.; Syn, W.K.; Choi, S.S.; Bradrick, S.; Karaca, G.F.; Agboola, K.M.; Jung, Y.; Omenetti, A.; Moylan, C.A.; et al. Viral factors induce Hedgehog pathway activation in humans with viral hepatitis, cirrhosis, and hepatocellular carcinoma. *Lab. Investig.* **2010**, *90*, 1690–1703. [CrossRef] [PubMed]
33. Granato, M.; Zompetta, C.; Vescarelli, E.; Rizzello, C.; Cardi, A.; Valia, S.; Antonelli, G.; Marchese, C.; Torrisi, M.R.; Faggioni, A.; et al. HCV derived from sera of HCV-infected patients induces pro-fibrotic effects in human primary fibroblasts by activating GLI2. *Sci. Rep.* **2016**, *6*, 30649. [CrossRef] [PubMed]
34. Kim, H.Y.; Cho, H.K.; Hong, S.P.; Cheong, J. Hepatitis B virus X protein stimulates the Hedgehog-Gli activation through protein stabilization and nuclear localization of Gli1 in liver cancer cells. *Cancer Lett.* **2011**, *309*, 176–184. [CrossRef] [PubMed]
35. Arzumanyan, A.; Sambandam, V.; Clayton, M.M.; Choi, S.S.; Xie, G.; Diehl, A.M.; Yu, D.Y.; Feitelson, M.A. Hedgehog signaling blockade delays hepatocarcinogenesis induced by hepatitis B virus X protein. *Cancer Res.* **2012**, *72*, 5912–5920. [CrossRef] [PubMed]
36. Jo, B.B.; Jeong, M.S.; Park, S.Y.; Cheong, J.; Jang, S.B. The Binding of Hepatitis B Virus X Protein to Glioma-Associated Oncogene Homologue 1 and its Biological Characterization In vitro. *Appl. Biochem. Biotechnol.* **2011**, *165*, 109. [CrossRef] [PubMed]
37. Choi, S.S.; Bradrick, S.; Qiang, G.; Mostafavi, A.; Chaturvedi, G.; Weinman, S.A.; Diehl, A.M.; Jhaveri, R. Up-regulation of Hedgehog pathway is associated with cellular permissiveness for hepatitis C virus replication. *Hepatology* **2011**, *54*, 1580–1590. [CrossRef] [PubMed]
38. Port, R.J.; Pinheiro-Maia, S.; Hu, C.; Arrand, J.R.; Wei, W.; Young, L.S.; Dawson, C.W. Epstein-Barr virus induction of the Hedgehog signalling pathway imposes a stem cell phenotype on human epithelial cells. *J. Pathol.* **2013**, *231*, 367–377. [CrossRef] [PubMed]

39. Pal, A.D.; Banerjee, S. Epstein-Barr virus latent membrane protein 2A mediated activation of Sonic Hedgehog pathway induces HLA class Ia downregulation in gastric cancer cells. *Virology* **2015**, *484*, 22–32.
40. Lan, X.; Wen, H.; Cheng, K.; Plagov, A.; Shoshtari, S.S.M.; Malhotra, A.; Singhal, P.C. Hedgehog pathway plays a vital role in HIV-induced epithelial-mesenchymal transition of podocyte. *Exp. Cell Res.* **2017**, *352*, 193–201. [CrossRef] [PubMed]
41. Fuccio, L.; Eusebi, L.H.; Bazzoli, F. Gastric cancer, Helicobacter pylori infection and other risk factors. *World J. Gastrointest. Oncol.* **2010**, *2*, 342–347. [CrossRef] [PubMed]
42. Suerbaum, S.; Michetti, P. Helicobacter pylori infection. *N. Engl. J. Med.* **2002**, *347*, 1175–1186. [CrossRef] [PubMed]
43. Schumacher, M.A.; Donnelly, J.M.; Engevik, A.C.; Xiao, C.; Yang, L.; Kenny, S.; Varro, A.; Hollande, F.; Samuelson, L.C.; Zavros, Y. Gastric Sonic Hedgehog acts as a macrophage chemoattractant during the immune response to Helicobacter pylori. *Gastroenterology* **2012**, *142*, 1150–1159. [CrossRef] [PubMed]
44. Schumacher, M.A.; Feng, R.; Aihara, E.; Engevik, A.C.; Montrose, M.H.; Ottemann, K.M.; Zavros, Y. Helicobacter pylori-induced Sonic Hedgehog expression is regulated by NFkappaB pathway activation: The use of a novel in vitro model to study epithelial response to infection. *Helicobacter* **2015**, *20*, 19–28. [CrossRef] [PubMed]
45. El-Zaatari, M.; Tobias, A.; Grabowska, A.M.; Kumari, R.; Scotting, P.J.; Kaye, P.; Atherton, J.; Clarke, P.A.; Powe, D.G.; Watson, S.A. De-regulation of the sonic hedgehog pathway in the InsGas mouse model of gastric carcinogenesis. *Br. J. Cancer* **2007**, *96*, 1855–1861. [CrossRef] [PubMed]
46. Suzuki, H.; Minegishi, Y.; Nomoto, Y.; Ota, T.; Masaoka, T.; van den Brink, G.R.; Hibi, T. Down-regulation of a morphogen (sonic hedgehog) gradient in the gastric epithelium of Helicobacter pylori-infected Mongolian gerbils. *J. Pathol.* **2005**, *206*, 186–197. [CrossRef] [PubMed]
47. Stasikowska-Kanicka, O.; Wagrowska-Danilewicz, M.; Bialek, I.; Danilewicz, M. The immunoexpression of Shh, Smo and Gli2 in Helicobacter pylori positive and negative gastric biopsies. *Pol. J. Pathol.* **2012**, *63*, 25–30. [PubMed]
48. Nishizawa, T.; Suzuki, H.; Nakagawa, I.; Minegishi, Y.; Masaoka, T.; Iwasaki, E.; Hibi, T. Early Helicobacter pylori eradication restores sonic hedgehog expression in the gastric mucosa of Mongolian gerbils. *Digestion* **2009**, *79*, 99–108. [CrossRef] [PubMed]
49. Shiotani, A.; Uedo, N.; Iishi, H.; Tatsuta, M.; Ishiguro, S.; Nakae, Y.; Kamada, T.; Haruma, K.; Merchant, J.L. Re-expression of sonic hedgehog and reduction of CDX2 after Helicobacter pylori eradication prior to incomplete intestinal metaplasia. *Int. J. Cancer* **2007**, *121*, 1182–1189. [CrossRef] [PubMed]
50. Nishizawa, T.; Suzuki, H.; Masaoka, T.; Minegishi, Y.; Iwasahi, E.; Hibi, T. Helicobacter pylori eradication restored sonic hedgehog expression in the stomach. *Hepato Gastroenterol.* **2007**, *54*, 697–700.
51. Jung, D.H.; Kim, J.H.; Chung, H.S.; Park, J.C.; Shin, S.K.; Lee, S.K.; Lee, Y.C. Helicobacter pylori Eradication on the Prevention of Metachronous Lesions after Endoscopic Resection of Gastric Neoplasm: A Meta-Analysis. *PLoS ONE* **2015**, *10*, e0124725. [CrossRef] [PubMed]
52. Mutoh, H.; Hayakawa, H.; Sashikawa, M.; Sakamoto, H.; Sugano, K. Direct repression of Sonic Hedgehog expression in the stomach by Cdx2 leads to intestinal transformation. *Biochem. J.* **2010**, *427*, 423–434. [CrossRef] [PubMed]
53. Barros, R.; Marcos, N.; Reis, C.A.; De Luca, A.; David, L.; Almeida, R. CDX2 expression is induced by Helicobacter pylori in AGS cells. *Scand. J. Gastroenterol.* **2009**, *44*, 124–125. [CrossRef] [PubMed]
54. Matsuda, K.; Yamauchi, K.; Matsumoto, T.; Sano, K.; Yamaoka, Y.; Ota, H. Quantitative analysis of the effect of Helicobacter pylori on the expressions of SOX2, CDX2, MUC2, MUC5AC, MUC6, TFF1, TFF2, and TFF3 mRNAs in human gastric carcinoma cells. *Scand. J. Gastroenterol.* **2008**, *43*, 25–33. [CrossRef] [PubMed]
55. Smelkinson, M.G.; Guichard, A.; Teijaro, J.R.; Malur, M.; Loureiro, M.E.; Jain, P.; Ganesan, S.; Zúñiga, E.I.; Krug, R.M.; Oldstone, M.B.; et al. Influenza NS1 directly modulates Hedgehog signaling during infection. *PLoS Pathog.* **2017**, *13*, e1006588. [CrossRef] [PubMed]
56. Bier, E.; Guichard, A. Deconstructing host-pathogen interactions in Drosophila. *Dis. Model. Mech.* **2012**, *5*, 48–61. [CrossRef] [PubMed]
57. Bier, E. Drosophila, the golden bug, emerges as a tool for human genetics. *Nat. Rev. Genet.* **2005**, *6*, 9–23. [CrossRef] [PubMed]
58. Reiter, L.T.; Potocki, L.; Chien, S.; Gribskov, M.; Bier, E. A systematic analysis of human disease-associated gene sequences in Drosophila melanogaster. *Genome Res.* **2001**, *11*, 1114–1125. [CrossRef] [PubMed]

59. Hughes, T.T.; Allen, A.L.; Bardin, J.E.; Christian, M.N.; Daimon, K.; Dozier, K.D.; Hansen, C.L.; Holcomb, L.M. Ahlander, J. Drosophila as a genetic model for studying pathogenic human viruses. *Virology* **2012**, *423*, 1–5. [CrossRef] [PubMed]
60. Lee, K.A.; Kim, B.; Bhin, J.; Kim, D.H.; You, H.; Kim, E.K.; Kim, S.H.; Ryu, J.H.; Hwang, D.; Lee, W.J. Bacterial uracil modulates Drosophila DUOX-dependent gut immunity via Hedgehog-induced signaling endosomes. *Cell Host Microbe* **2015**, *17*, 191–204. [CrossRef] [PubMed]
61. Biehs, B.; Sturtevant, M.A.; Bier, E. Boundaries in the Drosophila wing imaginal disc organize vein-specific genetic programs. *Development* **1998**, *125*, 4245–4257. [PubMed]
62. St Johnston, R.D.; Hoffmann, F.M.; Blackman, R.K.; Segal, D.; Grimaila, R.; Padgett, R.W.; Irick, H.A.; Gelbart, W.M. Molecular organization of the decapentaplegic gene in Drosophila melanogaster. *Genes Dev.* **1990**, *4*, 1114–1127. [CrossRef] [PubMed]
63. Lamb, R.A.; Krug, R.M. Othromyxoviridae: The viruses and their replication. In *Fundamental Virology*; Knipe, D.M., Howley, P.M., Eds.; Lippincott Williams & Wilkins: Philadelphia, PA, USA, 2001.
64. Hale, B.G.; Randall, R.E.; Ortin, J.; Jackson, D. The multifunctional NS1 protein of influenza A viruses. *J. Gen. Virol.* **2008**, *89*, 2359–2376. [CrossRef] [PubMed]
65. Marc, D. Influenza virus non-structural protein NS1: Interferon antagonism and beyond. *J. Gen. Virol.* **2014**, *95*, 2594–2611. [CrossRef] [PubMed]
66. El-Zaatari, M.; Kao, J.Y.; Tessier, A.; Bai, L.; Hayes, M.M.; Fontaine, C.; Eaton, K.A.; Merchant, J.L. Gli1 deletion prevents Helicobacter-induced gastric metaplasia and expansion of myeloid cell subsets. *PLoS ONE* **2013**, *8*, e58935. [CrossRef] [PubMed]
67. Zheng, L.L.; Li, C.; Ping, J.; Zhou, Y.; Li, Y.; Hao, P. The domain landscape of virus-host interactomes. *Biomed. Res. Int.* **2014**, *2014*, 867235. [CrossRef] [PubMed]
68. Watkins, D.N.; Berman, D.M.; Burkholder, S.G.; Wang, B.; Beachy, P.A.; Baylin, S.B. Hedgehog signalling within airway epithelial progenitors and in small-cell lung cancer. *Nature* **2003**, *422*, 313–317. [CrossRef] [PubMed]
69. Pogach, M.S.; Cao, Y.; Millien, G.; Ramirez, M.I.; Williams, M.C. Key developmental regulators change during hyperoxia-induced injury and recovery in adult mouse lung. *J. Cell. Biochem.* **2007**, *100*, 1415–1429. [CrossRef] [PubMed]
70. Moshai, E.F.; Wemeau-Stervinou, L.; Cigna, N.; Brayer, S.; Somme, J.M.; Crestani, B.; Mailleux, A.A. Targeting the hedgehog-glioma-associated oncogene homolog pathway inhibits bleomycin-induced lung fibrosis in mice. *Am. J. Respir. Cell Mol. Biol.* **2014**, *51*, 11–25. [CrossRef] [PubMed]
71. Fisher, C.E.; Ahmad, S.A.; Fitch, P.M.; Lamb, J.R.; Howie, S.E. FITC-induced murine pulmonary inflammation: CC10 up-regulation and concurrent Shh expression. *Cell Biol. Int.* **2005**, *29*, 868–876. [CrossRef] [PubMed]
72. Asai, J.; Takenaka, H.; Kusano, K.F.; Ii, M.; Luedemann, C.; Curry, C.; Eaton, E.; Iwakura, A.; Tsutsumi, Y.; Hamada, H.; et al. Topical sonic hedgehog gene therapy accelerates wound healing in diabetes by enhancing endothelial progenitor cell-mediated microvascular remodeling. *Circulation* **2006**, *113*, 2413–2424. [CrossRef] [PubMed]
73. Chen, J.K.; Taipale, J.; Cooper, M.K.; Beachy, P.A. Inhibition of Hedgehog signaling by direct binding of cyclopamine to Smoothened. *Genes Dev.* **2002**, *16*, 2743–2748. [CrossRef] [PubMed]
74. Luo, J.D.; Hu, T.P.; Wang, L.; Chen, M.S.; Liu, S.M.; Chen, A.F. Sonic hedgehog improves delayed wound healing via enhancing cutaneous nitric oxide function in diabetes. *Am. J. Physiol. Endocrinol. Metab.* **2009**, *297*, E525–E531. [CrossRef] [PubMed]
75. Mathew, E.; Collins, M.A.; Fernandez-Barrena, M.G.; Holtz, A.M.; Yan, W.; Hogan, J.O.; Tata, Z.; Allen, B.L.; Fernandez-Zapico, M.E.; Di Magliano, M.P. The transcription factor GLI1 modulates the inflammatory response during pancreatic tissue remodeling. *J. Biol. Chem.* **2014**, *289*, 27727–27743. [CrossRef] [PubMed]
76. Mills, L.D.; Zhang, Y.; Marler, R.J.; Herreros-Villanueva, M.; Zhang, L.; Almada, L.L.; Couch, F.; Wetmore, C.; Di Magliano, M.P.; Fernandez-Zapico, M.E. Loss of the transcription factor GLI1 identifies a signaling network in the tumor microenvironment mediating KRAS oncogene-induced transformation. *J. Biol. Chem.* **2013**, *288*, 11786–11794. [CrossRef] [PubMed]
77. Hu, L.; Lin, X.; Lu, H.; Chen, B.; Bai, Y. An overview of hedgehog signaling in fibrosis. *Mol. Pharmacol.* **2015**, *87*, 174–182. [CrossRef] [PubMed]

78. Bolanos, A.L.; Milla, C.M.; Lira, J.C.; Ramirez, R.; Checa, M.; Barrera, L.; García-Alvarez, J.; Carbajal, V.; Becerril, C.; Gaxiola, M.; et al. Role of Sonic Hedgehog in idiopathic pulmonary fibrosis. *Am. J. Physiol. Lung Cell. Mol. Physiol.* **2012**, *303*, L978–L990. [CrossRef] [PubMed]
79. Cigna, N.; Moshai, E.F.; Brayer, S.; Marchal-Somme, J.; Wemeau-Stervinou, L.; Fabre, A.; Mal, H.; Lesèche, G.; Dehoux, M.; Soler, P.; et al. The hedgehog system machinery controls transforming growth factor-beta-dependent myofibroblastic differentiation in humans: Involvement in idiopathic pulmonary fibrosis. *Am. J. Pathol.* **2012**, *181*, 2126–2137. [CrossRef] [PubMed]
80. Lowrey, J.A.; Stewart, G.A.; Lindey, S.; Hoyne, G.F.; Dallman, M.J.; Howie, S.E.; Lamb, J.R. Sonic hedgehog promotes cell cycle progression in activated peripheral CD4(+) T lymphocytes. *J. Immunol.* **2002**, *169*, 1869–1875. [CrossRef] [PubMed]
81. Stewart, G.A.; Lowrey, J.A.; Wakelin, S.J.; Fitch, P.M.; Lindey, S.; Dallman, M.J.; Lamb, J.R.; Howie, S.E. Sonic hedgehog signaling modulates activation of and cytokine production by human peripheral CD4+ T cells. *J. Immunol.* **2002**, *169*, 5451–5457. [CrossRef] [PubMed]
82. Fitch, P.M.; Howie, S.E.; Wallace, W.A. Oxidative damage and TGF-beta differentially induce lung epithelial cell sonic hedgehog and tenascin-C expression: Implications for the regulation of lung remodelling in idiopathic interstitial lung disease. *Int. J. Exp. Pathol.* **2011**, *92*, 8–17. [CrossRef] [PubMed]
83. Coon, D.R.; Roberts, D.J.; Loscertales, M.; Kradin, R. Differential epithelial expression of SHH and FOXF1 in usual and nonspecific interstitial pneumonia. *Exp. Mol. Pathol.* **2006**, *80*, 119–123. [CrossRef] [PubMed]
84. Liu, L.; Kugler, M.C.; Loomis, C.A.; Samdani, R.; Zhao, Z.; Chen, G.J.; Brandt, J.P.; Brownell, I.; Joyner, A.L.; Rom, W.N.; et al. Hedgehog signaling in neonatal and adult lung. *Am. J. Respir. Cell Mol. Biol.* **2013**, *48*, 703–710. [CrossRef] [PubMed]
85. Jung, Y.; Diehl, A.M. Non-alcoholic steatohepatitis pathogenesis: Role of repair in regulating the disease progression. *Dig. Dis.* **2010**, *28*, 225–228. [CrossRef] [PubMed]
86. Omenetti, A.; Diehl, A.M. Hedgehog signaling in cholangiocytes. *Curr. Opin. Gastroenterol.* **2011**, *27*, 268–275. [CrossRef] [PubMed]
87. Ochoa, B.; Syn, W.K.; Delgado, I.; Karaca, G.F.; Jung, Y.; Wang, J.; Zubiaga, A.M.; Fresnedo, O.; Omenetti, A.; Zdanowicz, M.; et al. Hedgehog signaling is critical for normal liver regeneration after partial hepatectomy in mice. *Hepatology* **2010**, *51*, 1712–1723. [CrossRef] [PubMed]
88. Syn, W.K.; Choi, S.S.; Liaskou, E.; Karaca, G.F.; Agboola, K.M.; Oo, Y.H.; Mi, Z.; Pereira, T.A.; Zdanowicz, M.; Malladi, P.; Chen, Y. Osteopontin is induced by hedgehog pathway activation and promotes fibrosis progression in nonalcoholic steatohepatitis. *Hepatology* **2011**, *53*, 106–115. [CrossRef] [PubMed]
89. Das, S.; Harris, L.G.; Metge, B.J.; Liu, S.; Riker, A.I.; Samant, R.S.; Shevde, L.A. The hedgehog pathway transcription factor GLI1 promotes malignant behavior of cancer cells by up-regulating osteopontin. *J. Biol. Chem.* **2009**, *284*, 22888–22897. [CrossRef] [PubMed]
90. Fabian, S.L.; Penchev, R.R.; Jacques, B.S.; Rao, A.N.; Sipila, P.; West, K.A.; McMahon, A.P.; Humphreys, B.D. Hedgehog-Gli Pathway Activation during Kidney Fibrosis. *Am. J. Pathol.* **2012**, *180*, 1441–1453. [CrossRef] [PubMed]
91. Ding, H.; Zhou, D.; Hao, S.; Zhou, L.; He, W.; Nie, J.; Hou, F.F.; Liu, Y. Sonic Hedgehog Signaling Mediates Epithelial-Mesenchymal Communication and Promotes Renal Fibrosis. *J. Am. Soc. Nephrol.* **2012**, *23*, 801–813. [CrossRef] [PubMed]
92. Kayed, H.; Kleeff, J.; Keleg, S.; Buchler, M.W.; Friess, H. Distribution of Indian hedgehog and its receptors patched and smoothened in human chronic pancreatitis. *J. Endocrinol.* **2003**, *178*, 467–478. [CrossRef] [PubMed]
93. Jung, I.H.; Jung, D.E.; Park, Y.N.; Song, S.Y.; Park, S.W. Aberrant Hedgehog ligands induce progressive pancreatic fibrosis by paracrine activation of myofibroblasts and ductular cells in transgenic zebrafish. *PLoS ONE* **2011**, *6*, e27941. [CrossRef] [PubMed]
94. Shinozaki, S.; Ohnishi, H.; Hama, K.; Kita, H.; Yamamoto, H.; Osawa, H.; Sato, K.; Tamada, K.; Mashima, H.; Sugano, K. Indian hedgehog promotes the migration of rat activated pancreatic stellate cells by increasing membrane type-1 matrix metalloproteinase on the plasma membrane. *J. Cell. Physiol.* **2008**, *216*, 38–46. [CrossRef] [PubMed]
95. Bier, E.; Nizet, V. Hedgehog: Linking uracil to innate defense. *Cell Host Microbe* **2015**, *17*, 146–148. [CrossRef] [PubMed]

96. Lee, K.A.; Kim, B.; You, H.; Lee, W.J. Uracil-induced signaling pathways for DUOX-dependent gut immunity. *Fly* **2015**, *9*, 115–120. [CrossRef] [PubMed]
97. Crompton, T.; Outram, S.V.; Hager-Theodorides, A.L. Sonic hedgehog signalling in T-cell development and activation. *Nat. Rev. Immunol.* **2007**, *7*, 726–735. [CrossRef] [PubMed]
98. De la Roche, M.; Ritter, A.T.; Angus, K.L.; Dinsmore, C.; Earnshaw, C.H.; Reiter, J.F.; Griffiths, G.M. Hedgehog signaling controls T cell killing at the immunological synapse. *Science* **2013**, *342*, 1247–1250. [CrossRef] [PubMed]
99. Godfrey, D.I.; Zlotnik, A. Control points in early T-cell development. *Immunol. Today* **1993**, *14*, 547–553. [CrossRef]
100. Haks, M.C.; Oosterwegel, M.A.; Blom, B.; Spits, H.M.; Kruisbeek, A.M. Cell-fate decisions in early T cell development: Regulation by cytokine receptors and the pre-TCR. *Semin. Immunol.* **1999**, *11*, 23–37. [CrossRef] [PubMed]
101. Borowski, C.; Martin, C.; Gounari, F.; Haughn, L.; Aifantis, I.; Grassi, F.; von Boehmer, H. On the brink of becoming a T cell. *Curr. Opin. Immunol.* **2002**, *14*, 200–206. [CrossRef]
102. Shah, D.K.; Hager-Theodorides, A.L.; Outram, S.V.; Ross, S.E.; Varas, A.; Crompton, T. Reduced thymocyte development in sonic hedgehog knockout embryos. *J. Immunol.* **2004**, *172*, 2296–2306. [CrossRef] [PubMed]
103. Hager-Theodorides, A.L.; Dessens, J.T.; Outram, S.V.; Crompton, T. The transcription factor Gli3 regulates differentiation of fetal CD4–CD8–double-negative thymocytes. *Blood* **2005**, *106*, 1296–1304. [CrossRef] [PubMed]
104. Outram, S.V.; Hager-Theodorides, A.L.; Shah, D.K.; Rowbotham, N.J.; Drakopoulou, E.; Ross, S.E.; Lanske, B.; Dessens, J.T.; Crompton, T. Indian hedgehog (Ihh) both promotes and restricts thymocyte differentiation. *Blood* **2009**, *113*, 2217–2228. [CrossRef] [PubMed]
105. Wakelin, S.J.; Forsythe, J.L.; Garden, O.J.; Howie, S.E. Commercially available recombinant sonic hedgehog up-regulates Ptc and modulates the cytokine and chemokine expression of human macrophages: An effect mediated by endotoxin contamination? *Immunobiology* **2008**, *213*, 25–38. [CrossRef] [PubMed]
106. Marwaha, S.; Schumacher, M.A.; Zavros, Y.; Eghbalnia, H.R. Crosstalks between cytokines and Sonic Hedgehog in Helicobacter pylori infection: A mathematical model. *PLoS ONE* **2014**, *9*, e111338. [CrossRef] [PubMed]
107. Chiaramonte, M.G.; Schopf, L.R.; Neben, T.Y.; Cheever, A.W.; Donaldson, D.D.; Wynn, T.A. IL-13 is a key regulatory cytokine for Th2 cell-mediated pulmonary granuloma formation and IgE responses induced by Schistosoma mansoni eggs. *J. Immunol.* **1999**, *162*, 920–930. [PubMed]
108. Fichtner-Feigl, S.; Strober, W.; Kawakami, K.; Puri, R.K.; Kitani, A. IL-13 signaling through the IL-13alpha2 receptor is involved in induction of TGF-beta1 production and fibrosis. *Nat. Med.* **2006**, *12*, 99–106. [CrossRef] [PubMed]
109. Syn, W.K.; Witek, R.P.; Curbishley, S.M.; Jung, Y.; Choi, S.S.; Enrich, B.; Omenetti, A.; Agboola, K.M.; Fearing, C.M.; Tilg, H. Role for hedgehog pathway in regulating growth and function of invariant NKT cells. *Eur. J. Immunol.* **2009**, *39*, 1879–1892. [CrossRef] [PubMed]
110. Standing, A.S.; Yanez, D.C.; Ross, R.; Crompton, T.; Furmanski, A.L. Frontline Science: Shh production and Gli signaling is activated in vivo in lung, enhancing the Th2 response during a murine model of allergic asthma. *J. Leukoc. Biol.* **2017**. [CrossRef] [PubMed]
111. Qin, S.; Sun, D.; Li, H.; Li, X.; Pan, W.; Yan, C.; Tang, R.; Liu, X. The Effect of SHH-Gli Signaling Pathway on the Synovial Fibroblast Proliferation in Rheumatoid Arthritis. *Inflammation* **2016**, *39*, 503–512. [CrossRef] [PubMed]
112. Samanta, J.; Grund, E.M.; Silva, H.M.; Lafaille, J.J.; Fishell, G.; Salzer, J.L. Inhibition of Gli1 mobilizes endogenous neural stem cells for remyelination. *Nature* **2015**, *526*, 448–452. [CrossRef] [PubMed]
113. Seifert, T.; Bauer, J.; Weissert, R.; Fazekas, F.; Storch, M.K. Differential expression of sonic hedgehog immunoreactivity during lesion evolution in autoimmune encephalomyelitis. *J. Neuropathol. Exp. Neurol.* **2005**, *64*, 404–411. [CrossRef] [PubMed]
114. Kobune, M.; Takimoto, R.; Murase, K.; Iyama, S.; Sato, T.; Kikuchi, S.; Kawano, Y.; Miyanishi, K.; Sato, Y.; Niitsu, Y.; et al. Drug resistance is dramatically restored by hedgehog inhibitors in CD34+ leukemic cells. *Cancer Sci.* **2009**, *100*, 948–955. [CrossRef] [PubMed]

115. Dagklis, A.; Pauwels, D.; Lahortiga, I.; Geerdens, E.; Bittoun, E.; Cauwelier, B.; Tousseyn, T.; Uyttebroeck, A.; Maertens, J.; Verhoef, G.; et al. Hedgehog pathway mutations in T-cell acute lymphoblastic leukemia. *Haematologica* **2015**, *100*, e102–e105. [CrossRef] [PubMed]
116. Pomes, A.; Chapman, M.D.; Wunschmann, S. Indoor Allergens and Allergic Respiratory Disease. *Curr. Allergy Asthma Rep.* **2016**, *16*, 43. [CrossRef] [PubMed]
117. Nhu, Q.M.; Aceves, S.S. Tissue Remodeling in Chronic Eosinophilic Esophageal Inflammation: Parallels in Asthma and Therapeutic Perspectives. *Front. Med.* **2017**, *4*, 128. [CrossRef] [PubMed]
118. Potter, C.W.; Jennings, R. A definition for influenza pandemics based on historical records. *J. Infect.* **2011**, *63*, 252–259. [CrossRef] [PubMed]
119. Aditya, S.; Rattan, A. Vismodegib: A smoothened inhibitor for the treatment of advanced basal cell carcinoma. *Indian Dermatol. Online J.* **2013**, *4*, 365–368. [CrossRef] [PubMed]
120. Rimkus, T.K.; Carpenter, R.L.; Qasem, S.; Chan, M.; Lo, H.W. Targeting the Sonic Hedgehog Signaling Pathway: Review of Smoothened and GLI Inhibitors. *Cancers* **2016**, *8*, 22. [CrossRef] [PubMed]
121. Philips, G.M.; Chan, I.S.; Swiderska, M.; Schroder, V.T.; Guy, C.; Karaca, G.F.; Moylan, C.; Venkatraman, T.; Feuerlein, S.; Syn, W.K.; et al. Hedgehog signaling antagonist promotes regression of both liver fibrosis and hepatocellular carcinoma in a murine model of primary liver cancer. *PLoS ONE* **2011**, *6*, e23943. [CrossRef] [PubMed]
122. Pratap, A.; Panakanti, R.; Yang, N.; Eason, J.D.; Mahato, R.I. Inhibition of endogenous hedgehog signaling protects against acute liver injury after ischemia reperfusion. *Pharm. Res.* **2010**, *27*, 2492–2504. [CrossRef] [PubMed]
123. Samarzija, I.; Beard, P. Hedgehog pathway regulators influence cervical cancer cell proliferation, survival and migration. *Biochem. Biophys. Res. Commun.* **2012**, *425*, 64–69. [CrossRef] [PubMed]
124. Cai, X.; Yu, K.; Zhang, L.; Li, Y.; Li, Q.; Yang, Z.; Shen, T.; Duan, L.; Xiong, W.; Wang, W. Synergistic inhibition of colon carcinoma cell growth by Hedgehog-Gli1 inhibitor arsenic trioxide and phosphoinositide 3-kinase inhibitor LY294002. *OncoTargets Ther.* **2015**, *8*, 877–883.
125. Kerl, K.; Moreno, N.; Holsten, T.; Ahlfeld, J.; Mertins, J.; Hotfilder, M.; Kool, M.; Bartelheim, K.; Schleicher, S.; Handgretinger, R.; et al. Arsenic trioxide inhibits tumor cell growth in malignant rhabdoid tumors in vitro and in vivo by targeting overexpressed Gli1. *Int. J. Cancer* **2014**, *135*, 989–995. [CrossRef] [PubMed]
126. Nakamura, S.; Nagano, S.; Nagao, H.; Ishidou, Y.; Yokouchi, M.; Abematsu, M.; Yamamoto, T.; Komiya, S.; Setoguchi, T. Arsenic trioxide prevents osteosarcoma growth by inhibition of GLI transcription via DNA damage accumulation. *PLoS ONE* **2013**, *8*, e69466. [CrossRef] [PubMed]
127. Yang, D.; Cao, F.; Ye, X.; Zhao, H.; Liu, X.; Li, Y.; Shi, C.; Wang, H.; Zhou, J. Arsenic trioxide inhibits the Hedgehog pathway which is aberrantly activated in acute promyelocytic leukemia. *Acta Haematol.* **2013**, *130*, 260–267. [CrossRef] [PubMed]
128. You, M.; Varona-Santos, J.; Singh, S.; Robbins, D.J.; Savaraj, N.; Nguyen, D.M. Targeting of the Hedgehog signal transduction pathway suppresses survival of malignant pleural mesothelioma cells in vitro. *J. Thorac. Cardiovasc. Surg.* **2014**, *147*, 508–516. [CrossRef] [PubMed]
129. Bansal, N.; Farley, N.J.; Wu, L.; Lewis, J.; Youssoufian, H.; Bertino, J.R. Darinaparsin inhibits prostate tumor-initiating cells and Du145 xenografts and is an inhibitor of hedgehog signaling. *Mol. Cancer Ther.* **2015**, *14*, 23–30. [CrossRef] [PubMed]
130. Singh, V.B.; Singh, M.V.; Gorantla, S.; Poluektova, L.Y.; Maggirwar, S.B. Smoothened Agonist Reduces Human Immunodeficiency Virus Type-1-Induced Blood-Brain Barrier Breakdown in Humanized Mice. *Sci. Rep.* **2016**, *6*, 26876. [CrossRef] [PubMed]
131. Singh, V.B.; Singh, M.V.; Piekna-Przybylska, D.; Gorantla, S.; Poluektova, L.Y.; Maggirwar, S.B. Sonic Hedgehog mimetic prevents leukocyte infiltration into the CNS during acute HIV infection. *Sci. Rep.* **2017**, *7*, 9578. [CrossRef] [PubMed]

7

Location, Location, Location: Signals in Muscle Specification

Chih-Ning Chang [1,2] and Chrissa Kioussi [1,2,*]

[1] Department of Pharmaceutical Sciences, College of Pharmacy, Oregon State University, Corvallis, OR 97331, USA; chanchih@oregonstate.edu
[2] Molecular Cell Biology Graduate Program, Oregon State University, Corvallis, OR 97331, USA
* Correspondence: chrissa.kioussi@oregonstate.edu

Academic Editor: Samuel J. Pleasure

Abstract: Muscles control body movement and locomotion, posture and body position and soft tissue support. Mesoderm derived cells gives rise to 700 unique muscles in humans as a result of well-orchestrated signaling and transcriptional networks in specific time and space. Although the anatomical structure of skeletal muscles is similar, their functions and locations are specialized. This is the result of specific signaling as the embryo grows and cells migrate to form different structures and organs. As cells progress to their next state, they suppress current sequence specific transcription factors (SSTF) and construct new networks to establish new myogenic features. In this review, we provide an overview of signaling pathways and gene regulatory networks during formation of the craniofacial, cardiac, vascular, trunk, and limb skeletal muscles.

Keywords: myogenesis; mesoderm; WNT; FGF; BMP; SHH; RA; NOTCH; ephrins; sequence specific transcription factor

---

1. Introduction

In mammals, skeletal muscles begin to develop shortly after gastrulation and potentially can continue developing throughout their life. During gastrulation, epiblast cells ingress through the primitive streak to give rise to the three basic germ layers: ectoderm, mesoderm, and endoderm. During neurulation, as the primitive streak recedes posteriorly, ectodermal cells at the cranial end of the embryo proliferate, migrate, elevate, fold, and form the notochord and neural tube. Ectodermal cells shed and form the mesoderm, which further subdivides into cranial and trunk mesoderm. Early ingressing mesodermal cells migrate and populate underneath the ectoderm layer to form the lateral plate mesoderm (LPM), whereas the others subdivide into the paraxial (PM) (presomitic) and intermediate mesoderm. The skeletal muscle progenitor cells arise from the paraxial mesoderm, which flanks the axial mesoderm, notochord, and neural tube. Craniofacial muscles and cardiac muscle originate from the unsegmented PM, whereas, the skeletal muscles of body and limbs originate from the segmented PM that gives rise to somites (Figure 1a) [1].

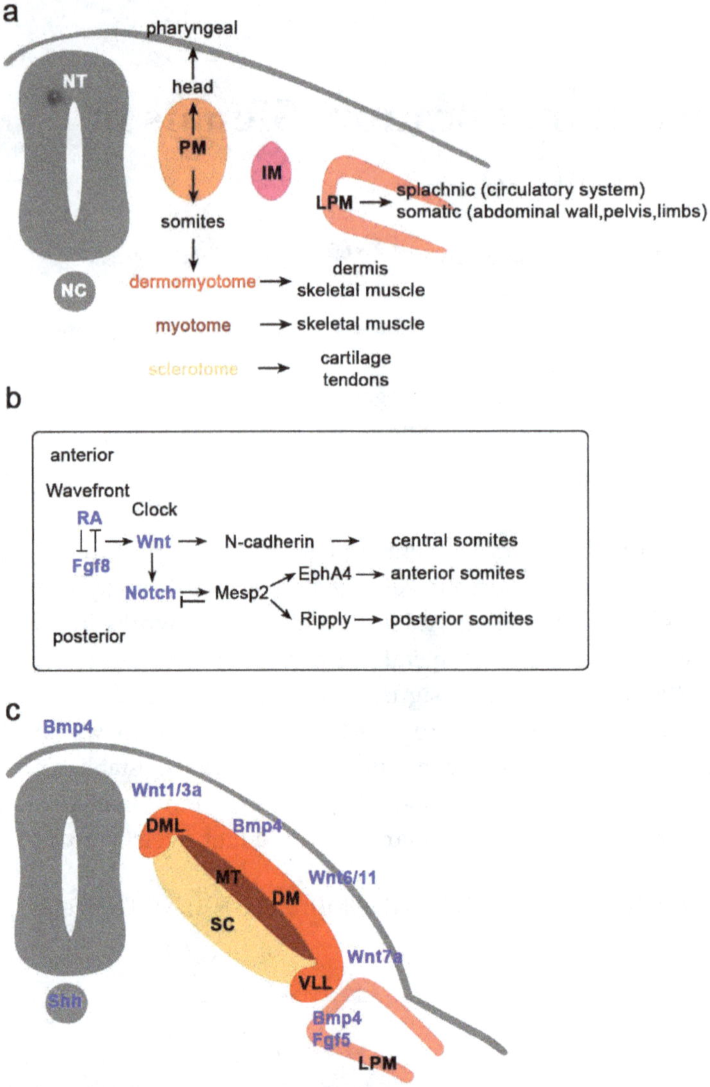

Figure 1. Signaling Molecules in Myogenesis (**a**) Paraxial mesoderm (PM) located next to the neural tube (NT). Dorsal PM gives rise to head and pharyngeal muscles while dorsal PM gives rise to somites. Intermediate mesoderm (IM) is located between the PM and the lateral plate mesoderm (LPM). LPM gives rise to the vascular system and skeletal muscles of the abdomen pelvis and some limb muscle. NC, notochord; (**b**) Anterior-posterior pattern of skeletal muscle is initiated by the RA/FGF8 inhibitory network, with AR promoting the anterior muscles and FGF8 the posterior. Activation of *Wnt* promotes the central somites and NOTCH signaling that will support formation of the anterior and posterior somites via EPH4 and RIPPLY, respectively; (**c**) Segmentation of somites into dermomyotome (DM), myotome (MT) and sclerotome (SC) is the result of signaling molecules secreted from the dorsal NT (BMP, WNT1/3a), notochord (SHH), ectoderm (WNT6/11, WNT7a, BMP4), and LPM (BMP4, FGF5). DML, dorsomedial lip; VLL, Ventrolateral lip.

The development of myogenic progenitors has been extensively investigated in vertebrate embryos since the 1990s using grafting [2], lineage tracing [3], in situ hybridization [4] and reporter (*LacZ*, *EGFP*) gene replacement [5–7] in both avian and mammalian embryos. These studies along with recent genomic wide association studies have established that the craniofacial, trunk, and skeletal musculature have unique anatomical origins and molecular networks [8–11]. Signals like retinoic acid (RA), sonic hedgehog homolog (SHH), bone morphogenetic proteins (BMP), WNT, fibroblast growth factor (FGF) are emitted from the neural tube, notochord, ectoderm, dorsal laminae, and neural crest [11]. RA and FGF set up the anterior-posterior cues respectively in early gastrulation. High concentration of RA triggers the formation of cranial PM and initiates the expression of cardiac (*Pitx2, Tbx1, Pax7*) and cranial (*Nkx2.5, Tbx5, Gata4*) SSTF networks [12,13]. FGFs from the tail end of embryo antagonize RA activity and initiate the formation of trunk PM, segmentation, and somite formation. BMP, SHH, WNT, and NOGGIN promote dorsal-ventral and medial-lateral cues that mediate trunk and abdominal SSTF networks (*Six1/4, Pax3, Pax7, Pitx2*) that in turn regulate the expression of basic helix-loop-helix (bHLH) myogenic regulatory factors (MRF; *Myf5, Myf6, Myod1, Myog*) [14]. MRFs regulate the commitment, determination, and differentiation of muscle progenitor cells [15]. MRFs act redundantly but control the cell-autonomous activation of myogenesis and regulate the expression of genes involved in contraction [16–20]. MYF5 and MYOD define the committed myoblasts, whereas MYF6 and MYOG define the differentiated myocytes. Differentiated myocytes exit the cell cycle and undergo morphological changes. In this review, we discuss the signaling pathways and gene networks that control skeletal muscle development in head and trunk, in addition to vascular and cardiac muscle (Table 1).

Table 1. Signaling molecules during myogenesis.

Gene	Anatomical Location	Model	Function	Reference
		Bone Morphogenetic Factor (BMP)		
Bmp2		murine fibroblasts	prevention of myogenesis	[21,22]
		murine myoblasts	inhibition of *MyoG* expression	[23,24]
		fish	delayed muscle differentiation	[25]
Bmp4		human fetal myoblasts	regulation of proliferation	[26]
		chick	induction of embryonic and fetal myogenesis	[27]
		frog	induction of ventral mesoderm	[28]
		chick	inhibition of *Myod* and *Myf5*	[29]
		axolotl	somite development	[30]
colloid-like 1 BMP4 regulator	anterior somites	chick	neck muscle formation	[31]
Noggin BMP4 antagonist	Neural Tube; Notochord	mouse	differentiation	[32,33]
		chick	growth inhibition	[34]
		mouse	fetal myogenesis; migration of muscle progenitor cells	[35,36]
Bmp7	dorsal Neural Tube	chick	muscle growth, apoptosis	[34]
		Fibroblast Growth Factor (FGF)		
Fgf2	Central Nervous System	fish	inhibition of muscle pioneer cells	[19,24,25]
Fgf4	myotubes	mouse	mesoderm formation	[37]
		chick	inhibition of terminal differentiation of limb	[38]
		chick	induction of tendon-specific markers	[39]
		chick	patterning during gastrulation	[40]
		frog	activation of *Myod*	[41,42]
Fgf5	Lateral Plate Mesoderm	mouse	formation of germ layers	[43]
		chick	inhibition of myogenesis	[44]
Fgf8	caudal end of embryo	fish	terminal muscle differentiation	[45,46]
		chick	tendon formation	[39]
		fish	somitic muscle formation	[47]
		murine myoblasts	myotubes formation	[48]
		fly	mesodermal cell migration	[49]
		chick	mesodermal cell migration	[40]
		chick; mouse	myoblast proliferation	[50]

Table 1. *Cont.*

Gene	Anatomical Location	Model	Function	Reference
		Notch Pathway		
Notch	Ectoderm	mouse	terminal muscle differentiation	[32,33]
		frog	inhibition of cell fusion; myotube formation	[51]
		mouse	cell fate decisions	[52]
		murine myoblasts	inhibition of muscle differentiation	[53]
rbp-j NOTCH ligand	trunk somites	frog	determination of dorsolateral and ventral mesoderm	[28]
		axolotl	formation of the dermomyotome	[30]
		myeloma cells	inhibition of differentiation	[54]
		mouse	differentiation	[32,33]
		murine myoblasts	inhibition of differentiation	[55]
Jagged-1 NOTCH ligand	trunk mesenchyme, splachnopleure	mouse	differentiation	[32,33]
		murine myoblasts	inhibition of differentiation	[56,57]
Delta1 NOTCH ligand	Neural Crest cells	mouse	terminal differentiation	[32,33]
		chick	inhibition of terminal differentiation	[58]
		chick	terminal differentiation, inhibition of exiting the cell cycle	[59]
Mesp2 NOTCH ligand	Presomitic Mesoderm	mouse	somitic boundaries	[60]
		mouse	cellular epithelialization	[61]
		mouse	somitogenesis; rostro-caudal polarity	[62]
		mouse	inhibition of NOTCH targets	[63]
		Ephrin Ligands		
EphA4	Presomitic Mesoderm	mouse	somitogenesis	[61]
		chick	somitogenesis	[64]
ephinb2	somites	chick	somitogenesis, expression of N-cadherin	[64]
Ripply	Presomitic Mesoderm	mouse	rostro-caudal polarity	[62]
		mouse	*Mesp2* expression	[65]
neurotrophin 3	Neural Tube	chick	*Pax3* expression induction, somitic myogenesis regulation	[66]
		chick	dermatome dissociation; epithelial-mesenchymal transition	[67]
		Retinoic Acid (RA)		
Ra	cranial end of the embryo	mouse	*Myf5* expression	[68]
		fish	somitogenesis; fast muscle differentiation	[69]
		murine myoblasts	inhibition of myoblast proliferation; differentiation	[69]
		mouse	somitogenesis	[70]
		Sonic Hedgehog (SHH)		
Shh	Neural Tube, Notochord	chick	somitogenesis	[66]
		fish	formation of slow muscle precursor cells	[71]
		chick	induction of myogenesis	[72]
		fish	inhibition of *pax3/7*; activation of *myf5* and *myod*	[47]
		chick	induction of *pax3* and *myod1*	[73]
		chick	activation of slow MyHC	[74]
		mouse	epaxial and hypaxial myogenesis	[75]

Table 1. *Cont.*

Gene	Anatomical Location	Model	Function	Reference
		Wnt		
Wnt1	Neural Tube	chick	*pax3* activation; somitogenesis	[66]
		murine myoblasts	formation of slow fiber types promotion	[76]
		chick	formation of paraxial mesoderm	[72]
		mouse	Formation of medial and dorsal portion of somites	[77]
		mouse	*Myf5* activation	[15]
Wnt3a	Neural Tube, Apical Ectodermal Ridge (AER)	murine myoblasts	*Bmp4* activation; formation of slow fibers	[76]
		mouse	Formation of dorsomedial part of somites	[77]
		murine myoblasts	myotube formation	[78]
		human embryonic stem cells	myogenic commitment	[79]
		murine myoblasts	myotube formation	[80]
		pluripotent stem cells	cardiomyocytes proliferation	[81]
Wnt4	Neural Tube	chickmurine myoblasts	activation of *pax7* and *myod1*; formation of fast myofibers	[82]
		mouse embryos	activation of *Myf5* and *Myod1*	[15]
		murine myoblasts	myotube formation	[78]
		murine satellite cells and myoblasts	activation of myogenesis	[83]
Wnt6	Paraxial Ectoderm	mouse	activation of *Myf5* and *Myod1* in paraxial mesoderm	[15]
		chick	activation of *Pax3*, *Paraxis*, *Myf5*, *Myog*, *Desmin* and MyHC	[84]
Wnt7a	Dorsal Ectoderm	mouse	activation of *Myod1*	[15]
Wnt11	somites	chick	formation of fast myofibers	[85]
		chick	elongation of myocytes	[86]

2. Craniofacial Myogenesis

Sixty muscle groups provide support for seeing, feeding, breathing, and moving of the head and neck. These muscles originate from the cranial mesoderm as a part of the paraxial mesoderm. Cranial mesoderm later specifies into the prechordal, pharyngeal, and lateral mesoderm. Prechordal mesoderm gives rise to the extraocular muscles (EOM) which encompass six muscles that control the eye movement [87]. Pharyngeal mesoderm forms the branchial arches (BA) that give rise to the masticatory, facial, pharyngeal and laryngeal muscles, and the secondary heart field [88,89]. Part of the pharyngeal mesoderm contributes to splanchnic mesoderm, that populates the tongue, axial neck, and heart (Figure 1a).

Craniofacial myogenesis is influenced by RA, BMP, FGF, and WNT and BMP inhibitors. RA activates *Pitx2* and inhibits *Tbx1* expression in the anterior head mesoderm. As RA is weakened in posterior head mesoderm, RA antagonist, FGF8, activates *Tbx1* expression [90–92]. FGF and BMP reinforce TBX1 activity and activate *Alx4* and *Msc* expression. FGF signaling spreads posteriorly and establishes the mature head mesoderm patterns, drives *MyoR* and *Tbx1* expression in pharynx, BA muscles, and heart [93]. Grafting experiments in chick embryos have indicated that WNT1 and WNT3A from the dorsal neural tube, WNT13 from the surface ectoderm, and BMP4 and BMP7 from the dorsal neural tube and ectoderm inhibit head muscle myogenesis [94,95]. However, the WNT antagonists, FRZB, and the BMP inhibitors, NOGGIN and GREMLIN, from the cranial neural crest and other surrounding tissues, induce craniofacial myogenesis [95]. Lineage tracing approaches in both avian and mouse models demonstrated that EOM and BA cells contribute to distinct muscles [96]. Homeodomain transcription factors *Tbx1*, *Pitx2*, *Tcf21* and *Lhx2* are expressed in the pharyngeal mesoderm and regulate cranial myogenesis. PITX2 specifies EOM by activating *Myf5*, *Myf6*, and *Myod1* [14]. During BA myogenesis,

PITX2 activates *Tbx1*, *Msc*, *Tcf21*, and *Six2* [97,98]. TBX1 induces the expression of *Fgf* receptors and their ligands. Perturbation of *Fgf* in mice or fish leads to hypoplastic and asymmetric muscles [99,100]. TBX1 and FGF activate *en2* in myogenic cells in the dorsal mandibular pharyngeal arch and contribute to jaw development. Although function of EN2 is still unclear, studies in chick embryos implied that signaling pathways prevent prematurely specification. TBX1, MSC, and TCF21 together activate MYF5 and lead to MYOD1 and MYOG expression (Figure 2) [101]. PAX7 is expressed in MRF^+ cells [88,94,102] and controls the formation of head satellite cells [103]. Head satellite cells derive from the $MESP1^+$ cells [96] but only the EOM and pharyngeal satellite cells express ISL1, ALX4 PITX1, PITX2, TCF21, cytokines, and chemokines [14,96,104].

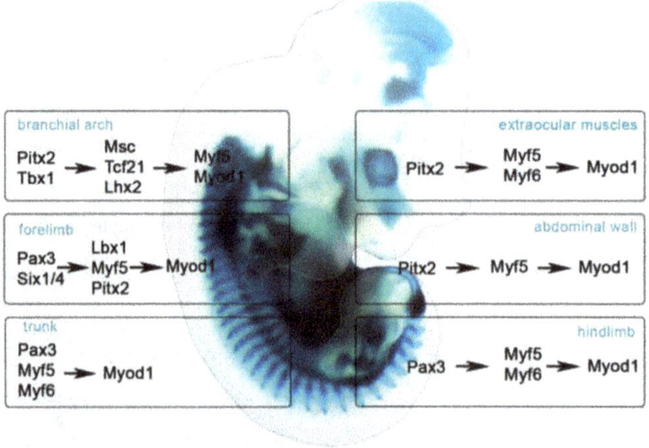

Figure 2. Gene Networks Involved in Muscle Specification. LacZ staining of E12.5 *Pitx2*$^{LacZ/+}$ mouse. Extraocular muscles (EOM) and abdominal wall muscles are specified by PITX2. Branchial arch muscles are specified by PITX2 and TBX1. Forelimb, trunk and hindlimb muscles are specified by PAX3.

Tongue [105] and neck muscles arise from the medial-dorsal and lateral-ventral domains of the occipital and cervical somites [106]. WNT signaling regulates tongue myogenesis [107] and differentially activation of MYF5 and MYOD1 [108]. WNT from the neural tube induces the expression of *Colloid-like1*, and reinforces BMP activity in the most anterior medial dermomyotome (DM) and regulates neck myogenesis [31].

3. Cardiac Myogenesis

Cranial mesoderm cells ingress laterally and contribute to the formation of the myocardium and endocardium [109], while medial and lateral splanchnic mesoderm give rise to the chambers of the heart [97]. RA establishes the posterior boundary of the secondary heart field [110], influences the development of the posterior and sinoatrial region [111] and transforms the cranial cells into cardiomyocytes [112]. BMP is expressed at the posterior and lateral region of the embryo and recruits head mesoderm cells to the heart [109,112]. WNT1, WNT3a, and WNT8c from the posterior primitive streak and lateral mesoderm inhibit cardiogenesis [113] and define the migratory limits of the cardiomyocytes. BMP antagonist NOGGIN from the notochord provides timely control to initiate early heart formation [12,13]. Collectively, the above signals promote expression of NKX2.5, TBX5, MEF2C and GATA4 [12,13] and the chromatin remodeling subunit SMARCD3 [114] and force the transition of mesodermal cells into cardiocytes [115].

Just after gastrulation, the cardiac crescent or 1st heart field, forms a tubular structure, with an outer myocardium and an inner endocardium. The linear heart tube connects with the artery at the anterior pole which forms the outflow tract and with a vein at the venous pole, which forms the inflow tract [116]. BMP2 activates NKX2.5 expression in the cardiac crescent. NKX2.5 interacts with GATA4 and TBX5 [117] to promote cell differentiation [118] and suppress the expression of FGF10 to prevent prenatal development [119]. BA-derived cells, 2nd heart field, express FGF8, FGF10, ISL1 [120] and TBX1 [121] proliferate, contribute to the extension of the tubular heart, and the formation of outflow tract and right ventricle. PITX2 activates MSC [97,98] and TBX*1* [122], which triggers FGF10 expression to stimulate cardiac muscle progenitors towards migration and proliferation [123]. Cranial neural crest (NC) cells contribute to the formation of valve cushions of the outflow tract [124,125]. Histone deacetylases (HDACS) [126] and TGFβ modulate their migration [127] throughout the dorso-ventral axis [128,129]. TGFβ represses cardiomyocyte specification to prevent premature differentiation [130].

4. Smooth Muscle Formation

Smooth muscle cells (SMC) give rise to gastrointestinal, urogenital and respiratory tract, and blood vessels [131]. The specific mechanisms responsible for SMC determination and differentiation are largely unknown. A heterogeneous population of cells derive from somites [132], secondary heart field [133], splanchnic mesoderm [134], and NC cells [135] contribute to the vascular SMC formation. BMP, NOTCH, SHH, and TGFβ1 promote vascular SMC specification and differentiation by activating serum response factors (SRF) and their cofactors. Activation of SRF leads to expression of ACTA), SM myosin heavy chain, and SM22 (review [136,137]). BMP and NOTCH initiate myogenic patterning in vascular SMC development. BMP2 stimulates their migration and the expression of the Va myosin [138], whereas NOTCH is required for differentiation of the vascular SMC progenitors [139]. SHH regulates vascular SMC proliferation and the formation of coronary vessels [140]. TGFβ1 inhibits vascular SMC growth, increases proliferation of the NC-derived vascular SMC via plasminogen activator inhibitor (PAI) [141]. TGFβ1 also stimulates vascular SMC differentiation by activating ACTA2 [142], SM22 [143], SMAD3 [144], and RhoA [145]. TGFβ/BMP co-receptor ENDOGLIN, a descendent of the PAX3$^+$ vascular SMCs is required for angiogenesis [146]. More recent studies also indicated the PAX3$^+$ DM might act as a stem cell niche for vascular SMCs [147].

5. Trunk Myogenesis

Skeletal muscles of the trunk originate from somites, which are derived from the segmented PM and located on the either side of the neural tube. Somitogenesis occurs progressively in an anterior to posterior sequence, simultaneous with regression of the primitive streak soon after neurulation (Figure 1a). After somites are formed and segmented, cells start to differentiate along the dorsal-ventral axis [1]. Ventral cells undergo an epithelial to mesenchymal transition to form the sclerotome, which will generate cartilage and associated connective tissues of the vertebrate and ribs. Dorsal cells remain epithelial and form the DM, which will generate the dermis, skeletal muscle of the truck and limbs, and brown fat [148]. The first myogenic tissue, myotome, arises later from the dorsomedial (DML) and ventrolateral (VLL) lips of the DM [149]. The DM is subdivided into the lateral hypaxial and medial epaxial DM. Cells of the VLL form the hypaxial myotome that gives rise to the lateral trunk and limb musculature, and cells from the epaxial myotome give rise to the deep muscles of the back (Figure 1c) [150,151].

5.1. Somitogenesis

Cells from the anterior tip of presomitic mesoderm undergo different morphogenetic changes and form as an epithelial ball, termed somitomeres [152,153]. Somitic cells are temporally controlled by the cyclin expression of intrinsic oscillating genes ("clock") that set the pace [154] and signals secreted from nearby tissues ("wave front") that define the position of the posterior border of each new somite [155,156]. Clock genes are expressed cyclically to define the time interval. Combinatorial signals (RA, FGF8, WNT3, and SHH) regulate the region where the clock genes are segregated and gives rise to a new pair of somites. Gradient signals provide the spatial clue and define the boundaries of segmentation of newly formed somites. FGF8 and WNT3 are produced in the caudal region of the embryo, RA is secreted by the cranial region of newly formed somites [152,153,157], and SHH is released from the notochord [155]. SHH signaling is required to preserve the FGF8 gradient. FGF8 signaling counteracts an opposing RA gradient [158]. The interaction of WNT and FGF8 regulates not only the separation, but also the structure of somites. At the anterior end of PM, WNT induces the expression of *ß-catenin* and *N-cadherin* in the center of somite. N-cadherin regulates the adhesion of the epithelial surface of the somite by forming tight junctions and a basal lamina that separates it from nearby tissues [64]. At the same time, WNT induces its antagonist, NOTCH, at the posterior end of somite, while NOTCH induces *Epha4* and either *ephrinB2* (chick) or *Ripply* (mouse). Together, they define the posterior and anterior end of the somite and promote their separation (Figure 1b) [64,159,160].

Mature somites differentiate into different compartments and determine the fates of cells (Figure 1c). Compartmentalization is induced by SHH and NOGGIN from the floor plate of neural tube and notochord, and by WNTs and BMPs from the ectoderm and dorsal laminae of the neural tube. Noggin binds and inactivates Bmp4 and creates morphogenic gradients [161,162]. Cells at the ventromedial part of the somite, located closer to the notochord and the ventral part of neural tube, receive a higher concentration of NOGGIN and SHH. As they proliferate, they lose the expression of N-cadherin and later become mesenchymal cells and develop into sclerotome. They express PAX1 and SOX9 that constitute the cartilage, vertebrae, and ribs and I-MF that inhibits muscle differentiation. Cells located at the dorsomedial part of the somite receive a low concentration of NOGGIN and SHH, remain epithelial, and give rise to DM [163–165]. The epithelial DM is divided into dorsomedial (epaxial), central (dermatome), and ventromedial (hypaxial) DM. The epithelial DM is influenced by different signals to activate distinct regulatory gene networks that contribute to specific muscle formation in the body. PAX3 marks the myogenic lineage in the DM. The $PAX3^+$ cells mark the DM-derived cells that give rise to skeletal [166,167] and vascular muscles [168,169]. Ablation of *Pax3* leads to the absence of forelimb muscles [170,171]. SIX1, SIX4, and their cofactors EYA1 and EYA2 directly bind on the promoter of *Pax3* and activate its expression [15,172,173]. Double mutants of *Six1/4* and *Eya1/2* exhibit loss of PAX3 expression that leads to migratory defects in hypaxial muscles [172,173].

PAX3 plays a key role in activating the MRF expression. MYF5 is expressed earlier in the dorsomedial lip of DM and then in the hypaxial myotome under the influenced of WNT1 and SHH to give rise to the epaxial muscles [174,175]. MYOD1 expression is induced by WNT7a and inhibited by BMP4 in the ventrolateral DM. Expression of MYF6 and MYOG coincides with the innervation and myofiber formation [18,19].

5.2. Epaxial Muscle Formation

Cells in the epaxial DM receive WNT1 and WNT3a from the dorsal neural tube and low levels of SHH from the floor plate of the neural tube. WNT1 and WNT3 activate *Myf5* [176] and *Myod1* expression through the WNT canonical pathway [177–181]. Low level of SHH contributes to *Myf5* expression through the GLI2 and ZIC1 [182,183]. Epaxial DM receives NOTCH/DELTA1 signaling from migrating NC cells and induces MYF5 expression [184]. PAX3 also regulates *Myf5* expression through direct binding to *Dmrt2* motifs [15,181,185]. PAX3 and MYF5 trigger MYOd1 expression, which drives myoblast differentiation [175]. MYF6 is expressed in the somites simultaneously with MYF5, although what activates its expression is still unknown [20].

5.3. Dermatome Formation

The dermatome is heterogenic and distinct, marks the boundary between the epaxial and hypaxial DM, and gives rise to the dermis of head and neck, satellite cells, and brown adipocytes [9,151,186–188]. The combined influence of SHH from the notochord, WNT1 from the dorsal neural tube, WNT11 from surface ectoderm and dermatome, and the negative influenced of BMP4 from the LPM marks the epaxial-hypaxial border. Homeobox gene *engrailed1* (*en1*) blocks the epaxial-hypaxial interface. EN1 is expressed on the border of the hypaxial maker SIM1, which marks the territories of motor neurons and regulates muscle patterning [189]. The EN1-SIM1 expression boundary marks the epaxial-hypaxial DM boundary and the dermatome. The specification of cells in the dermatome is diverse. A subset of cells in the medial part of dermatome receives BMP from the roof plate of the neural tube and migrates dorsally into the sub-ectodermal space to form cartilage and the medial margin of scapula in birds and mammals [190–193]. Another subset of cells, which is under the influence of *Wnt6* from the dorsal ectoderm, expresses PAX3 and PAX7 and forms the satellite cells [9,151,186–188]. PAX3 induces the expression of PAX7 in the central DM. The $PAX3^+/PAX7^+$ muscle progenitor cells populate the myotome [167,187,188] and receive *neurogulin1* (*nrg1*) from the migrating NC cells through NRG1 receptor ERBB3. The NRG1-ERBB3 signaling maintains PAX7 expression and prevents premature myogenic differentiation [194]. In addition, DERMO-1 also modulates the development of mesenchymal cell lineages including muscle and dermis [195]. DERMO-1 represses transcriptional activity of MEF2 and MYOD in a dose-dependent fashion [196].

5.4. Hypaxial Muscle Formation

The hypaxial DM gives rise to vertebral muscle, the diaphragm, abdominal muscles and muscles of the limb. Development of hypaxial DM is highly influenced by the LPM [165,197]. Upon receiving WNT7a from the dorsal ectoderm and Bmp4 from the LPM, cells give rise to hypaxial myotome [165,197]. BMP signaling induces somite-derived endothelial cell differentiation and migration via VEGFR2 expression [198,199], whereas the WNT7a/ PKC activates Pax3 in hypaxial DM [200]. PAX3 is critical for the onset of embryonic myogenesis by regulating delamination via C-MET, migration via LBX, and determination of muscle progenitors [166,167] in the limb [201–205]. PITX2 lies genetically downstream of PAX3 in the hypaxial DM [206,207]. Activation of *Pitx2* regulates expression of MYOD1 and MYOG, myoblast motility, and skeletal muscle maintenance [208,209]. PAX3 regulates the expression of ITM2a in DM, limb buds, adult muscle fibers, and satellite cells [210]. In turn, ITM2a regulates muscle creatine kinase (CK-M) expression and muscle differentiation [211].

5.5. Myotome Formation

Myotome is the first myogenic structure located between the DM dorsally, and the sclerotome ventrally. Cells of the DM, lose expression of N-cadherin, delaminate, and translocate underneath the DM to form the new structure, the myotome. Surrounding signals trigger the expression of *MRFs* and establish their myogenic identity as they populate the myotome. Upon arrival, the cells start to differentiate and fuse to become myofibers.

During the first wave of myogenesis, myogenic progenitors from the DM develop into mononucleated myocytes. Cells from the dorsal part of myotome originate from the dorsomedial lip (DML), migrate ventrally as they receive WNT1 and WNT3a from the dorsal neural tube, WNT7a from dorsal ectoderm, and SHH from the ventral region of the neural tube and notochord. The WNT and SHH signaling defines the myogenic properties of migrating muscle precursors by activating the expression of MYF5 [72,180,182] and MYOD1 [148]. Although the cells from the central DM contribute to the myotome, they are less significant than cells from the DML and ventrolateral lip (VLL). Cells in the central DM receive neurophrin-3 and WNT1 from the neural tube and become part of the myotome [212]. Cells from the VLL are essential for lateral extension of the myotome [213]. They receive WNT7a from the dorsal ectoderm and BMP4 and FGF5 from the LPM, inducing the expression of MYOD1, but not MYF5 [214].

After entering the myotome, MYF5 and SHH together activate the expression of *Fgf4*, which regulates the proliferation and differentiation of the *MRF* expressing myocytes [215]. Once the myocyte identity is established, they express PAX7 before undergoing terminal differentiation. SHH is crucial to maintain the expression of PAX7 in migrating $MYF5^+/MYOD1^+$ cells. Interfered by SHH results in accumulation of $PAX7^+$ cells and a small myotome in chick and mouse embryos [216,217]. Myocytes receive WNT11 as they elongate along the somites and express both slow- and fast-type cytoskeletal proteins, including MYH (slow type I), MYH3 (embryonic), ACTC1, ACTA1, DESMIN, JAG2, ß-ENOLASE, and CA2 [8,217,218]. Myocytes fuse and form the first multinucleated $MyHC^+$ myofibers spreading from the ventral to the dorsal myotome [217]. *Six1* and *Six4* expression activates the fast muscle program [218].

5.6. Limb Muscle Formation

Limb muscle development is a triphasic process. During embryonic myogenesis in mice (E9-E12), the first multinucleated muscle fibers are formed from $PAX3^+$ embryonic myoblasts, followed by $PAX3^+LBX1^+C\text{-}MET^+$ migratory myoblasts that provide the basic pattern and primary myofibers throughout the limbs. During fetal myogenesis (E13-E16), the $PAX3^+/PAX7^+$ fetal muscle progenitors use embryonic fibers as a scaffold to expand the muscle mass. During perinatal myogenesis (E17-P5), $PAX7^+$ muscle progenitors (satellite cells) are located between the basal lamina and the fiber plasma membrane and remain mononucleated and quiescent. After muscle damage, satellite cells can activate, fuse with existing myofibers, and repair damaged muscle (Figure 2) [10,219].

Embryonic (primary) myogenesis begins when the first embryonic myoblast progenitors of the hypaxial DM start to delaminate and migrate to forelimb buds [220]. PAX3 activates expression of c-MET and LBX1, and the $PAX3^+LBX1^+c\text{-}MET^+$ cells delaminate, become mobile, and start to migrate distally [203–205]. LBX1 is crucial for the migrating embryonic myoblast progenitors (EMP) [205,221]. The migratory directions are controlled by signals from the surface of limb buds, zone of polarizing activity (ZPA). Misrouted EMPs either migrate slowly or accumulate in the mesoderm of ventral body wall and form smaller or no muscles [222,223]. SHH signals from ZPA regulate the migratory patterns of EMPs, maintenance of limb bug outgrowth, dorso-ventral limb patterning, and development of skeleton, cartilage, and tendons of digits [224,225]. SHH and FGFs from the ectodermal ridge control the expression of HGF/SF, the only known ligand of c-MET [226], which directs the migration of EMPs [203,204]. Migratory EMPs also express the chemokine receptor CXCR4. This receptor responds to chemo-attractant SDF1 from the limb buds, which also directs the migratory routes [227]. SF/HGF, BMP, WNT and SFRP-2 inhibit prematurely differentiating EMPs and allow them to multiply and populate the limb bud parallel to establishing the myogenic program [198,228]. The Wnt/ß-catenin/PITX2 pathway controls motility and proliferation of the EMPs by regulating expression of the growth control genes *Ccnd1*, *Ccnd2* and *c-myc*, [209]. IGF contributes to the proliferation and determination of EMPs [229]. IGF-1 and IGF-2 play a role in PAX3 and MYOG expression via PI3K and MYOD through FGF18 [230]. Upon their entry into the limb bud, EMPs express high levels of MRFs and become embryonic myoblasts and myofibers. At the same time, SHH initiates the expression of MYF5 though GLIA [231]. The spindle-shaped EMPs receive FGF and TGFβ, and increase in cell number while they align to each other. FGFs maintain the expression of MSX1 to prevent their premature differentiation. The newly formed $MYOD1^+$embryonic myoblasts exit the cell cycle and express MYOG and MEF2 [232]. During differentiation, embryonic myoblasts undergo several changes, including cell aggregation, elongation, metabolic changes, cell membrane fragmentation and cytoskeleton assembly. They start to express MyHC and myosin light chain 1 (MYL1), secrete fibronectin, RGD-binding integrins, cadherins, and cell adhesion molecules that promote the alignment of migratory cells [233–235], and fuse to each other to become multinucleated primary myofibers [236].

Fetal myogenesis begins when a subset of $PAX3^+$ myogenic progenitors start to express PAX7 while their PAX3 expression is decreased in the central DM [167,188,237,238]. Mouse *Pax7* mutants exhibit defects in fetal myogenesis with smaller muscles, fewer myofibers, and impaired satellite cells, leading to defective muscle regeneration [239–243]. $PAX7^+$ cells constitute the fetal myogenic progenitors (FMP). FMP receive TGFβ, BMP, and WNT/β-catenin signaling from the ectoderm and the dorsal laminae to block the premature differentiation. TGFβ2 from the ventral region of the limb buds and later from the muscle mass [173] represses MYOD1 and MYOG activity [244], induces FMP migration and inhibits their differentiation [245,246]. They increase their number and contribute to all muscles [237] as they migrate and fuse to the primary fibers and each other to form the secondary fibers. PAX7 directly binds and activates *Nfix*, which marks the fetal myoblasts. NFIX represses the embryonic muscle genes *Sox6* and *MyHCI*, and activates the fetal muscle genes *α7-integrin*, *β-enolase*, muscle creatine kinase, and muscle sarcomeric proteins [247]. PITX2 and PITX3 are also expressed in fetal myoblasts [248] as they commit to the fetal program and become myocytes by subsequent expression of MYOD1, MYF5, and MYOG [167]. When differentiated, fetal myocytes fuse to embryonic myofibers and give rise to doughnut-shaped, multinucleated fetal myofibers [188,249].

During birth and the first few weeks after birth, the third wave of myogenesis begins with an increase in number and size of myofibers. $PAX7^+$satellite cells express myogenic (c-MET, HGF/SF, MSX1) and endothelial (CD34) markers [250,251]. Expression of *c-met* and *Hgf/Sf* promotes delamination and migration from the basal lamina once activated in response to stress. Activated satellite cells turn off *Pax7* and begin to express *Myod1* and cell-cycle markers and undergo multiple rounds of asymmetric cell division. A small number of activated satellite cells returns to an undifferentiated state ($PAX7^+MYOD1^-$) for self-renewal. PITX2 regulates satellite cell division by stimulating the expression of *Ccnd1*, *Ccnd2* and *Myf5*, and downregulates *miR-15b*, *miR-106b*, *miR-23b*, and *miR-503* [252]. Daughter satellite cells express MYF5 and MYOD, and fuse to pre-existing myofibers to form new fibers [238]. MSTN, a member of TGF family, regulates satellite cell renewal and muscle growth [253]. MSTN inhibits *Pax7* expression and binds to activin. Activin activates ALK4/5, mediates SMA2/3 phosphorylation [254,255], and recruits SMAD4. MSTN/activin/SMAD signaling pathway inhibits MRF expression [256], and limits proliferation and differentiation into myofibers via AKT/TORc1/P70S6K pathway [257–259]. MSTN antagonist, Follistatin (*Fst*), promotes PAX7 activity [260–262], regulates MRFs expression and muscle fiber formation [263]. Bone morphogenetic protein and activin membrane-bound inhibitor (BAMBI), another member of TGF family, responds to Wnt/β-catenin pathway and plays a role in regulating SMAD activity [264].

6. Abdominal Myogenesis

Abdominal muscles encompass three layers of skeletal muscles, which merge toward the midline and form a sheath to assist breathing and protect the inner organs. Abdominal muscles derive from LPM and somites. LPM gives rise to somatopleure (lateral) and splanchnopleure (medial). The somatopleure and the surface epithelial endoderm give rise to the body wall in three waves similar to trunk myogenesis (Figure 1a) [265,266]. Myogenic cells within a thin sheet of LPM-derived cells proliferate without delamination, remain within the DM epithelium, and migrate ventrally into the somatopleure to cover the abdomen and form the primary body wall [266,267]. The primary body wall is composed of myogenic and epithelial cells that provide the first tissue to cover the endoderm. After several days, a second and third wave of myogenic progenitor cells move into the somatopleure and all layers are joined at the midline [268,269]. The secondary body wall forms muscle, skin, ribs, and sternum. Cells of the secondary body wall are heterogeneous, composed of myogenic, mesenchymal and NC cells. WNT/PITX2 signaling plays a role in the abdominal myogenesis. WNT/β-catenin signaling instructs mesodermal specification of somites and LPM-derived mesenchymal cells in ventral body wall as a paracrine signal [270]. WNT7a from the dorsal ectoderm and BMP from the LPM activate expression of PAX3 and its targets. WNT signaling activates *Pitx2* expression in LPM. $PAX3^+PITX2^+$ cells proliferate, migrate to the midline, and express *MRFs* [271]. Ablation of *Pitx2* results in repression of *T-box* and activation of *Hox9-11* genes in the abdominal wall [266,272]. AP-2a regulates the closure of all four body folds by regulating the epithelial-mesenchymal interactions and cell-cell communication from the ectoderm to the abdominal mesoderm [268,273]. Defects in these pathways lead to hypoplasia in the abdominal wall or organ exposure. HOXb2 and HOXb4 activate ALX3/4 [274,275] and regulate the primary ventral body formation [276–278]. Independent from the PITX2 pathway [279], BMP1 might enhance the activity or availability of $TFGß5^+$ myofibroblasts during the first wave [280] and initiate the migration in the second wave [281]. TGFβ2/3 plays a fundamental role in activating cytoskeletal components (*Tagln*, *α-Sma* and *desmin*) that support myofibroblast migration towards the midline [282]. The signaling pathways contribute to ventral wall closing by involving the assembly and function of extracellular matrix and collagen fibrils [283]. NC cells migrate to the midline and contribute to the second body wall formation. AP-2 is expressed in migrating NC cells [284].

7. Conclusions

Skeletal muscle progenitor cells arise from the PM and LPM that are heavily influenced by signals from the neural tube, notochord, ectoderm, and dorsal laminae. Combinatorial networks of signaling molecules regulate the expression of SSTFs that further regulate the commitment and differentiation to the myogenic lineage. The PM of the cranial domain forms early and gives rise to craniofacial, cardiac and vascular smooth muscles. $PITX2^{+}MSC^{+}TBX1^{+}$ cells from the paraxial head develop the cardiogenic and head muscle cells. Head myogenesis is inhibited by WNT signaling. The segmented trunk PM forms the somites, in which the $PAX3^{+}$ cells are influenced by WNT and SHH to commit to myogenic $MYOD1^{+}$ cells and $PAX7^{+}$ satellite stem cells.

Acknowledgments: We thank colleagues for inspiring discussions, and Jason Lattier for editing the manuscript. We apologize for not including many pivotal studies due to space limitations. This work was supported by the College of Pharmacy, Oregon State University.

Conflicts of Interest: The authors declare no conflict of interest.

References

1. Burke, A.C.; Nowicki, J.L. A new view of patterning domains in the vertebrate mesoderm. *Dev. Cell* **2003**, *4*, 159–165. [CrossRef]
2. Borycki, A.G.; Emerson, C.P., Jr. Multiple tissue interactions and signal transduction pathways control somite myogenesis. *Curr. Top. Dev. Biol.* **2000**, *48*, 165–224. [PubMed]
3. Ordahl, C.P.; Le Douarin, N.M. Two myogenic lineages within the developing somite. *Development* **1992**, *114*, 339–353. [PubMed]
4. Pownall, M.E.; Emerson, C.P., Jr. Sequential activation of three myogenic regulatory genes during somite morphogenesis in quail embryos. *Dev. Biol.* **1992**, *151*, 67–79. [CrossRef]
5. Goldhamer, D.J.; Faerman, A.; Shani, M.; Emerson, C.P. Regulatory elements that control the lineage-specific expression of myoD. *Science* **1992**, *256*, 538–542. [CrossRef] [PubMed]
6. Pinney, D.F.; de la Brousse, F.C.; Faerman, A.; Shani, M.; Maruyama, K.; Emerson, C.P., Jr. Quail myoD is regulated by a complex array of cis-acting control sequences. *Dev. Biol.* **1995**, *170*, 21–38. [CrossRef] [PubMed]
7. Tajbakhsh, S.; Rocancourt, D.; Buckingham, M. Muscle progenitor cells failing to respond to positional cues adopt non-myogenic fates in myf-5 null mice. *Nature* **1996**, *384*, 266–270. [CrossRef] [PubMed]
8. Chal, J.; Pourquie, O. Making muscle: Skeletal myogenesis in vivo and in vitro. *Development* **2017**, *144*, 2104–2122. [CrossRef] [PubMed]
9. Christ, B.; Huang, R.; Scaal, M. Amniote somite derivatives. *Dev. Dyn.* **2007**, *236*, 2382–2396. [CrossRef] [PubMed]
10. Biressi, S.; Molinaro, M.; Cossu, G. Cellular heterogeneity during vertebrate skeletal muscle development. *Dev. Biol.* **2007**, *308*, 281–293. [CrossRef] [PubMed]
11. Placzek, M.; Tessier-Lavigne, M.; Yamada, T.; Jessell, T.; Dodd, J. Mesodermal control of neural cell identity: Floor plate induction by the notochord. *Science* **1990**, *250*, 985–988. [CrossRef] [PubMed]
12. Schultheiss, T.M.; Lassar, A.B. Induction of chick cardiac myogenesis by bone morphogenetic proteins. *Cold Spring Harb. Symp. Quant. Biol.* **1997**, *62*, 413–419. [PubMed]
13. Schlange, T.; Andree, B.; Arnold, H.H.; Brand, T. BMP2 is required for early heart development during a distinct time period. *Mech. Dev.* **2000**, *91*, 259–270. [CrossRef]
14. Sambasivan, R.; Gayraud-Morel, B.; Dumas, G.; Cimper, C.; Paisant, S.; Kelly, R.G.; Tajbakhsh, S. Distinct regulatory cascades govern extraocular and pharyngeal arch muscle progenitor cell fates. *Dev. Cell* **2009**, *16*, 810–821. [CrossRef] [PubMed]
15. Tajbakhsh, S.; Borello, U.; Vivarelli, E.; Kelly, R.; Papkoff, J.; Duprez, D.; Buckingham, M.; Cossu, G. Differential activation of Myf5 and MyoD by different Wnts in explants of mouse paraxial mesoderm and the later activation of myogenesis in the absence of Myf5. *Development* **1998**, *125*, 4155–4162. [PubMed]
16. Weintraub, H.; Davis, R.; Tapscott, S.; Thayer, M.; Krause, M.; Benezra, R.; Blackwell, T.K.; Turner, D.; Rupp, R.; Hollenberg, S.; et al. The myoD gene family: Nodal point during specification of the muscle cell lineage. *Science* **1991**, *251*, 761–766. [CrossRef] [PubMed]

17. Delfini, M.C.; Duprez, D. Ectopic Myf5 or MyoD prevents the neuronal differentiation program in addition to inducing skeletal muscle differentiation, in the chick neural tube. *Development* **2004**, *131*, 713–723. [CrossRef] [PubMed]
18. Olson, E.N.; Arnold, H.H.; Rigby, P.W.; Wold, B.J. Know your neighbors: Three phenotypes in null mutants of the myogenic bHLH gene MRF4. *Cell* **1996**, *85*, 1–4. [CrossRef]
19. Hasty, P.; Bradley, A.; Morris, J.H.; Edmondson, D.G.; Venuti, J.M.; Olson, E.N.; Klein, W.H. Muscle deficiency and neonatal death in mice with a targeted mutation in the myogenin gene. *Nature* **1993**, *364*, 501–506. [CrossRef] [PubMed]
20. Kassar-Duchossoy, L.; Gayraud-Morel, B.; Gomes, D.; Rocancourt, D.; Buckingham, M.; Shinin, V.; Tajbakhsh, S. Mrf4 determines skeletal muscle identity in Myf5:Myod double-mutant mice. *Nature* **2004**, *431*, 466–471. [CrossRef] [PubMed]
21. Wang, E.A.; Israel, D.I.; Kelly, S.; Luxenberg, D.P. Bone morphogenetic protein-2 causes commitment and differentiation in C3H10T1/2 and 3T3 cells. *Growth Factors* **1993**, *9*, 57–71. [CrossRef] [PubMed]
22. Denker, A.E.; Haas, A.R.; Nicoll, S.B.; Tuan, R.S. Chondrogenic differentiation of murine C3H10T1/2 multipotential mesenchymal cells: I. Stimulation by bone morphogenetic protein-2 in high-density micromass cultures. *Differentiation* **1999**, *64*, 67–76. [CrossRef] [PubMed]
23. Katagiri, T.; Yamaguchi, A.; Komaki, M.; Abe, E.; Takahashi, N.; Ikeda, T.; Rosen, V.; Wozney, J.M.; Fujisawa-Sehara, A.; Suda, T. Bone morphogenetic protein-2 converts the differentiation pathway of C2C12 myoblasts into the osteoblast lineage. *J. Cell Biol.* **1994**, *127*, 1755–1766. [CrossRef] [PubMed]
24. Okubo, Y.; Bessho, K.; Fujimura, K.; Iizuka, T.; Miyatake, S. Expression of bone morphogenetic protein-2 via adenoviral vector in C2C12 myoblasts induces differentiation into the osteoblast lineage. *Biochem. Biophys. Res. Commun.* **1999**, *262*, 739–743. [CrossRef] [PubMed]
25. Patterson, S.E.; Bird, N.C.; Devoto, S.H. BMP regulation of myogenesis in zebrafish. *Dev. Dyn.* **2010**, *239*, 806–817. [CrossRef] [PubMed]
26. Frank, N.Y.; Kho, A.T.; Schatton, T.; Murphy, G.F.; Molloy, M.J.; Zhan, Q.; Ramoni, M.F.; Frank, M.H.; Kohane, I.S.; Gussoni, E. Regulation of myogenic progenitor proliferation in human fetal skeletal muscle by BMP4 and its antagonist Gremlin. *J. Cell Biol.* **2006**, *175*, 99–110. [CrossRef] [PubMed]
27. Wang, H.; Noulet, F.; Edom-Vovard, F.; Tozer, S.; Le Grand, F.; Duprez, D. Bmp signaling at the tips of skeletal muscles regulates the number of fetal muscle progenitors and satellite cells during development. *Dev. Cell* **2010**, *18*, 643–654. [CrossRef] [PubMed]
28. Re'em-Kalma, Y.; Lamb, T.; Frank, D. Competition between noggin and bone morphogenetic protein 4 activities may regulate dorsalization during Xenopus development. *Proc. Natl. Acad. Sci. USA* **1995**, *92*, 12141–12145. [CrossRef] [PubMed]
29. Reshef, R.; Maroto, M.; Lassar, A.B. Regulation of dorsal somitic cell fates: BMPs and Noggin control the timing and pattern of myogenic regulator expression. *Genes Dev.* **1998**, *12*, 290–303. [CrossRef] [PubMed]
30. Epperlein, H.H.; Vichev, K.; Heidrich, F.M.; Kurth, T. BMP-4 and Noggin signaling modulate dorsal fin and somite development in the axolotl trunk. *Dev. Dyn.* **2007**, *236*, 2464–2474. [CrossRef] [PubMed]
31. Pais de Azevedo, T.; Zuzarte-Luis, V.; Goncalves, L.; Marques, C.; Palmeirim, I. Tissue regulation of somitic colloid-like1 gene expression. *Biochem. Biophys. Res. Commun.* **2012**, *424*, 295–300. [CrossRef] [PubMed]
32. Schuster-Gossler, K.; Cordes, R.; Gossler, A. Premature myogenic differentiation and depletion of progenitor cells cause severe muscle hypotrophy in Delta1 mutants. *Proc. Natl. Acad. Sci. USA* **2007**, *104*, 537–542. [CrossRef] [PubMed]
33. Hrabe de Angelis, M.; McIntyre, J., II; Gossler, A. Maintenance of somite borders in mice requires the Delta homologue DII1. *Nature* **1997**, *386*, 717–721. [CrossRef] [PubMed]
34. Amthor, H.; Christ, B.; Rashid-Doubell, F.; Kemp, C.F.; Lang, E.; Patel, K. Follistatin regulates bone morphogenetic protein-7 (BMP-7) activity to stimulate embryonic muscle growth. *Dev. Biol.* **2002**, *243*, 115–127. [CrossRef] [PubMed]
35. Costamagna, D.; Mommaerts, H.; Sampaolesi, M.; Tylzanowski, P. Noggin inactivation affects the number and differentiation potential of muscle progenitor cells in vivo. *Sci. Rep.* **2016**, *6*, 31949. [CrossRef] [PubMed]
36. Tylzanowski, P.; Mebis, L.; Luyten, F.P. The Noggin null mouse phenotype is strain dependent and haploinsufficiency leads to skeletal defects. *Dev. Dyn.* **2006**, *235*, 1599–1607. [CrossRef] [PubMed]
37. Niswander, L.; Martin, G.R. Fgf-4 expression during gastrulation, myogenesis, limb and tooth development in the mouse. *Development* **1992**, *114*, 755–768. [PubMed]

38. Edom-Vovard, F.; Bonnin, M.A.; Duprez, D. Misexpression of Fgf-4 in the chick limb inhibits myogenesis by down-regulating Frek expression. *Dev. Biol.* **2001**, *233*, 56–71. [CrossRef] [PubMed]
39. Edom-Vovard, F.; Schuler, B.; Bonnin, M.A.; Teillet, M.A.; Duprez, D. Fgf4 positively regulates scleraxis and tenascin expression in chick limb tendons. *Dev. Biol.* **2002**, *247*, 351–366. [CrossRef] [PubMed]
40. Yang, X.; Dormann, D.; Munsterberg, A.E.; Weijer, C.J. Cell movement patterns during gastrulation in the chick are controlled by positive and negative chemotaxis mediated by FGF4 and FGF8. *Dev. Cell* **2002**, *3*, 425–437. [CrossRef]
41. Burks, P.J.; Isaacs, H.V.; Pownall, M.E. FGF signalling modulates transcriptional repression by Xenopus groucho-related-4. *Biol. Cell* **2009**, *101*, 301–308. [CrossRef] [PubMed]
42. Fisher, M.E.; Isaacs, H.V.; Pownall, M.E. eFGF is required for activation of XmyoD expression in the myogenic cell lineage of Xenopus laevis. *Development* **2002**, *129*, 1307–1315. [PubMed]
43. Hebert, J.M.; Boyle, M.; Martin, G.R. mRNA localization studies suggest that murine FGF-5 plays a role in gastrulation. *Development* **1991**, *112*, 407–415. [PubMed]
44. Clase, K.L.; Mitchell, P.J.; Ward, P.J.; Dorman, C.M.; Johnson, S.E.; Hannon, K. FGF5 stimulates expansion of connective tissue fibroblasts and inhibits skeletal muscle development in the limb. *Dev. Dyn.* **2000**, *219*, 368–380. [CrossRef]
45. Groves, J.A.; Hammond, C.L.; Hughes, S.M. Fgf8 drives myogenic progression of a novel lateral fast muscle fibre population in zebrafish. *Development* **2005**, *132*, 4211–4222. [CrossRef] [PubMed]
46. Tu, C.F.; Tsao, K.C.; Lee, S.J.; Yang, R.B. SCUBE3 (signal peptide-CUB-EGF domain-containing protein 3) modulates fibroblast growth factor signaling during fast muscle development. *J. Biol. Chem.* **2014**, *289*, 18928–18942. [CrossRef] [PubMed]
47. Hammond, C.L.; Hinits, Y.; Osborn, D.P.; Minchin, J.E.; Tettamanti, G.; Hughes, S.M. Signals and myogenic regulatory factors restrict pax3 and pax7 expression to dermomyotome-like tissue in zebrafish. *Dev. Biol.* **2007**, *302*, 504–521. [CrossRef] [PubMed]
48. Omoteyama, K.; Takagi, M. FGF8 regulates myogenesis and induces Runx2 expression and osteoblast differentiation in cultured cells. *J. Cell Biochem.* **2009**, *106*, 546–552. [CrossRef] [PubMed]
49. Gryzik, T.; Muller, H.A. FGF8-like1 and FGF8-like2 encode putative ligands of the FGF receptor Htl and are required for mesoderm migration in the Drosophila gastrula. *Curr. Biol.* **2004**, *14*, 659–667. [CrossRef] [PubMed]
50. Rinon, A.; Lazar, S.; Marshall, H.; Buchmann-Moller, S.; Neufeld, A.; Elhanany-Tamir, H.; Taketo, M.M.; Sommer, L.; Krumlauf, R.; Tzahor, E. Cranial neural crest cells regulate head muscle patterning and differentiation during vertebrate embryogenesis. *Development* **2007**, *134*, 3065–3075. [CrossRef] [PubMed]
51. Kopan, R.; Nye, J.S.; Weintraub, H. The intracellular domain of mouse Notch: A constitutively activated repressor of myogenesis directed at the basic helix-loop-helix region of MyoD. *Development* **1994**, *120*, 2385–2396. [PubMed]
52. Vasyutina, E.; Lenhard, D.C.; Wende, H.; Erdmann, B.; Epstein, J.A.; Birchmeier, C. RBP-J (Rbpsuh) is essential to maintain muscle progenitor cells and to generate satellite cells. *Proc. Natl. Acad. Sci. USA* **2007**, *104*, 4443–4448. [CrossRef] [PubMed]
53. Nofziger, D.; Miyamoto, A.; Lyons, K.M.; Weinmaster, G. Notch signaling imposes two distinct blocks in the differentiation of C2C12 myoblasts. *Development* **1999**, *126*, 1689–1702. [PubMed]
54. Kuroda, K.; Tani, S.; Tamura, K.; Minoguchi, S.; Kurooka, H.; Honjo, T. Delta-induced Notch signaling mediated by RBP-J inhibits MyoD expression and myogenesis. *J. Biol. Chem.* **1999**, *274*, 7238–7244. [CrossRef] [PubMed]
55. Kato, H.; Taniguchi, Y.; Kurooka, H.; Minoguchi, S.; Sakai, T.; Nomura-Okazaki, S.; Tamura, K.; Honjo, T. Involvement of RBP-J in biological functions of mouse Notch1 and its derivatives. *Development* **1997**, *124*, 4133–4141. [PubMed]
56. Lindsell, C.E.; Shawber, C.J.; Boulter, J.; Weinmaster, G. Jagged: A mammalian ligand that activates Notch1. *Cell* **1995**, *80*, 909–917. [CrossRef]
57. Shawber, C.; Nofziger, D.; Hsieh, J.J.; Lindsell, C.; Bogler, O.; Hayward, D.; Weinmaster, G. Notch signaling inhibits muscle cell differentiation through a CBF1-independent pathway. *Development* **1996**, *122*, 3765–3773. [PubMed]
58. Delfini, M.C.; Hirsinger, E.; Pourquie, O.; Duprez, D. Delta 1-activated notch inhibits muscle differentiation without affecting Myf5 and Pax3 expression in chick limb myogenesis. *Development* **2000**, *127*, 5213–5224. [PubMed]

59. Hirsinger, E.; Malapert, P.; Dubrulle, J.; Delfini, M.C.; Duprez, D.; Henrique, D.; Ish-Horowicz, D.; Pourquie, O. Notch signalling acts in postmitotic avian myogenic cells to control MyoD activation. *Development* **2001**, *128*, 107–116. [PubMed]
60. Morimoto, M.; Takahashi, Y.; Endo, M.; Saga, Y. The Mesp2 transcription factor establishes segmental borders by suppressing Notch activity. *Nature* **2005**, *435*, 354–359. [CrossRef] [PubMed]
61. Nakajima, Y.; Morimoto, M.; Takahashi, Y.; Koseki, H.; Saga, Y. Identification of Epha4 enhancer required for segmental expression and the regulation by Mesp2. *Development* **2006**, *133*, 2517–2525. [CrossRef] [PubMed]
62. Morimoto, M.; Sasaki, N.; Oginuma, M.; Kiso, M.; Igarashi, K.; Aizaki, K.; Kanno, J.; Saga, Y. The negative regulation of Mesp2 by mouse Ripply2 is required to establish the rostro-caudal patterning within a somite. *Development* **2007**, *134*, 1561–1569. [CrossRef] [PubMed]
63. Sasaki, N.; Kiso, M.; Kitagawa, M.; Saga, Y. The repression of Notch signaling occurs via the destabilization of mastermind-like 1 by Mesp2 and is essential for somitogenesis. *Development* **2011**, *138*, 55–64. [CrossRef] [PubMed]
64. Glazier, J.A.; Zhang, Y.; Swat, M.; Zaitlen, B.; Schnell, S. Coordinated action of N-CAM, N-cadherin, EphA4, and ephrinB2 translates genetic prepatterns into structure during somitogenesis in chick. *Curr. Top. Dev. Biol.* **2008**, *81*, 205–247. [PubMed]
65. Takahashi, J.; Ohbayashi, A.; Oginuma, M.; Saito, D.; Mochizuki, A.; Saga, Y.; Takada, S. Analysis of Ripply1/2-deficient mouse embryos reveals a mechanism underlying the rostro-caudal patterning within a somite. *Dev. Biol.* **2010**, *342*, 134–145. [CrossRef] [PubMed]
66. Munsterberg, A.E.; Kitajewski, J.; Bumcrot, D.A.; McMahon, A.P.; Lassar, A.B. Combinatorial signaling by Sonic hedgehog and Wnt family members induces myogenic bHLH gene expression in the somite. *Genes Dev.* **1995**, *9*, 2911–2922. [CrossRef] [PubMed]
67. Brill, G.; Kahane, N.; Carmeli, C.; von Schack, D.; Barde, Y.A.; Kalcheim, C. Epithelial-mesenchymal conversion of dermatome progenitors requires neural tube-derived signals: Characterization of the role of Neurotrophin-3. *Development* **1995**, *121*, 2583–2594. [PubMed]
68. Chiang, C.; Litingtung, Y.; Lee, E.; Young, K.E.; Corden, J.L.; Westphal, H.; Beachy, P.A. Cyclopia and defective axial patterning in mice lacking Sonic hedgehog gene function. *Nature* **1996**, *383*, 407–413. [CrossRef] [PubMed]
69. Hamade, A.; Deries, M.; Begemann, G.; Bally-Cuif, L.; Genet, C.; Sabatier, F.; Bonnieu, A.; Cousin, X. Retinoic acid activates myogenesis in vivo through Fgf8 signalling. *Dev. Biol.* **2006**, *289*, 127–140. [CrossRef] [PubMed]
70. Cunningham, T.J.; Brade, T.; Sandell, L.L.; Lewandoski, M.; Trainor, P.A.; Colas, A.; Mercola, M.; Duester, G. Retinoic Acid Activity in Undifferentiated Neural Progenitors Is Sufficient to Fulfill Its Role in Restricting Fgf8 Expression for Somitogenesis. *PLoS ONE* **2015**, *10*, e0137894. [CrossRef] [PubMed]
71. Du, S.J.; Devoto, S.H.; Westerfield, M.; Moon, R.T. Positive and negative regulation of muscle cell identity by members of the hedgehog and TGF-beta gene families. *J. Cell Biol.* **1997**, *139*, 145–156. [CrossRef] [PubMed]
72. Stern, H.M.; Brown, A.M.; Hauschka, S.D. Myogenesis in paraxial mesoderm: Preferential induction by dorsal neural tube and by cells expressing Wnt-1. *Development* **1995**, *121*, 3675–3686. [PubMed]
73. Duprez, D.; Fournier-Thibault, C.; Le Douarin, N. Sonic Hedgehog induces proliferation of committed skeletal muscle cells in the chick limb. *Development* **1998**, *125*, 495–505. [PubMed]
74. Bren-Mattison, Y.; Olwin, B.B. Sonic hedgehog inhibits the terminal differentiation of limb myoblasts committed to the slow muscle lineage. *Dev. Biol.* **2002**, *242*, 130–148. [CrossRef] [PubMed]
75. Kruger, M.; Mennerich, D.; Fees, S.; Schafer, R.; Mundlos, S.; Braun, T. Sonic hedgehog is a survival factor for hypaxial muscles during mouse development. *Development* **2001**, *128*, 743–752. [PubMed]
76. Kuroda, K.; Kuang, S.; Taketo, M.M.; Rudnicki, M.A. Canonical Wnt signaling induces BMP-4 to specify slow myofibrogenesis of fetal myoblasts. *Skelet Muscle* **2013**, *3*, 5. [CrossRef] [PubMed]
77. Ikeya, M.; Takada, S. Wnt signaling from the dorsal neural tube is required for the formation of the medial dermomyotome. *Development* **1998**, *125*, 4969–4976. [PubMed]
78. Tanaka, S.; Terada, K.; Nohno, T. Canonical Wnt signaling is involved in switching from cell proliferation to myogenic differentiation of mouse myoblast cells. *J. Mol. Signal.* **2011**, *6*, 12. [CrossRef] [PubMed]
79. Hwang, Y.; Suk, S.; Shih, Y.R.; Seo, T.; Du, B.; Xie, Y.; Li, Z.; Varghese, S. WNT3A promotes myogenesis of human embryonic stem cells and enhances in vivo engraftment. *Sci. Rep.* **2014**, *4*, 5916. [CrossRef] [PubMed]
80. Abraham, S.T. A role for the Wnt3a/beta-catenin signaling pathway in the myogenic program of C2C12 cells. *In Vitro Cell. Dev. Biol. Anim.* **2016**, *52*, 935–941. [CrossRef] [PubMed]

81. Buikema, J.W.; Mady, A.S.; Mittal, N.V.; Atmanli, A.; Caron, L.; Doevendans, P.A.; Sluijter, J.P.; Domian, I.J. Wnt/beta-catenin signaling directs the regional expansion of first and second heart field-derived ventricular cardiomyocytes. *Development* **2013**, *140*, 4165–4176. [CrossRef] [PubMed]
82. Takata, H.; Terada, K.; Oka, H.; Sunada, Y.; Moriguchi, T.; Nohno, T. Involvement of Wnt4 signaling during myogenic proliferation and differentiation of skeletal muscle. *Dev. Dyn.* **2007**, *236*, 2800–2807. [CrossRef] [PubMed]
83. Bernardi, H.; Gay, S.; Fedon, Y.; Vernus, B.; Bonnieu, A.; Bacou, F. Wnt4 activates the canonical beta-catenin pathway and regulates negatively myostatin: Functional implication in myogenesis. *Am. J. Physiol. Cell Physiol.* **2011**, *300*, C1122–C1138. [CrossRef] [PubMed]
84. Geetha-Loganathan, P.; Nimmagadda, S.; Prols, F.; Patel, K.; Scaal, M.; Huang, R.; Christ, B. Ectodermal Wnt-6 promotes Myf5-dependent avian limb myogenesis. *Dev. Biol.* **2005**, *288*, 221–233. [CrossRef] [PubMed]
85. Anakwe, K.; Robson, L.; Hadley, J.; Buxton, P.; Church, V.; Allen, S.; Hartmann, C.; Harfe, B.; Nohno, T.; Brown, A.M.; et al. Wnt signalling regulates myogenic differentiation in the developing avian wing. *Development* **2003**, *130*, 3503–3514. [CrossRef] [PubMed]
86. Gros, J.; Serralbo, O.; Marcelle, C. WNT11 acts as a directional cue to organize the elongation of early muscle fibres. *Nature* **2009**, *457*, 589–593. [CrossRef] [PubMed]
87. Noden, D.M.; Francis-West, P. The differentiation and morphogenesis of craniofacial muscles. *Dev. Dyn.* **2006**, *235*, 1194–1218. [CrossRef] [PubMed]
88. Hacker, A.; Guthrie, S. A distinct developmental programme for the cranial paraxial mesoderm in the chick embryo. *Development* **1998**, *125*, 3461–3472. [PubMed]
89. Rochais, F.; Mesbah, K.; Kelly, R.G. Signaling pathways controlling second heart field development. *Circ. Res.* **2009**, *104*, 933–942. [CrossRef] [PubMed]
90. Abu-Issa, R.; Smyth, G.; Smoak, I.; Yamamura, K.; Meyers, E.N. Fgf8 is required for pharyngeal arch and cardiovascular development in the mouse. *Development* **2002**, *129*, 4613–4625. [PubMed]
91. Mitsiadis, T.A.; Tucker, A.S.; De Bari, C.; Cobourne, M.T.; Rice, D.P. A regulatory relationship between Tbx1 and FGF signaling during tooth morphogenesis and ameloblast lineage determination. *Dev. Biol.* **2008**, *320*, 39–48. [CrossRef] [PubMed]
92. Von Scheven, G.; Bothe, I.; Ahmed, M.U.; Alvares, L.E.; Dietrich, S. Protein and genomic organisation of vertebrate MyoR and Capsulin genes and their expression during avian development. *Gene Expr. Patterns* **2006**, *6*, 383–393. [CrossRef] [PubMed]
93. Bothe, I.; Tenin, G.; Oseni, A.; Dietrich, S. Dynamic control of head mesoderm patterning. *Development* **2011**, *138*, 2807–2821. [CrossRef] [PubMed]
94. Mootoosamy, R.C.; Dietrich, S. Distinct regulatory cascades for head and trunk myogenesis. *Development* **2002**, *129*, 573–583. [PubMed]
95. Tzahor, E.; Kempf, H.; Mootoosamy, R.C.; Poon, A.C.; Abzhanov, A.; Tabin, C.J.; Dietrich, S.; Lassar, A.B. Antagonists of Wnt and BMP signaling promote the formation of vertebrate head muscle. *Genes Dev.* **2003**, *17*, 3087–3099. [CrossRef] [PubMed]
96. Harel, I.; Nathan, E.; Tirosh-Finkel, L.; Zigdon, H.; Guimaraes-Camboa, N.; Evans, S.M.; Tzahor, E. Distinct origins and genetic programs of head muscle satellite cells. *Dev. Cell* **2009**, *16*, 822–832. [CrossRef] [PubMed]
97. Dong, F.; Sun, X.; Liu, W.; Ai, D.; Klysik, E.; Lu, M.F.; Hadley, J.; Antoni, L.; Chen, L.; Baldini, A.; et al. Pitx2 promotes development of splanchnic mesoderm-derived branchiomeric muscle. *Development* **2006**, *133*, 4891–4899. [CrossRef] [PubMed]
98. Shih, H.P.; Gross, M.K.; Kioussi, C. Cranial muscle defects of Pitx2 mutants result from specification defects in the first branchial arch. *Proc. Natl. Acad. Sci. USA* **2007**, *104*, 5907–5912. [CrossRef] [PubMed]
99. Kelly, R.G.; Jerome-Majewska, L.A.; Papaioannou, V.E. The del22q11.2 candidate gene Tbx1 regulates branchiomeric myogenesis. *Hum. Mol. Genet.* **2004**, *13*, 2829–2840. [CrossRef] [PubMed]
100. Hu, T.; Yamagishi, H.; Maeda, J.; McAnally, J.; Yamagishi, C.; Srivastava, D. Tbx1 regulates fibroblast growth factors in the anterior heart field through a reinforcing autoregulatory loop involving forkhead transcription factors. *Development* **2004**, *131*, 5491–5502. [CrossRef] [PubMed]
101. Schubert, F.R.; Singh, A.J.; Afoyalan, O.; Kioussi, C.; Dietrich, S. To roll the eyes and snap a bite—Function, development and evolution of craniofacial muscles. *Semin. Cell Dev. Biol.* **2018**. [CrossRef] [PubMed]
102. Bothe, I.; Dietrich, S. The molecular setup of the avian head mesoderm and its implication for craniofacial myogenesis. *Dev. Dyn.* **2006**, *235*, 2845–2860. [CrossRef] [PubMed]

103. Nogueira, J.M.; Hawrot, K.; Sharpe, C.; Noble, A.; Wood, W.M.; Jorge, E.C.; Goldhamer, D.J.; Kardon, G.; Dietrich, S. The emergence of Pax7-expressing muscle stem cells during vertebrate head muscle development. *Front. Aging Neurosci.* **2015**, *7*, 62. [CrossRef] [PubMed]
104. Randolph, M.E.; Phillips, B.L.; Choo, H.J.; Vest, K.E.; Vera, Y.; Pavlath, G.K. Pharyngeal Satellite Cells Undergo Myogenesis Under Basal Conditions and Are Required for Pharyngeal Muscle Maintenance. *Stem. Cells* **2015**, *33*, 3581–3595. [CrossRef] [PubMed]
105. Sambasivan, R.; Kuratani, S.; Tajbakhsh, S. An eye on the head: The development and evolution of craniofacial muscles. *Development* **2011**, *138*, 2401–2415. [CrossRef] [PubMed]
106. Matsuoka, T.; Ahlberg, P.E.; Kessaris, N.; Iannarelli, P.; Dennehy, U.; Richardson, W.D.; McMahon, A.P.; Koentges, G. Neural crest origins of the neck and shoulder. *Nature* **2005**, *436*, 347–355. [CrossRef] [PubMed]
107. Suzuki, A.; Pelikan, R.C.; Iwata, J. WNT/beta-Catenin Signaling Regulates Multiple Steps of Myogenesis by Regulating Step-Specific Targets. *Mol. Cell Biol.* **2015**, *35*, 1763–1776. [CrossRef] [PubMed]
108. Zhong, Z.; Zhao, H.; Mayo, J.; Chai, Y. Different requirements for Wnt signaling in tongue myogenic subpopulations. *J. Dent. Res.* **2015**, *94*, 421–429. [CrossRef] [PubMed]
109. Tirosh-Finkel, L.; Elhanany, H.; Rinon, A.; Tzahor, E. Mesoderm progenitor cells of common origin contribute to the head musculature and the cardiac outflow tract. *Development* **2006**, *133*, 1943–1953. [CrossRef] [PubMed]
110. Ryckebusch, L.; Wang, Z.; Bertrand, N.; Lin, S.C.; Chi, X.; Schwartz, R.; Zaffran, S.; Niederreither, K. Retinoic acid deficiency alters second heart field formation. *Proc. Natl. Acad. Sci. USA* **2008**, *105*, 2913–2918. [CrossRef] [PubMed]
111. Hochgreb, T.; Linhares, V.L.; Menezes, D.C.; Sampaio, A.C.; Yan, C.Y.; Cardoso, W.V.; Rosenthal, N.; Xavier-Neto, J. A caudorostral wave of RALDH2 conveys anteroposterior information to the cardiac field. *Development* **2003**, *130*, 5363–5374. [CrossRef] [PubMed]
112. Daughters, R.S.; Keirstead, S.A.; Slack, J.M. Transformation of jaw muscle satellite cells to cardiomyocytes. *Differentiation* **2017**, *93*, 58–65. [CrossRef] [PubMed]
113. Marvin, M.J.; Di Rocco, G.; Gardiner, A.; Bush, S.M.; Lassar, A.B. Inhibition of Wnt activity induces heart formation from posterior mesoderm. *Genes Dev.* **2001**, *15*, 316–327. [CrossRef] [PubMed]
114. Lopez-Sanchez, C.; Garcia-Martinez, V. Molecular determinants of cardiac specification. *Cardiovasc. Res.* **2011**, *91*, 185–195. [CrossRef] [PubMed]
115. Takeuchi, J.K.; Bruneau, B.G. Directed transdifferentiation of mouse mesoderm to heart tissue by defined factors. *Nature* **2009**, *459*, 708–711. [CrossRef] [PubMed]
116. Jacobson, A.G.; Sater, A.K. Features of embryonic induction. *Development* **1988**, *104*, 341–359. [PubMed]
117. Garg, V.; Kathiriya, I.S.; Barnes, R.; Schluterman, M.K.; King, I.N.; Butler, C.A.; Rothrock, C.R.; Eapen, R.S.; Hirayama-Yamada, K.; Joo, K.; et al. GATA4 mutations cause human congenital heart defects and reveal an interaction with TBX5. *Nature* **2003**, *424*, 443–447. [CrossRef] [PubMed]
118. Bruneau, B.G.; Nemer, G.; Schmitt, J.P.; Charron, F.; Robitaille, L.; Caron, S.; Conner, D.A.; Gessler, M.; Nemer, M.; Seidman, C.E.; et al. A murine model of Holt-Oram syndrome defines roles of the T-box transcription factor Tbx5 in cardiogenesis and disease. *Cell* **2001**, *106*, 709–721. [CrossRef]
119. Prall, O.W.; Menon, M.K.; Solloway, M.J.; Watanabe, Y.; Zaffran, S.; Bajolle, F.; Biben, C.; McBride, J.J.; Robertson, B.R.; Chaulet, H.; et al. An Nkx2-5/Bmp2/Smad1 negative feedback loop controls heart progenitor specification and proliferation. *Cell* **2007**, *128*, 947–959. [CrossRef] [PubMed]
120. Cai, C.L.; Liang, X.; Shi, Y.; Chu, P.H.; Pfaff, S.L.; Chen, J.; Evans, S. Isl1 identifies a cardiac progenitor population that proliferates prior to differentiation and contributes a majority of cells to the heart. *Dev. Cell* **2003**, *5*, 877–889. [CrossRef]
121. Theveniau-Ruissy, M.; Dandonneau, M.; Mesbah, K.; Ghez, O.; Mattei, M.G.; Miquerol, L.; Kelly, R.G. The del22q11.2 candidate gene Tbx1 controls regional outflow tract identity and coronary artery patterning. *Circ. Res.* **2008**, *103*, 142–148. [CrossRef] [PubMed]
122. Harel, I.; Maezawa, Y.; Avraham, R.; Rinon, A.; Ma, H.Y.; Cross, J.W.; Leviatan, N.; Hegesh, J.; Roy, A.; Jacob-Hirsch, J. Pharyngeal mesoderm regulatory network controls cardiac and head muscle morphogenesis. *Proc. Natl. Acad. Sci. USA* **2012**, *109*, 18839–18844. [CrossRef] [PubMed]
123. Watanabe, Y.; Zaffran, S.; Kuroiwa, A.; Higuchi, H.; Ogura, T.; Harvey, R.P.; Kelly, R.G.; Buckingham, M. Fibroblast growth factor 10 gene regulation in the second heart field by Tbx1, Nkx2-5, and Islet1 reveals a genetic switch for down-regulation in the myocardium. *Proc. Natl. Acad. Sci. USA* **2012**, *109*, 18273–18280. [CrossRef] [PubMed]

124. Doetschman, T.; Barnett, J.V.; Runyan, R.B.; Camenisch, T.D.; Heimark, R.L.; Granzier, H.L.; Conway, S.J.; Azhar, M. Transforming growth factor beta signaling in adult cardiovascular diseases and repair. *Cell Tissue Res.* **2012**, *347*, 203–223. [CrossRef] [PubMed]
125. Bartram, U.; Molin, D.G.; Wisse, L.J.; Mohamad, A.; Sanford, L.P.; Doetschman, T.; Speer, C.P.; Poelmann, R.E.; Gittenberger-de Groot, A.C. Double-outlet right ventricle and overriding tricuspid valve reflect disturbances of looping, myocardialization, endocardial cushion differentiation, and apoptosis in TGF-beta(2)-knockout mice. *Circulation* **2001**, *103*, 2745–2752. [CrossRef] [PubMed]
126. Lewandowski, S.L.; Janardhan, H.P.; Trivedi, C.M. Histone Deacetylase 3 Coordinates Deacetylase-independent Epigenetic Silencing of Transforming Growth Factor-beta1 (TGF-beta1) to Orchestrate Second Heart Field Development. *J. Biol. Chem.* **2015**, *290*, 27067–27089. [CrossRef] [PubMed]
127. Dickson, M.C.; Martin, J.S.; Cousins, F.M.; Kulkarni, A.B.; Karlsson, S.; Akhurst, R.J. Defective haematopoiesis and vasculogenesis in transforming growth factor-beta 1 knock out mice. *Development* **1995**, *121*, 1845–1854. [PubMed]
128. Trainor, P.A.; Tam, P.P. Cranial paraxial mesoderm and neural crest cells of the mouse embryo: co-distribution in the craniofacial mesenchyme but distinct segregation in branchial arches. *Development* **1995**, *121*, 2569–2582. [PubMed]
129. Kirby, M.L.; Gale, T.F.; Stewart, D.E. Neural crest cells contribute to normal aorticopulmonary septation. *Science* **1983**, *220*, 1059–1061. [CrossRef] [PubMed]
130. Willems, E.; Cabral-Teixeira, J.; Schade, D.; Cai, W.; Reeves, P.; Bushway, P.J.; Lanier, M.; Walsh, C.; Kirchhausen, T.; Izpisua Belmonte, J.C.; et al. Small molecule-mediated TGF-beta type II receptor degradation promotes cardiomyogenesis in embryonic stem cells. *Cell Stem Cell* **2012**, *11*, 242–252. [CrossRef] [PubMed]
131. Gays, D.; Hess, C.; Camporeale, A.; Ala, U.; Provero, P.; Mosimann, C.; Santoro, M.M. An exclusive cellular and molecular network governs intestinal smooth muscle cell differentiation in vertebrates. *Development* **2017**, *144*, 464–478. [CrossRef] [PubMed]
132. Pouget, C.; Gautier, R.; Teillet, M.A.; Jaffredo, T. Somite-derived cells replace ventral aortic hemangioblasts and provide aortic smooth muscle cells of the trunk. *Development* **2006**, *133*, 1013–1022. [CrossRef] [PubMed]
133. Waldo, K.L.; Kumiski, D.H.; Wallis, K.T.; Stadt, H.A.; Hutson, M.R.; Platt, D.H.; Kirby, M.L. Conotruncal myocardium arises from a secondary heart field. *Development* **2001**, *128*, 3179–3188. [PubMed]
134. Rosenquist, T.H.; Kirby, M.L.; van Mierop, L.H. Solitary aortic arch artery. A result of surgical ablation of cardiac neural crest and nodose placode in the avian embryo. *Circulation* **1989**, *80*, 1469–1475. [CrossRef] [PubMed]
135. Le Lievre, C.S.; Le Douarin, N.M. Mesenchymal derivatives of the neural crest: Analysis of chimaeric quail and chick embryos. *J. Embryol. Exp. Morphol.* **1975**, *34*, 125–154. [PubMed]
136. Owens, G.K.; Wise, G. Regulation of differentiation/maturation in vascular smooth muscle cells by hormones and growth factors. *Agents Actions Suppl.* **1997**, *48*, 3–24. [PubMed]
137. Miano, J.M.; Long, X.; Fujiwara, K. Serum response factor: Master regulator of the actin cytoskeleton and contractile apparatus. *Am. J. Physiol. Cell Physiol.* **2007**, *292*, C70–C81. [CrossRef] [PubMed]
138. Zhang, M.; Yang, M.; Liu, L.P.; Lau, W.B.; Gao, H.; Xin, M.K.; Su, L.X.; Wang, J.; Cheng, S.J.; Fan, Q.; et al. BMP-2 overexpression augments vascular smooth muscle cell motility by upregulating myosin Va via Erk signaling. *Oxid. Med. Cell. Longev.* **2014**, *2014*, 294150. [CrossRef] [PubMed]
139. Chang, L.; Noseda, M.; Higginson, M.; Ly, M.; Patenaude, A.; Fuller, M.; Kyle, A.H.; Minchinton, A.I.; Puri, M.C.; Dumont, D.J.; et al. Differentiation of vascular smooth muscle cells from local precursors during embryonic and adult arteriogenesis requires Notch signaling. *Proc. Natl. Acad. Sci. USA* **2012**, *109*, 6993–6998. [CrossRef] [PubMed]
140. Lavine, K.J.; Long, F.; Choi, K.; Smith, C.; Ornitz, D.M. Hedgehog signaling to distinct cell types differentially regulates coronary artery and vein development. *Development* **2008**, *135*, 3161–3171. [CrossRef] [PubMed]
141. Topouzis, S.; Majesky, M.W. Smooth muscle lineage diversity in the chick embryo. Two types of aortic smooth muscle cell differ in growth and receptor-mediated transcriptional responses to transforming growth factor-beta. *Dev. Biol.* **1996**, *178*, 430–445. [CrossRef] [PubMed]
142. Hautmann, M.B.; Madsen, C.S.; Owens, G.K. A transforming growth factor beta (TGFbeta) control element drives TGFbeta-induced stimulation of smooth muscle alpha-actin gene expression in concert with two CArG elements. *J. Biol. Chem.* **1997**, *272*, 10948–10956. [CrossRef] [PubMed]

143. Hirschi, K.K.; Rohovsky, S.A.; D'Amore, P.A. PDGF, TGF-beta, and heterotypic cell-cell interactions mediate endothelial cell-induced recruitment of 10T1/2 cells and their differentiation to a smooth muscle fate. *J. Cell Biol.* **1998**, *141*, 805–814. [CrossRef] [PubMed]
144. Qin, B.Y.; Lam, S.S.; Correia, J.J.; Lin, K. Smad3 allostery links TGF-beta receptor kinase activation to transcriptional control. *Genes Dev.* **2002**, *16*, 1950–1963. [CrossRef] [PubMed]
145. Tsapara, A.; Luthert, P.; Greenwood, J.; Hill, C.S.; Matter, K.; Balda, M.S. The RhoA activator GEF-H1/Lfc is a transforming growth factor-beta target gene and effector that regulates alpha-smooth muscle actin expression and cell migration. *Mol. Biol. Cell* **2010**, *21*, 860–870. [CrossRef] [PubMed]
146. Young, K.; Krebs, L.T.; Tweedie, E.; Conley, B.; Mancini, M.; Arthur, H.M.; Liaw, L.; Gridley, T.; Vary, C. Endoglin is required in Pax3-derived cells for embryonic blood vessel formation. *Dev. Biol.* **2016**, *409*, 95–105. [CrossRef] [PubMed]
147. Singh, A.J.; Chang, C.N.; Ma, H.Y.; Ramsey, S.A.; Flitz, T.M.; Kioussi, C. FACS-Seq analysis of Pax3-derived cells identifies non-myogenic lineages in the embryonic forelimb. *Sci. Rep.* **2018**, in press. [CrossRef] [PubMed]
148. Cossu, G.; Tajbakhsh, S.; Buckingham, M. How is myogenesis initiated in the embryo? *Trends Genet.* **1996**, *12*, 218–223. [CrossRef]
149. Brent, A.E.; Schweitzer, R.; Tabin, C.J. A somitic compartment of tendon progenitors. *Cell* **2003**, *113*, 235–248. [CrossRef]
150. Tajbakhsh, S.; Bober, E.; Babinet, C.; Pournin, S.; Arnold, H.; Buckingham, M. Gene targeting the myf-5 locus with nlacZ reveals expression of this myogenic factor in mature skeletal muscle fibres as well as early embryonic muscle. *Dev. Dyn.* **1996**, *206*, 291–300. [CrossRef]
151. Gros, J.; Manceau, M.; Thome, V.; Marcelle, C. A common somitic origin for embryonic muscle progenitors and satellite cells. *Nature* **2005**, *435*, 954–958. [CrossRef] [PubMed]
152. Oates, A.C.; Morelli, L.G.; Ares, S. Patterning embryos with oscillations: Structure, function and dynamics of the vertebrate segmentation clock. *Development* **2012**, *139*, 625–639. [CrossRef] [PubMed]
153. Mallo, M. Revisiting the involvement of signaling gradients in somitogenesis. *FEBS J.* **2016**, *283*, 1430–1437. [CrossRef] [PubMed]
154. Palmeirim, I.; Henrique, D.; Ish-Horowicz, D.; Pourquie, O. Avian hairy gene expression identifies a molecular clock linked to vertebrate segmentation and somitogenesis. *Cell* **1997**, *91*, 639–648. [CrossRef]
155. Resende, T.P.; Ferreira, M.; Teillet, M.A.; Tavares, A.T.; Andrade, R.P.; Palmeirim, I. Sonic hedgehog in temporal control of somite formation. *Proc. Natl. Acad. Sci. USA* **2010**, *107*, 12907–12912. [CrossRef] [PubMed]
156. Aulehla, A.; Herrmann, B.G. Segmentation in vertebrates: Clock and gradient finally joined. *Genes Dev.* **2004**, *18*, 2060–2067. [CrossRef] [PubMed]
157. Ferjentsik, Z.; Hayashi, S.; Dale, J.K.; Bessho, Y.; Herreman, A.; De Strooper, B.; del Monte, G.; de la Pompa, J.L.; Maroto, M. Notch is a critical component of the mouse somitogenesis oscillator and is essential for the formation of the somites. *PLoS Genet.* **2009**, *5*, e1000662. [CrossRef] [PubMed]
158. Diez del Corral, R.; Olivera-Martinez, I.; Goriely, A.; Gale, E.; Maden, M.; Storey, K. Opposing FGF and retinoid pathways control ventral neural pattern, neuronal differentiation, and segmentation during body axis extension. *Neuron* **2003**, *40*, 65–79. [CrossRef]
159. Watanabe, T.; Takahashi, Y. Tissue morphogenesis coupled with cell shape changes. *Curr. Opin. Genet. Dev.* **2010**, *20*, 443–447. [CrossRef] [PubMed]
160. Yabe, T.; Hoshijima, K.; Yamamoto, T.; Takada, S. Quadruple zebrafish mutant reveals different roles of Mesp genes in somite segmentation between mouse and zebrafish. *Development* **2016**, *143*, 2842–2852. [CrossRef] [PubMed]
161. Marcelino, J.; Sciortino, C.M.; Romero, M.F.; Ulatowski, L.M.; Ballock, R.T.; Economides, A.N.; Eimon, P.M.; Harland, R.M.; Warman, M.L. Human disease-causing NOG missense mutations: Effects on noggin secretion, dimer formation, and bone morphogenetic protein binding. *Proc. Natl. Acad. Sci. USA* **2001**, *98*, 11353–11358. [CrossRef] [PubMed]
162. Bok, J.; Brunet, L.J.; Howard, O.; Burton, Q.; Wu, D.K. Role of hindbrain in inner ear morphogenesis: Analysis of Noggin knockout mice. *Dev. Biol.* **2007**, *311*, 69–78. [CrossRef] [PubMed]
163. Fan, C.M.; Tessier-Lavigne, M. Patterning of mammalian somites by surface ectoderm and notochord: Evidence for sclerotome induction by a hedgehog homolog. *Cell* **1994**, *79*, 1175–1186. [CrossRef]
164. Johnson, R.L.; Laufer, E.; Riddle, R.D.; Tabin, C. Ectopic expression of Sonic hedgehog alters dorsal-ventral patterning of somites. *Cell* **1994**, *79*, 1165–1173. [CrossRef]

165. Chen, C.M.; Kraut, N.; Groudine, M.; Weintraub, H. I-mf, a novel myogenic repressor, interacts with members of the MyoD family. *Cell* **1996**, *86*, 731–741. [CrossRef]
166. Williams, B.A.; Ordahl, C.P. Pax-3 expression in segmental mesoderm marks early stages in myogenic cell specification. *Development* **1994**, *120*, 785–796. [PubMed]
167. Kassar-Duchossoy, L.; Giacone, E.; Gayraud-Morel, B.; Jory, A.; Gomes, D.; Tajbakhsh, S. Pax3/Pax7 mark a novel population of primitive myogenic cells during development. *Genes Dev.* **2005**, *19*, 1426–1431. [CrossRef] [PubMed]
168. Etchevers, H.C.; Vincent, C.; Le Douarin, N.M.; Couly, G.F. The cephalic neural crest provides pericytes and smooth muscle cells to all blood vessels of the face and forebrain. *Development* **2001**, *128*, 1059–1068. [PubMed]
169. Goupille, O.; Pallafacchina, G.; Relaix, F.; Conway, S.J.; Cumano, A.; Robert, B.; Montarras, D.; Buckingham, M. Characterization of Pax3-expressing cells from adult blood vessels. *J. Cell Sci.* **2011**, *124*, 3980–3988. [CrossRef] [PubMed]
170. Goulding, M.; Lumsden, A.; Paquette, A.J. Regulation of Pax-3 expression in the dermomyotome and its role in muscle development. *Development* **1994**, *120*, 957–971. [PubMed]
171. Bober, E.; Franz, T.; Arnold, H.H.; Gruss, P.; Tremblay, P. Pax-3 is required for the development of limb muscles: A possible role for the migration of dermomyotomal muscle progenitor cells. *Development* **1994**, *120*, 603–612. [PubMed]
172. Grifone, R.; Demignon, J.; Houbron, C.; Souil, E.; Niro, C.; Seller, M.J.; Hamard, G.; Maire, P. Six1 and Six4 homeoproteins are required for Pax3 and Mrf expression during myogenesis in the mouse embryo. *Development* **2005**, *132*, 2235–2249. [CrossRef] [PubMed]
173. Grifone, R.; Kelly, R.G. Heartening news for head muscle development. *Trends Genet.* **2007**, *23*, 365–369. [CrossRef] [PubMed]
174. Ott, M.O.; Bober, E.; Lyons, G.; Arnold, H.; Buckingham, M. Early expression of the myogenic regulatory gene, myf-5, in precursor cells of skeletal muscle in the mouse embryo. *Development* **1991**, *111*, 1097–1107. [PubMed]
175. Tajbakhsh, S.; Rocancourt, D.; Cossu, G.; Buckingham, M. Redefining the genetic hierarchies controlling skeletal myogenesis: Pax-3 and Myf-5 act upstream of MyoD. *Cell* **1997**, *89*, 127–138. [CrossRef]
176. Borello, U.; Berarducci, B.; Murphy, P.; Bajard, L.; Buffa, V.; Piccolo, S.; Buckingham, M.; Cossu, G. The Wnt/beta-catenin pathway regulates Gli-mediated Myf5 expression during somitogenesis. *Development* **2006**, *133*, 3723–3732. [CrossRef] [PubMed]
177. Rong, P.M.; Teillet, M.A.; Ziller, C.; Le Douarin, N.M. The neural tube/notochord complex is necessary for vertebral but not limb and body wall striated muscle differentiation. *Development* **1992**, *115*, 657–672. [PubMed]
178. Fontaine-Perus, J.; Halgand, P.; Cheraud, Y.; Rouaud, T.; Velasco, M.E.; Cifuentes Diaz, C.; Rieger, F. Mouse-chick chimera: A developmental model of murine neurogenic cells. *Development* **1997**, *124*, 3025–3036. [PubMed]
179. Pownall, M.E.; Strunk, K.E.; Emerson, C.P., Jr. Notochord signals control the transcriptional cascade of myogenic bHLH genes in somites of quail embryos. *Development* **1996**, *122*, 1475–1488. [PubMed]
180. Borycki, A.G.; Mendham, L.; Emerson, C.P., Jr. Control of somite patterning by Sonic hedgehog and its downstream signal response genes. *Development* **1998**, *125*, 777–790. [PubMed]
181. Dietrich, S.; Schubert, F.R.; Lumsden, A. Control of dorsoventral pattern in the chick paraxial mesoderm. *Development* **1997**, *124*, 3895–3908. [PubMed]
182. Gustafsson, M.K.; Pan, H.; Pinney, D.F.; Liu, Y.; Lewandowski, A.; Epstein, D.J.; Emerson, C.P., Jr. Myf5 is a direct target of long-range Shh signaling and Gli regulation for muscle specification. *Genes Dev.* **2002**, *16*, 114–126. [CrossRef] [PubMed]
183. Teboul, L.; Summerbell, D.; Rigby, P.W. The initial somitic phase of Myf5 expression requires neither Shh signaling nor Gli regulation. *Genes Dev.* **2003**, *17*, 2870–2874. [CrossRef] [PubMed]
184. Rios, A.C.; Serralbo, O.; Salgado, D.; Marcelle, C. Neural crest regulates myogenesis through the transient activation of NOTCH. *Nature* **2011**, *473*, 532–535. [CrossRef] [PubMed]
185. Sato, T.; Rocancourt, D.; Marques, L.; Thorsteinsdottir, S.; Buckingham, M. A Pax3/Dmrt2/Myf5 regulatory cascade functions at the onset of myogenesis. *PLoS Genet.* **2010**, *6*, e1000897. [CrossRef] [PubMed]

186. Relaix, F.; Montarras, D.; Zaffran, S.; Gayraud-Morel, B.; Rocancourt, D.; Tajbakhsh, S.; Mansouri, A.; Cumano, A.; Buckingham, M. Pax3 and Pax7 have distinct and overlapping functions in adult muscle progenitor cells. *J. Cell Biol.* **2006**, *172*, 91–102. [CrossRef] [PubMed]
187. Ben-Yair, R.; Kalcheim, C. Lineage analysis of the avian dermomyotome sheet reveals the existence of single cells with both dermal and muscle progenitor fates. *Development* **2005**, *132*, 689–701. [CrossRef] [PubMed]
188. Relaix, F.; Rocancourt, D.; Mansouri, A.; Buckingham, M. A Pax3/Pax7-dependent population of skeletal muscle progenitor cells. *Nature* **2005**, *435*, 948–953. [CrossRef] [PubMed]
189. Ahmed, M.U.; Maurya, A.K.; Cheng, L.; Jorge, E.C.; Schubert, F.R.; Maire, P.; Basson, M.A.; Ingham, P.W.; Dietrich, S. Engrailed controls epaxial-hypaxial muscle innervation and the establishment of vertebrate three-dimensional mobility. *Dev. Biol.* **2017**, *430*, 90–104. [CrossRef] [PubMed]
190. Huang, R.; Zhi, Q.; Patel, K.; Wilting, J.; Christ, B. Dual origin and segmental organisation of the avian scapula. *Development* **2000**, *127*, 3789–3794. [PubMed]
191. Huang, R.; Zhi, Q.; Schmidt, C.; Wilting, J.; Brand-Saberi, B.; Christ, B. Sclerotomal origin of the ribs. *Development* **2000**, *127*, 527–532. [PubMed]
192. Ehehalt, F.; Wang, B.; Christ, B.; Patel, K.; Huang, R. Intrinsic cartilage-forming potential of dermomyotomal cells requires ectodermal signals for the development of the scapula blade. *Anat. Embryol.* **2004**, *208*, 431–437. [CrossRef] [PubMed]
193. Valasek, P.; Theis, S.; DeLaurier, A.; Hinits, Y.; Luke, G.N.; Otto, A.M.; Minchin, J.; He, L.; Christ, B.; Brooks, G.; et al. Cellular and molecular investigations into the development of the pectoral girdle. *Dev. Biol.* **2011**, *357*, 108–116. [CrossRef] [PubMed]
194. Van Ho, A.T.; Hayashi, S.; Brohl, D.; Aurade, F.; Rattenbach, R.; Relaix, F. Neural crest cell lineage restricts skeletal muscle progenitor cell differentiation through Neuregulin1-ErbB3 signaling. *Dev. Cell* **2011**, *21*, 273–287. [PubMed]
195. Li, L.; Cserjesi, P.; Olson, E.N. Dermo-1: A novel twist-related bHLH protein expressed in the developing dermis. *Dev. Biol.* **1995**, *172*, 280–292. [CrossRef] [PubMed]
196. Gong, X.Q.; Li, L. Dermo-1, a multifunctional basic helix-loop-helix protein, represses MyoD transactivation via the HLH domain, MEF2 interaction, and chromatin deacetylation. *J. Biol. Chem.* **2002**, *277*, 12310–12317. [CrossRef] [PubMed]
197. Dietrich, S.; Schubert, F.R.; Healy, C.; Sharpe, P.T.; Lumsden, A. Specification of the hypaxial musculature. *Development* **1998**, *125*, 2235–2249. [PubMed]
198. Pourquie, O.; Fan, C.M.; Coltey, M.; Hirsinger, E.; Watanabe, Y.; Breant, C.; Francis-West, P.; Brickell, P.; Tessier-Lavigne, M.; Le Douarin, N.M. Lateral and axial signals involved in avian somite patterning: A role for BMP4. *Cell* **1996**, *84*, 461–471. [CrossRef]
199. Ben-Yair, R.; Kalcheim, C. Notch and bone morphogenetic protein differentially act on dermomyotome cells to generate endothelium, smooth, and striated muscle. *J. Cell Biol.* **2008**, *180*, 607–618. [CrossRef] [PubMed]
200. Brunelli, S.; Relaix, F.; Baesso, S.; Buckingham, M.; Cossu, G. Beta catenin-independent activation of MyoD in presomitic mesoderm requires PKC and depends on Pax3 transcriptional activity. *Dev. Biol.* **2007**, *304*, 604–614. [CrossRef] [PubMed]
201. Giordani, J.; Bajard, L.; Demignon, J.; Daubas, P.; Buckingham, M.; Maire, P. Six proteins regulate the activation of Myf5 expression in embryonic mouse limbs. *Proc. Natl. Acad. Sci. USA* **2007**, *104*, 11310–11315. [CrossRef] [PubMed]
202. Bajard, L.; Relaix, F.; Lagha, M.; Rocancourt, D.; Daubas, P.; Buckingham, M.E. A novel genetic hierarchy functions during hypaxial myogenesis: Pax3 directly activates Myf5 in muscle progenitor cells in the limb. *Genes Dev.* **2006**, *20*, 2450–2464. [CrossRef] [PubMed]
203. Bladt, F.; Riethmacher, D.; Isenmann, S.; Aguzzi, A.; Birchmeier, C. Essential role for the c-met receptor in the migration of myogenic precursor cells into the limb bud. *Nature* **1995**, *376*, 768–771. [CrossRef] [PubMed]
204. Yang, X.M.; Vogan, K.; Gros, P.; Park, M. Expression of the met receptor tyrosine kinase in muscle progenitor cells in somites and limbs is absent in Splotch mice. *Development* **1996**, *122*, 2163–2171. [PubMed]
205. Gross, M.K.; Moran-Rivard, L.; Velasquez, T.; Nakatsu, M.N.; Jagla, K.; Goulding, M. Lbx1 is required for muscle precursor migration along a lateral pathway into the limb. *Development* **2000**, *127*, 413–424. [PubMed]
206. Kioussi, C.; Briata, P.; Baek, S.H.; Rose, D.W.; Hamblet, N.S.; Herman, T.; Ohgi, K.A.; Lin, C.; Gleiberman, A.; Wang, J.; et al. Identification of a Wnt/Dvl/beta-Catenin –> Pitx2 pathway mediating cell-type-specific proliferation during development. *Cell* **2002**, *111*, 673–685. [CrossRef]

207. Shih, H.P.; Gross, M.K.; Kioussi, C. Expression pattern of the homeodomain transcription factor Pitx2 during muscle development. *Gene Expr. Patterns* **2007**, *7*, 441–451. [CrossRef] [PubMed]
208. Lagha, M.; Sato, T.; Regnault, B.; Cumano, A.; Zuniga, A.; Licht, J.; Relaix, F.; Buckingham, M. Transcriptome analyses based on genetic screens for Pax3 myogenic targets in the mouse embryo. *BMC Genom.* **2010**, *11*, 696. [CrossRef] [PubMed]
209. Campbell, A.L.; Shih, H.P.; Xu, J.; Gross, M.K.; Kioussi, C. Regulation of motility of myogenic cells in filling limb muscle anlagen by Pitx2. *PLoS ONE* **2012**, *7*, e35822. [CrossRef] [PubMed]
210. Lagha, M.; Mayeuf-Louchart, A.; Chang, T.; Montarras, D.; Rocancourt, D.; Zalc, A.; Kormish, J.; Zaret, K.S.; Buckingham, M.E.; Relaix, F. Itm2a is a Pax3 target gene, expressed at sites of skeletal muscle formation in vivo. *PLoS ONE* **2013**, *8*, e63143. [CrossRef] [PubMed]
211. Van den Plas, D.; Merregaert, J. Constitutive overexpression of the integral membrane protein Itm2A enhances myogenic differentiation of C2C12 cells. *Cell Biol. Int.* **2004**, *28*, 199–207. [CrossRef] [PubMed]
212. Delfini, M.C.; De La Celle, M.; Gros, J.; Serralbo, O.; Marics, I.; Seux, M.; Scaal, M.; Marcelle, C. The timing of emergence of muscle progenitors is controlled by an FGF/ERK/SNAIL1 pathway. *Dev. Biol.* **2009**, *333*, 229–237. [CrossRef] [PubMed]
213. Pu, Q.; Abduelmula, A.; Masyuk, M.; Theiss, C.; Schwandulla, D.; Hans, M.; Patel, K.; Brand-Saberi, B.; Huang, R. The dermomyotome ventrolateral lip is essential for the hypaxial myotome formation. *BMC Dev. Biol.* **2013**, *13*, 37. [CrossRef] [PubMed]
214. Deries, M.; Thorsteinsdottir, S. Axial and limb muscle development: Dialogue with the neighbourhood. *Cell. Mol. Life Sci.* **2016**, *73*, 4415–4431. [CrossRef] [PubMed]
215. Fraidenraich, D.; Iwahori, A.; Rudnicki, M.; Basilico, C. Activation of fgf4 gene expression in the myotomes is regulated by myogenic bHLH factors and by sonic hedgehog. *Dev. Biol.* **2000**, *225*, 392–406. [CrossRef] [PubMed]
216. Kahane, N.; Ribes, V.; Kicheva, A.; Briscoe, J.; Kalcheim, C. The transition from differentiation to growth during dermomyotome-derived myogenesis depends on temporally restricted hedgehog signaling. *Development* **2013**, *140*, 1740–1750. [CrossRef] [PubMed]
217. Sacks, L.D.; Cann, G.M.; Nikovits, W., Jr.; Conlon, S.; Espinoza, N.R.; Stockdale, F.E. Regulation of myosin expression during myotome formation. *Development* **2003**, *130*, 3391–3402. [CrossRef] [PubMed]
218. Niro, C.; Demignon, J.; Vincent, S.; Liu, Y.; Giordani, J.; Sgarioto, N.; Favier, M.; Guillet-Deniau, I.; Blais, A.; Maire, P. Six1 and Six4 gene expression is necessary to activate the fast-type muscle gene program in the mouse primary myotome. *Dev. Biol.* **2010**, *338*, 168–182. [CrossRef] [PubMed]
219. Bryson-Richardson, R.J.; Currie, P.D. The genetics of vertebrate myogenesis. *Nat. Rev. Genet.* **2008**, *9*, 632–646. [CrossRef] [PubMed]
220. Relaix, F.; Rocancourt, D.; Mansouri, A.; Buckingham, M. Divergent functions of murine Pax3 and Pax7 in limb muscle development. *Genes Dev.* **2004**, *18*, 1088–1105. [CrossRef] [PubMed]
221. Brohmann, H.; Jagla, K.; Birchmeier, C. The role of Lbx1 in migration of muscle precursor cells. *Development* **2000**, *127*, 437–445. [PubMed]
222. Amthor, H.; Christ, B.; Patel, K. A molecular mechanism enabling continuous embryonic muscle growth—A balance between proliferation and differentiation. *Development* **1999**, *126*, 1041–1053. [PubMed]
223. Bendall, A.J.; Ding, J.; Hu, G.; Shen, M.M.; Abate-Shen, C. Msx1 antagonizes the myogenic activity of Pax3 in migrating limb muscle precursors. *Development* **1999**, *126*, 4965–4976. [PubMed]
224. Riddle, R.D.; Johnson, R.L.; Laufer, E.; Tabin, C. Sonic hedgehog mediates the polarizing activity of the ZPA. *Cell* **1993**, *75*, 1401–1416. [CrossRef]
225. Niswander, L.; Jeffrey, S.; Martin, G.R.; Tickle, C. A positive feedback loop coordinates growth and patterning in the vertebrate limb. *Nature* **1994**, *371*, 609–612. [CrossRef] [PubMed]
226. Scaal, M.; Bonafede, A.; Dathe, V.; Sachs, M.; Cann, G.; Christ, B.; Brand-Saberi, B. SF/HGF is a mediator between limb patterning and muscle development. *Development* **1999**, *126*, 4885–4893. [PubMed]
227. Vasyutina, E.; Stebler, J.; Brand-Saberi, B.; Schulz, S.; Raz, E.; Birchmeier, C. CXCR4 and Gab1 cooperate to control the development of migrating muscle progenitor cells. *Genes Dev.* **2005**, *19*, 2187–2198. [CrossRef] [PubMed]
228. Linker, C.; Lesbros, C.; Stark, M.R.; Marcelle, C. Intrinsic signals regulate the initial steps of myogenesis in vertebrates. *Development* **2003**, *130*, 4797–4807. [CrossRef] [PubMed]

229. Mitchell, P.J.; Johnson, S.E.; Hannon, K. Insulin-like growth factor I stimulates myoblast expansion and myofiber development in the limb. *Dev. Dyn.* **2002**, *223*, 12–23. [CrossRef] [PubMed]
230. Mohammed, R.H.; Anderton, H.; Brameld, J.M.; Sweetman, D. Correction: Effects of insulin like growth factors on early embryonic chick limb myogenesis. *PLoS ONE* **2017**, *12*, e0189395. [CrossRef] [PubMed]
231. Anderson, C.; Williams, V.C.; Moyon, B.; Daubas, P.; Tajbakhsh, S.; Buckingham, M.E.; Shiroishi, T.; Hughes, S.M.; Borycki, A.G. Sonic hedgehog acts cell-autonomously on muscle precursor cells to generate limb muscle diversity. *Genes Dev.* **2012**, *26*, 2103–2117. [CrossRef] [PubMed]
232. Sassoon, D.; Lyons, G.; Wright, W.E.; Lin, V.; Lassar, A.; Weintraub, H.; Buckingham, M. Expression of two myogenic regulatory factors myogenin and MyoD1 during mouse embryogenesis. *Nature* **1989**, *341*, 303–307. [CrossRef] [PubMed]
233. Menko, A.S.; Boettiger, D. Occupation of the extracellular matrix receptor, integrin, is a control point for myogenic differentiation. *Cell* **1987**, *51*, 51–57. [CrossRef]
234. Boettiger, D.; Enomoto-Iwamoto, M.; Yoon, H.Y.; Hofer, U.; Menko, A.S.; Chiquet-Ehrismann, R. Regulation of integrin alpha 5 beta 1 affinity during myogenic differentiation. *Dev. Biol.* **1995**, *169*, 261–272. [CrossRef] [PubMed]
235. Knudsen, K.A. The calcium-dependent myoblast adhesion that precedes cell fusion is mediated by glycoproteins. *J. Cell Biol.* **1985**, *101*, 891–897. [CrossRef] [PubMed]
236. Spitz, F.; Demignon, J.; Porteu, A.; Kahn, A.; Concordet, J.P.; Daegelen, D.; Maire, P. Expression of myogenin during embryogenesis is controlled by Six/sine oculis homeoproteins through a conserved MEF3 binding site. *Proc. Natl. Acad. Sci. USA* **1998**, *95*, 14220–14225. [CrossRef] [PubMed]
237. Hutcheson, D.A.; Zhao, J.; Merrell, A.; Haldar, M.; Kardon, G. Embryonic and fetal limb myogenic cells are derived from developmentally distinct progenitors and have different requirements for beta-catenin. *Genes Dev.* **2009**, *23*, 997–1013. [CrossRef] [PubMed]
238. Seale, P.; Sabourin, L.A.; Girgis-Gabardo, A.; Mansouri, A.; Gruss, P.; Rudnicki, M.A. Pax7 is required for the specification of myogenic satellite cells. *Cell* **2000**, *102*, 777–786. [CrossRef]
239. Gunther, S.; Kim, J.; Kostin, S.; Lepper, C.; Fan, C.M.; Braun, T. Myf5-positive satellite cells contribute to Pax7-dependent long-term maintenance of adult muscle stem cells. *Cell Stem Cell* **2013**, *13*, 590–601. [CrossRef] [PubMed]
240. Soleimani, V.D.; Punch, V.G.; Kawabe, Y.; Jones, A.E.; Palidwor, G.A.; Porter, C.J.; Cross, J.W.; Carvajal, J.J.; Kockx, C.E.; van IJcken, W.F.; et al. Transcriptional dominance of Pax7 in adult myogenesis is due to high-affinity recognition of homeodomain motifs. *Dev. Cell* **2012**, *22*, 1208–1220. [CrossRef] [PubMed]
241. Von Maltzahn, J.; Jones, A.E.; Parks, R.J.; Rudnicki, M.A. Pax7 is critical for the normal function of satellite cells in adult skeletal muscle. *Proc. Natl. Acad. Sci. USA* **2013**, *110*, 16474–16479. [CrossRef] [PubMed]
242. Oustanina, S.; Hause, G.; Braun, T. Pax7 directs postnatal renewal and propagation of myogenic satellite cells but not their specification. *EMBO J.* **2004**, *23*, 3430–3439. [CrossRef] [PubMed]
243. Kuang, S.; Charge, S.B.; Seale, P.; Huh, M.; Rudnicki, M.A. Distinct roles for Pax7 and Pax3 in adult regenerative myogenesis. *J. Cell Biol.* **2006**, *172*, 103–113. [CrossRef] [PubMed]
244. Martin, J.F.; Li, L.; Olson, E.N. Repression of myogenin function by TGF-beta 1 is targeted at the basic helix-loop-helix motif and is independent of E2A products. *J. Biol. Chem.* **1992**, *267*, 10956–10960. [PubMed]
245. Bischoff, R. Chemotaxis of skeletal muscle satellite cells. *Dev. Dyn.* **1997**, *208*, 505–515. [CrossRef]
246. Yamagishi, T.; Ando, K.; Nakamura, H.; Nakajima, Y. Expression of the Tgfbeta2 gene during chick embryogenesis. *Anat. Rec.* **2012**, *295*, 257–267. [CrossRef] [PubMed]
247. Messina, G.; Biressi, S.; Monteverde, S.; Magli, A.; Cassano, M.; Perani, L.; Roncaglia, E.; Tagliafico, E.; Starnes, L.; Campbell, C.E.; et al. Nfix regulates fetal-specific transcription in developing skeletal muscle. *Cell* **2010**, *140*, 554–566. [CrossRef] [PubMed]
248. L'Honore, A.; Coulon, V.; Marcil, A.; Lebel, M.; Lafrance-Vanasse, J.; Gage, P.; Camper, S.; Drouin, J. Sequential expression and redundancy of Pitx2 and Pitx3 genes during muscle development. *Dev. Biol.* **2007**, *307*, 421–433. [CrossRef] [PubMed]
249. Duxson, M.J.; Usson, Y.; Harris, A.J. The origin of secondary myotubes in mammalian skeletal muscles: Ultrastructural studies. *Development* **1989**, *107*, 743–750. [PubMed]
250. De Angelis, L.; Berghella, L.; Coletta, M.; Lattanzi, L.; Zanchi, M.; Cusella-De Angelis, M.G.; Ponzetto, C.; Cossu, G. Skeletal myogenic progenitors originating from embryonic dorsal aorta coexpress endothelial and myogenic

markers and contribute to postnatal muscle growth and regeneration. *J. Cell Biol.* **1999**, *147*, 869–878. [CrossRef] [PubMed]

251. Beauchamp, J.R.; Heslop, L.; Yu, D.S.; Tajbakhsh, S.; Kelly, R.G.; Wernig, A.; Buckingham, M.E.; Partridge, T.A.; Zammit, P.S. Expression of CD34 and Myf5 defines the majority of quiescent adult skeletal muscle satellite cells. *J. Cell Biol.* **2000**, *151*, 1221–1234. [CrossRef] [PubMed]
252. Lozano-Velasco, E.; Vallejo, D.; Esteban, F.J.; Doherty, C.; Hernandez-Torres, F.; Franco, D.; Aranega, A.E. A Pitx2-MicroRNA Pathway Modulates Cell Proliferation in Myoblasts and Skeletal-Muscle Satellite Cells and Promotes Their Commitment to a Myogenic Cell Fate. *Mol. Cell Biol.* **2015**, *35*, 2892–2909. [CrossRef] [PubMed]
253. McFarlane, C.; Hennebry, A.; Thomas, M.; Plummer, E.; Ling, N.; Sharma, M.; Kambadur, R. Myostatin signals through Pax7 to regulate satellite cell self-renewal. *Exp. Cell Res.* **2008**, *314*, 317–329. [CrossRef] [PubMed]
254. Rebbapragada, A.; Benchabane, H.; Wrana, J.L.; Celeste, A.J.; Attisano, L. Myostatin signals through a transforming growth factor beta-like signaling pathway to block adipogenesis. *Mol. Cell. Biol.* **2003**, *23*, 7230–7242. [CrossRef] [PubMed]
255. Massague, J.; Seoane, J.; Wotton, D. Smad transcription factors. *Genes Dev.* **2005**, *19*, 2783–2810. [CrossRef] [PubMed]
256. Liu, D.; Black, B.L.; Derynck, R. TGF-beta inhibits muscle differentiation through functional repression of myogenic transcription factors by Smad3. *Genes Dev.* **2001**, *15*, 2950–2966. [CrossRef] [PubMed]
257. McPherron, A.C.; Lawler, A.M.; Lee, S.J. Regulation of skeletal muscle mass in mice by a new TGF-beta superfamily member. *Nature* **1997**, *387*, 83–90. [CrossRef] [PubMed]
258. McPherron, A.C.; Lee, S.J. Double muscling in cattle due to mutations in the myostatin gene. *Proc. Natl. Acad. Sci. USA* **1997**, *94*, 12457–12461. [CrossRef] [PubMed]
259. Sartori, R.; Milan, G.; Patron, M.; Mammucari, C.; Blaauw, B.; Abraham, R.; Sandri, M. Smad2 and 3 transcription factors control muscle mass in adulthood. *Am. J. Physiol. Cell Physiol.* **2009**, *296*, C1248–C1257. [CrossRef] [PubMed]
260. Iezzi, S.; Di Padova, M.; Serra, C.; Caretti, G.; Simone, C.; Maklan, E.; Minetti, G.; Zhao, P.; Hoffman, E.P.; Puri, P.L.; et al. Deacetylase inhibitors increase muscle cell size by promoting myoblast recruitment and fusion through induction of follistatin. *Dev. Cell* **2004**, *6*, 673–684. [CrossRef]
261. Pisconti, A.; Brunelli, S.; Di Padova, M.; De Palma, C.; Deponti, D.; Baesso, S.; Sartorelli, V.; Cossu, G.; Clementi, E. Follistatin induction by nitric oxide through cyclic GMP: A tightly regulated signaling pathway that controls myoblast fusion. *J. Cell Biol.* **2006**, *172*, 233–244. [CrossRef] [PubMed]
262. Fainsod, A.; Deissler, K.; Yelin, R.; Marom, K.; Epstein, M.; Pillemer, G.; Steinbeisser, H.; Blum, M. The dorsalizing and neural inducing gene follistatin is an antagonist of BMP-4. *Mech. Dev.* **1997**, *63*, 39–50. [CrossRef]
263. Amthor, H.; Nicholas, G.; McKinnell, I.; Kemp, C.F.; Sharma, M.; Kambadur, R.; Patel, K. Follistatin complexes Myostatin and antagonises Myostatin-mediated inhibition of myogenesis. *Dev. Biol.* **2004**, *270*, 19–30. [CrossRef] [PubMed]
264. Zhang, Q.; Shi, X.E.; Song, C.; Sun, S.; Yang, G.; Li, X. BAMBI Promotes C2C12 Myogenic Differentiation by Enhancing Wnt/beta-Catenin Signaling. *Int. J. Mol. Sci.* **2015**, *16*, 17734–17745. [CrossRef] [PubMed]
265. Christ, B.; Jacob, M.; Jacob, H.J. On the origin and development of the ventrolateral abdominal muscles in the avian embryo. An experimental and ultrastructural study. *Anat. Embryol.* **1983**, *166*, 87–101. [CrossRef] [PubMed]
266. Eng, D.; Ma, H.Y.; Xu, J.; Shih, H.P.; Gross, M.K.; Kioussi, C. Loss of abdominal muscle in Pitx2 mutants associated with altered axial specification of lateral plate mesoderm. *PLoS ONE* **2012**, *7*, e42228. [CrossRef]
267. Sadler, T.W. The embryologic origin of ventral body wall defects. *Semin. Pediatr. Surg.* **2010**, *19*, 209–214. [CrossRef] [PubMed]
268. Brewer, S.; Williams, T. Finally, a sense of closure? Animal models of human ventral body wall defects. *Bioessays* **2004**, *26*, 1307–1321. [CrossRef] [PubMed]
269. Sadler, T.W.; Feldkamp, M.L. The embryology of body wall closure: Relevance to gastroschisis and other ventral body wall defects. *Am. J. Med. Genet. C Semin. Med. Genet.* **2008**, *148*, 180–185. [CrossRef] [PubMed]
270. Zhang, L.; Li, H.; Yu, J.; Cao, J.; Chen, H.; Zhao, H.; Zhao, J.; Yao, Y.; Cheng, H.; Wang, L.; et al. Ectodermal Wnt signaling regulates abdominal myogenesis during ventral body wall development. *Dev. Biol.* **2014**, *387*, 64–72. [CrossRef] [PubMed]
271. L'Honore, A.; Ouimette, J.F.; Lavertu-Jolin, M.; Drouin, J. Pitx2 defines alternate pathways acting through MyoD during limb and somitic myogenesis. *Development* **2010**, *137*, 3847–3856. [CrossRef] [PubMed]

272. Hilton, T.; Gross, M.K.; Kioussi, C. Pitx2-dependent occupancy by histone deacetylases is associated with T-box gene regulation in mammalian abdominal tissue. *J. Biol. Chem.* **2010**, *285*, 11129–11142. [CrossRef] [PubMed]
273. Brewer, S.; Williams, T. Loss of AP-2alpha impacts multiple aspects of ventral body wall development and closure. *Dev. Biol.* **2004**, *267*, 399–417. [CrossRef] [PubMed]
274. Ten Berge, D.; Brouwer, A.; el Bahi, S.; Guenet, J.L.; Robert, B.; Meijlink, F. Mouse Alx3: An aristaless-like homeobox gene expressed during embryogenesis in ectomesenchyme and lateral plate mesoderm. *Dev. Biol.* **1998**, *199*, 11–25. [CrossRef] [PubMed]
275. Qu, S.; Niswender, K.D.; Ji, Q.; van der Meer, R.; Keeney, D.; Magnuson, M.A.; Wisdom, R. Polydactyly and ectopic ZPA formation in Alx-4 mutant mice. *Development* **1997**, *124*, 3999–4008. [PubMed]
276. Doi, T.; Puri, P.; Bannigan, J.; Thompson, J. HoxB2, HoxB4 and Alx4 genes are downregulated in the cadmium-induced omphalocele in the chick model. *Pediatr. Surg. Int.* **2010**, *26*, 1017–1023. [CrossRef] [PubMed]
277. Manley, N.R.; Barrow, J.R.; Zhang, T.; Capecchi, M.R. Hoxb2 and hoxb4 act together to specify ventral body wall formation. *Dev. Biol.* **2001**, *237*, 130–144. [CrossRef] [PubMed]
278. Barrow, J.R.; Capecchi, M.R. Targeted disruption of the Hoxb-2 locus in mice interferes with expression of Hoxb-1 and Hoxb-4. *Development* **1996**, *122*, 3817–3828. [PubMed]
279. Kitamura, K.; Miura, H.; Miyagawa-Tomita, S.; Yanazawa, M.; Katoh-Fukui, Y.; Suzuki, R.; Ohuchi, H.; Suehiro, A.; Motegi, Y.; Nakahara, Y.; et al. Mouse Pitx2 deficiency leads to anomalies of the ventral body wall, heart, extra- and periocular mesoderm and right pulmonary isomerism. *Development* **1999**, *126*, 5749–5758. [PubMed]
280. Hogan, B.L. Bone morphogenetic proteins: Multifunctional regulators of vertebrate development. *Genes Dev.* **1996**, *10*, 1580–1594. [CrossRef] [PubMed]
281. Dickson, M.C.; Slager, H.G.; Duffie, E.; Mummery, G.; Mummery, C.L.; Akhurst, R.J. RNA and protein localisations of TGF beta 2 in the early mouse embryo suggest an involvement in cardiac development. *Development* **1993**, *117*, 625–639. [PubMed]
282. Aldeiri, B.; Roostalu, U.; Albertini, A.; Wong, J.; Morabito, A.; Cossu, G. Transgelin-expressing myofibroblasts orchestrate ventral midline closure through TGFbeta signalling. *Development* **2017**, *144*, 3336–3348. [CrossRef] [PubMed]
283. Suzuki, N.; Labosky, P.A.; Furuta, Y.; Hargett, L.; Dunn, R.; Fogo, A.B.; Takahara, K.; Peters, D.M.; Greenspan, D.S.; Hogan, B.L. Failure of ventral body wall closure in mouse embryos lacking a procollagen C-proteinase encoded by Bmp1, a mammalian gene related to Drosophila tolloid. *Development* **1996**, *122*, 3587–3595. [PubMed]
284. Zhang, J.; Hagopian-Donaldson, S.; Serbedzija, G.; Elsemore, J.; Plehn-Dujowich, D.; McMahon, A.P.; Flavell, R.A.; Williams, T. Neural tube, skeletal and body wall defects in mice lacking transcription factor AP-2. *Nature* **1996**, *381*, 238–241. [CrossRef] [PubMed]

8

LHX2 Mediates the FGF-to-SHH Regulatory Loop during Limb Development

Billy A. Watson [1,2], Jennifer M. Feenstra [1], Jonathan M. Van Arsdale [1], Karndeep S. Rai-Bhatti [1], Diana J. H. Kim [1], Ashley S. Coggins [1], Gennaya L. Mattison [1], Stephen Yoo [1], Eric D. Steinman [1], Charmaine U. Pira [1], Brendan R. Gongol [3] and Kerby C. Oberg [1,*]

[1] Department of Pathology and Human Anatomy, School of Medicine, Loma Linda University, Loma Linda, CA 92354, USA; bawatson@llu.edu (B.A.W.); jennifer.feenstra@ki.se (J.M.F.); Jonvan83@gmail.com (J.M.V.A.); kraib001@ucr.edu (K.S.R.-B.); djhkim@llu.edu (D.J.H.K.); ashley.coggins.1@gmail.com (A.S.C.); gennaya.mattison@gmail.com (G.L.M.); Yoomanchu@gmail.com (S.Y.); esteinman@gmail.com (E.D.S.); cpira@llu.edu (C.U.P.)

[2] Division of Microbiology and Molecular Genetics, Department of Basic Sciences, School of Medicine, Loma Linda University, Loma Linda, CA 92354, USA

[3] Department of Cardiopulmonary Sciences, School of Allied Health Professions, Loma Linda University, Loma Linda, CA 92354, USA; bgongol@llu.edu

* Correspondence: koberg@llu.edu

Abstract: During limb development, fibroblast growth factors (Fgfs) govern proximal–distal outgrowth and patterning. FGFs also synchronize developmental patterning between the proximal–distal and anterior–posterior axes by maintaining Sonic hedgehog (Shh) expression in cells of the zone of polarizing activity (ZPA) in the distal posterior mesoderm. Shh, in turn, maintains Fgfs in the apical ectodermal ridge (AER) that caps the distal tip of the limb bud. Crosstalk between Fgf and Shh signaling is critical for patterned limb development, but the mechanisms underlying this feedback loop are not well-characterized. Implantation of Fgf beads in the proximal posterior limb bud can maintain *SHH* expression in the former ZPA domain (evident 3 h after application), while prolonged exposure (24 h) can induce *SHH* outside of this domain. Although temporally and spatially disparate, comparative analysis of transcriptome data from these different populations accentuated genes involved in SHH regulation. Comparative analysis identified 25 candidates common to both treatments, with eight linked to *SHH* expression or function. Furthermore, we demonstrated that LHX2, a LIM Homeodomain transcription factor, is an intermediate in the FGF-mediated regulation of SHH. Our data suggest that LHX2 acts as a competency factor maintaining distal posterior *SHH* expression subjacent to the AER.

Keywords: limb development; fibroblast growth factor (FGF); sonic hedgehog (SHH); LIM Homeobox 2 (LHX2)

---

1. Introduction

Limb development generates a structure that has asymmetry along three coordinate axes: proximal–distal (PD), anterior–posterior (AP), and dorsal–ventral (DV). Each axis has its own signaling center that mediates patterning. The apical ectodermal ridge (AER), a thickening of ectoderm located along the distal rim of the limb, is a signaling center from which secreted fibroblast growth factors (Fgfs) regulate PD patterning and outgrowth. Sonic hedgehog (Shh) is secreted from a cluster of cells in the distal posterior limb bud mesoderm called the zone of polarizing activity (ZPA) and regulates AP expansion and patterning. The dorsal ectoderm secretes Wnt7a, which dorsalizes the developing limb. During development, crosstalk among these axes is required to coordinate proper limb patterning [1].

Along the PD and AP axes, development is coordinated through crosstalk between Fgf and Shh in a reciprocal feedback loop [2,3]. The mechanism by which Shh maintains Fgf in the AER is relatively well-characterized and includes Shh directing interactions between Formin, Gremlin, and Bone morphogenic proteins (Bmps) [4–6]. In the absence of Shh, the AER regresses and posterior elements of the developing limb are lost [7,8]. In contrast, the mechanism by which AER-secreted Fgfs regulate *Shh* is less clear. As the limb bud elongates, the ZPA (and *Shh* expression) persists distally subjacent to the AER. *Shh* expression in the more proximal posterior mesodermal cells, which we refer to as the former ZPA domain, wanes as the cells move beyond the influence of the AER-Fgfs [9,10].

In the chicken, FGF2 [11,12], FGF4 [2,13,14], or FGF8 [15,16] is sufficient to recover *SHH* expression in the absence of the AER. When applied to the mid-posterior half of the limb bud, nearly 24 h is required to upregulate *SHH* in cells outside the ZPA or former ZPA domain (Figure 1A). Interestingly, in cells of the former ZPA domain (proximal posterior limb margin), *SHH* could be maintained in the presence of proximal FGF, evident within 3 h of exposure. The former ZPA domain and the non-ZPA domain differ in context (maintenance versus induction, respectively); nevertheless, we suspected that the final pathway to *SHH* expression is common in both domains.

FGF2 is endogenously expressed in the chicken [11,17] and human AER [18], has broader effects than FGF4 and FGF8 [19], and interacts with greater affinity to the primary ZPA-related Fgf receptor (FGFr1cIII) [20,21]. Thus, we applied FGF2-laden beads to chicken limb buds to generate ZPA domain and the non-ZPA domain transcriptomes. Using comparative analysis, we identified genes that were common to both transcriptomes, genes we suspected would accentuate the FGF-to-SHH pathway. From our analysis, we were able to identify a number of candidates in the FGF-to-SHH pathway and confirm a role for LIM Homeobox 2 (LHX2) as an intermediate in FGF-regulated *SHH* expression.

2. Materials and Methods

2.1. FGF Bead Implants, WNT5a Cell Implants, and Harvesting Embryos

Heparin acrylic beads (H5263 Sigma, St. Louis, MO, USA), about 150 μm in diameter, were soaked in 4 μL of 0.5 mg/mL recombinant human FGF2 (233-FB), FGF4 (235-F4), or FGF8 (423-F8) (R&D Systems, Minneapolis, MN, USA) overnight at 4 °C or for at least 1 h at room temperature. For the control samples, the heparin acrylic beads were soaked in 4 μL 1× phosphate-buffered saline (PBS) before implantation. Using tungsten needle and forceps, FGF2 or PBS-soaked beads were implanted (*in ovo*) into the posterior forelimb bud mesoderm of 23 Hamburger–Hamilton stage (HH) chicken embryos about 500 μm from the distal tip. Cells secreting Wnt5a (CRL-2814, ATCC®, Manassas, VA, USA) were grown to confluence and then stained with neutral red. Strips of Wnt5a-secreting cells were then harvested and implanted into the posterior margin of HH23 right limb bud mesoderm as previously described [22]. Embryos were incubated at 39 °C in a humidified chamber for the experimental time (3–24 h) and quickly harvested in PBS. Tissue from embryos for transcriptome and RT-qPCR analyses was removed using a tungsten needle, flash frozen using liquid nitrogen, and stored at −80 °C until further processing. Embryos for whole-mount in situ hybridization (WMISH) were fixed in MEMFA (1 M MOPS, 20 mM EGTA, 10 mM $MgSO_4$, 38% formaldehyde) overnight at 4 °C as described previously [23], then stored in 90% methanol at −20 °C until further processing.

2.2. RNA-Seq Analysis

After 3 h of incubation, the tissue directly posterior to the FGF2 or PBS bead was extracted, and RNA was isolated using the RNeasy Plus Micro Kit (Qiagen, Valencia, CA, USA). Three independent experiments were conducted and RNA within each sample group (FGF or PBS) from each experiment was pooled to decrease genetic variability (n = 15–20 embryos per sample). RNA-seq was conducted by Norris Comprehensive Cancer Center, University of Southern California. RNA-seq differential gene expression and statistical analysis was conducted in R coding language with the systempipeR package tools. Reads were aligned with Hisat2 alignment algorithm to release-87 of *Gallus gallus* genome build 5. Fold changes were calculated, and statistical comparisons made with EdgeR. Genes exhibiting differential expression between FGF2- and PBS-treated embryos with $p < 0.05$ were included in further analyses. RNA-seq data were submitted to Gene Expression Omnibus (http://www.ncbi.nlm.nih.gov/geo/) [24] and can be located under series accession number GES114663.

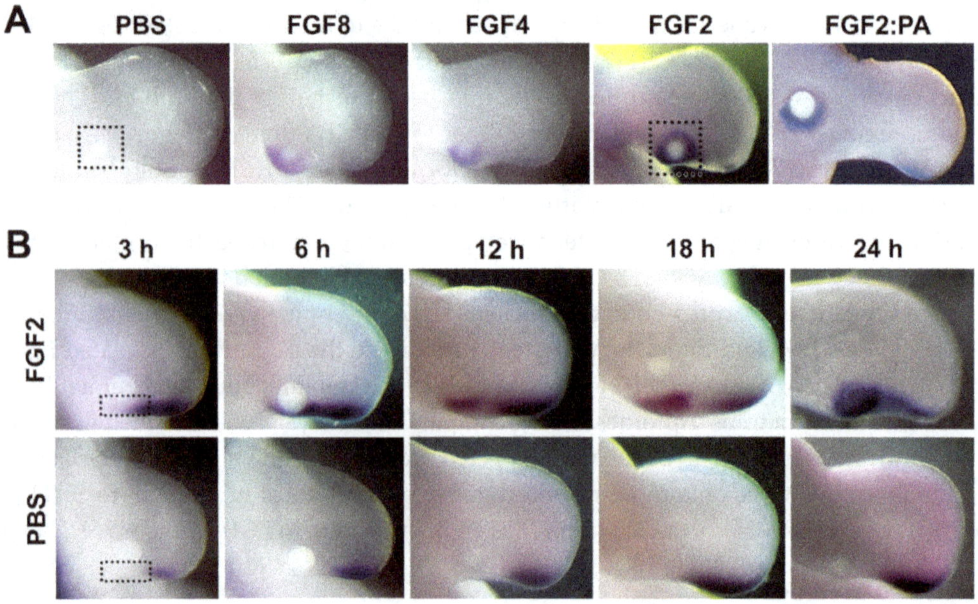

Figure 1. FGF has the capacity to maintain or ectopically induce SHH expression. (**A**) SHH expression following 24 h of exposure to FGF8, FGF4, or FGF2-soaked beads implanted into the posterior HH23 chicken wing mesoderm proximal to the ZPA (500 μm from the AER). Ectopic SHH expression is present around the FGF bead, but not in the PBS (vehicle control). This ectopic expression was noted in all embryos tested (n = 3 for each FGF). The expression was present in cells of the former ZPA domain (along the posterior margin) and in cells that likely lacked prior ZPA exposure (anterior aspect of the bead). Induction of non-ZPA cells was accentuated when initial bead placement was more proximal (600 μm from the AER) and anterior (500 μm from posterior edge) (FGF2:PA). (**B**) Persistent SHH expression is evident in former ZPA domains over a 3–24 h time course post-FGF bead implant. In addition, robust ectopic SHH expression is present around the FGF bead at 24 h (top panel) as compared to the PBS control (bottom panel). For each time point, 5 embryos were assayed for response to FGF, while 3 embryos were for expression in the presence of PBS (vehicle control). Two independent experiments were used to collect embryos, and the pattern of ectopic SHH expression was consistent in all time-matched embryos. Tissue harvested for transcriptome analysis and RT-qPCR validation is indicated by black dotted lines.

2.3. *DNA Microarray Analysis*

RNA was isolated from tissue surrounding an FGF2 or PBS bead after 24 h of incubation using the RNeasy kit (Qiagen). The RNA from 6 embryos was pooled for each treatment group to decrease genetic variability, and the experiment was repeated 3 times on separate days. Microarray analysis was performed by University of California at Irvine Genomics High Throughput Facility as previously described [23]. Samples were hybridized to the Affymetrix GeneChip® Chicken Genome Array (ThermoFisher Scientific, Chino, CA, USA). The data was normalized using Robust Multi-array Averaging (RMA) and analyzed using the Comprehensive R and Bioconductor-based web service for microarray data analysis. Only genes exhibiting differential expression between FGF- and PBS-treated embryos with $p < 0.05$ were included in further analyses. Microarray data were submitted to Gene Expression Omnibus and can be located under series accession number GES114663.

2.4. Gene Ontology

Differentially expressed genes at 3 and 24 h post-FGF bead implantation were classified according to gene ontology biological pathways using Ingenuity Pathway Analysis (IPA) software and database (Qiagen).

2.5. Gene Expression Analysis via Whole-Mount In Situ Hybridization (WMISH)

To validate differential gene expression in response to 3 and 24 h of FGF treatment, digoxigenin-labeled probes were generated, and embryos were processed for WMISH as previously described [23,25]. Primers used for probe generation are listed in Table S2; alternatively, probes were generated from plasmids received as gifts: *DKK1, WNT5a, PTCH2,* (Dr. Clifford Tabin); *FZD4* (Dr. Philippa Francis-West); *FGF8* (Dr. Eric Swindell); *PYST1/DUSP6* (Dr. Stephen Keyse). For all WMISH procedures, 3–5 embryos were examined for each gene per experiment and at least two independent experiments were performed.

2.6. RT-qPCR Validation of Transcriptome Data and Quantitation of LHX2 Overexpression/Knockdown Analysis

Further validation of the differential expression pattern of selected genes following 3 and 24 h of FGF treatment was performed via RT-qPCR. RNA was extracted from FGF2- or PBS-treated tissue 3 and 24 h after bead implantation. In order to minimize biological variation, 7–10 embryos from each treatment group were pooled. Primers used for RT-qPCR validation are listed in Table S2.

RT-qPCR was also used to detect changes in *SHH* and *PTCH2* expression following overexpression and knockdown of LHX2. For all samples, RNA extraction and on-column DNA digestion were performed using the RNeasy plus micro kit (Qiagen). The extracted RNA was converted to cDNA by reverse transcription using the iScript Advanced cDNA Synthesis kit (Bio-Rad, Irvine, CA, USA). Primers used are listed in Table S2.

The RT-qPCR experiments were performed using the SsoAdvanced Universal SYBR Green Supermix (Bio-Rad, Irvine, CA, USA) on the CFX96 ThermoCycler and analyzed using the CFX Manager 3.0 software (Bio-Rad). All RT-qPCR reactions were performed in triplicate for at least 2 independent experiments. Gene expression levels relative to a PBS-treated control were calculated using the $2^{-\Delta\Delta CT}$ method after values were normalized to the housekeeping gene *PGK1*. The resulting fold changes are plotted with error bars representing the standard deviation. The significance (p-value) was calculated using the Student's paired t-test.

2.7. Comparative Transcriptome Analysis

Integration of the RNA-seq and microarray data into R coding platforms facilitated categorization of genes according to their fold change and p-values then visualized by a volcano plot and table. We ranked the genes by dividing the fold change of each gene by their corresponding p-value. This narrowed down our target selection to LHX2.

2.8. Gene Overexpression and Knockdown via Electroporation

A plasmid encoding the mouse Lhx2 gene (pCGC-Lhx2; used at a final concentration of 1 μg/μL) or the translation-blocking anti-*LHX2* morpholino (1 mM) was injected into HH23 chicken embryo forelimbs around the FGF bead or into the ZPA. Confined micro-electroporation (CMEP) as previously described [22] was used to introduce the plasmid or morpholino into the limb cells surrounding the injection site. A β-actin promoter-driven red fluorescent protein (RFP) vector (pCAGGS-RFP) was used as an empty vector control. Sterile mineral oil is used in our syringe and needle to facilitate accurate DNA delivery. After injection, a small amount of oil is typically introduced to occlude the site of injection and confine the DNA to the target region.

The LHX2-GFP expression plasmid (pCGC-*Lhx2*) was received as a gift from Dr. Shubha Tole and used with the permission of Dr. Toshi Ohshima. The design and efficacy of the vector has been previously reported [26]. Representative data from three independent experiments is shown in the results section with number of embryos assayed included in the figure legends.

A translation-blocking anti-*LHX2* morpholino (GeneTools, LLC, Philomath, OR, USA) was designed against chicken *LHX2* (NCBI Reference Sequence: NM_204889.1). A negative-control morpholino was generated from the anti-*LHX2* sequence incorporating 5 base mismatches to alter the sequence. A positive-control anti-*SHH* morpholino was also generated against chicken *SHH* (NCBI Reference Sequence: NM_204821.1) A 3′ carboxyfluorescein (green) tag was incorporated into the anti-c*LHX2* and anti-c*SHH* morpholinos, while a lissamine (red) tag was incorporated on the 3′ end of the negative-control morpholino (anti-c*LHX2*-5mis). The fluorescent tags were used to assess targeting of the injection and electroporation efficiency prior to harvesting embryos for WMISH or RT-qPCR analysis. Morpholino (MO) sequences are as follows: anti-c*SHH* MO: 5′-TTGTCAACAGCAGCATTTCGACCAT-3′; anti-c*LHX2* MO: 5′-ACAGGCTGTGGAAAAGCATCGCT-3′; anti-c*LHX2*-5mis: 5′-AgAGcCTGTcGAAAAcCATCGgT-3′. In the 5-base mismatch morpholino (anti-c*LHX2*-5mis), nucleotides in lower case specify the mismatches. All morpholinos were used at a concentration of 1 millimolar.

Controlling for Off-Target Effects Using Morpholinos

To control for target specificity, we used 3 previously described approaches [27,28], including: (1) The use of a mismatched negative-control morpholino that differed from the anti-target morpholino by 5 nucleotides; (2) comparison of phenotypic changes with published reports. Our LHX2 knockdown was consistent with published reports of *Lhx2* mutants [29,30]; (3) we used "rescue" experiments that showed that overexpression of *Lhx2* was able to recover the expression of its downstream target, *SHH*.

3. Results

3.1. FGF Can Maintain and Induce SHH Expression in the Posterior Limb Bud

SHH expression persists in the former ZPA domain following implantation of an FGF-laden bead. *SHH* is detectable as early as 3 h after application; at 24 h, robust *SHH* expression is also induced around the bead (non-ZPA domain) (Figure 1B). This spatial and temporal difference in *SHH* upregulation suggests that induction or reactivation of *SHH* expression outside the former ZPA domain (24 h) requires additional FGF-mediated factors besides those required to maintain *SHH* in the former ZPA domain (3 h). We analyzed the transcriptome along the posterior margin proximal to the ZPA 3 h after FGF application and surrounding the FGF-laden bead after 24 h to uncover molecules involved in FGF-mediated SHH expression (areas outlined by dotted lines in Figure 1).

3.2. Brief (3 h) FGF Exposure Affects Biological Processes Associated with Its Role in Gene Expression

We found 150 genes differentially regulated 3 h after FGF exposure by RNA-seq analysis, with 102 targets upregulated and 48 downregulated ($p < 0.05$). Four of the top biological processes affected by FGF exposure were identified in our 3 h dataset, but not our 24 h dataset: "Gene Expression", "Cell death and survival", "Cellular development", and "Tissue morphology" (Figure 2A). This collection of biological processes was predicted to be downstream of FGF2 and FGF8, in addition to some other growth factors. Moreover, limb-related functions formed subsets of many of these biological processes (listed in Table S1).

Approximately 30% of the targets (n = 42 out of 150) were associated with "Gene Expression", with 36 categorized as transcription/translation regulators. Transcription factors Early growth response 1 (*EGR1*), LIM Homeobox 2 (*LHX2*), and Transcription factor AP2 gamma (*TFAP2C*) were upregulated

12.5, 3.9, and 3.4-fold, respectively. Their regulation in response to 3 h FGF exposure was confirmed by WMISH and quantified by RT-qPCR (Figure 2B,C).

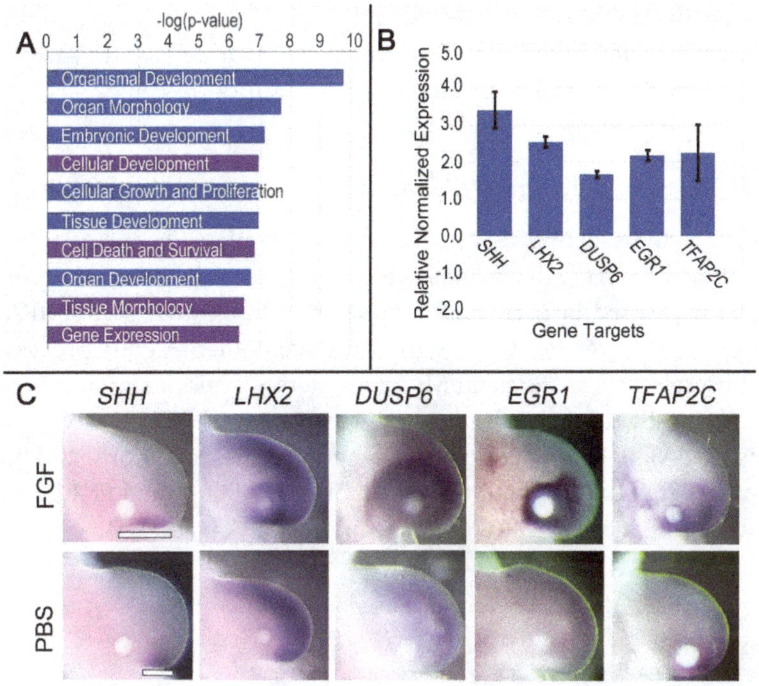

Figure 2. Brief (3 h) FGF treatment promotes transcription of early developmental genes and maintains *SHH* expression in the former ZPA domain. (**A**) Top pathways and biological processes affected 3 h after FGF treatment. Purple bars indicate top pathways affected that differed from the 3 h transcriptome analysis. (**B**) Fold change of selected targets treated with FGF for 3 h compared to PBS treatment via RT-qPCR ($p < 0.05$). The y-axis shows the relative normalized gene expression. Assays were performed in triplicate and at least two independent experiments were performed. Expression of each gene was compared to a PBS control ($p < 0.05$ for each comparison). (**C**) WMISH validation of selected targets regulated by FGF after 3 h exposure. Top panel shows upregulation of respective genes, while the bottom panel consists of PBS-treated limbs. n = 6 embryos per gene for FGF treatment and 3 embryos per gene for PBS. White bars highlight the proximal extension of the *SHH* expression domain in the FGF-treated limb.

3.3. Brief (3 h) FGF Exposure Regulates Genes Involved in FGF Feedback Inhibition, Distalization, and SHH Expression

After 3 h of FGF treatment, Dual specificity phosphatase 6 (*DUSP6*) and Sprouty 2 (*SPRY2*), both negative regulators of FGF signaling, were upregulated 2.6-fold and 3.4-fold, respectively, consistent with FGF feedback inhibition. The upregulation of *DUSP6* was validated by WMISH and RT-qPCR (Figure 2B,C). Coupled with *DUSP6* and *SPRY2*, several other distally restricted genes were upregulated in the former ZPA domain 3 h after FGF exposure, including Bone morphogenetic protein 4 (*BMP4*; 1.8-fold). BMP4 plays a role in AER maintenance and regression and cooperates with SHH in determining digit identity [31,32]. In contrast, the proximally restricted Pre-B-cell leukemia homeobox 1 (*PBX1*) was downregulated 1.8-fold. *Pbx1*-deficient mice have malformations in proximal limb elements, suggesting a role in proximal limb development [33]. Downregulation of proximal markers by brief FGF exposure and the upregulation of distal factors support a distal respecification of the mid-proximal limb, which may be necessary for *SHH* expression.

Several targets known to be upstream of *SHH* expression, including Homeobox D9 [34,35], *ELF2* (a member of the ETS family), and *HAND2* (a basic helix–loop–helix transcription factor) were up-regulated 2.3-fold, 1.9-fold, and 1.6-fold, respectively. A complex interaction between HOX,

PBX, ETS, and HAND2 has been implicated in the activation and localization of *SHH* expression to the ZPA [36]. These transcription factors coordinate their interactions on *SHH* expression through a conserved long range *cis*-regulatory module, the ZPA regulatory sequence (*ZRS*) [37]. *LHX2* and *TFAP2C* are two additional distally restricted transcription factors with potential binding sites in the *ZRS* that were upregulated after 3 h of FGF treatment. Our dataset revealed a 2.6-fold increase in *SHH* expression after 3 h of FGF exposure, which is supported by in situ and RT-qPCR data (Figure 2B,C).

3.4. Prolonged (24 h) FGF Exposure Affects Cell Processes Related to Organ and Organismal Development

DNA microarray analysis identified 3434 differentially regulated targets after 24 h of FGF exposure ($p < 0.05$). In addition to *SHH*, 2152 targets were upregulated, while 1281 were downregulated. Of the differentially expressed targets, 1085 mapped to IPA-curated genes. IPA revealed the top 10 biological processes were related to growth and development. Four processes were found at 24 h that were not present at 3 h: "Skeletal Muscle System Development and Function", "Nervous Tissue Development and Function", "Organismal Survival", and "Connective Tissue Development and Function" (Figure 3A). IPA also predicted that the molecules, functions, and biological processes identified in the microarray data were downstream of FGF2, FGF4, and FGF8. Several limb-related functions form subsets of the biological processes mined by IPA (Table S1).

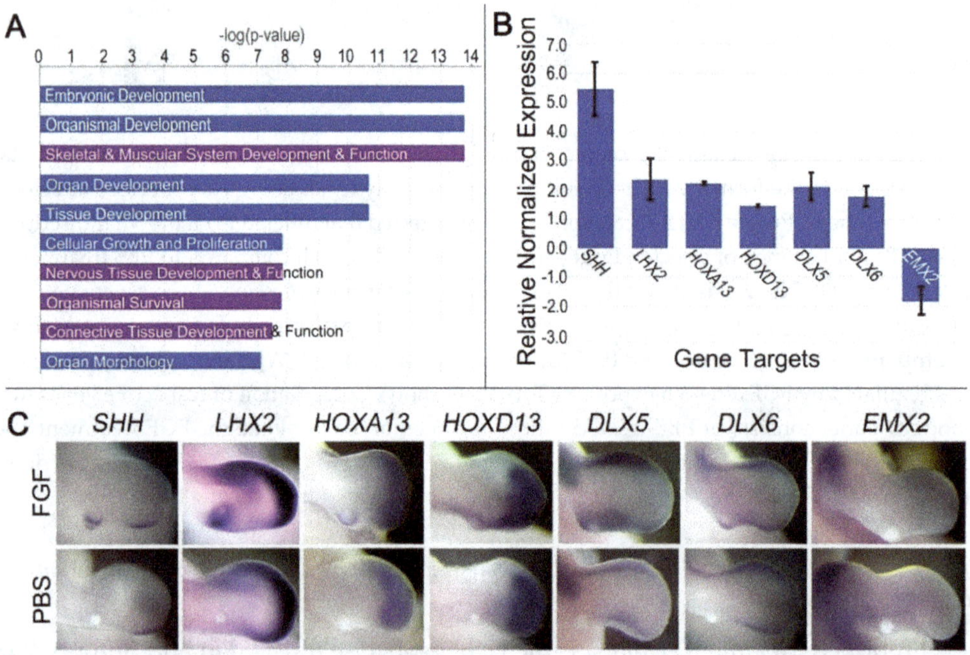

Figure 3. Prolonged (24 h) FGF treatment in the mid-proximal limb bud promotes distalization, supports processes involved in embryonic development, and leads to SHH induction. (**A**) Top pathways and biological processes affected by 24 h of FGF treatment. Purple bars indicate top pathways affected that differed from the 3 h transcriptome analysis. (**B**) RT-qPCR validation of FGF-mediated expression of target genes. The y-axis shows the relative normalized expression of selected target genes. Assays were performed in triplicate and at least two independent experiments were performed. Expression of each gene was compared to the same region on the contralateral limb buds ($p < 0.05$ for each comparison). (**C**) WMISH of distally restricted transcription factors confirm their regulation in response to 24 h of FGF exposure. Top panel shows upregulation/downregulation in response to ectopic FGF when compared to a PBS-treated limb (control; bottom panel). n = 6 embryos per gene for FGF treatment and 3 embryos per gene for PBS. Note that *EMX2* is strongly expressed proximally, but has a weak distal expression that was reduced by FGF treatment after 24 h.

Of note, an abundance of differentially regulated genes encoding Wnt signaling proteins were detected (n = 46). Wnt proteins are involved in limb initiation, outgrowth, cell migration, cell differentiation, and patterning [38,39]. This led us to question whether a Wnt pathway might be involved in the upregulation of *SHH*. However, overexpression of the highest upregulated Wnt protein (*WNT5a*; 1.6-fold) in the posterior limb bud did not lead to ectopic *SHH* expression (Figure S1).

3.5. Prolonged FGF (24 h) Regulates Genes Associated with Dedifferentiation, Distalization, and SHH Regulation

Several transcription factors associated with cell dedifferentiation were upregulated in response to prolonged FGF exposure: Spalt-Like Transcription Factor 4 (*SALL4*; 1.8-fold), Muscle segment homeobox genes 1 and 2 (*MSX1*; 4.4-fold and *MSX2*; 3.0-fold), *LHX2* (5.6-fold), and Matrix metallopeptidase 11 and 17 (*MMP11* and *MMP17*, each 1.4-fold). FGF treatment also downregulated members of the collagen family, which are associated with limb differentiation and chondrocyte maturation (*COL12A1, COL6A1, COL6A3, COL21A1*; >1.5-fold). Dedifferentiation of the mid-proximal limb bud may be a crucial step in reprogramming this region to express *SHH*. Alternatively, FGF may be functioning to prevent differentiation and promote the maintenance of cells in an undifferentiated state. Another step could be the distal respecification of the mid-proximal limb supported by the upregulation of several distally restricted genes, including *HOXA13* (13.0-fold), *HOXD13* (1.7-fold), *DLX5* (2.2-fold), *DLX6* (3.6-fold), and *LHX2* (5.6-fold) (Figure 3B,C); while proximally restricted genes such as Empty spiracles homeobox 2 (*EMX2*) and *MEIS2* were downregulated 1.7- and 1.5-fold, respectively. Downregulation of *EMX2* was validated by WMISH and RT-qPCR (Figure 3B,C). Together, the data suggest that dedifferentiation and the distal respecification of the mid-proximal limb bud mesoderm play a role in FGF-mediated *SHH* expression.

After 24 h of FGF exposure, *SHH* is upregulated 2.1-fold around the FGF bead. This correlated with RT-qPCR data (upregulated 5.5-fold) (Figure 3C). Genes associated with *SHH* expression include the distally restricted *HOXA13*, *HOXD13*, and *LHX2*. Other upregulated *SHH*-associated genes include *HOXA10* (1.4-fold), *HOXA11* (1.7-fold), Gap junction protein alpha 1 (*GJA1*; 1.3-fold), *ETS2* (1.3-fold), and *ETV5* (1.1-fold). Although the fold change for *ETV5* might be less than typical cutoffs, it has a p value < 0.05 and an established functional role in linking FGF signaling to the *ZRS* and *Shh* expression [40,41]. The *SHH* antagonist Aristaless-like homeobox 4 (*ALX4*) was downregulated 1.7-fold. ETS2, ETV5, and ALX4 contribute to the spatial localization of *SHH* expression [41–43]. Notably, our 24 h dataset includes upregulation of factors downstream of SHH signaling such as *PTCH1* (1.8-fold) and *PTCH2* (1.1-fold). Because the 24 h transcriptome may include a number of downstream targets of SHH, we compared it with the 3 h transcriptome to highlight common genes most likely to be upstream of *SHH*.

3.6. Common Pathways Accentuated by Comparative Transcriptome Analysis

Six of the top IPA-curated pathways affected by FGF treatment were common between the 3 and 24 h datasets, including "Embryonic/Organismal Development" and "Growth and Proliferation": hallmarks of FGF signaling (Figures 2A and 3A). We also detected mutual limb-related functions associated with limb development and digit morphogenesis (see Table S1). Twenty-five genes were common to both transcriptomes, with 19 being upregulated, five downregulated, and one gene downregulated at 3 h but upregulated at 24 h (Figure 4).

3.6.1. Common Targets Involved in Wnt Signaling

We identified five genes associated with the Wnt pathway within the shared 25 FGF-regulated genes. *APCDD1* [44] and *RANBP3* [45] are Wnt signaling inhibitors that were upregulated by FGF, while R-spondin 3 (*RSPO3*), a secreted ligand that binds cell-surface receptors and activates Wnt/β-catenin or Wnt/planar cell polarity signaling [46,47], was downregulated. One downstream target of Wnt signaling, SNAI2, was upregulated, while another, *TUSC3* [48], was downregulated.

Taken together, these data support a role for Fgf in the complicated regulation of Wnt signaling during limb development [39,49].

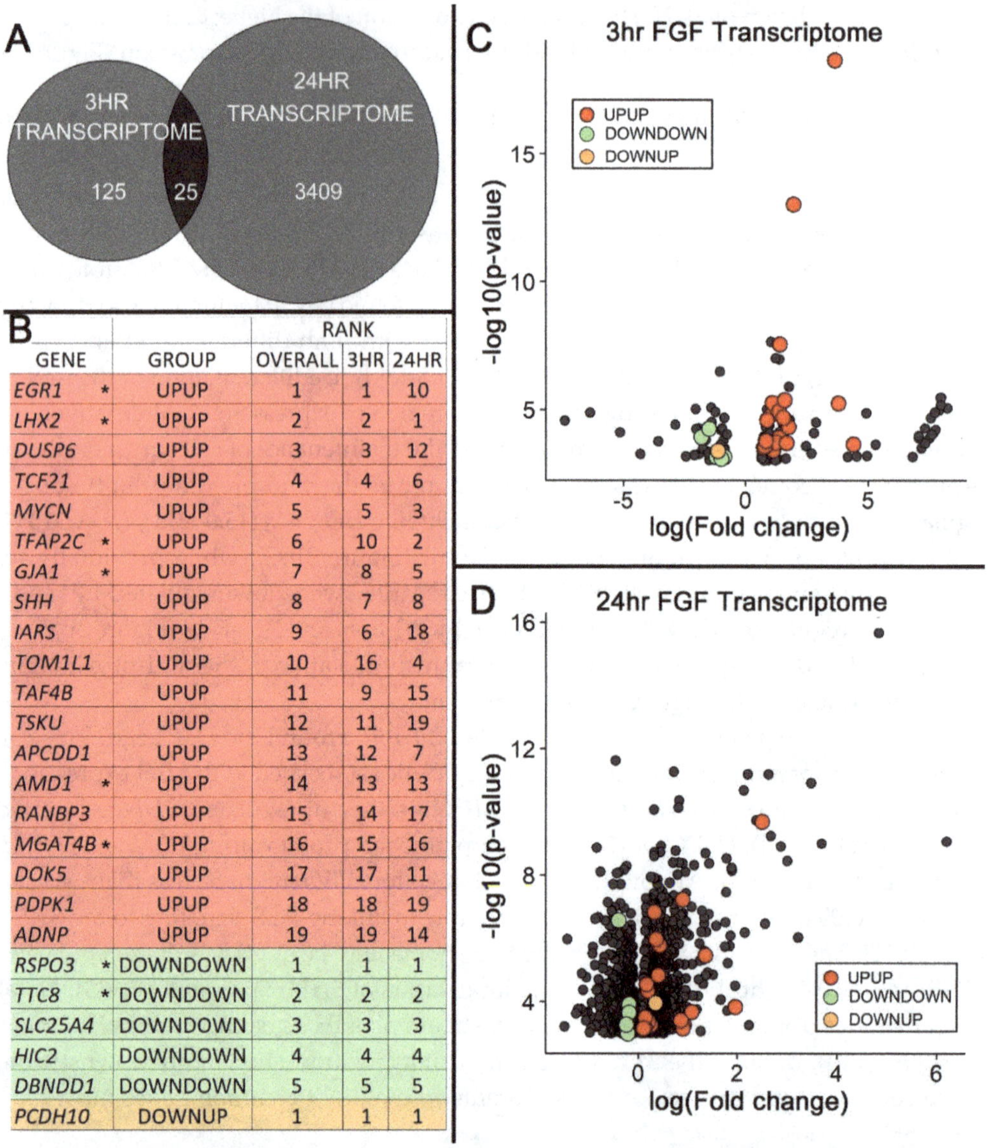

B

GENE		GROUP	RANK OVERALL	RANK 3HR	RANK 24HR
EGR1	*	UPUP	1	1	10
LHX2	*	UPUP	2	2	1
DUSP6		UPUP	3	3	12
TCF21		UPUP	4	4	6
MYCN		UPUP	5	5	3
TFAP2C	*	UPUP	6	10	2
GJA1	*	UPUP	7	8	5
SHH		UPUP	8	7	8
IARS		UPUP	9	6	18
TOM1L1		UPUP	10	16	4
TAF4B		UPUP	11	9	15
TSKU		UPUP	12	11	19
APCDD1		UPUP	13	12	7
AMD1	*	UPUP	14	13	13
RANBP3		UPUP	15	14	17
MGAT4B	*	UPUP	16	15	16
DOK5		UPUP	17	17	11
PDPK1		UPUP	18	18	19
ADNP		UPUP	19	19	14
RSPO3	*	DOWNDOWN	1	1	1
TTC8	*	DOWNDOWN	2	2	2
SLC25A4		DOWNDOWN	3	3	3
HIC2		DOWNDOWN	4	4	4
DBNDD1		DOWNDOWN	5	5	5
PCDH10		DOWNUP	1	1	1

Figure 4. Common transcripts regulated by FGF2 during maintenance and induction of *SHH* expression are likely candidates for FGF-mediated SHH expression. (**A**) The number of FGF-regulated transcripts common to both 3 and 24 h datasets is shown in the overlapping region of the Venn diagram (n = 25). (**B**) Table showing the 25 common targets from the 3 and 24 h transcriptome that are likely candidates for FGF-mediated SHH expression. Targets are ranked based on differential expression (fold change) divided by significance (p-value). Asterisks (*) denote genes associated to *SHH* expression/function. Genes upregulated at both 3 and 24 h are designated with Group ID "UPUP": red (n = 19). Genes downregulated at both time points are designated "DOWNDOWN": green (n = 5), while genes downregulated at 3 h and upregulated at 24 h are designated "DOWNUP": orange (n = 1). (**C**) Distribution (by log10 p-value and log2 fold change) of the 150 genes differentially regulated after 3 h FGF treatment, with the 25 common targets highlighted in color ($p < 0.05$). (**D**) Distribution (by log10 p-value and log2 fold change) of the 3434 genes differentially regulated by 24 h FGF treatment, with the 25 common targets highlighted in color ($p < 0.05$).

3.6.2. Common Targets Associated with SHH Expression

Of the 25 common targets, eight relate to *SHH* expression, function, or signaling (highlighted by asterisks in Figure 4B). Four genes associated with *SHH* expression were upregulated: *EGR1*, *GJA1* (also known as Connexin 43), *LHX2*, and *TFAP2C*. *TTC8*, which is associated with ciliary-associated GLI-Kruppel family member (GLI) processing, was downregulated. Collectively, the regulation of these genes is expected to enhance SHH function/signaling. Paradoxically, *RSPO3*, a gene associated with *SHH* upregulation, was downregulated. *AMD1* and *MGAT4B* were both upregulated and are decreased in Shh-deficient mice [50]. They could therefore be downstream of Shh signaling. Neither *AMD1* nor *MGAT4B* has been well-characterized, however, and they require further investigation.

3.7. LHX2 as an Intermediate in FGF-Regulated SHH Expression

Our screening identified LHX2 as a candidate for mediating FGF regulation of SHH. In the limb, *LHX2* is robustly upregulated by FGF following 3 and 24 h of exposure (Figures 2 and 3). Additionally, *LHX2* is distally restricted subjacent to the AER overlapping the ZPA (Figure 5A). Since Lhx9 appears to play a redundant role in regulating *Shh* expression in mice, we evaluated its expression pattern in the chicken model. In chicken wings, *LHX9* is restricted to the anterior margin of the developing limb mesoderm and does not overlap the ZPA. Moreover, *LHX9* was not significantly upregulated in our transcriptome data and not convincingly upregulated around applied FGF beads (Figure 5B). Thus, in chickens, LHX2, but not LHX9, could function as an intermediate in the FGF-to-SHH pathway.

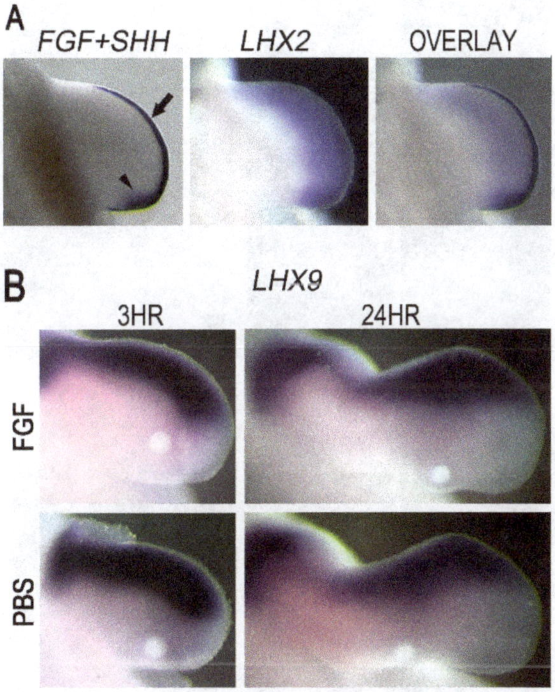

Figure 5. The *LHX2* expression pattern supports a role for maintaining distal posterior *SHH* expression during progressive limb outgrowth. (**A**) Left panel shows endogenous *FGF8* (arrow) and *SHH* (arrowhead) expression in a HH23 limb bud. Middle panel shows endogenous *LHX2* expression. Right panel is a composite of all three expression patterns. *LHX2* expression is restricted to the distal mesoderm in the developing limb subjacent to the *FGF8*-expressing AER. Importantly, *SHH* is expressed only within the posterior boundary of the *LHX2* expression domain. The overlapping expression pattern is consistent with a role for LHX2 in facilitating SHH expression in the developing limb. (**B**) After 3 and 24 h of FGF2 exposure, no appreciable ectopic *LHX9* expression is observed (n = 3 embryos per treatment group for each time point). PBS beads were used as controls. Note that endogenous *LHX9* expression is restricted to the anterior mesoderm.

To determine the role of LHX2 in FGF-regulated *SHH* expression, we electroporated a mouse Lhx2 expression vector into the posterior limb mesoderm adjacent to an FGF- or PBS-soaked bead. Lhx2 was not sufficient to induce *SHH* independently. However, when combined with FGF, Lhx2 increased the expression of *SHH* (2.7-fold by RT-qPCR) and *PTCH2*, a downstream target of SHH signaling (Figure 6). We also examined the loss of LHX2 function using an anti-*LHX2* morpholino (anti-c*LHX2* MO) designed to block translation of the chicken *LHX2* transcript. Electroporation of this morpholino around an FGF bead implant resulted in decreased *SHH* expression (35% by RT-qPCR) and a decrease in *PTCH2*. Targeted SHH knockdown gives a similar decrease in *PTCH2* expression (Figure 6). Additionally, a decrease in LHX2 at the ZPA resulted in decreased limb outgrowth and reduced endogenous *SHH* expression after 24 h (Figure 7). This is consistent with a previous report that suggested that competitive inhibition of Lhx2 activity inhibits limb outgrowth in chicken as well as interfering with the expression of distally restricted genes such as *SHH* [29].

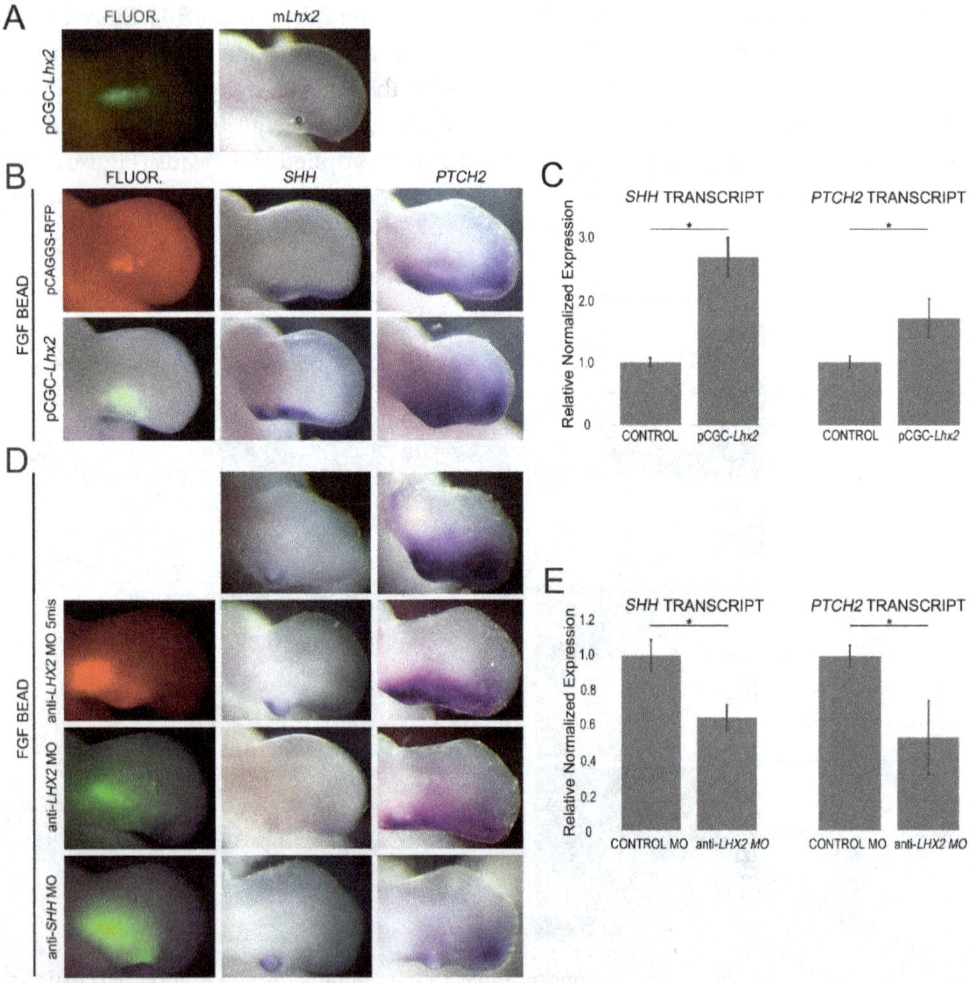

Figure 6. LHX2 regulates SHH expression and function. (**A**) Fluorescence (FLUOR.) and whole mount in situ hybridization (WMISH) pictures confirming ectopic mouse *Lhx2* following electroporation of an Lhx2 expression vector containing a GFP reporter (pCGC-*Lhx2*). "o" is used to identify the oil bubble which is occasionally present after the injection of a plasmid. (**B**) Increasing the level of local LHX2 by electroporation of pCGC-*Lhx2* around an FGF bead leads to an increase in the expression of *SHH* (n = 12) and its downstream target *PTCH2* (n = 8) when compared to an empty vector control (pCAGGS-red fluorescent protein (RFP)). (**C**) RT-qPCR data from two independent experiments reveals a 2.7-fold upregulation of FGF-induced *SHH* and a 1.7-fold upregulation of *PTCH2* transcripts in the presence of additional LHX2 when compared to FGF alone ($p < 0.05$). Tissue samples from 7–10 embryos were pooled

for each treatment group and assayed in triplicate. (**D**) Transfection of an anti-*LHX2* morpholino decreases FGF-induced *SHH* expression (third row; n = 7) when compared to limbs treated with FGF beads only (top row; n = 5) or FGF with a negative control (five-base mismatch anti-*LHX2* MO; second row; n = 5). *PTCH2* expression was also decreased following treatment with the anti-*LHX2* morpholino (n = 10) and mimics the *PTCH2* reduction observed with a knockdown of SHH using an anti-*SHH* MO (n = 5). *SHH* expression is not affected by the electroporation of an anti-*SHH* morpholino designed to block translation of *SHH* transcript (bottom panel; n = 3). (**E**) Following a knockdown of LHX2, *SHH* and *PTCH2* transcript levels were reduced by 35% and 53%, respectively, when compared to the anti-*LHX2* MO five-base mismatch control ($p < 0.05$). Tissue samples from 7–10 embryos were pooled for each treatment group and assayed in triplicate for two independent RT-qPCR runs. * $p < 0.05$.

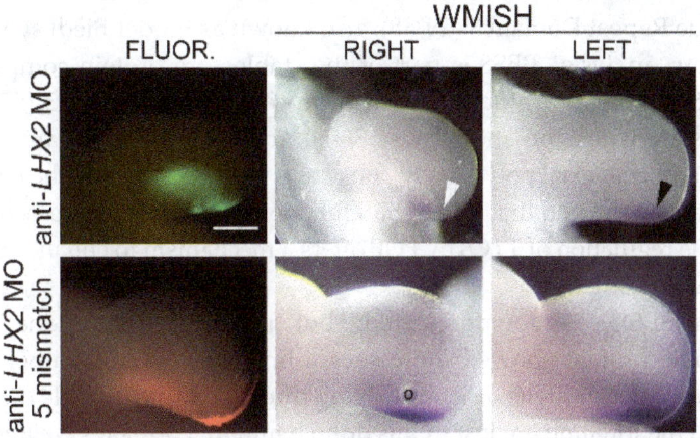

Figure 7. LHX2 knockdown at the ZPA decreases *SHH* expression. Twenty-four hours after electroporation of anti-*LHX2* MO at the ZPA, there is reduced *SHH* expression and retarded limb outgrowth when compared to the contralateral limb (LEFT) and to the negative control (anti-*LHX2* MO five-base mismatch). Reduced *SHH* expression and retarded limb outgrowth were observed in nine out of 10 embryos from two independent experiments. Please note the oil bubble "o" in the bottom row (middle panel). The images of the left contralateral control limbs have been reversed for comparison. Scalebar = 0.25 inches.

4. Discussion

One of the roles of apical ectodermal ridge (AER)-secreted FGFs is the maintenance of *SHH* expression in the subjacent ZPA during limb outgrowth [15]. We were able to maintain the expression of *SHH* in the former ZPA domain by application of FGF proximal to the ZPA. Persistent *SHH* expression was evident within as little as 3 h. FGF was also able to induce *SHH* in non-ZPA domain-related mesoderm, although this required prolonged exposure (24 h) (Figure 1). Cells in the population responding to prolonged FGF are likely to include some prior SHH-expressing cells. Most of the prior SHH-expressing cells, however, are present within the autopod and prospective digits at a comparable stage of development [9]. Irrespective of their prior history, induction or reactivation of cells in the non-ZPA domain requires prolonged FGF exposure to induce *SHH* expression and a mechanism disparate from that of cells residing within the former ZPA domain. By comparing transcriptome data from these two populations of *SHH*-expressing cells, we were able to identify 25 common genes, including 19 synexpressed with *SHH* and five downregulated.

4.1. FGF Regulates Genes that Support SHH Expression and Signaling/Function

Eight of the 25 common targets differentially regulated by FGF exposure were associated with *SHH* expression, signaling, or function (Figure 4). Consistent with our data, EGR1 has been reported to act downstream of FGF signaling [51,52]. Additionally, it serves as a direct transcriptional regulator of *SHH* in glioma cells [53]. In the limb, EGR1 participates in FGF-induced tendon differentiation [52], but its role in FGF-mediated *Shh* expression has yet to be evaluated. FGF upregulated *GJA1*, which is a gene that encodes Connexin 43, a gap junction protein. Gap junction proteins have been reported to relay FGF signals to neighboring cells; moreover, the conditional knockout of Connexin 43 in mice leads to reduced *Shh* expression, limb truncation, and patterning defects [54,55]. Thus, an increase in *GJA1* in both of our datasets suggests that FGF supports *SHH* expression by upregulating intercellular communication. FGF also upregulated *TFAP2C*. TFAP2C is an activating, enhancer-binding protein-2 (AP2) transcription factor expressed subjacent to the AER in normal limbs (Figure 2C). The ZPA regulatory sequence (*ZRS*) contains a potential AP2 binding site. Interestingly, a mutation generating an extra AP2 binding site in the *ZRS* is associated with anterior ectopic *SHH* expression and preaxial polydactyly [56]. Although its expression pattern and role in misregulation of the *ZRS* are intriguing, a role for AP2 transcription factors in normal *SHH* expression has yet to be determined.

Tetratricopeptide Repeat Domain 8 (TTC8), also known as Bardet Biedl syndrome 8 (BBS8), was downregulated in our analysis. BBS8 is part of the stable core protein complex of cilia involved with Smoothened (Smo) ciliary trafficking, which processes Gli2 and Gli3 [57,58]. Processed GLI transcription factors are truncated and repress SHH targets [59]. In humans, TTC8 mutations are associated with pre- and post-axial polydactyly, consistent with activation of the hedgehog pathway [60, 61]. Tayeh and colleagues showed that loss of BBS function in zebrafish increased fin/limb expression of *SHH* [62]. Thus, the regulation of TTC8 by FGF offers a mechanism to enhance the activation of SHH that warrants further investigation.

R-spondin 3 (RSPO3) is downregulated, but potentially functions upstream of SHH. A *Rspo2*/*Rspo3* double mutant showed more severe limb defects than either single mutant with the most anterior and posterior digits missing, leaving three shortened middle digits [46]. The fact that posterior elements in the forelimb are lost in this double mutant correlates well with the phenotype of *SHH*-deficient limbs [7,8,63] and suggests that R-spondin genes could affect *SHH* expression/function. No limb expression patterns were found for *AMD1* and *MGAT4B*, but our data indicate that they are upregulated by FGF signaling. Additionally, *Amd1* and the Mgat4b paralog *Mgat4a* are reported to be decreased in the limbs of Shh-deficient mice [50], indicating that they may be downstream of SHH signaling.

4.2. FGF Regulates Factors that Localize SHH Expression

The *ZRS* is a conserved *cis*-acting regulatory element responsible for limb-specific *SHH* expression [7,37,64,65]. The *ZRS* houses multiple binding sites for the ETS/ETV family of transcription factors. At 3 h, *ELF2* (E74-like ETS transcription factor 2) and *ETV6* were upregulated, while *ETS2* and *ETV5* were upregulated at 24 h. Lettice et al. reported that a balance between occupancy of the ETS/ETV binding sites within the *ZRS* contributes to the expression and location of *Shh* in the limb bud [42]. The upregulation of members of the ETS family in our data supports the notion that FGF regulates ETS/ETV transcription factors to modify *SHH* expression in the developing limb and highlights the possibility that family members other than ETS2 and ETV4/5 [41,42] may be involved.

The TAATTA binding motif for LHX2 [66] is found in the *ZRS* and is conserved across 16 vertebrate species including the human, mouse, and chicken [67]. Of note, other Hox proteins share this binding motif and 5′ Hox genes have been reported to bind the ZRS [33]; therefore, LHX2 binding to this region needs to be confirmed.

4.3. LHX2, but Not LHX9, Regulates FGF-Mediated SHH Expression during Chicken Limb Development

Reports have suggested that LHX2 and a homolog LHX9 may be functionally redundant due to their overlapping expression patterns [30,68,69]. Simultaneous knockout of *Lhx2* and *Lhx9* causes a marked reduction in *Shh* expression and altered limb growth and patterning characterized by oligodactyly, loss of digit morphology, and a shortened limb [30]. In chickens, *LHX2* is expressed in the distal mesoderm subjacent to the AER in a pattern that overlaps the ZPA, while *LHX9* is restricted to the anterior and distal rim of the limb mesoderm distant from the ZPA (Figure 5) [29,70]. Unlike the mouse model, where an *Lhx2/Lhx9* double knockout was necessary to perturb limb outgrowth and *Shh* expression, Rodriguez-Esteban and coworkers showed that a retroviral LHX2-targeted repressor caused limb truncations [29]. We further show that targeted knockdown of LHX2 within the ZPA is sufficient to decrease *SHH* expression and disrupt limb outgrowth (Figure 7). These collective findings indicate a species-specific difference in the function of Lhx family members, but highlight Lhx2 as a common mediator of *Shh* expression.

Although the pattern of *Lhx2* and *Lhx9* expression in mice is consistent with regulation by FGF, Tzchori et al. suggested that Fgf4/8 did not control their expression [30]. In contrast, Yang and colleagues showed that Fgf signaling was required for *Lhx9* expression in mouse limb explants [71]. In chicken, we demonstrate by transcriptome, WMISH, and RT-qPCR that FGF robustly upregulates *LHX2*, but not *LHX9*. *LHX2* overexpression or knockdown in the context of ectopic FGF-bead application also resulted in a robust increase or decrease in *SHH* expression, respectively. Our data indicate that LHX2 is a target of FGF signaling and an intermediate in the FGF to SHH regulatory loop.

Interestingly, Tzchori and colleagues demonstrated that Ldb1, a cofactor of LIM transcription factors, was required for FGF-mediated induction of *Shh*. *Ldb1* is ubiquitously expressed in the limb and is known to associate with a variety of LIM-domain genes [72,73]. The widespread expression of Ldb1 and its required presence to permit Fgf-mediated *Shh* expression points to a cofactor, such as Lhx2, as an intermediate in upregulating *Shh*.

4.4. LHX2 as a Competency Factor for SHH Expression in the Limb

Our data corroborates other reports that indicate LHX2 is necessary for *SHH* expression [30]. Additionally, *LHX2* expression overlaps the ZPA and the proximal extent of *LHX2* expression corresponds to the proximal boundary of the ZPA; beyond this boundary, *SHH* expression wanes. Taken together, LHX2 is likely a competence factor for *SHH*, keeping its expression juxtaposed to the AER during progressive limb outgrowth. Other known competence factors such as *HoxB8* [74], *Hox9* paralogs [34], and *Hand2* [75,76] display expression domains larger than, but inclusive of, the ZPA to permit *SHH* expression, indicating a collective cooperation among competence factors.

Ectopic apical *Shh* expression subjacent to the AER in an *Lhx2*-like pattern has been reported [77,78]. Interestingly, in both reports, there was a reduction in *Gli3*. *Gli3* is expressed throughout the limb, except within the distal posterior mesoderm [79], and together with Alx4 [43], Twist [80], and the Etv proteins [40,80], plays a role in restricting *Shh* expression to this limited posterior domain. Reduction in the repressive activity of Gli3 results in the anterior expansion of Hand2, while *Lhx2* expression in *Gli3*-deficient limb buds remains unchanged [81]. The ectopic apical *SHH* expression pattern overlapped the expanded *Hand2* domain, but was more distally restricted, suggesting that a distally restricted factor, such as Lhx2, was required. We suspect that Lhx2 is a transcription partner with Hand2 and other competency factors that function to regulate and maintain *Shh* transcription.

In conclusion, we identified a common set of genes regulated by FGF with potential to function as intermediates in limb-related FGF-mediated *SHH* expression. Additionally, we have extended the role of LHX2 from previous reports, providing evidence that LHX2 mediates the FGF-to-SHH regulatory loop during limb development.

Supplementary Materials: The Supplementary Materials are available online at http://www.mdpi.com/2221-3759/6/2/13/s1.

Author Contributions: Conceptualization: B.A.W., K.C.O.; investigation: B.A.W., C.U.P., J.M.F., J.M.V.A., K.S.R.-B., D.J.H.K., A.S.C., G.L.M., S.Y., E.D.S.; data analysis: B.A.W., J.M.F., B.R.G., K.C.O.; manuscript preparation: B.A.W., K.C.O., C.U.P.

Funding: This work was funded in part by the Loma Linda University Pathology Endowment Fund.

Acknowledgments: The authors would like to thank Charles Nicolet for his advice and initial alignment of the RNA-seq data; Shubha Tole and Toshi Ohshima for the kind gift and permission to use the LHX2 expression vector (pCGC-LHX2); Clifford Tabin, Philippa Francis-West, Eric Swindell, and Stephen Keyse for their kind gift of plasmids used to generate in situ hybridization probes; and GeneTools, LLC for their aid in the design of the morpholinos.

Conflicts of Interest: The authors declare no conflict of interest.

References

1. Delgado, I.; Torres, M. Coordination of limb development by crosstalk among axial patterning pathways. *Dev. Biol.* **2017**, *429*, 382–386. [CrossRef] [PubMed]
2. Laufer, E.; Nelson, C.E.; Johnson, R.L.; Morgan, B.A.; Tabin, C. Sonic hedgehog and fgf-4 act through a signaling cascade and feedback loop to integrate growth and patterning of the developing limb bud. *Cell* **1994**, *79*, 993–1003. [CrossRef]
3. Niswander, L.; Jeffrey, S.; Martin, G.R.; Tickle, C. A positive feedback loop coordinates growth and patterning in the vertebrate limb. *Nature* **1994**, *371*, 609–612. [CrossRef] [PubMed]
4. Zeller, R.; Haramis, A.G.; Zuniga, A.; McGuigan, C.; Dono, R.; Davidson, G.; Chabanis, S.; Gibson, T. Formin defines a large family of morphoregulatory genes and functions in establishment of the polarising region. *Cell Tissue Res.* **1999**, *296*, 85–93. [CrossRef] [PubMed]
5. Capdevila, J.; Tsukui, T.; Rodriquez Esteban, C.; Zappavigna, V.; Izpisua Belmonte, J.C. Control of vertebrate limb outgrowth by the proximal factor meis2 and distal antagonism of bmps by gremlin. *Mol. Cell* **1999**, *4*, 839–849. [CrossRef]
6. Zuniga, A.; Haramis, A.P.; McMahon, A.P.; Zeller, R. Signal relay by bmp antagonism controls the shh/fgf4 feedback loop in vertebrate limb buds. *Nature* **1999**, *401*, 598–602. [CrossRef] [PubMed]
7. Ros, M.A.; Dahn, R.D.; Fernandez-Teran, M.; Rashka, K.; Caruccio, N.C.; Hasso, S.M.; Bitgood, J.J.; Lancman, J.J.; Fallon, J.F. The chick oligozeugodactyly (ozd) mutant lacks sonic hedgehog function in the limb. *Development* **2003**, *130*, 527–537. [CrossRef] [PubMed]
8. Chiang, C.; Litingtung, Y.; Harris, M.P.; Simandl, B.K.; Li, Y.; Beachy, P.A.; Fallon, J.F. Manifestation of the limb prepattern: Limb development in the absence of sonic hedgehog function. *Dev. Biol.* **2001**, *236*, 421–435. [CrossRef] [PubMed]
9. Harfe, B.D.; Scherz, P.J.; Nissim, S.; Tian, H.; McMahon, A.P.; Tabin, C.J. Evidence for an expansion-based temporal shh gradient in specifying vertebrate digit identities. *Cell* **2004**, *118*, 517–528. [CrossRef] [PubMed]
10. Ingham, P.W.; Placzek, M. Orchestrating ontogenesis: Variations on a theme by sonic hedgehog. *Nat. Rev. Genet.* **2006**, *7*, 841–850. [CrossRef] [PubMed]
11. Fallon, J.F.; Lopez, A.; Ros, M.A.; Savage, M.P.; Olwin, B.B.; Simandl, B.K. Fgf-2: Apical ectodermal ridge growth signal for chick limb development. *Science* **1994**, *264*, 104–107. [CrossRef] [PubMed]
12. Li, S.; Anderson, R.; Reginelli, A.D.; Muneoka, K. Fgf-2 influences cell movements and gene expression during limb development. *J. Exp. Zool.* **1996**, *274*, 234–247. [CrossRef]
13. Niswander, L.; Tickle, C.; Vogel, A.; Booth, I.; Martin, G.R. Fgf-4 replaces the apical ectodermal ridge and directs outgrowth and patterning of the limb. *Cell* **1993**, *75*, 579–587. [CrossRef]
14. Vogel, A.; Tickle, C. Fgf-4 maintains polarizing activity of posterior limb bud cells in vivo and in vitro. *Development* **1993**, *119*, 199–206. [PubMed]
15. Crossley, P.H.; Minowada, G.; MacArthur, C.A.; Martin, G.R. Roles for fgf8 in the induction, initiation, and maintenance of chick limb development. *Cell* **1996**, *84*, 127–136. [CrossRef]
16. Vogel, A.; Rodriguez, C.; Izpisua-Belmonte, J.C. Involvement of fgf-8 in initiation, outgrowth and patterning of the vertebrate limb. *Development* **1996**, *122*, 1737–1750. [PubMed]
17. Savage, M.P.; Hart, C.E.; Riley, B.B.; Sasse, J.; Olwin, B.B.; Fallon, J.F. Distribution of fgf-2 suggests it has a role in chick limb bud growth. *Dev. Dyn.* **1993**, *198*, 159–170. [CrossRef] [PubMed]
18. Becic, T.; Kero, D.; Vukojevic, K.; Mardesic, S.; Saraga-Babic, M. Growth factors fgf8 and fgf2 and their receptor fgfr1, transcriptional factors msx-1 and msx-2, and apoptotic factors p19 and rip5 participate in the early human limb development. *Acta Histochem.* **2018**, *120*, 205–214. [CrossRef] [PubMed]
19. Harduf, H.; Halperin, E.; Reshef, R.; Ron, D. Sef is synexpressed with fgfs during chick embryogenesis and its expression is differentially regulated by fgfs in the developing limb. *Dev. Dyn.* **2005**, *233*, 301–312. [CrossRef] [PubMed]

20. Verheyden, J.M.; Lewandoski, M.; Deng, C.; Harfe, B.D.; Sun, X. Conditional inactivation of fgfr1 in mouse defines its role in limb bud establishment, outgrowth and digit patterning. *Development* **2005**, *132*, 4235–4245. [CrossRef] [PubMed]

21. Sheeba, C.J.; Andrade, R.P.; Duprez, D.; Palmeirim, I. Comprehensive analysis of fibroblast growth factor receptor expression patterns during chick forelimb development. *Int. J. Dev. Biol.* **2010**, *54*, 1517–1526. [CrossRef] [PubMed]

22. Oberg, K.C.; Pira, C.U.; Revelli, J.P.; Ratz, B.; Aguilar-Cordova, E.; Eichele, G. Efficient ectopic gene expression targeting chick mesoderm. *Dev. Dyn.* **2002**, *224*, 291–302. [CrossRef] [PubMed]

23. Feenstra, J.M.; Kanaya, K.; Pira, C.U.; Hoffman, S.E.; Eppey, R.J.; Oberg, K.C. Detection of genes regulated by lmx1b during limb dorsalization. *Dev. Growth Differ.* **2012**, *54*, 451–462. [CrossRef] [PubMed]

24. Barrett, T.; Wilhite, S.E.; Ledoux, P.; Evangelista, C.; Kim, I.F.; Tomashevsky, M.; Marshall, K.A.; Phillippy, K.H.; Sherman, P.M.; Holko, M.; et al. Ncbi geo: Archive for functional genomics data sets–update. *Nucl. Acids Res.* **2013**, *41*, D991–D995. [CrossRef] [PubMed]

25. Yamada, M.; Szendro, P.I.; Prokscha, A.; Schwartz, R.J.; Eichele, G. Evidence for a role of smad6 in chick cardiac development. *Dev. Biol.* **1999**, *215*, 48–61. [CrossRef] [PubMed]

26. Subramanian, L.; Sarkar, A.; Shetty, A.S.; Muralidharan, B.; Padmanabhan, H.; Piper, M.; Monuki, E.S.; Bach, I.; Gronostajski, R.M.; Richards, L.J.; et al. Transcription factor lhx2 is necessary and sufficient to suppress astrogliogenesis and promote neurogenesis in the developing hippocampus. *Proc. Natl. Acad. Sci. USA* **2011**, *108*, E265–E274. [CrossRef] [PubMed]

27. Moulton, J.D.; Yan, Y.L. Using Morpholinos to control gene expression. *Curr. Protoc. Mol. Biol.* **2008**, *83*, 26.8.1–26.8.29.

28. Eisen, J.S.; Smith, J.C. Controlling morpholino experiments: Don't stop making antisense. *Development* **2008**, *135*, 1735–1743. [CrossRef] [PubMed]

29. Rodriguez-Esteban, C.; Schwabe, J.W.; Pena, J.D.; Rincon-Limas, D.E.; Magallon, J.; Botas, J.; Izpisua Belmonte, J.C. Lhx2, a vertebrate homologue of apterous, regulates vertebrate limb outgrowth. *Development* **1998**, *125*, 3925–3934. [PubMed]

30. Tzchori, I.; Day, T.F.; Carolan, P.J.; Zhao, Y.; Wassif, C.A.; Li, L.; Lewandoski, M.; Gorivodsky, M.; Love, P.E.; Porter, F.D.; et al. Lim homeobox transcription factors integrate signaling events that control three-dimensional limb patterning and growth. *Development* **2009**, *136*, 1375–1385. [CrossRef] [PubMed]

31. Bastida, M.F.; Sheth, R.; Ros, M.A. A bmp-shh negative-feedback loop restricts shh expression during limb development. *Development* **2009**, *136*, 3779–3789. [CrossRef] [PubMed]

32. Norrie, J.L.; Lewandowski, J.P.; Bouldin, C.M.; Amarnath, S.; Li, Q.; Vokes, M.S.; Ehrlich, L.I.R.; Harfe, B.D.; Vokes, S.A. Dynamics of bmp signaling in limb bud mesenchyme and polydactyly. *Dev. Biol.* **2014**, *393*, 270–281. [CrossRef] [PubMed]

33. Capellini, T.D.; Di Giacomo, G.; Salsi, V.; Brendolan, A.; Ferretti, E.; Srivastava, D.; Zappavigna, V.; Selleri, L. Pbx1/pbx2 requirement for distal limb patterning is mediated by the hierarchical control of hox gene spatial distribution and shh expression. *Development* **2006**, *133*, 2263–2273. [CrossRef] [PubMed]

34. Xu, B.; Wellik, D.M. Axial hox9 activity establishes the posterior field in the developing forelimb. *Proc. Natl. Acad. Sci. USA* **2011**, *108*, 4888–4891. [CrossRef] [PubMed]

35. Matsubara, H.; Saito, D.; Abe, G.; Yokoyama, H.; Suzuki, T.; Tamura, K. Upstream regulation for initiation of restricted shh expression in the chick limb bud. *Dev. Dyn.* **2017**, *246*, 417–430. [CrossRef] [PubMed]

36. Osterwalder, M.; Speziale, D.; Shoukry, M.; Mohan, R.; Ivanek, R.; Kohler, M.; Beisel, C.; Wen, X.; Scales, S.J.; Christoffels, V.M.; et al. Hand2 targets define a network of transcriptional regulators that compartmentalize the early limb bud mesenchyme. *Dev. Cell* **2014**, *31*, 345–357. [CrossRef] [PubMed]

37. Lettice, L.; Heaney, S.; Hill, R. 2 Preaxial polydactyly in human and mouse: Regulatory anomalies in digit patterning. *J. Anat.* **2002**, *201*, 417. [PubMed]

38. Church, V.L.; Francis-West, P. Wnt signalling during limb development. *Int. J. Dev. Biol.* **2002**, *46*, 927–936. [PubMed]

39. Geetha-Loganathan, P.; Nimmagadda, S.; Scaal, M. Wnt signaling in limb organogenesis. *Organogenesis* **2008**, *4*, 109–115. [CrossRef] [PubMed]

40. Mao, J.; McGlinn, E.; Huang, P.; Tabin, C.J.; McMahon, A.P. Fgf-dependent etv4/5 activity is required for posterior restriction of sonic hedgehog and promoting outgrowth of the vertebrate limb. *Dev. Cell* **2009**, *16*, 600–606. [CrossRef] [PubMed]

41. Zhang, Z.; Verheyden, J.M.; Hassell, J.A.; Sun, X. Fgf-regulated etv genes are essential for repressing shh expression in mouse limb buds. *Dev. Cell* **2009**, *16*, 607–613. [CrossRef] [PubMed]
42. Lettice, L.A.; Williamson, I.; Wiltshire, J.H.; Peluso, S.; Devenney, P.S.; Hill, A.E.; Essafi, A.; Hagman, J.; Mort, R.; Grimes, G.; et al. Opposing functions of the ets factor family define shh spatial expression in limb buds and underlie polydactyly. *Dev. Cell* **2012**, *22*, 459–467. [CrossRef] [PubMed]
43. Qu, S.; Niswender, K.D.; Ji, Q.; van der Meer, R.; Keeney, D.; Magnuson, M.A.; Wisdom, R. Polydactyly and ectopic zpa formation in alx-4 mutant mice. *Development* **1997**, *124*, 3999–4008. [PubMed]
44. Shimomura, Y.; Agalliu, D.; Vonica, A.; Luria, V.; Wajid, M.; Baumer, A.; Belli, S.; Petukhova, L.; Schinzel, A.; Brivanlou, A.H.; et al. Apcdd1 is a novel wnt inhibitor mutated in hereditary hypotrichosis simplex. *Nature* **2010**, *464*, 1043–1047. [CrossRef] [PubMed]
45. Hendriksen, J.; Fagotto, F.; van der Velde, H.; van Schie, M.; Noordermeer, J.; Fornerod, M. Ranbp3 enhances nuclear export of active (beta)-catenin independently of crm1. *J. Cell Biol.* **2005**, *171*, 785–797. [CrossRef] [PubMed]
46. Neufeld, S.; Rosin, J.M.; Ambasta, A.; Hui, K.; Shaneman, V.; Crowder, R.; Vickerman, L.; Cobb, J. A conditional allele of rspo3 reveals redundant function of r-spondins during mouse limb development. *Genesis* **2012**, *50*, 741–749. [CrossRef] [PubMed]
47. Zhang, M.; Zhang, P.; Liu, Y.; Lv, L.; Zhang, X.; Liu, H.; Zhou, Y. Rspo3-lgr4 regulates osteogenic differentiation of human adipose-derived stem cells via erk/fgf signalling. *Sci. Rep.* **2017**, *7*, 42841. [CrossRef] [PubMed]
48. Gu, Y.; Wang, Q.; Guo, K.; Qin, W.; Liao, W.; Wang, S.; Ding, Y.; Lin, J. Tusc3 promotes colorectal cancer progression and epithelial-mesenchymal transition (emt) through wnt/beta-catenin and mapk signalling. *J. Pathol.* **2016**, *239*, 60–71. [CrossRef] [PubMed]
49. Ten Berge, D.; Brugmann, S.A.; Helms, J.A.; Nusse, R. Wnt and fgf signals interact to coordinate growth with cell fate specification during limb development. *Development* **2008**, *135*, 3247–3257. [CrossRef] [PubMed]
50. Probst, S.; Kraemer, C.; Demougin, P.; Sheth, R.; Martin, G.R.; Shiratori, H.; Hamada, H.; Iber, D.; Zeller, R.; Zuniga, A. Shh propagates distal limb bud development by enhancing cyp26b1-mediated retinoic acid clearance via aer-fgf signalling. *Development* **2011**, *138*, 1913–1923. [CrossRef] [PubMed]
51. Han, P.; Guerrero-Netro, H.; Estienne, A.; Cao, B.; Price, C.A. Regulation and action of early growth response 1 in bovine granulosa cells. *Reproduction* **2017**, *154*, 547–557. [CrossRef] [PubMed]
52. Lejard, V.; Blais, F.; Guerquin, M.J.; Bonnet, A.; Bonnin, M.A.; Havis, E.; Malbouyres, M.; Bidaud, C.B.; Maro, G.; Gilardi-Hebenstreit, P.; et al. Egr1 and egr2 involvement in vertebrate tendon differentiation. *J. Biol. Chem.* **2011**, *286*, 5855–5867. [CrossRef] [PubMed]
53. Sakakini, N.; Turchi, L.; Bergon, A.; Holota, H.; Rekima, S.; Lopez, F.; Paquis, P.; Almairac, F.; Fontaine, D.; Baeza-Kallee, N.; et al. A positive feed-forward loop associating egr1 and pdgfa promotes proliferation and self-renewal in glioblastoma stem cells. *J. Biol. Chem.* **2016**, *291*, 10684–10699. [CrossRef] [PubMed]
54. Law, L.Y.; Lin, J.S.; Becker, D.L.; Green, C.R. Knockdown of connexin43-mediated regulation of the zone of polarizing activity in the developing chick limb leads to digit truncation. *Dev. Growth Differ.* **2002**, *44*, 537–547. [CrossRef] [PubMed]
55. Dobrowolski, R.; Hertig, G.; Lechner, H.; Worsdorfer, P.; Wulf, V.; Dicke, N.; Eckert, D.; Bauer, R.; Schorle, H.; Willecke, K. Loss of connexin43-mediated gap junctional coupling in the mesenchyme of limb buds leads to altered expression of morphogens in mice. *Hum. Mol. Genet.* **2009**, *18*, 2899–2911. [CrossRef] [PubMed]
56. Fuxman Bass, J.I.; Sahni, N.; Shrestha, S.; Garcia-Gonzalez, A.; Mori, A.; Bhat, N.; Yi, S.; Hill, D.E.; Vidal, M.; Walhout, A.J.M. Human gene-centered transcription factor networks for enhancers and disease variants. *Cell* **2015**, *161*, 661–673. [CrossRef] [PubMed]
57. Zhang, Q.; Seo, S.; Bugge, K.; Stone, E.M.; Sheffield, V.C. Bbs proteins interact genetically with the ift pathway to influence shh-related phenotypes. *Hum. Mol. Genet.* **2012**, *21*, 1945–1953. [CrossRef] [PubMed]
58. Haycraft, C.J.; Banizs, B.; Aydin-Son, Y.; Zhang, Q.; Michaud, E.J.; Yoder, B.K. Gli2 and gli3 localize to cilia and require the intraflagellar transport protein polaris for processing and function. *PLoS Genet.* **2005**, *1*, e53. [CrossRef] [PubMed]
59. Sheeba, C.J.; Andrade, R.P.; Palmeirim, I. Getting a handle on embryo limb development: Molecular interactions driving limb outgrowth and patterning. *Semin. Cell Dev. Biol.* **2016**, *49*, 92–101. [CrossRef] [PubMed]

60. Manouvrier-Hanu, S.; Moerman, A.; Lefevre, J. Bardet-biedl syndrome with preaxial polydactyly. *Am. J. Med. Genet.* **1999**, *84*, 75. [CrossRef]
61. M'Hamdi, O.; Ouertani, I.; Chaabouni-Bouhamed, H. Update on the genetics of bardet-biedl syndrome. *Mol. Syndromol.* **2014**, *5*, 51–56. [CrossRef] [PubMed]
62. Tayeh, M.K.; Yen, H.J.; Beck, J.S.; Searby, C.C.; Westfall, T.A.; Griesbach, H.; Sheffield, V.C.; Slusarski, D.C. Genetic interaction between bardet-biedl syndrome genes and implications for limb patterning. *Hum. Mol. Genet.* **2008**, *17*, 1956–1967. [CrossRef] [PubMed]
63. Kraus, P.; Fraidenraich, D.; Loomis, C.A. Some distal limb structures develop in mice lacking sonic hedgehog signaling. *Mech. Dev.* **2001**, *100*, 45–58. [CrossRef]
64. Smyth, J.R., Jr.; Sreekumar, G.P.; Coyle, C.A.; Bitgood, J.J. A new recessive ametapodia mutation in the chicken (gallus domesticus). *J. Hered.* **2000**, *91*, 340–342. [CrossRef] [PubMed]
65. Sharpe, J.; Lettice, L.; Hecksher-Sorensen, J.; Fox, M.; Hill, R.; Krumlauf, R. Identification of sonic hedgehog as a candidate gene responsible for the polydactylous mouse mutant sasquatch. *Curr. Biol.* **1999**, *9*, 97–100. [CrossRef]
66. Roberson, M.S.; Schoderbek, W.E.; Tremml, G.; Maurer, R.A. Activation of the glycoprotein hormone alpha-subunit promoter by a lim-homeodomain transcription factor. *Mol. Cell Biol.* **1994**, *14*, 2985–2993. [CrossRef] [PubMed]
67. Kvon, E.Z.; Kamneva, O.K.; Melo, U.S.; Barozzi, I.; Osterwalder, M.; Mannion, B.J.; Tissieres, V.; Pickle, C.S.; Plajzer-Frick, I.; Lee, E.A.; et al. Progressive loss of function in a limb enhancer during snake evolution. *Cell* **2016**, *167*, 633–642. [CrossRef] [PubMed]
68. Bertuzzi, S.; Porter, F.D.; Pitts, A.; Kumar, M.; Agulnick, A.; Wassif, C.; Westphal, H. Characterization of lhx9, a novel lim/homeobox gene expressed by the pioneer neurons in the mouse cerebral cortex. *Mech. Dev.* **1999**, *81*, 193–198. [CrossRef]
69. Peukert, D.; Weber, S.; Lumsden, A.; Scholpp, S. Lhx2 and lhx9 determine neuronal differentiation and compartition in the caudal forebrain by regulating wnt signaling. *PLoS Biol.* **2011**, *9*, e1001218. [CrossRef] [PubMed]
70. Nohno, T.; Kawakami, Y.; Wada, N.; Ishikawa, T.; Ohuchi, H.; Noji, S. Differential expression of the two closely related lim-class homeobox genes lh-2a and lh-2b during limb development. *Biochem. Biophys. Res. Commun.* **1997**, *238*, 506–511. [CrossRef] [PubMed]
71. Yang, Y.; Wilson, M.J. Lhx9 gene expression during early limb development in mice requires the fgf signalling pathway. *Gene Expr. Patterns* **2015**, *19*, 45–51. [CrossRef] [PubMed]
72. Tucker, A.S.; Al Khamis, A.; Ferguson, C.A.; Bach, I.; Rosenfeld, M.G.; Sharpe, P.T. Conserved regulation of mesenchymal gene expression by fgf-8 in face and limb development. *Development* **1999**, *126*, 221–228. [PubMed]
73. Jurata, L.W.; Pfaff, S.L.; Gill, G.N. The nuclear lim domain interactor nli mediates homo- and heterodimerization of lim domain transcription factors. *J. Biol. Chem.* **1998**, *273*, 3152–3157. [CrossRef] [PubMed]
74. Lu, H.C.; Revelli, J.P.; Goering, L.; Thaller, C.; Eichele, G. Retinoid signaling is required for the establishment of a zpa and for the expression of hoxb-8, a mediator of zpa formation. *Development* **1997**, *124*, 1643–1651. [PubMed]
75. Charite, J.; McFadden, D.G.; Olson, E.N. The bhlh transcription factor dhand controls sonic hedgehog expression and establishment of the zone of polarizing activity during limb development. *Development* **2000**, *127*, 2461–2470. [PubMed]
76. Fernandez-Teran, M.; Piedra, M.E.; Kathiriya, I.S.; Srivastava, D.; Rodriguez-Rey, J.C.; Ros, M.A. Role of dhand in the anterior-posterior polarization of the limb bud: Implications for the sonic hedgehog pathway. *Development* **2000**, *127*, 2133–2142. [PubMed]
77. Chen, Y.; Knezevic, V.; Ervin, V.; Hutson, R.; Ward, Y.; Mackem, S. Direct interaction with hoxd proteins reverses gli3-repressor function to promote digit formation downstream of shh. *Development* **2004**, *131*, 2339–2347. [CrossRef] [PubMed]
78. Zhulyn, O.; Li, D.; Deimling, S.; Vakili, N.A.; Mo, R.; Puviindran, V.; Chen, M.H.; Chuang, P.T.; Hopyan, S.; Hui, C.C. A switch from low to high shh activity regulates establishment of limb progenitors and signaling centers. *Dev. Cell* **2014**, *29*, 241–249. [CrossRef] [PubMed]

79. Abbasi, A.A.; Paparidis, Z.; Malik, S.; Bangs, F.; Schmidt, A.; Koch, S.; Lopez-Rios, J.; Grzeschik, K.H. Human intronic enhancers control distinct sub-domains of gli3 expression during mouse cns and limb development. *BMC Dev. Biol.* **2010**, *10*, 44. [CrossRef] [PubMed]
80. Zhang, Z.; Sui, P.; Dong, A.; Hassell, J.; Cserjesi, P.; Chen, Y.T.; Behringer, R.R.; Sun, X. Preaxial polydactyly: Interactions among etv, twist1 and hand2 control anterior-posterior patterning of the limb. *Development* **2010**, *137*, 3417–3426. [CrossRef] [PubMed]
81. Yokoyama, S.; Furukawa, S.; Kitada, S.; Mori, M.; Saito, T.; Kawakami, K.; Belmonte, J.C.I.; Kawakami, Y.; Ito, Y.; Sato, T.; et al. Analysis of transcription factors expressed at the anterior mouse limb bud. *PLoS ONE* **2017**, *12*, e0175673. [CrossRef] [PubMed]

9

The Impact of Two Different Cold-Extruded Feeds and Feeding Regimens on Zebrafish Survival, Growth and Reproductive Performance

Joana F. Monteiro *,†, Sandra Martins †, Matheus Farias, Telma Costa and Ana Catarina Certal *

Fish Platform, Champalimaud Center for the Unknown, 1400-038 Lisbon, Portugal;
sandra.martins@neuro.fchampalimaud.org (S.M.); matheus.farias@neuro.fchampalimaud.org (M.F.);
telma.costa@iol.pt (T.C.)
* Correspondence: joana.monteiro@neuro.fchampalimaud.org (J.F.M.);
catarina.certal@fundacaochampalimaud.pt (A.C.C.)
† These authors contributed equally to this work.

Academic Editors: Przemko Tylzanowski, Lisa Maves and Mark W. Majesky

Abstract: Zebrafish (*Danio rerio*) is one of the top model organisms used in biomedical research. Therefore, it is fundamental that zebrafish facilities continuously improve husbandry methods to provide fish with the best physiological and welfare conditions that suit each experimental purpose. Nutrition is a husbandry aspect that needs further optimization, as it greatly affects growth, reproduction, health and behaviour. Here, we have compared the impact of different feeding regimens on zebrafish survival, growth and reproductive performance. Mutant and wild-type zebrafish were raised using several combinations of two cold-extruded processed feeds—Skretting®GemmaMicro and Sparos®Zebrafeed—and one live feed (rotifers). Zebrafeed® outperformed GemmaMicro® in terms of survival rate, and embryo viability was also higher when the spawners were fed with Zebrafeed® either from larval stage or upon sexual maturation. In contrast, GemmaMicro® favoured growth, both in size and weight. The use of rotifers until 60 days post-fertilization improved survival of fish co-fed with GemmaMicro®, while delaying their growth. Zebrafeed® performance was not affected by co-feeding rotifers. Overall, we showed that different nutritional formulas affect physiological parameters, allowing for the establishment of feeding protocols adapted to the objectives of each facility. At the same time, we validated Skretting®GemmaMicro and Sparos®Zebrafeed as two commercially available feeds that are well suited for zebrafish nutrition in a laboratory environment.

Keywords: zebrafish; dietary regimen; rotifers; feeds; survival; growth; reproductive performance

1. Introduction

Research in different models has historically promoted the importance of nutrition in outcomes related to development, health, disease and responses to toxic compounds in food and the environment [1–3]. Yet, there is a striking absence of literature showing the role of nutrition as an important regulator of health and disease in zebrafish. If nutrition is to be accepted as a critical part of health management, it is imperative to consider the typical end goal of fish production in a research setting: to provide fish that are clinically healthy and representative of animals exhibiting normal physiology [4]. Although the known nutritional requirements of other cyprinid fish can be an acceptable approach to zebrafish requirements [5,6], this does not take into account some particularities of the species, such as high growth rate, infinite growth, continuous egg production after maturation. The feeds and feeding regimens implemented by research laboratories for rearing zebrafish are varied [7]. In some cases, feeds and feeding regimens are implemented without a formal evaluation [8]. While the world aquaculture relies on fish meal (FM) as the main protein source in aquafeeds [9], research over the last decade has resulted in a significant decrease in the portion of FM in fish feeds, and there are species-specific limits in the FM replacement rate by, for example, soybean (SBM), above which irreversible damages occur affecting fish growth performance and welfare [10,11]. Therefore, meeting the specific nutritional requirements of zebrafish will potentiate growth and promote an optimal physiological status, minimizing undesirable effects and allowing for more consistent results [12]. The use of purified diets with defined ingredients facilitates the determination of the precise nutritional requirements, promoting optimal performance in fish stocks [13].

Zebrafish is an established model species and a powerful tool to study the modulation of reproductive processes through broodstock nutrition [14]. Proper management of broodstock nutrition encompasses providing the right amounts of the proper specialized nutrients, such as protein, essential fatty acids, vitamin E, vitamin C, carotenoids and phosphoglycerides during the different developmental stages, optimizing gonadal development and fecundity [15–23].

Having alternative suppliers of live and processed feeds, which results in fish with high, optimized and standardized growth, survival and breeding rates, is a management requisite for zebrafish facilities. Comparing different feeds tested in previous studies [14,24] that have demonstrated good impacts on growth and reproductive performance is the first step towards achieving this. This study aimed to understand which combination of commercial processed feeds, Skretting®GemmaMicro® and/or Sparos®Zebrafeed®, complemented with Type "L" saltwater live rotifers *Brachionus plicatilis*, is more suitable for improved survival, growth and reproductive performance.

2. Materials and Methods

2.1. Zebrafish Strains and Embryo Generation

In this study, we used two strains of zebrafish that are widely used in biomedical research: wild-type AB and non-pigmented Nacre mutant (*mitfa* −/−). Embryos were obtained by overnight crossing 8–10 females with 4–6 males of each strain stock in 1.7 L sloping breeding tanks (Tecniplast®). The housing and husbandry of all animals were performed as published [25], and dietary adaptations were in accordance with the specifications of the present study (Table 1). Briefly, all embryos were collected, placed in 90 mm diameter petri dishes with standard embryo medium [26] and kept at 28 °C in an incubator with a photoperiod of 14 h:10 h/light:dark up to 5 days post-fertilization (dpf). Early 6 dpf larvae were transferred to 3.5 L tanks at a density of 35 fish/tank, each containing 400 mL of fish water and 100 mL of type "L" saltwater rotifers (*Brachionus plicatilis*) solution (0.5 mL of RG Complete™, ~800–900 rotifers/mL, conductivity ~3 ppt). At 8 dpf, water flow was started through the 3.5 L tanks, and larvae began to be fed according to their experimental group (Table 1).

Table 1. Experimental dietary groups, starting at 8 days post-fertilization (dpf). The number of daily meals of each feed is listed as well as the size of the grains of processed feeds.

	Age (dpf)	Skretting®GemmaMicro	Sparos®Zebrafeed	Live Feed
Experimental Dietary Group 1	8–15	-	1× 100–200	3× Rotifers
	15–30	-	1× 100–200	3× Rotifers
	30–60	-	1× 200–400	3× Rotifers
	60–90	-	3× 200–400	-
	>90	-	2× 400–600	-
Experimental Dietary Group 2	8–15	1× 150	-	3× Rotifers
	15–30	1× 150	-	3× Rotifers
	30–60	1× 300	-	3× Rotifers
	60–90	3× 300	-	-
	>90	2× 500	-	-

Table 1. *Cont.*

	Age (dpf)	Skretting®GemmaMicro	Sparos®Zebrafeed	Live Feed
Experimental Dietary Group 3	8–15	-	1× 100–200	3× Rotifers
	15–30	-	4× 100–200	-
	30–60	-	4× 200–400	-
	60–90	-	3× 200–400	-
	>90	-	2× 400–600	-
Experimental Dietary Group 4	8–15	1× 150		3× Rotifers
	15–30	4× 150		-
	30–60	4× 300		-
	60–90	3× 300		-
	>90	2× 500		-
Experimental Dietary Group 5	8–15	1× 150	-	3× Rotifers
	15–30	1× 150	-	3× Rotifers
	30–60	1× 300	-	3× Rotifers
	60–90	-	3× 200–400	-
	>90	-	2× 400–600	-

2.2. *Experimental Dietary Groups*

Five experimental dietary groups were designed using different combinations of processed feeds (Skretting®GemmaMicro®; Sparos®Zebrafeed®) and a live feed (type "L" saltwater rotifers, *B. plicatilis*) (Table 1). The nutritional compositions of the processed feeds are described in Table 2. For each dietary group, there were 4 replicate tanks per strain. The size of the processed feed particles varied according to the fish developmental stage—using a seed dispenser, we provided Skretting®Gemma Micro® 150, 300 or 500 or Sparos®Zebrafeed® 100–200, 200–400 or 400–600 to animals at ages <30 dpf, 30–90 dpf and >90 dpf, respectively. Both dry feeds are easily distinguished by their colour—GemmaMicro® grains are green, whereas Zebrafeed® particles are orange red. Rotifer cultures were maintained as described [25], including feeding with refrigerated algae (RGcomplete®; Reed Mariculture). Regardless of the dietary regimen, all fish were fed 4×/day between 8 dpf and 60 dpf, 3×/day from 60 to 90 dpf and 2×/day from 90 dpf onwards. On weekends and holidays, animals were fed 1×/day with the processed feed of the corresponding experimental group. All tanks were given a similar volume of rotifer solution and a similar amount of processed feed, always in excess so that the feed quantity would not be a limiting factor. In order to avoid the impact of survival on growth, fish density/tank was readjusted at 30 dpf within each dietary group. To achieve this, fish from tanks with higher density (higher survival) were transferred to tanks with lower density (lower survival) resulting in a similar number of fish/tank.

Table 2. Nutritional composition of the processed feeds tested. The nutritional content was the same for all grain sizes. Information was gathered from the products' labels.

	Skretting® Gemma Micro (%)	Sparos® Zebrafeed (%)
Protein	59	63
Fat and Oils	14	14
Fiber	0.2	1.8
Ash	14	12
Phosphorus	1.3	2.3
Main ingredients	Fish meal, lecithin, wheat gluten, dried seaweed, fish oil, maize starch, vitamins and minerals	Squid meal, wheat gluten, fish meal, fish protein concentrate, krill meal, pea protein concentrate, starch, lecithin, krill oil, fish oil.

2.3. *Fish Survival Rate*

The total number of fish in each tank was counted at the end of larval stage (30 dpf) and during adulthood (90 dpf), by manually transferring animals one by one to new tanks with a fishing net. The survival rates for each dietary treatment were calculated at 30 and 90 dpf as the total number of living fish in each dietary treatment per total number at the beginning of the experimental assay.

Fish welfare was checked twice/day by visual inspection for behavioural and/or physical abnormalities [27]. Carcasses were removed immediately upon detection and discarded as described [25].

2.4. Growth

Fish growth was examined using fork length (linear measurement) and dry weight (mass measurement). Both parameters were evaluated at 30 dpf and 90 dpf, using one tank with Nacre fish and another with AB fish for each dietary group. To determine the fork length, all fish were anesthetized with 0.16 g/L Tricaine (MS-222), photographed with a SONY® SteadyShot (DSC-W510) and measured from snout to tail split using ImageJ software (v. 1.45s, Wayne Rasband, National Institutes of Health, Bethesda, MD, USA). To estimate the dry weight, fish were euthanized with 4 g/L Tricaine and individually stored in 1.5 mL microcentrifuge plastic tubes at −20 °C. They were then lyophilized for 48 h, and the dried residue was weighed on a precision scale.

2.5. Fish Sex Ratio

Fish were individually sexed at 90 dpf based on body shape and colour [28,29]. Briefly, males are slimmer and have pinky-blue streaks (in AB strain) or a yellowish colour (in Nacre strain). On the other hand, females have a protruding belly and silvery-blue streaks (in AB strain) or a pink colour (in Nacre strain). Only animals with clear sexual dimorphism were considered. The sex ratio for each dietary group was calculated as the percentage of males/females in each tank.

2.6. Reproductive Performance

To evaluate the impact of the five dietary regimens on embryo viability, all fish from each treatment were crossed at ≥90 dpf, in groups of 8–10 females and 4–6 males from each dietary treatment, and placed overnight in 1.7 L sloping breeding tanks (Tecniplast®). This procedure was repeated three times, on a weekly basis. Upon spawning, embryos from each cross were collected and distributed in 90 mm diameter Petri dishes at a density of ~50 embryos/plate. All unfertilized eggs were counted and removed, and plates were incubated at 28 °C. After 24 h, the numbers of live and non-viable embryos were counted. The embryo viability rate was calculated as the percentage of live embryos per total number of embryos.

Both Zebrafeed® and GemmaMicro® have been associated to good reproductive performances [14,30], but the two had never been compared. To assess if the use of Zebrafeed® or GemmaMicro® around the time of sexual maturation was enough to produce an effect on reproduction, we designed our experimental protocol to include two dietary groups where the only difference was the dry feed from 60 dpf onwards—both consumed GemmaMicro® as the primary feed up to 60 dpf, but from then onwards group 2 was fed GemmaMicro® and group 5 was fed Zebrafeed®. The fertilization rate and sperm viability were assessed for both groups. For the former, a new set of weekly incrosses was performed. Embryos and unfertilized eggs were counted at 7 hpf, and live embryos were gently transferred to a clean embryo medium-filled plate. The fertilization rate was calculated as the percentage of embryos per total number of eggs. At 24 hpf, live and non-viable embryos were counted and used to calculate embryo viability.

Sperm viability is often evaluated using several parameters, such as morphology, membrane integrity, ability to bind oocytes and motility, the latter being the most widely used assay. However, fertilization is considered to be the most informative parameter [31]. Therefore, sperm viability was assessed through the rate of in vitro fertilization using wild type oocytes and sperm from males of dietary groups 2 and 5. In particular, 8 rounds of in vitro fertilization (1 round every 2 weeks) were performed. Briefly, oocytes from 5–10 month old AB and Tübingen (TU) females under the normal feeding regimen [31] were collected by gentle squeezing. Each egg clutch was divided into two petri dishes and held in Aquaboost® Ovacoat extender (Cryogenetics, USA). Sperm collected from males of dietary group 2 or 5 was added to each half of the egg clutch. After mixing, fish water was added to activate gametes for fertilization. At 24 hpf, the fertilization rate was assessed as described above. The survival and malformation rates (defined as larvae viability rate) were evaluated between 6 and 7 dpf.

2.7. Statistical Analysis

All data was analysed with the IBM SPSS Statistics software (v. 23, IBM Corp., Chicago, IL, USA), and all statistical analysis were based on widely established rules [24]. Specifically, after assumption verification (normality and homoscedasticity), Student *t*-tests and one-way ANOVA tests were used. When significant differences were observed, they were analysed with the Tukey's HSD and DSM test in order to perceive which groups differed. For groups where assumptions were not verified, the Wilcoxon–Mann–Whitney and Kruskall–Wallis tests were used followed by pairwise comparison. Results were considered statistically significant for *p*-values < 0.05. Data sets where assumptions were verified were graphically presented in histograms with means and standard deviations. The remaining data sets were presented in boxplots with medians, interquartile ranges, maximums and minimums. Groups with statistically significant differences were graphically identified with a horizontal bar with small, perpendicular dashes on the top of each group, and the *p*-value level was represented on the top of the group that differs from the others, as follows: * $p < 0.05$, ** $p < 0.01$ and *** $p < 0.0001$.

3. Results

The initial comparisons between the AB and Nacre fish within each experimental treatment group returned no statistically significant differences between the two strains (*p*-value > 0.05, data not shown), suggesting that fish with different genetic backgrounds respond in the same way to each dietary regimen. Therefore, data from both strains were pooled and treated together to test for comparisons between the five dietary groups.

3.1. Survival Rate

In this study, the larval median survival rate at 30 dpf ranged from 77.1% to 87.1%, and there were no significant differences among the dietary groups tested (Figure 1A). Nevertheless, at 90 dpf, fish fed exclusively with GemmaMicro® from 15 dpf onwards (dietary group 4) had statistically significant lower survival rates than the remaining dietary groups (Figure 1B). On the other hand, fish that ate only Zebrafeed® from 15 dpf (dietary group 3) had survival rates similar to dietary regimens that included rotifers up to later developmental stages (dietary groups 1, 2 and 5).

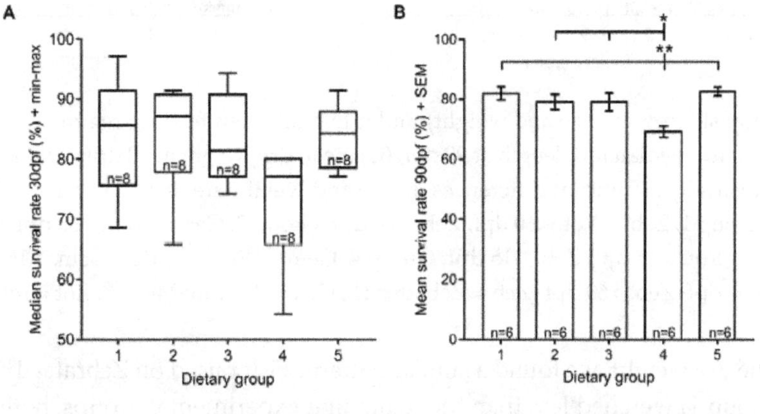

Figure 1. Zebrafish survival rates under the five different dietary regimens. (**A**) Median survival rate at 30 dpf; (**B**) mean survival rate at 90 dpf. Statistically significant differences in survival rates depicted: * $p < 0.05$, ** $p < 0.01$. Dietary Groups (Zeb = Zebrafeed®; Gem = GemmaMicro®; Rot = rotifers): group 1, Zeb. + Rot. <60 dpf, Zeb. >60 dpf; group 2, Gem. + Rot. <60 dpf, Gem. >60 dpf; group 3, Zeb. + Rot. <15 dpf, Zeb. >15 dpf; group 4, Gem. + Rot. <15 dpf, Gem. >15 dpf; group 5, Gem. + Rot. <60 dpf, Zeb. >60 dpf.

3.2. Growth

Organismal growth is the most commonly used metric to estimate nutrient requirements. Here, we analysed two growth measurements: fork length and dry weight. Regarding the fork length, all fish exhibited similar sizes at 30 dpf (Figure 2A). However, at 90 dpf, fish fed with both Zebrafeed® and rotifers until 60 dpf (dietary group 1) were smaller than all others (fork length: 2.77 ± 0.15 cm). On the contrary, a diet composed exclusively of GemmaMicro® from 15 dpf (dietary group 4) was associated with fish with higher fork length than all diets that included Zebrafeed® (dietary groups 1, 3 and 5) (Figure 2B).

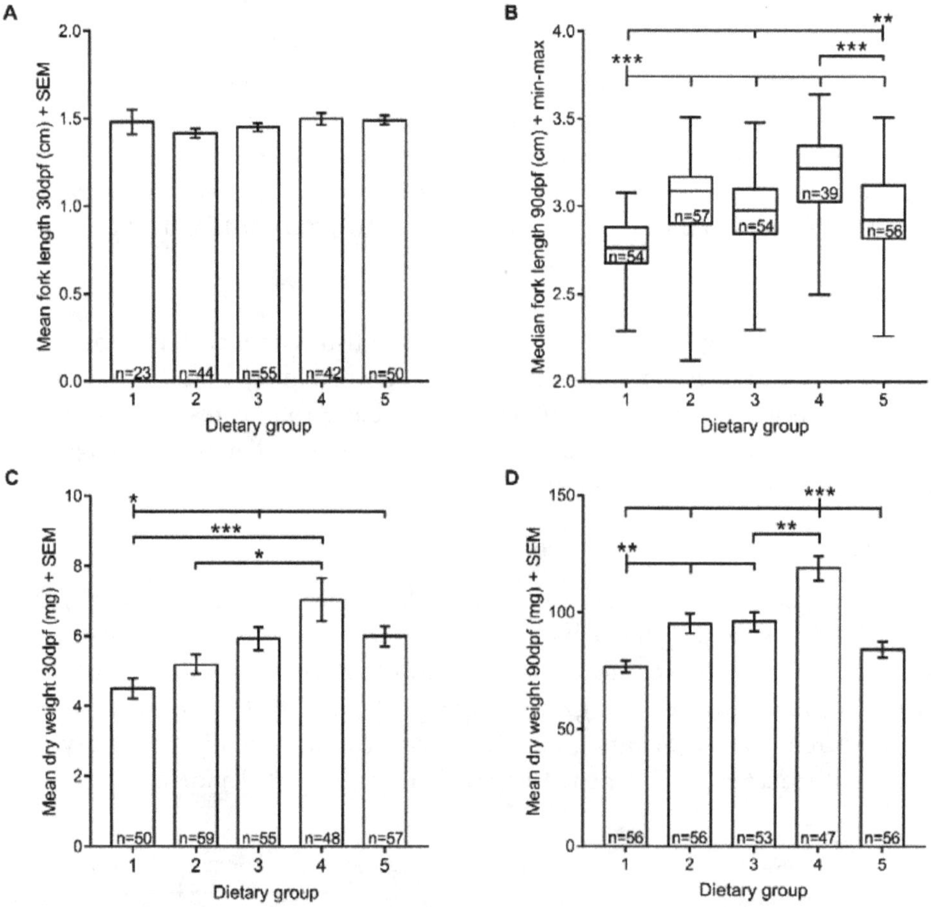

Figure 2. Zebrafish growth (size and weight) under five different feeding regimens. (**A**) Mean fork length at 30 dpf; (**B**) median fork length at 90 dpf; (**C**) mean dry weight at 30 dpf; (**D**) mean dry weight at 90 dpf. Statistically significant differences in size and weight are depicted: * $p < 0.05$, ** $p < 0.01$, *** $p < 0.001$. Group 1, Zeb. + Rot. <60 dpf, Zeb. >60 dpf; Group 2, Gem. + Rot. <60 dpf, Gem. >60 dpf; Group 3, Zeb. + Rot. <15 dpf, Zeb. >15 dpf; Group 4, Gem. + Rot. <15 dpf, Gem. >15 dpf; Group 5, Gem. + Rot. <60 dpf, Zeb. >60 dpf (Zeb = Zebrafeed®; Gem = GemmaMicro®; Rot = rotifers).

Regarding the dry weight, we found a similar pattern. Fish raised on Zebrafeed® and rotifers until 60 dpf (dietary group 1) weighed less than the remaining experimental groups, both at 30 and 90 dpf, with a few exceptions (dietary group 2 at 30 dpf; dietary group 5 at 90 dpf) (Figure 2C,D). Conversely, at 30 dpf, animals raised exclusively using GemmaMicro® from 15 dpf onwards (dietary group 4) weighed more than when rotifers were still provided (dietary groups 1 and 2), and at 90 dpf, the same group had heavier fish than all others (Figure 2C,D). In fact, dietary group 1 weighed significantly less than dietary group 5 at 30 dpf when the two were provided with different processed feeds (Figure 2C), but they were not significantly different at 90 dpf, a timepoint at which both groups had been fed with the same dry feed (Zebrafeed®) for at least one month (Figure 2C). This not only agrees with GemmaMicro® yielding higher growth rates than Zebrafeed®, but it also suggests that changing processed feds at late developmental stages produces significant effects in a time period as short as one month.

3.3. Sex Ratio

To determine if the different feeds affected sex differentiation, we compared the numbers of males and females in the different experimental groups at 90 dpf and did not observe significant differences ($U = 357.2$; $p = 0.328$) (Figure 3).

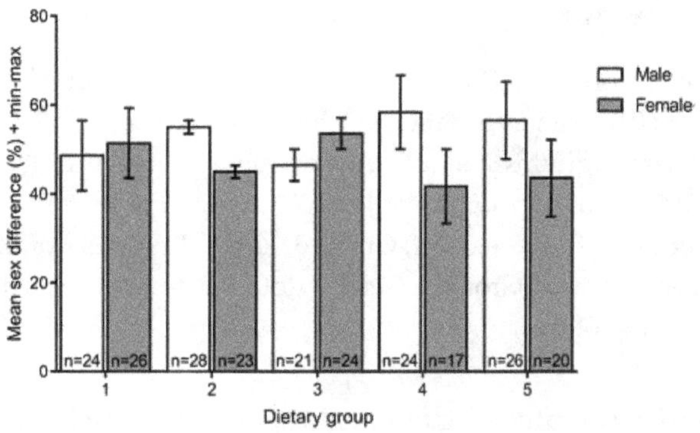

Figure 3. Mean sex ratio for each dietary group at 90 dpf. No statistically significant differences were found among the five dietary regimens. Group 1, Zeb. + Rot. <60 dpf, Zeb. >60 dpf; Group 2: Gem. + Rot. <60 dpf, Gem. >60 dpf; Group 3, Zeb. + Rot. <15 dpf, Zeb. >15 dpf; Group 4, Gem. + Rot. <15 dpf, Gem. >15 dpf; Group 5, Gem. + Rot. <60 dpf, Zeb. >60 dpf (Zeb = Zebrafeed®; Gem = GemmaMicro®; Rot = rotifers).

3.4. Reproductive Performance

A positive correlation between Zebrafeed® and reproductive performance has been previously reported [14]. Therefore, we tested several reproductive parameters, starting with the viability of embryos from all five experimental dietary regimens. Our results showed that extending the period of rotifers co-feeding from 15 dpf to 60 dpf did not improve embryo viability (Figure 4A), yet processed feeds affected zebrafish reproductive function. Specifically, embryo viability was higher when spawners were fed exclusively with Zebrafeed® from the larval stage (15 dpf) or when Zebrafeed® replaced GemmaMicro® at 60 dpf, which is around the time that fish reach sexual maturation. Moreover, in dietary groups fed with GemmaMicro® and/or with more rotifer meals than processed meals (dietary groups 1, 2 and 4), embryo viability was lower. All these results suggest that Zebrafeed® outperforms GemmaMicro® in terms of embryo viability, as long as it is provided to the fish in sufficient amounts.

To further confirm the positive effect of Zebrafeed® on zebrafish reproduction, we focused on dietary groups 2 and 5, because both groups were first raised on GemmaMicro® from the larval stage, but upon sexual maturation (60 dpf), dietary group 2 was kept under GemmaMicro®, whereas group 5 was switched to Zebrafeed®. After a new round of crosses using these fish, the fertilization and embryo viability rates were analysed. Our results showed no differences between fertilization rates of the two groups (Figure 4C). In contrast, embryo viability was higher in dietary group 5, where Zebrafeed® replaced GemmaMicro® at 60 dpf (Figure 4B), reinforcing that Zebrafeed® outperforms GemmaMicro® in terms of reproductive performance.

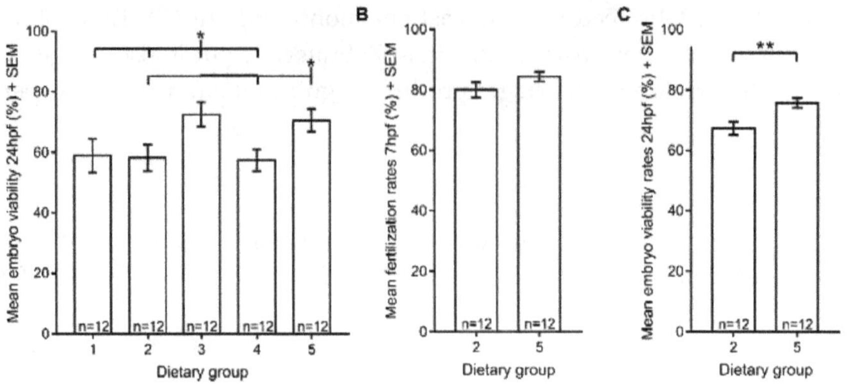

Figure 4. Reproductive performance is affected by processed feeds but not by rotifers. (**A**) Mean embryo viability at 24 hpf; (**B**) mean fertilization at 7 hpf for dietary groups 2 and 5; (**C**) mean embryo viability at 24 hpf for dietary groups 2 and 5. Statistically significant differences in reproductive parameters are depicted: *$p < 0.05$, ** $p < 0.01$, *** $p < 0.001$. Group 1, Zeb. + Rot. <60 dpf, Zeb. >60 dpf; Group 2, Gem. + Rot. <60 dpf, Gem. >60 dpf; Group 3, Zeb. + Rot. <15 dpf, Zeb. >15 dpf; Group 4, Gem. + Rot. <15 dpf, Gem. >15 dpf; Group 5, Gem. + Rot. <60 dpf, Zeb. >60 dpf (Zeb = Zebrafeed®; Gem = GemmaMicro®; Rot = rotifers).

It has been reported that Zebrafeed® increments sperm motility [14]. Taking that into account, we evaluated the sperm quality using sperm from males from dietary groups 2 and 5 in in vitro fertilization assays, and using the fertilization rate at 24 hpf and larvae viability rate at 6–7 dpf as proxies for sperm viability. We did not find any significant differences between the two processed feeds tested (Figure 5). Taken together, our results could not confirm that Zebrafeed® improves sperm quality.

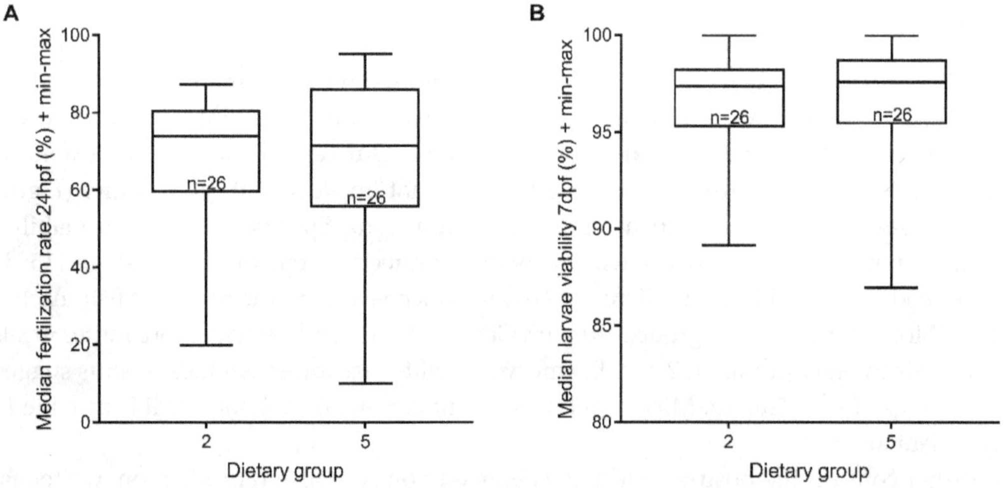

Figure 5. Sperm viability was not affected by processed feeds. (**A**) Median in vitro fertilization rate using sperm from males under GemmaMicro® (dietary group 2) or Zebrafeed® (dietary group 5) at 24 hpf; (**B**) corresponding larvae viability rate (6–7 dpf). There were no statistically significant differences, for either parameters. Group 1, Gem. + Rot. <60 dpf, Gem. >60 dpf; Group 5, Gem. + Rot. <60 dpf, Zeb. >60 dpf (Zeb = Zebrafeed®; Gem = GemmaMicro®; Rot = rotifers).

4. Discussion

In the Champalimaud Fish Platform, a feeding program comprising rotifers, *Artemia* and GemmaMicro® (Skretting®) has been optimized and implemented [25] based on studies that suggested this is the best diet in terms of both larval rearing and breeding performance [30,32,33]. Importantly, the success of GemmaMicro® compared to other dry feeds seems to be more associated with its production method [25], named "cold extrusion spheronizer agglomeration", which better preserves nutrients, rendering them more digestible and stable in water [34–36]. Recently, a new processed feed using the same production method has become commercially available: Zebrafeed® (Sparos®). As facilities need to constantly seek for new and improved solutions to maintain and breed fish, the present study was devised to compare the survival, growth and reproductive performance of zebrafish raised on either GemmaMicro® and/or Zebrafeed®. In addition, we wanted to understand to which extent fish benefit from the use of rotifers as a co-feed to these processed feeds beyond the early larval stage (>15 dpf). Our results validate both GemmaMicro® and Zebrafeed® as two commercially available feeds that are well suited for laboratory zebrafish and show that the benefits of co-feeding rotifers after early larval stage vary depending on the processed diet used and considering the main research interests of each facility.

The first parameter assessed was survival rate, because it generally reflects a response to an extreme limitation or toxicity of a specific nutrient [37]. At 30 dpf, we obtained similar survival rates in all dietary groups, ranging from 77.1% to 87.1%. Even though this indicates that there is room for further nutritional improvements to meet zebrafish needs, one should keep in mind that some mortality can be associated with genetic factors inherent to the strain [38]. Thus, our data suggest that all of the dietary regimens tested seem to respect the nutritional requirements of zebrafish during early development to a similar and satisfactory extent.

However, as fish develop, their nutritional requirements change, and by 90 dpf, some of the dietary regimens had different efficiencies in terms of survival. In a previous study using GemmaMicro® [30], survival rates under similar feeding conditions in our facility did not go beyond 41% by 90 dpf. In this study, by adding rotifers as a co-feed until 15 dpf before switching to a GemmaMicro®-only diet, we were able to increase survival rates from 41% to more than 65% at 90 dpf, demonstrating that the use of live feeds, in particular rotifers, in combination with this dry feed during early larval development greatly improves survival rate, even in later developmental stages, as previously suggested [30,32,39,40]. Moreover, extending the period of GemmaMicro® and rotifer co-feeding to 60 dpf increased the survival rate even further which could be related to the fact that larvae and juveniles feed most efficiently when they are given the opportunity to capture small, slow moving prey that move within their field of vision [33]. On the other hand, Zebrafeed® alone accounted for higher survival than GemmaMicro® and was able to replace the use of rotifers as a co-feed from as early as 15 dpf, revealing that Zebrafeed® could be a better choice than GemmaMicro® for facilities that need to rely only on processed feeds [32,39].

When it comes to growth performance, comparisons with previous studies are not easy due to the lack of uniformity in rearing conditions and evaluated parameters, such as growth measurements and developmental time points analysed. Even so, many studies have suggested that a mixed diet composed of live and processed feeds or live feeds alone accounts for higher growth [40–43]. Accordingly, at 90 dpf, we obtained higher fork lengths for fish raised on GemmaMicro® and also rotifers up to 15 dpf (dietary group 4) than what has been described for fish reared on GemmaMicro® alone [30]. Nevertheless, increasing the period of rotifer feeding from 15 to 60 dpf did not increase growth. This could have been caused by ingestion of non-equivalent amounts of rotifers and dry feed and consequently, different nutritional intakes [44], considering that we provided live and dry feeds until apparent satisfaction. In the future, it would be interesting to measure the biomass of the live and dry feeds and provide equivalent quantities. Another explanation for our results is that the first feeding stage (~8–15 dpf) could be a crucial time window for rotifer feeding, during which, nutrition may condition behaviour and physiological processes to such an extent that could affect growth in later developmental stages [32]. After that initial stage, a diet composed exclusively of processed feeds accounts for the increased growth, likely because it allows for greater control in terms of quantity and nutritional quality [45].

Comparing the growth performances of the dry feeds, GemmaMicro® outperformed Zebrafeed® both in length and weight. In addition, we detected a decrease in growth one month after switching feeds from GemmaMicro® to Zebrafeed®, further supporting the proposal that GemmaMicro® promotes higher growth than Zebrafeed®. Accordingly, previous studies comparing the growth performances of different dry feeds all agreed that GemmaMicro® sustains faster growth rates [14,30,45]. However, one should keep in mind that in our study, the faster growth rates of fish fed only with GemmaMicro® from 15 dpf (dietary group 4) may have been influenced by their lower survival rate and consequently, lower fish density, as the latter is a factor that is known to increase growth [30,37]. Furthermore, we noticed that, although not statistically significant, the dietary groups that included GemmaMicro® were slightly male-biased, whereas those that ate Zebrafeed® appeared to be more female-biased, which is an apparent benefit for facility management purposes where a higher female/male ratio is usually required. What determines the sexual ratio is still under debate, but it has been related to different fish stocking densities [38]. Our sex ratio results are in contrast with the larger sizes and weights obtained with GemmaMicro® which would be more likely to be related with more females, so it would be interesting to perform a relative growth analysis for fish raised with each feed as it could be the case that Zebrafeed® has a slower growth curve but would match the size of GemmaMicro®-fed fish within a few days [46,47]. In fact, size is usually more associated with sexual maturity than with age [48], and no delay in sexual maturation in fish raised with Zebrafeed® as compared to GemmaMicro® was observed.

Regarding reproductive performance, both sperm viability and fecundity were similar in all experimental dietary groups. Even though we could not confirm a direct benefit of Zebrafeed® on sperm quality, as previously described [14], it should be taken into account that the parameters used to analyse sperm quality on each study were not the same. In addition, we found that fish fed with Zebrafeed® more than once a day produced embryos with higher viability, which could be a long-term effect of diet-related sperm quality [49]. This higher embryo viability was observed not only for spawners that ate exclusively Zebrafeed® since larval stages (15 dpf) but also for fish fed with Zebrafeed® only from 60 dpf onwards, i.e., around the time they became sexually mature. This suggests that Zebrafeed® is capable of improving reproductive performance in a short time frame after changing feeds which we propose could be related to the specific lipid composition of Zebrafeed® [14]. Even though the crude lipid percentages of GemmaMicro® and Zebrafeed® are similar (Table 2) and a detailed lipid composition is not available for the two feeds, it has been shown that feeding zebrafish spawners with a diet enriched in essential polyunsaturated fatty acids (PUFA) is enough to increase the quality of gametes and larvae in as little as 14 days of change in feeding [50].

Several studies have made it clear that the presence of essential fatty acids (especially PUFAs) as well as several vitamins (vitamin A, C, E among others) in adequate proportions is required to ensure normal egg and embryo viability and normal larval development [15,19,42,49–51], even though the exact values for zebrafish are yet to be determined. In fact, adequate nutrition is fundamental for most physiological functions, not only affecting directly survival, growth and reproduction, but also influencing several other processes, including endocrine, neurological and immunological functions [39]. The two processed feeds compared here were manufactured through the same production method and have similar gross nutritional compositions (Table 2). Therefore, the dichotomy obtained here—higher growth for fish raised with GemmaMicro® and better survival and embryo viability for fish fed with Zebrafeed®—should be related to different micronutrient proportions or to the way they become available to cells. The answer may lie in the ingredients used in each feed: GemmaMicro® is based on fish meal and fish oil, whereas Zebrafeed® is based on fish, squid and krill meal and on fish and krill oil. It would be interesting to investigate the exact effects of krill meal and oil on zebrafish rearing, as these small crustaceans have a very high nutritional level and represent the primary feed for many fish and marine mammals in the wild [44]. Another possible factor worth assessing is the role of visual attractiveness of these processed feeds related to their colour (Zebrafeed® is orange red, GemmaMicro® is green), as visual appeal is known to affect prey-capture behaviour [52].

In conclusion, we have validated different dietary regimens that can be successfully implemented in a research zebrafish facility setting, adapted to the research objectives. For instance, Zebrafeed® seems to meet zebrafish demands more closely than GemmaMicro® in terms of survival and reproductive performance which could benefit facilities that prioritize the capacity of maximizing the number of animals and embryo production (e.g., for drug or mutagenesis screens). On the other hand, GemmaMicro® promotes higher growth rates, which could be useful when the biggest concern is to minimize the time animals take to reach a certain size of interest or to reduce the life cycle (e.g., for quick high-throughput transgenic screens or juvenile studies). Interestingly, other dietary regimens tested in previous studies have also concluded that the feeding regimes that are most conductive to growth do not necessarily maximize reproductive success in this species [45]. Regarding the use of rotifers, we confirmed that their use during the first stages of exogenous feeding improves larval rearing. However, the benefits of extending the use of rotifers until later developmental stages are not so straightforward, and their use should be considered in light of other factors, including the processed feeds used and the scientific goals of the fish. For instance, rotifers can be an important co-feed in facilities that serve research communities focused on behavioural studies, because live feeds ensure the maintenance of zebrafish natural prey-capture behaviour.

We believe this study brings new data and insights that will be useful for designing and implementing feeding programs in zebrafish facilities in accordance with their research programs.

Author Contributions: The different authors contributed to this study in the following way: Conceptualization, Methodology, Writing of Original Draft Preparation, Writing of Review & Editing, and Visualization, J.F.M., S.M. and A.C.C.; Formal Analysis, M.F.; Investigation, J.F.M., S.M. and T.C.; Resources, A.C.C.; Validation, A.C.C.; Supervision, A.C.C.; Project Administration, J.F.M. and A.C.C.; Funding Acquisition, A.C.C.

Funding: This research was funded by the Champalimaud Foundation and developed with the support of the research infrastructure Congento, project LISBOA-01-0145-FEDER-022170 co-financed by Lisboa Regional Operational Programme (Lisboa 2020), under the Portugal 2020 Partnership Agreement, through the European Regional Development Fund (ERDF) and Foundation for Science and Technology (Portugal).

Acknowledgments: The authors wish to acknowledge all researchers using the Champalimaud Fish Platform for their support of facility-based research programs. The authors also wish to thank the Vivarium staff for support during weekend feedings.

Conflicts of Interest: The authors declare no conflict of interest. The founding sponsors had no role in the design of the study; in the collection, analyses, or interpretation of data; in the writing of the manuscript, and in the decision to publish the results.

References

1. Barnard, D.E.; Lewis, S.M.; Teter, B.B.; Thigpen, J.E. Open- and closed-formula laboratory animal diets and their importance to research. *J. Am. Assoc. Lab. Anim. Sci.* **2009**, *48*, 709–713. [PubMed]
2. Harding, J.D.; Van Hoosier, J.L., Jr.; Grieder, F.B. The contribution of laboratory animals to medical progress—past, present and future. In *Handbook of Laboratory Animal Science—Essential Principles and Practices*, 3rd ed.; Hau, J., Schapiro, S.J., Eds.; CRC Press LLC: London, UK, 2011; Volume 1, pp. 1–20, ISBN 978-1-4200-8456-6.
3. Hau, J. Animal models. In *Handbook of Laboratory Animal Science—Animal Models*, 2nd ed.; Hau, J., Gerald, L., Eds.; CRC Press: Boca Raton, FL, USA, 2005; Volume 2, pp. 1–11, ISBN 0-8493-1086-5.
4. Watts, S.A.; Lawrence, C.; Powell, M.; D'Abramo, R.L. The Vital Relationship between Nutrition and Health in Zebrafish. *Zebrafish* **2016**, *13*, S72–S76. [CrossRef] [PubMed]
5. Lawrence, C. The husbandry of zebrafish (*Danio rerio*): A review. *Aquaculture* **2007**, *269*, 1–20. [CrossRef]
6. Kaushik, S.; Georgia, I.; Koumoundouros, G. Growth and body composition of zebrafish (*Danio rerio*) larvae fed a compound feed from first feeding onward: toward implications on nutrient requirements. *Zebrafish* **2011**, *8*, 87–95. [CrossRef] [PubMed]
7. Castranova, D.; Lawton, A.; Lawrence, C.; Baumann, D.; Best, J.; Coscolla, J.; Doherty, A.; Ramos, J.; Hakkesteeg, J.; Wang, C.; et al. The effect of stocking densities on reproductive performance in laboratory zebrafish (*Danio rerio*). *Zebrafish* **2011**, *8*, 141–146. [CrossRef] [PubMed]

8. Gonzales, J. Preliminary evaluations on the growth and early reproductive performance of zebrafish (*Danio rerio*). *J. Am. Assoc. Lab. Anim. Sci.* **2012**, *51*, 412–417. [PubMed]
9. *SOFIA: The State of World Fisheries and Aquaculture*; FAO Fisheries and Aquaculture Department: Roma, Italy, 2010.
10. Geay, F.; Ferraresso, S.; Zambonino-Infante, J.L.; Bargelloni, L.; Quentel, C.; Vandeputte, M.; Kaushik, S.; Cahu, C.L.; Mazurais, D. Effects of the total replacement of fish based diet with plant-based diet on the hepatic transcriptome of two European sea bass (Dicentrarchus labrax) half-sibfamilies showing different growth rates with the plant-based diet. *BMC Genom.* **2011**, *12*, 522. [CrossRef] [PubMed]
11. Sahlmann, C.; Sutherland, B.J.G.; Kortner, T.M.; Koop, B.F.; Krogdahl, A.; Bakke, A.M. Early response of gene expression in the distal intestine of Atlantic salmon (*Salmo salar* L.) during the development of soybean meal induced enteritis. *Fish Shellfish Immunol.* **2013**, *34*, 599–609. [CrossRef] [PubMed]
12. Fernandes, H.; Peres, H.; Carvalho, A.P. Dietary Protein Requirement during Juvenile Growth of Zebrafish (*Danio rerio*). *Zebrafish* **2016**, *6*, 548–555. [CrossRef] [PubMed]
13. Lawrence, C. The Zebrafish: Genetics, Genomics and Informatics. In *Methods in Cell Biology*, 3rd ed.; William, D.H., Ed.; Academic Press: Waltham, MA, USA, 2011; Volume 104, ISBN 9780080888576.
14. Diogo, P.; Martins, G.; Gavaia, P.; Pinto, W.; Dias, J.; Cancela, L.; Martınez-Páramo, S. Assessment of nutritional supplementation in phospholipids on the reproductive performance of zebrafish, *Danio rerio* (Hamilton, 1822). *Appl. Ichthyol.* **2015**, *31*, 3–9. [CrossRef]
15. Rainuzzo, J.R.; Reitan, K.I.; Olsen, Y. The significance of lipids at early stages of marine fish: A review. *Aquaculture* **1997**, *155*, 103–115. [CrossRef]
16. Watanabe, T. Importance of the study of broodstock nutrition for further development of aquaculture. In *Nutrition and Feeding in Fish*; Cowley, C.B., Mackie, A.M., Bell, J.K., Eds.; Academic Press: London, UK, 1985; pp. 395–414.
17. Fernandez-Palacios, H.; Izquierdo, M.; Robaina, L.; Valencia, A.; Salhi, M.; Montero, D. The effect of dietary protein and lipid from squid and fish meals on egg quality of broodstock for gilthead seabream (*Sparus aurata*). *Aquaculture* **1997**, *148*, 233–246. [CrossRef]
18. Izquierdo, M.S.; Fernandez-Palacios, H.; Tacon, A.G.J. Effect of broodstock nutrition on reproductive performance of fish. *Aquaculture* **2001**, *197*, 25–42. [CrossRef]
19. Alsop, D.; Matsumoto, J.; Brown, S.; Van Der Kraak, G. Retinoid requirements in the reproduction of zebrafish. *Gen. Comp. Endocrinol.* **2008**, *156*, 51–62. [CrossRef] [PubMed]
20. Jaya-Ram, A.; Kuah, M.K.; Lim, P.S.; Kolkovski, S.; Shu-Chien, A.C. Influence of dietary HUFA levels on reproductive performance, tissue fatty acid profile and desaturase and elongase mRNAs expression in female zebrafish *Danio rerio*. *Aquaculture* **2008**, *277*, 275–281. [CrossRef]
21. Miller, G.W.; Labut, E.M.; Lebold, K.M.; Floeter, A.; Tanguay, R.L.; Traber, M.G. Zebrafish (*Danio rerio*) fed vitamin E-deficient diets produce embryos with increased morphologic abnormalities and mortality. *J. Nutr. Biochem.* **2012**, *23*, 478–486. [CrossRef] [PubMed]
22. Robinson, B.D.; Drew, R.E.; Murdoch, G.K.; Powell, M.; Rodnick, K.J.; Settles, M.; Stone, D.; Churchill, E.; Hill, R.A.; Papasani, R.A.; et al. Sexual dimorphism in hepatic gene expression and the response to dietary carbohydrate manipulation in the zebrafish (*Danio rerio*). *Comp. Biochem. Physiol. D-Genom. Proteom.* **2008**, *3*, 141–154. [CrossRef]
23. Cabrita, E.; Robles, V.; Sarasquete, C.; Herráez, M.P. New insights on sperm quality analysis for the improvement of broodstock. In *Cryopreservation of Aquatic Species*, 2nd ed.; Tiersch, T., Mazik, P.M., Eds.; World Aquaculture Society: Baton Rouge, LA, USA, 2011; pp. 146–161.
24. James, G.; Witten, D.; Hastie, T.; Tibshirani, R. Statistical learning. In *An Introduction to Statistical Learning: with Applications in R*; Springer Publishing Company: New York, NY, USA, 2013; pp. 21–24, ISBN 978-1-4614-7138-7.
25. Martins, S.; Monteiro, J.F.; Vito, M.; Weintraub, D.; Almeida, J.; Certal, A.C. Toward an Integrated Zebrafish Health Management Program Supporting Cancer and Neuroscience Research. *Zebrafish* **2016**, *13* (Suppl. 1), 47–55. [CrossRef] [PubMed]
26. Westerfield, M. General Methods for zebrafish care. In *The Zebrafish Book. A Guide for the Laboratory Use of Zebrafish (Danio rerio)*, 4th ed.; Univ. of Oregon Press: Eugene, OR, USA, 2000.

27. Zebrafish International Resource Center—Monitoring of Fish Morbidity and Mortality. Available online: http://zebrafish.org/documents/protocols/pdf/health_monitoring/daily_monitoring_fish_morbidity_07_2015.pdf (accessed on 12 April 2018).
28. Darrow, K.O.; Harris, W. Characterization and development of courtship in Zebrafish, *Danio rerio*. *Zebrafish* **2004**, *1*, 40–45. [CrossRef] [PubMed]
29. Zebrafish in the Classroom. Available online: http://www.zfic.org/common%20techniques/Gender%20identification%20guide.pdf (accessed on 26 May 2018).
30. Farias, M.; Certal, A.C. Different Feeds and Feeding Regimens have an Impact on Zebrafish Larval Rearing and Breeding Performance. *SOJ Aquat. Res.* **2016**, *1*, 1–8.
31. Yang, H.; Tiersch, T.R. Current status of sperm cryopreservation in biomedical research fish models: Zebrafish, Medaka, and Xiphophorus. *Comp. Biochem. Physiol. C Toxicol. Pharmacol.* **2009**, *149*, 224–232. [CrossRef] [PubMed]
32. Best, J.; Adatto, I.; Cockington, J.; James, A.; Lawrence, C. A novel method for rearing first-feeding larval zebrafish: polyculture with Type L saltwater rotifers (*Brachionus plicatilis*). *Zebrafish* **2010**, *7*, 289–295. [CrossRef] [PubMed]
33. Lawrence, C.; James, A.; Mobley, S. Successful Replacement of *Artemia salina nauplii* with Marine Rotifers (*Brachionus plicatilis*) in the Diet of Preadult Zebrafish (*Danio rerio*). *Zebrafish* **2015**, *12*, 366–371. [CrossRef] [PubMed]
34. Boyaci, B.B.; Han, J.; Masatcioglu, M.T.; Yalcin, E.; Celik, S.; Ryu, G.; Koksel, H. Analytical Methods—Effects of cold extrusion process on thiamine and riboflavin contents of fortified corn extrudates. *Food Chem.* **2012**, *132*, 2165–2170. [CrossRef]
35. Steel, C.J.; Leoro, M.G.V.; Schmiele, M.; Ferreira, R.E.; Chang, Y.K. Thermoplastic Extrusion in Food Processing. In *Thermoplastic Elastomers*; El-Sonbati, A., Ed.; InTech Europe; University Campus: Rijeka, Croatia, 2012; pp. 265–290, ISBN 978-953-51-0346-2.
36. New! TECH Feature: Cold Extrusion—Headline News. Available online: http://www.aquafeed.com/news/headline-news-article/601/NEW-TECH-FEATURE-COLD-EXTRUSION/ (accessed on 21 March 2018).
37. Watts, S.A.; Powell, M.; D'Abramo, L.R. Fundamental approaches to the study of zebrafish nutrition. *ILAR J.* **2012**, *53*, 144–160. [CrossRef] [PubMed]
38. Wilson, C. Aspects of larval rearing. *ILAR J.* **2012**, *53*, 169–178. [CrossRef] [PubMed]
39. Siccardi, A.J.; Garris, H.W.; Jones, W.T.; Moseley, D.B.; D'Abramo, L.R.; Watts, SA. Growth and survival of zebrafish (*Danio rerio*) fed different commercial and laboratory diets. *Zebrafish* **2009**, *6*, 275–280. [CrossRef] [PubMed]
40. Carvalho, A.P.; Araújo, L.; Santos, M.M. Rearing zebrafish (*Danio rerio*) larvae without live food: Evaluation of a commercial, a practical and a purified starter diet on larval performance. *Aquac. Res.* **2006**, *37*, 1107–1111. [CrossRef]
41. Goolish, E.M.; Okutake, K.; Lesure, S. Growth and Survivorship of Larval Zebrafish *Danio rerio* on Processed Diets. *N. Am. J. Aquac.* **1999**, *61*, 189–198. [CrossRef]
42. Gonzales, J.M., Jr.; Law, S.H. Feed and feeding regime affect growth rate and gonadosomatic index of adult zebrafish (*Danio rerio*). *Zebrafish* **2013**, *10*, 532–540. [CrossRef] [PubMed]
43. Singh, S.K.; Mishra, U.; Roy, S.D.; Chadha, N.K.; Venkateshwarlu, G. Effect of feeding enriched formulated diet and live feed on growth, survival and fatty acid profile of Deccan Mahseer, Tor khudree (Sykes) first feeding fry. *J. Aquac. Res. Dev.* **2012**, *3*. [CrossRef]
44. Schramm, M.J. "Tiny Krill: Giants in Marine Food Chain". NOAA National Marine Sanctuary Program. Available online: https://sanctuaries.noaa.gov/news/features/1007_krill.html (accessed on 1 June 2018).
45. Lawrence, C.; Best, J.; James, A.; Maloney, K. The effects of feeding frequency on growth and reproduction in zebrafish (*Danio rerio*). *Aquaculture* **2012**, *s368–s369*, 103–108. [CrossRef]
46. Hopkins, K.D. Reporting Fish Growth: A Review of the Basics. *J. World Aquac. Soc.* **1992**, *23*, 173–179. [CrossRef]
47. Gómez-Requeni, P.; Conceição, L.E.; Jordal, A.E.O.; Rønnestad, I. A reference growth curve for nutritional experiments in zebrafish (*Danio rerio*) and changes in whole body proteome during development. *Fish Physiol. Biochem.* **2010**, *36*, 1199–1215. [CrossRef] [PubMed]
48. Harper, C.; Lawrence, C. Husbandry. In *The Laboratory Zebrafish*; Suckow, M.A., Ed.; CRC Press: Boca Raton, FL, USA, 2011; pp. 47–59, ISBN 9781439807439.

49. Bobe, J.; Labbé, C. Egg and sperm quality in fish. *Gen. Comp. Endocrinol.* **2010**, *165*, 535–548. [CrossRef] [PubMed]
50. Nowosad, J.; Kucharczyk, D.; Targońska, K. Enrichment of Zebrafish *Danio rerio* (Hamilton, 1822) Diet with Polyunsaturated Fatty Acids Improves Fecundity and Larvae Quality. *Zebrafish* **2017**, *14*, 364–370. [CrossRef] [PubMed]
51. Traber, M.G.; Atkinson, J. Vitamin E antioxidant and nothing more. *Free Radic. Biol. Med.* **2007**, *43*, 4–15. [CrossRef] [PubMed]
52. Lawrence, C.; Sanders, E.; Henry, E. Methods for culturing saltwater rotifers (*Brachionus licatilis*) for rearing larval zebrafish. *Zebrafish* **2012**, *9*, 140–146. [CrossRef] [PubMed]

10

Protein Kinase A Signaling Inhibits Iridophore Differentiation in Zebrafish

Cynthia D. Cooper [1,2,*], Steve D. Erickson [2], Scott Yin [2], Trevor Moravec [2], Brian Peh [2] and Kevin Curran [3]

1 School of Molecular Biosciences, Washington State University Vancouver, Vancouver, WA 98686, USA
2 College of Arts and Sciences, Washington State University Vancouver, Vancouver, WA 98686, USA; stevederickson@gmail.com (S.D.E.); scottyin@yahoo.com (S.Y.); moravect@wsu.edu (T.M.); brianpeh@comcast.net (B.P.)
3 Department of Biology, University of San Diego, San Diego, CA 92110, USA; kevincurranuw@gmail.com
* Correspondence: cdcooper@wsu.edu

Abstract: In zebrafish (*Danio rerio*), iridophores are specified from neural crest cells and represent a tractable system for examining mechanisms of cell fate and differentiation. Using this system, we have investigated the role of cAMP protein kinase A (PKA) signaling in pigment cell differentiation. Activation of PKA with the adenylyl cyclase activator forskolin reduces the number of differentiated iridophores in wildtype larvae, with insignificant changes to melanophore number. Inhibition of PKA with H89 significantly increases iridophore number, supporting a specific role for PKA during iridophore development. To determine the effects of altering PKA activity on iridophore and melanophore gene expression, we examined expression of iridophore marker *pnp4a*, melanophore marker *mitfa*, and the *mitfa* repressor *foxd3*. Consistent with our cell counts, forskolin significantly decreased *pnp4a* expression as detected by in situ hybridization and quantification of *pnp4a*+ cells. Forskolin had the opposite effect on *mitfa* and *foxd3* gene activity, increasing the area of expression. As *mitfa/nacre* mutants have extra iridophores as compared to wildtype larvae, we examined the function of *mitfa* during PKA-sensitive iridophore development. Forskolin treatment of *mitfa/nacre* mutants did significantly reduce the number of iridophores but to a lesser extent than that observed in treated wildtype larvae. Taken together, our data suggests that PKA inhibits iridophore development in a subset of iridophore precursors, potentially via a *foxd3*-independent pathway.

Keywords: iridophores; PKA signaling; forskolin; melanophores; differentiation; *mitfa*; *pnp4a*

---

1. Introduction

Chromatophores, or pigment/light-reflecting cells, develop from neural crest cells, a population of cells that arise near the dorsal aspect of the developing vertebrate neural tube. Five types of chromatophores have been characterized in different organisms, including several species of fish. These cell types include black melanophores, yellow xanthophores, silver iridescent iridophores, white reflective leucophores, and red erythrophores (reviewed in [1]). Chromatophores promote temporary or permanent color change and are thus critical for an aquatic organism's ability to evade predators, to blend into their environments, or to find mates. Zebrafish have a subset of these chromatophores, including melanophores, xanthophores, and iridophores. Each cell type represents an excellent system for understanding cell fate decisions and subsequent differentiation. Melanophores are the best characterized and loss of function mutations in zebrafish microthalmia-associated transcription factor (*mitfa*), *kit*, oculocutaneous albinism 2 (*oca2*), and vacuolar protein sorting 11 (*vps*) genes promote phenotypes similar to those found in mammalian knockouts (i.e., loss of pigment cell precursors or under/hypopigmented melanocytes; [2–9]). Some pathways affecting melanocyte development also impact iridophore development, offering a separate system for understanding the activity or function of these pathways during cell differentiation. For example, the number of iridophores is impacted by several loss of function mutations in genes previously characterized in melanocytes, including *oca2* (increased iridophores), *vps11* (decreased iridophores), and *mitfa* (increased iridophores) as compared to age-matched wildtype larvae. Additionally, Forkhead transcription factor *foxd3* regulates the expression of *mitfa* and nucleoside phosphorylase gene, *pnp4a*, in a bipotent melanophore/iridophore precursor cell, suggesting that a population of cells exists with the capability of becoming either pigment cell type [10]. It remains unclear as to which intracellular signaling proteins control iridophore differentiation decisions. Here, we propose a role for protein kinase A (PKA) in inhibiting the iridophore differentiation program.

PKA signaling is tightly controlled by regulation of cAMP gradients. Adenylyl cyclases generate cAMP and phosphodiesterases—the latter degrades cAMP (reviewed in [11]). Upon activation by cAMP (via regulatory unit binding), protein kinase A catalytic units phosphorylate target proteins involved in several cellular processes, including cell differentiation and growth [12–15]. PKA signaling also functions in organelle trafficking within chromatophores. Melanophore melanin-producing organelles, melanosomes, are trafficked towards the cell periphery (dispersion) or nucleus (aggregation) in response to hormones such as cAMP-dependent melanocyte-stimulating and melanocyte-concentrating hormones, respectively [16]. Similarly, xanthophore pigment-producing organelles, xanthosomes, internalize their pigment granules in response to changes in light in a cAMP- and phosphodiesterase-dependent manner [17]. Last, PKA signaling also regulates transcriptional programming important for neural crest cell fate choice in quail cell cultures, resulting in the induction of the melanocyte differentiation gene *Mitf* while simultaneously inducing expression of the RE-1 silencing transcription factor, a neuron-specific transcriptional repressor. These results were duplicated in zebrafish models where involvement of the Wnt signaling responsive transcription factor, CtBP2, was required for increased melanophore differentiation [18], suggesting a cross-talk between Wnt and PKA signaling. Thus, there is previous evidence of PKA serving as a switch to encourage differentiation of one lineage at the expense of a second. We explored a function for PKA signaling during iridophore development. Our data suggests that PKA signaling inhibits iridophore differentiation while promoting differentiation of melanophores in zebrafish larvae.

2. Materials and Methods

2.1. Fish Rearing and Crosses

Zebrafish work is approved by Washington State University Institutional Animal Care and Use Committee, Animal Subject Approval Form 03848. Wildtype fish were of the AB (ZDBGENO-960809-7) strain. Adult zebrafish were maintained on a 14/10 h light/dark cycle at 28.5 °C. Adult fish were maintained on a recirculating water system, fed dried/live diets, and checked daily. Embryos were acquired from natural crosses and grown at 28.5 °C in embryo media until analysis. Embryos were staged according to characterized morphological criteria [19]. The following mutant alleles were used: *nacre* [7].

2.2. Production of BACfoxd3::GFP Transgenic Zebrafish

We modified a *foxd3*-containing bacterial artificial chromosome (BAC) clone by *Escherichia coli*-based homologous recombination [20]. BAC clone 137512 contains around 130 kb of sequence upstream and 40 kb downstream of *foxd3* (http://www.sanger.ac.uk/Projects/D_rerio/mapping.shtml). BAC clone 137512 was inserted into pTAR BAC 2.1. Following recombination, the modified BAC clone contained an *egfp* gene positioned at an endogenous start site. The accuracy of recombination was evaluated by PCR, sequencing, and by transient expression assays. *BACfoxd3*::EGFP faithfully recapitulated endogenous *foxd3* expression in neural crest and other organs. To obtain a germline, we linearized BAC DNA with *Not 1*, injected linearized BAC DNA into zebrafish embryos, raised injected fish to adulthood, and screened their progeny for reporter gene expression. The germline transmission rate was 2%.

2.3. Forskolin and H-89 Treatments

Stocks of forskolin and H-89 were prepared in dimethyl sulfoxide (DMSO) at concentrations of 2.4 and 9.6 mM, respectively, and stored at −20 °C until use. Zebrafish were treated with 5 μM forskolin (Sigma, St. Louis, MO, USA) diluted in embryo media beginning at 17–20 h post fertilization (hpf) unless otherwise indicated. At 48–50 hpf, forskolin-containing medium was removed, and fish were rinsed twice in fresh embryo media and incubated at 28.5 °C until fixation. At the indicated timepoint, fish were fixed in 4% paraformaldehyde (4% PFA, diluted in phosphate-buffered saline, PBS), rinsed in PBS +0.01% Tween (PBTw), and stored in PBTw (pre-in situ) until in situ hybridization/quantification or PBS +0.01% Triton X-100 (PBTx) until antibody staining. Samples were stored in 50% glycerol/PBS for analysis and to preserve fluorescence of antibody-stained samples. For H-89 (Sigma, St Louis, MO, USA) treatment, fish were treated with 25 μM or 50 μM H-89 in embryo media beginning at 17–20 hpf unless otherwise indicated. AT 48–50 hpf, H-89-containing media was removed, and fish were rinsed twice in fresh embryo media and incubated at 28.5 °C until fixation for quantification. H-89 treated fish were fixed using 4% PFA and stored in PBTw until in situ hybridization analysis or 50% glycerol/PBS until quantification.

2.4. Cell Quantification and Statistics

For iridophore quantification, a fiber optic light source was used to illuminate iridophores. Iridophores were counted using Nikon SMZ-1500 (Nikon Inc., Melville, NY, USA) or Leica M80 (Leica Microsystems, Inc., Buffalo Grove, IL, USA) stereomicroscopes. Unless otherwise indicated, cell counts include all (dorsal and ventral) trunk and tail iridophores. Fixed fish were used for quantification and 6% methyl cellulose was used to position fish for counting. Fish were rotated in methyl cellulose as necessary for viewing individual stripes/locations. For in situ hybridization analysis, fish were divided into "anterior body", "posterior body", and "tail" regions (see Figure D′ for additional information). Anterior regions include tissue dorsal to rostral and caudal yolk boundaries, while posterior regions include tissue posterior to the caudal yolk boundary, ending at the end of the yolk extension. The tail region included tissue posterior to the yolk extension. Statistical tests performed for each experiment were the Student's *t*-test (Microsoft Excel 2007) or ANOVA (one- or two-way ANOVA; GraphPad Prism, San Diego, CA, USA). Data reported include a minimum of 8–10 samples per time point, with at least three replicates per experiment.

2.5. In Situ Hybridization

Embryos were fixed in 4% paraformaldehyde in phosphate-buffered saline at pH 7.2, and processed using standard protocols [21]. Digoxigenin-labeled probes for *mitfa* [7], *foxd3* [22,23], and *pnp4a* [10] have been previously described. All imaging was done using a Nikon SMZ-1500 stereomicroscope equipped with a Digital Sight DS-Ri1 Digital Camera. All images were processed for contrast, brightness, and color using Adobe Photoshop CS3 Extended Version 10.0 (Adobe Corporate Headquarters, San Jose, CA, USA).

2.6. Antibody Staining and Confocal Imaging

The following antibodies were used for immunohistochemistry at the indicated dilutions: mouse monoclonal anti-Green Fluorescent Protein (Invitrogen/ThermoFisher Scientific, Waltham, MA, USA), 1:500; rabbit anti-beta catenin (Pharmingen, San Diego, CA, USA), 1:100; anti-mouse (Alexa 488) and anti-rabbit (Alexa 568) secondary antibodies (Molecular Probes, Eugene, OR, USA) 1:500 or 1:750. Brightfield or incident lighting (epi-illumination from a fiber optic/ectopic light source) images were obtained on a Nikon dissecting microscope with a Spot RT Slider digital camera (Diagnostic Instruments, Grand Island, NY, USA) or a Nikon SMZ-1500 stereomicroscope with a Digital Sight DS-Ri1 Digital camera. Fluorescent confocal images were obtained on LSM 5 Pascal (Zeiss, Thornwood, NY, USA) or Olympus FV1000 laser scanning confocal microscopes. Images were equivalently processed for color balancing and brightness/contrast using Photoshop CS3 or CS4 (Adobe, San Jose, CA, USA) and formatted with Illustrator CS4 (Adobe, San Jose, CA, USA).

3. Results and Discussion

3.1. Effects of Protein Kinase A Activity Alter the Number of Differentiated Iridophores

To examine the role of protein kinase A during iridophore development, we treated zebrafish embryos with forskolin, an adenylyl cyclase activator. Treatments commenced at 17–20 h post fertilization (hpf) when iridophores are specified from neural crest cells [10]. Following washout of forskolin at 2 days post fertilization (dpf), larvae were grown until 4 dpf when they were fixed for iridophore quantification. At this timepoint, iridophores are positioned ~one cell per somite, making them easy to quantify. Using an incident light source, we were able to illuminate and observe individual iridophores in the dorsal and ventral stripes, beginning at the anterior trunk level. Imaging analysis indicates fewer iridophores in forskolin-treated individuals, especially in the ventral stripes (Figure 1A,B; white arrowheads indicate the ventral stripe). Quantification of iridophores in dorsal and ventral stripes confirms a significant reduction in both stripes, and a 38.4% reduction in total iridophores (Figure 1C, $p < 0.05$ via Student's t-test). To ensure this reduction was not due to developmental delay of the larvae, we also quantified melanophores which develop at a similar time to iridophores in zebrafish (Figure 1D). Statistical analysis suggests that while melanophore number is slightly higher with forskolin treatment (as compared to controls), this increase is not significant (Student's t-test, $p > 0.05$; we also noted no significant differences in dorsal or ventral stripe numbers). Normal somite shape and yolk extension patterning/development further suggest minimal impact on other tissues in forskolin-treated larvae.

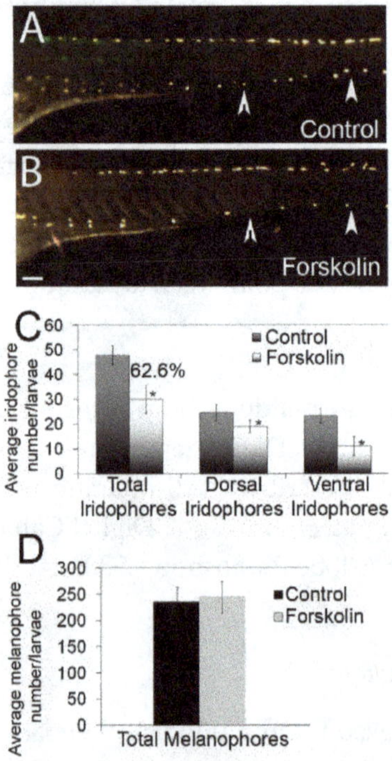

Figure 1. Forskolin treatment reduces iridophore number in zebrafish larvae at 4 days post fertilization (dpf). (**A**,**B**) Lateral images (using an incident lighting source) of larvae at 4 dpf treated with 0.1% dimethyl sulfoxide (DMSO) or 5 μM forskolin in embryo media. Forskolin-treated individuals have fewer iridophores, especially in the ventral stripe (arrowheads). Scale bar = 200 um and applies to panels A and BC) Quantification of total, dorsal stripe, and ventral stripe iridophores (n = 8–10 larvae per condition). Forskolin treatment leads to a 38.4% reduction in differentiated iridophores (62.6% of control number). (**D**) Quantification of melanophores following 0.1% DMSO or 5 μM forskolin treatment. Total melanophores include dorsal, lateral, and ventral stripes (n = 8–10 larvae per condition), showing no significant change following forskolin treatment.

To confirm that forskolin effects were specific to protein kinase A, we analyzed the effects of the protein kinase A inhibitor, H-89 dihydrochloride (H-89), on iridophore number. We tested several concentrations to best understand the effects of distinct doses on iridophore development (Figure 2). Inhibition of protein kinase A via 25 μM and 50 μM H-89 treatment significantly increases iridophore number (one-way ANOVA with Bonferroni multiple comparison (* $p < 0.05$ or ** $p < 0.01$)), consistent with our forskolin data. We do see a lower than expected number of iridophores at 100 μM, most likely due to developmental delay or toxicity with this dose. Taken together, this data suggests that protein kinase A functions to regulate the number of differentiated iridophores in developing zebrafish.

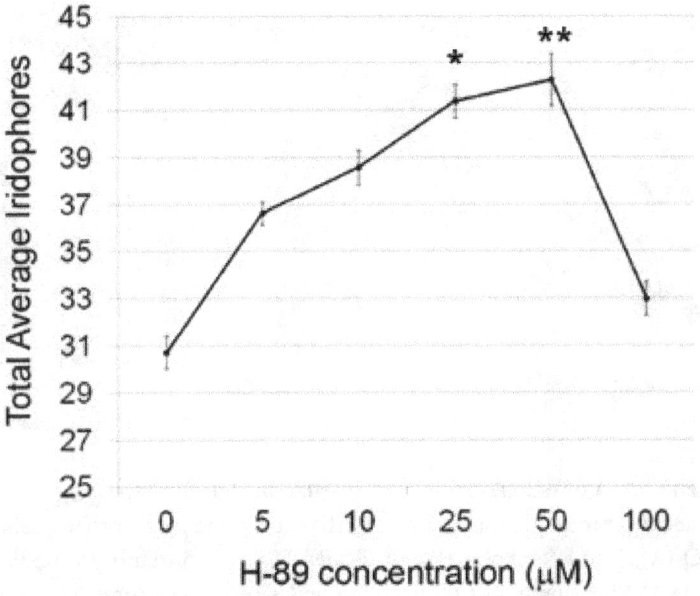

Figure 2. H89 treatment increases iridophore number in zebrafish larvae at 4 dpf. Quantification of iridophores following treatment of wildtype embryos with varying doses of the protein kinase A (PKA) inhibitor H89. One-way ANOVA with Bonferroni multiple comparison post-test indicates that doses of 25 μM and 50 μM are statistically significant (* $p < 0.05$ or ** $p < 0.01$).

3.2. PKA Modulation Inhibits Iridophore Gene Expression

Zebrafish iridophores are specified and then differentiate beginning at 20 h post fertilization (hpf), as detected by purine nucleoside phosphorylase (*pnp4a*) expression [10]. As PKA signaling functions during neuron differentiation [22], we hypothesized that PKA signaling regulates differentiation of the iridophore precursors, iridoblasts. To test this hypothesis, we examined *pnp4a* expression in forskolin and H89-treated larvae. Embryos were treated with forskolin, H89, or control (0.1% DMSO) embryo medium beginning at 22 hpf. Larvae were fixed at 2 dpf and analyzed via in situ hybridization for *pnp4a* expression. Forskolin treatment reduced the amount of *pnp4a*+ signal at 2 dpf (posterior body; Figure 3A,B). Conversely, H89 individuals showed an increase in *pnp4a* signal (compare Figure 3C to Figure 3A). To determine if the number of *pnp4a*+ cells had significantly changed, we quantified the number of *pnp4a*+ cells in the anterior body, posterior body, and tail. The number of *pnp4a*+ cells in posterior body/trunk regions of forskolin-treated larvae is significantly lower as compared to DMSO-treated controls (Figure 3D; *** $p < 0.0001$ via two way ANOVA with Bonferroni multiple comparisons analysis) and significantly higher with H89 treatment (** $p < 0.01$ via two way ANOVA with Bonferroni multiple comparisons analysis) at 2 dpf. This data suggests that PKA activity inhibits iridophore differentiation by reducing *pnp4a* expression.

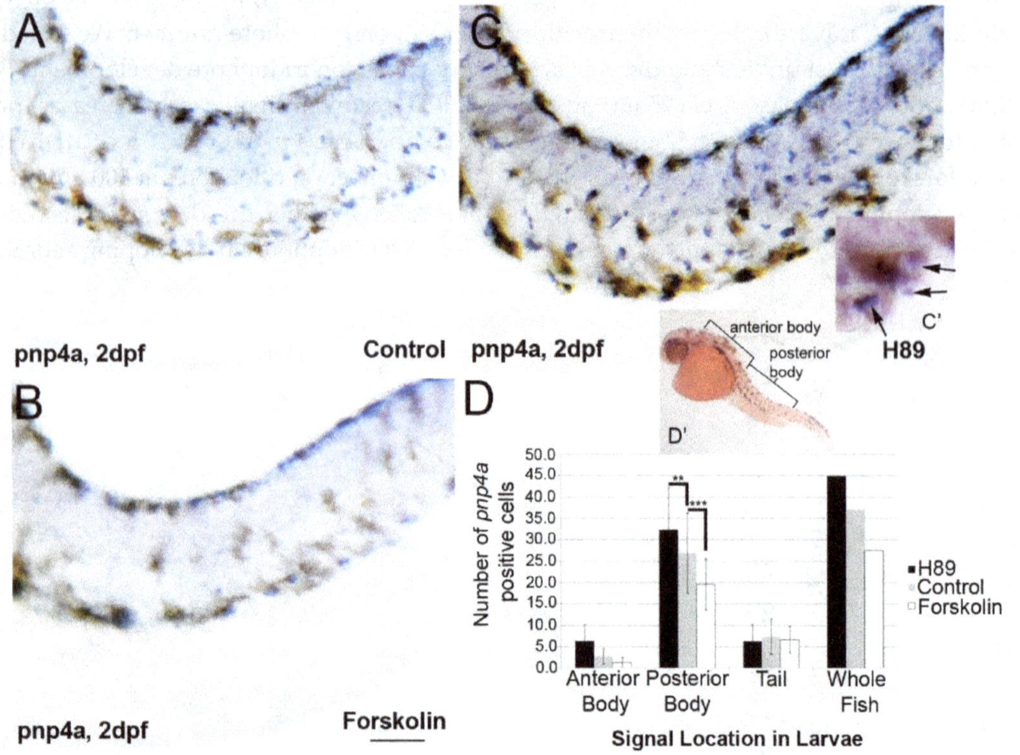

Figure 3. Forskolin treatment reduces iridoblast marker *pnp4a* expression. (**A**–**C**) Lateral brightfield images of larvae at 2 dpf processed for in situ hybridization using *pnp4a* probes, following treatment with 0.1% DMSO (**A**), 5 μM forskolin (**B**), or 25 μM H-89 (**C**) in embryo media. Activation of adenylyl cyclase (and PKA signaling) reduces the levels of *pnp4a* expression, whereas inhibition of PKA has the converse effect (increased *pnp4a* expression). Scale bar = 200 um andapplies to all images. (**C'**) Representative image with black arrows showing blue/purple *pnp4*+ cells included in panel D quantification. Brown melanophores were not included in counts. (**D**) Quantification of *pnp4a*+ cells confirms a significant reduction and increase in the number of *pnp4a*+ cells following forskolin and H-89 treatment, respectively (** $p < 0.01$ and *** $p < 0.0001$ via two-way ANOVA and Bonferonni multiple comparison analysis). (**D'**) The quantified regions are labeled.

3.3. mitfa Is Partially Required for the Forskolin-Dependent Reduction in Iridophore Number

mitfa is required and sufficient for melanophore specification as indicated by *nacre/mitfa* zebrafish mutants, which do not develop melanophores [7]. Additionally, iridophores are present in excess in *mitfa* mutants as compared to wildtype control zebrafish [7,10], suggesting that *mitfa* is critical for establishing correct iridophore number. Examination of simultaneous *pnp4a* and *mitfa* expression indicates co-expression of these gene products at 24 hpf (but loss of co-expression by 50 hpf [10]). To examine the requirement of *mitfa* during PKA-sensitive iridophore differentiation, we first observed *mitfa* expression with or without PKA activation (forskolin treatment). Forskolin-treated larvae showed larger regions of *mitfa* expression posterior to the eye and in the dorsal tail as compared to DMSO treated controls (Figure 4A,B).

We next examined the effects of forskolin treatment on *pnp4a* expression in the dorsal trunk region of *nacre/mitfa* mutants. Again, treatment of wildtype embryos with forskolin caused a reduction in *pnp4a* expression at 2 dpf as compared to controls (Figure 5A,B; consistent with previous results in Figure 3). Conversely, *pnp4a* expression was similar in DMSO/control and forskolin-treated *nacre* larvae at 2 dpf (Figure 5C,D). To confirm that iridophore number was less impacted by PKA activity in *nacre* mutants with forskolin treatment, we quantified the number of iridophores in untreated and forskolin treated *mitfa/nacre* larvae at 4 dpf (following a washout at 2 dpf). While wildtype/AB larvae at 4 dpf showed a 38.4% (See Figure 1) reduction in iridophores as compared to DMSO treated controls, *nacre* larvae showed a 22.2% reduction in iridophores (Figure 5E). Furthermore, only the forskolin-treated *nacre* mutant iridophore number was close to control numbers by 7 dpf (control, 46.6 ± 4.4; forskolin 43.2 ± 4.8). Taken together, this data suggests that zebrafish larvae have a population of iridophores that are not sensitive to PKA in the absence of *mitfa* expression.

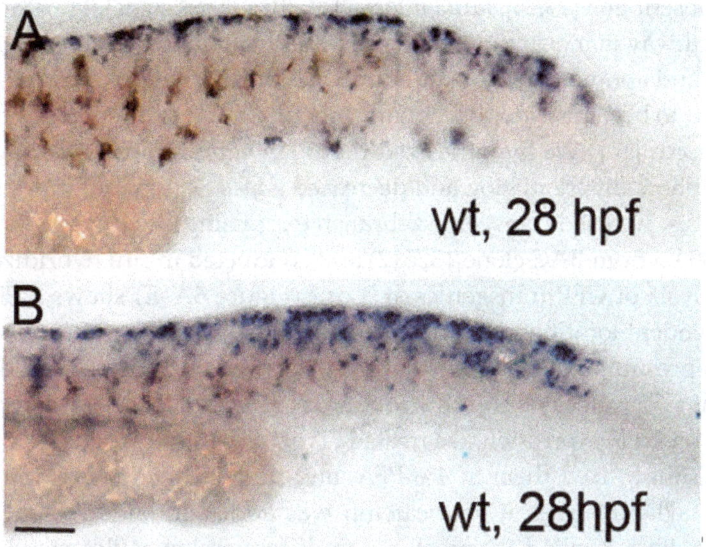

Figure 4. Forskolin treatment expands *mitfa* expression at 28 hpf. (**A**,**B**) Brightfield images of lateral tail regions of 28-h-old wildtype control and forskolin-treated embryos processed for in situ hybridization using *mitfa* probe. Note increased *mitfa* expression in the dorsal tail. Scale bar = 100 um and applies to both images.

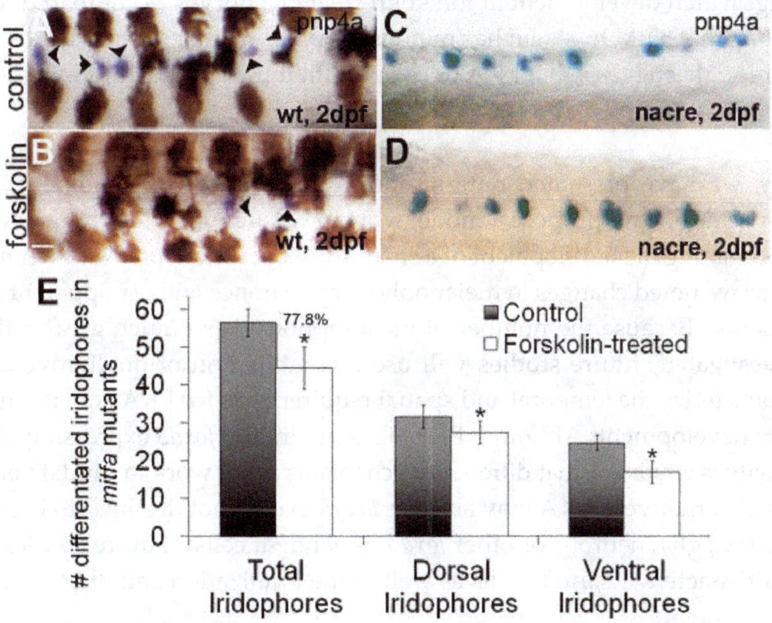

Figure 5. A subset of *mitfa* negative iridoblasts is resistant to PKA signaling. (**A**–**D**) Brightfield images of dorsal trunks of wildtype or *nacre/mitfa* mutant larvae at 2 dpf treated with 0.1% DMSO or 5 μM forskolin and processed for in situ hybridization using a *pnp4a* probe. A reduction in *pnp4a* is detected in wildtype (as previously observed) but not in *nacre/mitfa* larvae. Scale bar = 100 um and applies to all images. (**E**) Quantification of differentiated iridophores in 0.1% DMSO or 5 μM forskolin-treated *nacre/mitfa* larvae. We note a significant, yet attenuated reduction in iridophores with *mitfa* loss of function (22.2% reduction in iridophores or 77.8% of control). The reductions are significant via Student's *t*-test ($p < 0.05$).

3.4. Activation of PKA Expands Neural Crest Marker foxd3 Expression

Forkhead box transcriptional repressor *foxd3* activity is important for specification of specific neural crest derivatives, including iridophores [23,24] and melanophores [25]. *foxd3* is a well characterized marker for neural crest cells and presumptive iridophores. Consistently, *foxd3* is co-expressed with the iridophore marker *pnp4a* [10]. Additionally, evidence suggests that *foxd3* represses its own transcription [23,24]. *foxd3* intersects with several signaling pathways—including the ERBB3 signaling pathway in melanoma cells—via upregulation of specific signaling genes involved in cell proliferation and apoptosis [26,27] and directly represses *mitfa* expression—likely a necessary step in cells destined to become iridophores. Additionally, *foxd3* is required for *pnp4a* expression by developing iridophores [10]. We hypothesized that PKA inhibits *foxd3*, leading to increased *mitfa* expression/melanophore differentiation and decreased *pnp4a* expression/iridophore differentiation. To test this hypothesis, we used transgenic zebrafish expressing GFP under the control of the *foxd3* promoter (about 130 kb from BAC clone 137512) and conducted in situ hybridization using a probe against *foxd3*. Analysis of GFP transgenics at 3 dpf (Figure 6A–E) shows GFP expression in the expected *foxd3*-dependent locations, including the pineal gland (Figure 6B, arrowhead) and the skin (Figure 6B,C), with presumptive iridophores indicated by arrows [23,24,28]. Higher magnification (20×) confirms that GFP colocalizes with iridophores (Figure 6D–E). To determine the effects of forskolin treatment on *foxd3* expression, we treated embryos with forskolin or DMSO control embryo media from 19 to 25 hpf, fixed them in 4% PFA, and probed them using anti-GFP and anti-beta catenin antibodies. The analysis of beta catenin was added to better visualize cell boundaries. Unexpectedly, forskolin expanded the area(s) of *foxd3* expression as detected by the GFP reporter (Figure 6F–G; arrowheads indicate neural crest positive pharyngeal arches) and in situ hybridization (Figure 6H–I). Dorsal head and trunk skin also showed a wider domain of *foxd3* expression following forskolin treatment (arrows, Figure 6H–I). We also note similar locations of *foxd3* expression, eye size, and pharyngeal arch development in forskolin-treated embryos as compared to control treated embryos, suggesting that 5 μM forskolin has minimal impact on other developing tissues at this stage. Taken together, this data suggests that PKA does not inhibit iridophore differentiation by regulating *foxd3* expression.

In summary, we have presented data supporting a role of PKA signaling in pigment cell differentiation. Specifically, we propose a model where PKA inhibits iridophore development while simultaneously promoting the development of a small subset of melanophores. PKA may also regulate melanin content as we noted changes in melanophore appearance and/or apparent melanin content with both treatments. Because the number of melanophores is so much greater than iridophores at the stages investigated, future studies will use a combination of qualitative and quantitative approaches to characterize the temporal and spatial requirements for PKA activity during iridophore and melanophore development. Although PKA does not inhibit *foxd3* expression, it is possible that *foxd3* protein activity is altered or that different mechanisms are at work in caudal versus trunk *foxd3+* neural crest cells. Alternatively, PKA may activate *foxd3* in cells not destined to become iridophores or melanophores (i.e., glia, neurons or other *foxd3*-dependent cells). Future studies will investigate the fate of these PKA-activated *foxd3+* cells as well as the localization and phosphorylation status of Foxd3 protein.

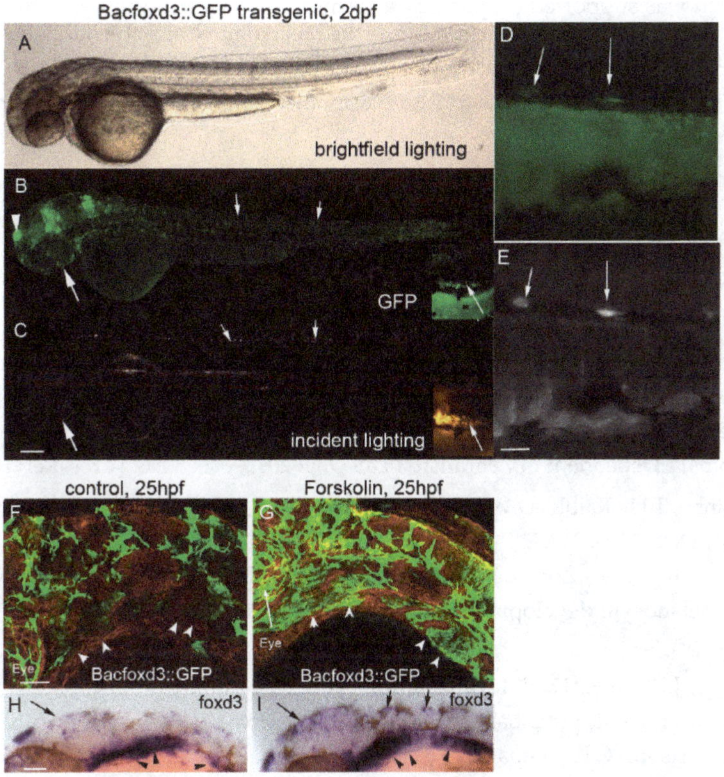

Figure 6. Forskolin increases expression of the *mitfa* transcriptional repressor *foxd3*. (**A**–**C**) 10× magnification of *Bacfoxd3::GFP* transgenic larvae at 2 dpf. Transgenic zebrafish express GFP in *foxd3*-dependent cells including the pineal gland (arrowhead) and presumptive iridophores (arrows). Scale bar in panel C = 200 um and applies to panels A–C. (**D**–**E**) 20× magnification of *BACfoxd3::GFP* zebrafish. Arrows indicate cells co-expressing GFP and iridescence typically observed in iridophores. The scale bar in panel E = 100 um and applies to panels D and E. (**F**,**G**) 20× magnification of 25 hpf 0.1% DMSO or 5 µM forskolin-treated *BACfoxd3::GFP* transgenic fish processed for immunocytochemistry using anti-GFP and anti-beta catenin antibodies (red). Forskolin treatment leads to an increase in GFP (arrowheads indicate pharyngeal arches), indicating expanded domains of *foxd3* expression. Scale bar in panel F = 100 um and applies to panels F and G. (**H**,**I**) 0.1% DMSO and 5 µM forskolin-treated wildtype embryos processed for in situ hybridization using the *foxd3* probe. Forskolin expands the areas of *foxd3* expression, notably in the dorsal head and trunk regions (arrows) which are locations of presumptive iridophores, and in pharyngeal arch neural crest streams (arrowheads). Scale bar in panel H = 100 um and applies to panels H and I.

4. Conclusions

Specification and differentiation of neural crest cells is a highly regulated process, relying on expression (or inhibition) of specific transcription factors and cell differentiation genes in response to environmental signals. Our work suggests that protein kinase A signaling turns on melanophore specific programming at the expense of iridophore programming in a subset of chromatophore precursor cells. This is the first study to implicate PKA signaling as an intracellular regulator of iridophore development.

Author Contributions: Conceptualization, C.D.C., S.D.E., and K.C.; Methodology, S.D.E., S.Y., T.M., B.P., K.C.; Formal Analysis, C.D.C., S.D.E., S.Y., T.M., B.P., K.C.; Investigation, S.D.E., S.Y., T.M., B.P., K.C.; Data Curation, C.D.C., S.D.E., S.Y., T.M., B.P., K.C.; Writing—Original Draft Preparation, C.D.C.; Writing—Review and Editing, C.D.C., S.D.E., S.Y., T.M., B.P., K.C.; Supervision, C.D.C.; Project Administration, C.D.C.; Funding Acquisition, C.D.C.; K.C.

Funding: This research was supported (in part) by an American Cancer Society Institutional Research Grant (IRG-77-003-32 awarded to C.D.C), a research grant from the Melanoma Research Foundation (awarded to C.D.C.) and the Washington State University Vancouver faculty mini grant program (awarded to C.D.C.).

Acknowledgments: The authors are grateful to Alex Nechiporuk and David Raible for donating the BAC*foxd3*::GFP clone.

Conflicts of Interest: The authors declare no conflict of interest.

References

1. Fujii, R. The regulation of motile activity in fish chromatophores. *Pigment Cell Res.* **2000**, *13*, 300–319. [CrossRef] [PubMed]
2. Beirl, A.J.; Linbo, T.H.; Cobb, M.J.; Cooper, C.D. oca2 regulation of chromatophore differentiation and number is cell type specific in zebrafish. *Pigment Cell Melanoma Res.* **2014**, *27*, 178–189. [CrossRef] [PubMed]
3. Clancey, L.F.; Beirl, A.J.; Linbo, T.H.; Cooper, C.D. Maintenance of Melanophore Morphology and Survival Is Cathepsin and vps11 Dependent in Zebrafish. *PLoS ONE* **2013**, *8*, e65096. [CrossRef] [PubMed]
4. Cooper, C.D.; Linbo, T.H.; Raible, D.W. Kit and foxd3 genetically interact to regulate melanophore survival in zebrafish. *Dev. Dyn.* **2009**, *238*, 875–886. [CrossRef] [PubMed]
5. Hultman, K.A.; Bahary, N.; Zon, L.I.; Johnson, S.L. Gene Duplication of the zebrafish kit ligand and partitioning of melanocyte development functions to kit ligand a. *PLoS Genet.* **2007**, *3*, e17. [CrossRef] [PubMed]
6. Lister, J.A.; Close, J.; Raible, D.W. Duplicate mitf genes in zebrafish: Complementary expression and conservation of melanogenic potential. *Dev. Biol.* **2001**, *237*, 333–344. [CrossRef] [PubMed]
7. Lister, J.A.; Robertson, C.P.; Lepage, T.; Johnson, S.L.; Raible, D.W. Nacre encodes a zebrafish microphthalmia-related protein that regulates neural-crest-derived pigment cell fate. *Development* **1999**, *126*, 3757–3767. [PubMed]
8. Parichy, D.M.; Rawls, J.F.; Pratt, S.J.; Whitfield, T.T.; Johnson, S.L. Zebrafish sparse corresponds to an orthologue of c-kit and is required for the morphogenesis of a subpopulation of melanocytes, but is not essential for hematopoiesis or primordial germ cell development. *Development* **1999**, *126*, 3425–3436. [PubMed]
9. Thomas, J.L.; Vihtelic, T.S.; denDekker, A.D.; Willer, G.; Luo, X.; Murphy, T.R.; Gregg, R.G.; Hyde, D.R.; Thummel, R. The loss of vacuolar protein sorting 11 (vps11) causes retinal pathogenesis in a vertebrate model of syndromic albinism. *Investig. Ophthalmol. Vis. Sci.* **2011**, *52*, 3119–3128. [CrossRef] [PubMed]
10. Curran, K.; Lister, J.A.; Kunkel, G.R.; Prendergast, A.; Parichy, D.M.; Raible, D.W. Interplay between Foxd3 and Mitf regulates cell fate plasticity in the zebrafish neural crest. *Dev. Biol.* **2010**, *344*, 107–118. [CrossRef] [PubMed]
11. Skroblin, P.; Grossmann, S.; Schafer, G.; Rosenthal, W.; Klussmann, E. Mechanisms of protein kinase A anchoring. *Int. Rev. Cell Mol. Biol.* **2010**, *283*, 235–330. [PubMed]
12. Ruehr, M.L.; Russell, M.A.; Bond, M. A-kinase anchoring protein targeting of protein kinase A in the heart. *J. Mol. Cell Cardiol.* **2004**, *37*, 653–665. [CrossRef] [PubMed]
13. Fimia, G.M.; Sassone-Corsi, P. Cyclic AMP signalling. *J. Cell Sci.* **2001**, *114*, 1971–1972. [PubMed]
14. Vandamme, J.; Castermans, D.; Thevelein, J.M. Molecular mechanisms of feedback inhibition of protein kinase A on intracellular cAMP accumulation. *Cell Signal.* **2012**, *24*, 1610–1618. [CrossRef] [PubMed]
15. Liang, W.; Ge, S.; Yang, L.; Yang, M.; Ye, Z.; Yan, M.; Du, J.; Luo, Z. Ginsenosides Rb1 and Rg1 promote proliferation and expression of neurotrophic factors in primary Schwann cell cultures. *Brain Res.* **2010**, *1357*, 19–25. [CrossRef] [PubMed]
16. Sheets, L.; Ransom, D.G.; Mellgren, E.M.; Johnson, S.L.; Schnapp, B.J. Zebrafish melanophilin facilitates melanosome dispersion by regulating dynein. *Curr. Biol.* **2007**, *17*, 1721–1734. [CrossRef] [PubMed]
17. Oshima, N.; Nakata, E.; Ohta, M.; Kamagata, S. Light-induced pigment aggregation in xanthophores of the medaka, Oryzias latipes. *Pigment Cell Res.* **1998**, *11*, 362–367. [CrossRef] [PubMed]
18. Liang, H.; Fekete, D.M.; Andrisani, O.M. CtBP2 downregulation during neural crest specification induces expression of Mitf and REST, resulting in melanocyte differentiation and sympathoadrenal lineage suppression. *Mol. Cell Biol.* **2011**, *31*, 955–970. [CrossRef] [PubMed]

19. Kimmel, C.B.; Ballard, W.W.; Kimmel, S.R.; Ullmann, B.; Schilling, T.F. Stages of Embryonic Development of the Zebrafish. *Dev. Dyn.* **1995**, *203*, 253–310. [CrossRef] [PubMed]
20. Zhang, Y.; Buchholz, F.; Muyrers, J.P.; Stewart, A.F. A new logic for DNA engineering using recombination in *Escherichia coli. Nat. Genet.* **1998**, *20*, 123–128. [CrossRef] [PubMed]
21. Thisse, C.; Thisse, B.; Schilling, T.F.; Postlethwait, J.H. Structure of the zebrafish snail1 gene and its expression in wild-type, spadetail and no tail mutant embryos. *Development* **1993**, *119*, 1203–1215. [PubMed]
22. Wang, S.; Li, Z.; Shen, H.; Zhang, Z.; Yin, Y.; Wang, Q.; Zhao, X.; Ji, J. Quantitative Phosphoproteomic Study Reveals that Protein Kinase A Regulates Neural Stem Cell Differentiation Through Phosphorylation of Catenin Beta-1 and Glycogen synthase Kinase 3β. *Stem Cells* **2016**, *34*, 2090–2101. [CrossRef] [PubMed]
23. Lister, J.A.; Cooper, C.; Nguyen, K.; Modrell, M.; Grant, K.; Raible, D.W. Zebrafish Foxd3 is required for development of a subset of neural crest derivatives. *Dev. Biol.* **2006**, *290*, 92–104. [CrossRef] [PubMed]
24. Stewart, R.A.; Arduini, B.L.; Berghmans, S.; George, R.E.; Kanki, J.P.; Henion, P.D.; Look, A.T. Zebrafish foxd3 is selectively required for neural crest specification, migration and survival. *Dev. Biol.* **2006**, *292*, 174–188. [CrossRef] [PubMed]
25. Curran, K.; Raible, D.W.; Lister, J.A. Foxd3 controls melanophore specification in the zebrafish neural crest by regulation of Mitf. *Dev. Biol.* **2009**, *332*, 408–417. [CrossRef] [PubMed]
26. Abel, E.V.; Aplin, A.E. FOXD3 is a mutant B-RAF-regulated inhibitor of G(1)-S progression in melanoma cells. *Cancer Res.* **2010**, *70*, 2891–2900. [CrossRef] [PubMed]
27. Abel, E.V.; Basile, K.J.; Kugel, C.H., 3rd; Witkiewicz, A.K.; Le, K.; Amaravadi, R.K.; Karakousis, G.C.; Xu, X.; Xu, W.; Schuchter, L.M.; et al. Melanoma adapts to RAF/MEK inhibitors through FOXD3-mediated upregulation of ERBB3. *J. Clin. Investig.* **2013**, *123*, 2155–2168. [CrossRef] [PubMed]
28. Gilmour, D.T.; Maischein, H.M.; Nusslein-Volhard, C. Migration and function of a glial subtype in the vertebrate peripheral nervous system. *Neuron* **2002**, *34*, 577–588. [CrossRef]

11

Drosophila as a Model for Developmental Biology: Stem Cell-Fate Decisions in the Developing Nervous System

Katherine Harding and Kristin White *

Massachusetts General Hospital Cutaneous Biology Research Center, Harvard Medical School, Boston, MA 02129, USA
* Correspondence: Kristin.white@mgh.harvard.edu

Abstract: Stem cells face a diversity of choices throughout their lives. At specific times, they may decide to initiate cell division, terminal differentiation, or apoptosis, or they may enter a quiescent non-proliferative state. Neural stem cells in the *Drosophila* central nervous system do all of these, at stereotypical times and anatomical positions during development. Distinct populations of neural stem cells offer a unique system to investigate the regulation of a particular stem cell behavior, while comparisons between populations can lead us to a broader understanding of stem cell identity. *Drosophila* is a well-described and genetically tractable model for studying fundamental stem cell behavior and the mechanisms that underlie cell-fate decisions. This review will focus on recent advances in our understanding of the factors that contribute to distinct stem cell-fate decisions within the context of the *Drosophila* nervous system.

Keywords: stem cell biology; neuroblast; cell cycle; apoptosis; *Drosophila*; development

---

1. Introduction

Control of stem cell divisions is critical for normal organismal development. The overall proliferative potential of a given tissue is defined by several parameters, including the total number of stem cells, temporal control of proliferation and capacity to divide, and longevity within the tissue. The generation of progeny cells with specific differentiation programs must be coordinated in both time and space for proper tissue organization. Accurate transmission of identity factors from stem cell to progeny is therefore also critically important for organogenesis. Unregulated proliferation of cells with stem-like behavior is commonly associated with tumorigenesis and can lead to metastases.

In this review, we will discuss the current understanding of the mechanisms that control stem cell-fate decisions in the *Drosophila melanogaster* central nervous system (CNS). Neural stem cells, called neuroblasts, are present throughout the *Drosophila* CNS, and generate both neurons and glia as progeny. The biology of these stem cells has been sufficiently characterized to begin to identify molecular differences between neuroblast populations, as well as heterogeneous behavior within a given population. Among these differences are proliferation patterns, progeny identities and terminal fate decisions (Figure 1). Stem cell populations are traditionally defined by their anatomical position, and a vast body of work has described spatial control of stem cell proliferation and terminal fate (Figure 2; [1–5]). In addition, neurogenesis occurs in several waves throughout the development of the fly, allowing temporal control of stem cell decisions to be examined. We summarize the strengths of the *Drosophila* CNS as an easily accessible, genetically tractable model to study fundamental processes of stem cell biology.

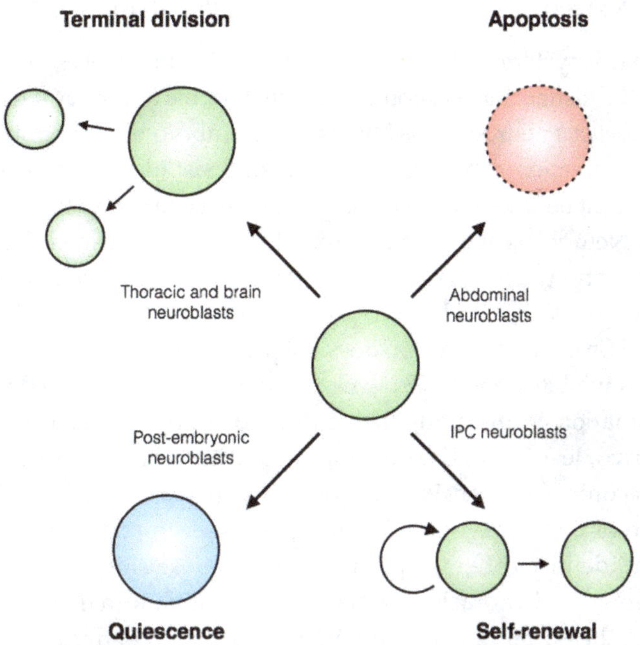

Figure 1. Stem cell-fate decisions in the *Drosophila* central nervous system. *Drosophila* neural stem cells, called neuroblasts, exhibit diverse stem cell behaviors. Thoracic and brain neuroblasts are eliminated by terminal division (top left), while abdominal neuroblasts undergo apoptosis (top right). Temporal controls regulate the period of neuroblast quiescence between embryonic and larval stages of development (bottom left). Certain larval neuroblasts within the inner proliferating center (IPC) of the optic lobe switch from renewing asymmetric to amplifying symmetric divisions (bottom right). Execution of these cell-fate decisions is intimately tied to neuroblast identity in time and space.

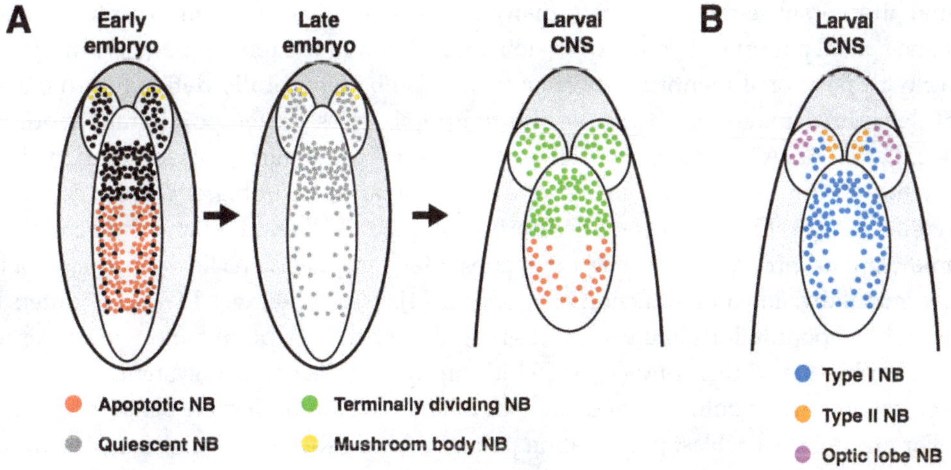

Figure 2. Neuroblast populations within the *Drosophila* central nervous system. (**A**) Cell-fate decisions are mapped onto anatomical populations of neuroblasts during embryonic and larval development. The majority of abdominal neuroblasts undergo apoptosis during embryogenesis, leaving 3 neuroblasts per hemisegment in the larval CNS. Neuroblasts stop proliferating between embryonic and larval development, except for 4 mushroom body neuroblasts. Following reactivation, neuroblasts in the thorax and brain undergo terminal division, while the remaining abdominal neuroblasts die through apoptosis. (**B**) Proliferation patterns of neuroblasts throughout the larval CNS. Most neuroblasts use the Type I division pattern, except a small population of Type II neuroblasts in the brain. The neuroblasts in the inner proliferating center of the optic lobe have been observed to divide symmetrically.

2. Overview of the Fly Nervous System and Its Stem Cell Populations

The *Drosophila* nervous system is shaped in stages: the distinct phases of neurogenesis and tissue sculpting correspond to the distinct innervation requirements of the animal as it develops from a crawling larva to an adult fly. *Drosophila* neurogenesis begins in the embryo with neuroblast delamination from the neuroectoderm. Delamination and acquisition of neuroblast identity depends on Notch signaling interactions between adjacent neuroectodermal cells, leading to stem cell-specific expression of pro-neural genes (reviewed in [6]). Neuroblast specification from the neuroectoderm is also dependent on the Sox genes *SoxN* and *Dichaete*, and members of the Snail family (*snail*, *worniu*, and *escargot*). Dichaete and SoxN act redundantly in the neuroectoderm to promote neuroblast identity through pro-neural gene expression [7–11]. Loss of Snail family function does not prevent neuroblast delamination, but expression of the pro-neural genes is inhibited and CNS development is severely impaired [12–14].

Neuroblast delamination occurs along the ventral side of the embryo and from neural placodes at the head of the embryo, leading to the development of the ventral nerve cord (VNC) and brain, respectively. The embryonic VNC consists of repeated structural units called segments, which are divided by the midline of the embryo into hemisegments, each containing 30 unique neuroblasts (Figure 3A). Segmental identity differs along the anterior-posterior axis of the embryo: there are 3 anterior gnathal segments, 3 thoracic segments and 10 abdominal segments, followed by the non-segmented telson at the posterior end of the VNC. In addition to the anterior-posterior positional information at the whole animal level, there is a positional map within each hemisegment of the VNC. Patterned expression of positional genes creates a Cartesian grid in which each of the 30 neuroblasts can be assigned a unique identity based on its molecular markers (Figure 3B, [15,16]). Transplantation and *in vitro* explant studies have shown that neuroblast molecular identity is defined early in embryogenesis and is retained cell-autonomously [17–21]. The spatial information within each hemisegment is superimposed upon the anterior-posterior axis, where positional information is provided by the Hox gene family (Figure 3C). The segmental pattern of molecular markers seems invariant along the anterior-posterior axis [22]. Birkholz et al. (2013) also determined that, while the terminal abdominal segments contain many fewer neuroblasts than more anterior abdominal segments, the spatial patterns of gene expression are retained in the remaining neuroblasts.

Along with positional identity, neuroblasts can also be temporally defined by the expression pattern of sequential markers, known as the temporal series or temporal transcription factors (TTFs; [23–27]). Figure 4A illustrates the TTF series for embryonic neuroblasts found in the thorax and abdomen, while Figure 4B shows the series for a subset of larval neuroblasts within the central brain. The TTFs confer temporal identity to neurons born within each TTF window, and their cross-regulation ensures unidirectional progression through the series. Many fate decisions lie downstream of the TTFs, including terminal division and induction of apoptosis [4]. While the exact TTF series differs between specific neuroblast populations, the use of cross-regulatory transcription factors to confer temporal identity to neuroblasts and their progeny is found throughout the nervous system.

Neuroblast identity, in both time and space, fundamentally underlies the behavior and fate of the stem cell. For example, neuroblast proliferation patterns differ between populations of stem cells, with three major division types occurring throughout the CNS (reviewed in [28]). Within the thorax and abdomen of the embryonic VNC and the central brain of the larval CNS, most stem cell divisions are asymmetric and produce a new stem cell and a smaller ganglion mother cell (GMC). The GMC will then divide to produce two daughter cells, which terminally differentiate into neurons or glia. This pattern of division is referred to as a Type I division (Figure 4C, left). A subset of neuroblasts in the larval central brain undergo Type II divisions, in which the neuroblast asymmetric division produces a stem cell and an intermediate neural progenitor (INP) that matures prior to dividing asymmetrically to produce another progenitor and a GMC (Figure 4C, middle; [29–31]). Type II neuroblasts have only been described within the brain lobes: their presence in this region may reflect the high proliferative capacity required to generate a complex CNS. If the central brain contained only Type I neuroblasts, it is tempting to speculate that the limited proliferative potential of this type of stem cell division would prevent the production of diverse types of neurons and supporting cells required for higher level processing.

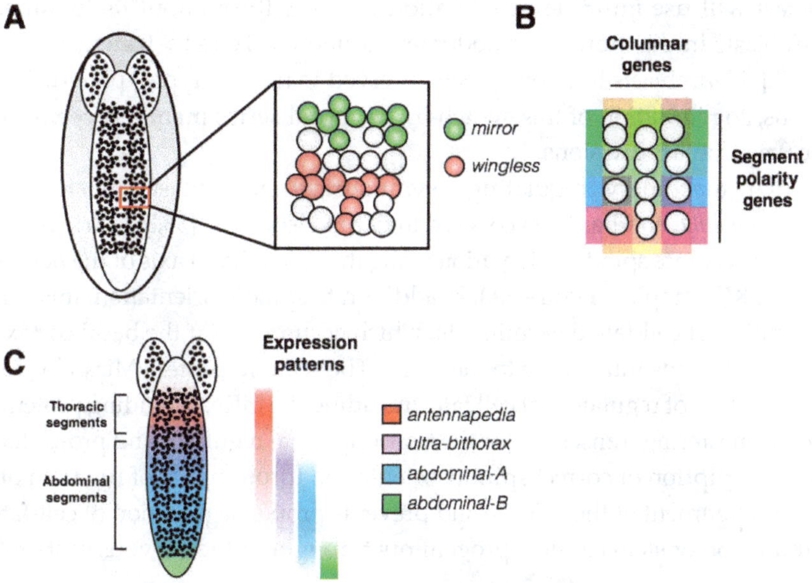

Figure 3. Spatial patterning of the *Drosophila* ventral nerve cord. (**A**) A single hemisegment containing approximately 30 neuroblasts is shown in the inset. Gene expression is patterned within the hemisegment: two examples of spatially restricted molecular markers are shown (*mirror* and *wingless*; from the Hyper-Neuroblast Map, Doe laboratory). (**B**) Upon delamination from the neuroectoderm, neuroblast identity is specified within a hemisegment by a Cartesian grid generated by expression patterns of the columnar and segment polarity genes (reviewed in [16]). (**C**) Along the anterior-posterior axis of the embryo, hemisegment identity is also subject to spatial regulation by the Hox gene family. All of these spatial patterns are superimposed *in vivo*, providing each neuroblast with a unique spatial identity.

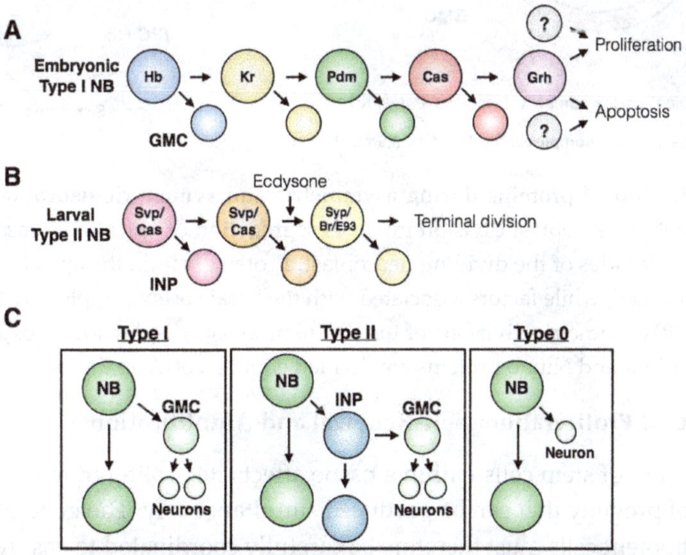

Figure 4. The temporal series and division patterns of *Drosophila* neuroblasts. (**A**) The temporal transcription factor (TTF) sequence expressed in embryonic Type I neuroblasts. The last member, *grainyhead*, has been shown to regulate both proliferation and apoptosis of neuroblasts, suggesting that context-dependent co-factors may dictate the cell-fate outcome for a given neuroblast. (**B**) The TTF sequence expressed in larval Type II neuroblasts. Progression to the late factors is dependent on the steroid hormone ecdysone, and leads to a terminal division [26,32]. (**C**) Three types of asymmetric neuroblast proliferation patterns observed in the nervous system. Type 0 divisions have only been observed following a period of Type I divisions in embryonic neuroblasts.

An alternative mode of division has also been described within the embryonic VNC, called a Type 0 division: the asymmetric division of the neuroblast produces a stem cell and a terminally differentiated neuron, without a GMC intermediate (Figure 4C, right; [33–35]). In certain cases, a single neuroblast will use multiple proliferation patterns throughout its lifetime. For example, embryonic neuroblasts in the thorax and abdomen undergo a Type I > 0 switch towards the end of embryogenesis [33]. Neuroblasts have only been observed to use the Type 0 pattern following a period of Type I divisions, so regulation of this switch by temporal series members seems to be the critical inducer of Type 0 neuroblast divisions.

The molecular machinery regulating asymmetrical neuroblast divisions has been well characterized in *Drosophila*, and has been covered in depth elsewhere [36,37]. Briefly, the asymmetrical division depends on accurate spindle orientation along the apical-basal axis of the neuroblast, mediated by the apical Par/aPKC complex (Figure 5A). In addition to spindle orientation, this complex regulates subcellular localization of cell-fate determinants, which accumulate at the basal cortex of the dividing neuroblast and are thus transmitted only to the GMC. The scaffold protein Miranda plays a key role in asymmetrical localization of regulators of cell fate, including the mRNA-binding protein Staufen, which is responsible for sequestering transcripts for the transcriptional regulator and pro-differentiation factor Prospero [38–44]. Disruption of correct spindle orientation through loss of function of the Par/aPCK complex causes misalignment of the spindle and prevents proper segregation of cell-fate determinants, leading to accumulation of stem cell-like progenitors throughout the nervous system [45–50].

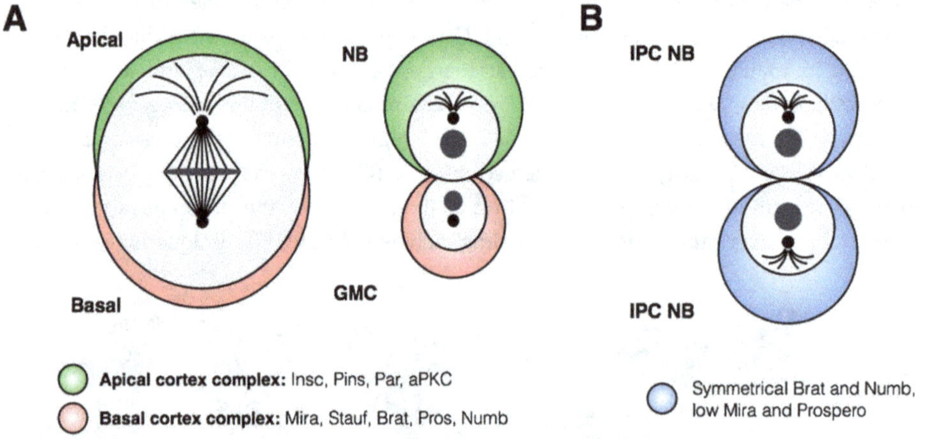

Figure 5. Localization of proteins during asymmetric and symmetric neuroblast divisions. (**A**) A typical asymmetric neuroblast division involves segregation of apical and basal cortex complex members to opposite sides of the dividing neuroblast. Proteins within the apical cortex complex are retained in the stem cell, while factors associated with the basal cortex complex are transmitted to the daughter GMC. (**B**) Symmetric divisions of the IPC neuroblasts display low Prospero and Miranda protein levels, but Brat and Numb proteins are divided equally between the two new stem cells [51].

3. Control of Stem Cell Proliferation: Self-Renewal and Amplification

The overall number of stem cells within a tissue affects its proliferative potential by limiting the maximum number of progeny that can be produced simultaneously. The generation of tissue-specific stem cells from multipotent cells must therefore be carefully coordinated to ensure proper downstream progeny numbers and organization. As described above, the production of stem cells from the embryonic neuroectoderm is limited by Notch-mediated lateral inhibition between a newly born neuroblast and its neighbors [52–54]. The embryonic phase of neuroblast delamination generates ~30 neuroblasts per thoracic or abdominal hemisegment in the VNC, and ~105 neuroblasts in each central brain lobe. Historically, the ~800 neuroblasts of the optic lobe were thought to be generated from the neuroepithelial layer during larval development [55]. However, a recent study observed the generation of embryonic optic neuroblasts (EONs) from the neuroepithelium overlying the developing embryonic brain [56]. This study indicates that optic lobe development begins much earlier than previously thought and provides an additional system for studying the coordination of embryonic and larval neurogenesis.

Until recently, the number of stem cells within the nervous system was thought to be defined by the number of delaminated neuroblasts, as all neuroblasts were only observed to divide asymmetrically. This model was challenged by a recent study that demonstrated that neuroblasts of the inner proliferating center (IPC) of the larval optic lobe undergo a temporal switch to symmetric amplifying divisions [51]. Unlike dividing Type I/II neuroblasts, the symmetric divisions of the larval IPC neuroblasts do not show asymmetric segregation of Prospero, Mira or Brat proteins (Figure 5B; [51]), indicating a fundamentally distinct mechanism of neuroblast division in these cells. This division pattern is induced by the pro-neural gene *atonal*, but the authors also observe that *atonal* maintains neuronal differentiation. It is therefore unclear which *atonal* targets contribute to the induction of symmetric neuroblast divisions in the IPC. As *atonal* is a pro-neural gene required for neuroblast identity, and symmetric divisions have only been observed in this limited population, context-dependent pathways are likely to play a role in promoting symmetric divisions.

With respect to the generation of Type I or Type II neuroblasts in the embryo, the upstream signals that direct stem cells to a particular division pattern remain unclear. Close characterization of TTF expression in each type of neuroblast demonstrates that they progress through distinct temporal identities (reviewed in [25]). Much of the work investigating TTF expression in neuroblasts has focused on the generation of neural diversity through transmission of temporal identity to progeny. It is unclear whether the TTF series themselves direct neuroblast proliferation patterns, or if expression of these transcription factors lies downstream of the Type I vs II decision. However, the switch during embryogenesis from Type I > 0 does lie downstream of TTFs: loss of function of *castor* or *grainyhead*, both late TTFs, results in extra neuroblast and daughter proliferation in the embryo [33]. Two recent studies showed that central brain Type II neuroblasts arise during embryogenesis, and identified the epidermal growth factor (EGF) pathway as an upstream determinant of the Type II identity [57,58]. EGF signaling pathway components, including the EGF receptor ligand Spitz, the Rhomboid protease required for ligand release and the downstream transcription factor Pointed P1 (PntP1) are required for embryonic specification of Type II neuroblasts [57,58]. Embryonic Type I neuroblasts are PntP1-negative [59,60], suggesting that EGF signaling is a positive driver of the Type II identity. Another outstanding question relates to whether neuroblasts can be induced to switch proliferation patterns following delamination, or if the Type I/II decision is an immutable quality determined by hemisegment identity and anterior-posterior position. It was recently shown that loss of function of *inscuteable* (Insc), a regulator of cell-fate determinant localization, results in INP-like progenitor cells within larval Type I lineages [61]. These findings begin to shed light on the question of how Type I and Type II neuroblasts are distinguished prior to the onset of their distinct TTF series and implicates the upstream regulators EGF and Insc in the Type I/II choice.

The previous examples described differences in the fundamental proliferation patterns of specific neuroblast populations. However, the division capacity and identity of progeny of individual neuroblasts within a population can differ widely across the anterior-posterior axis of the nervous system, even between stem cells that share the same identity in a hemisegment [18,20,62–66]. For example, during embryogenesis the thoracic neuroblast NB1-1 produces multiple neuronal types but no glia, while the abdominal NB1-1 produces fewer neurons and 3 glia [18]. Spatial information conferred by Hox gene expression is sufficient to induce changes in division capacity: ectopic expression of the Hox genes *ultrabithorax* or *abdominal-A* in the thoracic NB1-1 resulted in the production of glial progeny [18]. Berger et al. (2005) [67] demonstrated that transformation of thoracic NB6-4 by AbdA (as determined by progeny output) requires CycE, and that CycE expression was sufficient to induce the positional transformation. As other cell cycle manipulations did not produce this effect, the role of CycE in segmental identity was proposed to be independent of its function in the cell cycle. Indeed, the same group determined that these two functions of CycE are mediated through distinct protein domains, and that CycE both promotes the cortical localization of Prospero protein, and interferes with Pros-mediated gene regulation [68]. More recently, the upstream factors *glial cell missing* (gcm) and *apontic* (Apt) were found to be differentially expressed between thoracic and abdominal NB6-4, resulting in restriction of CycE expression to the thoracic NB6-4 [69]. These findings provide the most direct evidence for how Hox-mediated differences in segmental identity influence neuroblast proliferation behavior.

4. Regulation of Stem Cell Quiescence

As well as a model for stem cell proliferation, neuroblasts offer a tractable system to study the regulation and maintenance of endogenous stem cell quiescence. Embryonic neuroblast divisions were meticulously described by Hartenstein et al. (1987) [70], and the lull in neuroblast proliferation at the end of embryogenesis was first described thirty years ago (Figure 3A, middle; [1]). Since that time, a large body of work has characterized the timing and regulation of the reactivation of proliferation in the larval nervous system. With the exception of four mushroom body neuroblasts that do not enter quiescence ([71], Figure 3A), the neuroblasts in the central brain are the first to begin re-proliferating, within 24 h after larval hatching (ALH; [1,71,72]). This is followed by reactivation of the thoracic neuroblasts at ~30h ALH, and then the abdominal neuroblasts after ~50h ALH. The mechanisms regulating the anterior-posterior gradient of neuroblast proliferation have been recently investigated by the Thor lab. They observe that the lineage size of a neuroblast is regulated by Polycomb group proteins (PcG), whose members inhibit Hox gene expression in the brain and thereby promote continued neuroblast proliferation in this region of the nervous system [65,66]. While these studies establish a role for PcG and Hox genes in the regulation of neuroblast proliferation, the mechanism by which they control neuroblast reactivation following quiescence remains unclear.

Despite three decades having passed since the first description of neuroblast quiescence, it was only recently recognized that three quarters of post-embryonic neuroblasts exit the cell cycle in G2 (Figure 6A; [73]), while the remaining 25% display a more standard G0 quiescence. The authors identified the pseudokinase *tribbles* as a major regulator of G2 quiescence entry and stem cell reactivation (Figure 6A). Tribbles has been shown previously to mediate degradation of Cdc25/String [74–76], leading to maintenance of inhibitory phosphorylation on Cdk1 and preventing cell cycle progression. In addition to its role in inducing quiescence, Tribbles maintains neuroblast quiescence into larval life by inhibiting Akt activation [73]. Interestingly, the authors demonstrate that G2-arrested stem cells are reactivated more quickly in the larval nervous system compared to G0-arrested stem cells. This finding opens many questions regarding the significance and regulation of G2 quiescence: how broadly is G2 quiescence used in other stem cells in flies or mammals? What molecular signatures distinguish G2-quiescent cells from the G2 phase of actively cycling cells? Further work will be needed to link these newest observations with the previous body of work describing spatial and temporal regulation of neuroblast entry into quiescence.

The transcriptional regulator Prospero has been well characterized as an inhibitor of proliferation in the nervous system. Prospero is suggested to act as a "binary switch" between neural stem cells and differentiated neural progeny: targets of Prospero include genes involved in cell cycle regulation, neuroblast identity, the temporal series and neuronal differentiation [77]. In addition to its role in differentiation, Prospero has also been implicated in neuroblast quiescence, as embryonic neuroblasts express a pulse of transient nuclear Prospero that is required for their quiescence (Figure 6B; [78]). Inhibition of embryonic temporal series progression with *pdm* and *castor* mutants did not prevent quiescence, indicating that, while the timing of quiescence is shifted, the mechanism leading to accumulation of nuclear Prospero is independent of the temporal series [78]. In light of the recent work from the Brand lab [73], it is of strong interest to determine how nuclear Prospero could lead to reversible G2 quiescence in neuroblasts at the end of embryogenesis, while promoting differentiation in neuronal progeny (Figure 6B). Differences in target gene sensitivity to levels in Prospero protein may be involved: a threshold model was introduced by Lai & Doe (2014) [78] to explain differential repression of target genes. The threshold model could encompass both contexts in which target gene expression is solely dictated by Prospero levels, and situations in which Prospero cooperates with co-factors whose expression patterns dictate the identity and amplitude of target gene expression. An additional layer to this model is the possible regulation of co-factors by Prospero itself.

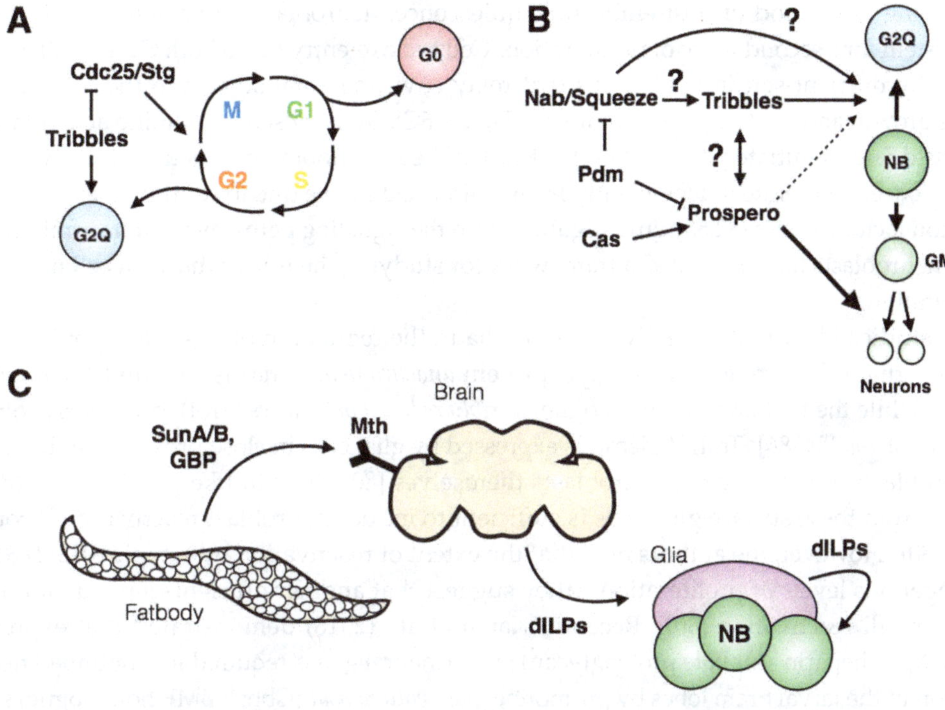

Figure 6. Regulation of neuroblast quiescence and reactivation. (**A**) The majority of post-embryonic neuroblasts exit the cell cycle at G2 phase (G2Q), while the remaining ~25% display a canonical G0 arrest [73]. Tribbles promotes and maintains quiescence through degradation of Cdc25/Stg and inhibition of Akt signaling (not shown). (**B**) Prospero regulates neural differentiation and is required for neuroblast quiescence, but the mechanism underlying its ability to induce two different cell fates is unclear. The connections between G2Q and the Prospero regulatory network are also unknown. (**C**) Secreted factors from the larval fatbody are sensed by the brain, leading to secretion of insulin-like peptides (dILPs). dILPs are also expressed by stellate surface glia, and these act on the neuroblast to promote growth and proliferation.

Context-specific regulation of quiescence entry can be examined by comparing neuroblasts across anatomical regions of the VNC. The timing of quiescence entry differs for a given embryonic neuroblast between the thorax and abdomen [79], indicating that both temporal and spatial signals influence the mechanisms regulating entry into quiescence. The Hox gene *antennapedia* (Antp) is required for quiescence of the thoracic NB3-3, while *abdA* expression is sufficient to induce extended proliferation of thoracic NB3-3, copying the endogenous division pattern of the abdominal NB3-3 [79]. Perhaps intuitively, the quiescent state itself prevents progression through the temporal series: continued expression of the late temporal series members Castor and Grainyhead ensures a retained memory of the temporal identity of the neuroblast upon cell cycle re-entry [79]. However, the molecular details of this inhibition are unclear. The regulation of embryonic neuroblast quiescence at the spatial and temporal level should ultimately coincide with the induction of a transient Prospero pulse and the expression of Tribbles in post-embryonic neuroblasts (Figure 6B). Tsuji et al. (2008) placed the transcription factors Nab and Squeeze downstream of the temporal series and showed their necessity for quiescence entry. It remains to be tested whether Nab and Squeeze represent the intermediaries to Prospero and Tribbles induction (Figure 6B), and whether their own expression is under control of spatial factors as well as the temporal regulation described in Tsuji et al. (2008). The reversible quiescence of post-embryonic neuroblasts represents a mechanism of cell cycle exit fundamentally distinct from Prospero-mediated terminal differentiation of neuronal progeny, and much remains to be uncovered with respect to its regulation and molecular signatures.

5. Induction of Neuroblast Cell Cycle Re-Entry

Following the period of post-embryonic quiescence, neuroblasts are reactivated in the larval nervous system for a second wave of proliferation. Cell cycle re-entry (from both G2 and G0 quiescence) is induced by nutrient-sensing pathways that relay environmental signals to the nervous system, promoting growth and continued development (Figure 6C). In the absence of amino acid intake, larval neuroblasts do not reinitiate proliferation [72,80,81]. The mushroom body neuroblasts, which do not enter quiescence, proliferate independently from amino acid intake due to continued expression of the transcription factor Eyeless [82,83]. Investigations into the signaling pathways that transmit nutritional signals to neuroblasts have provided a framework for studying the role of the stem cell niche on stem cell proliferation.

Early studies identified a positive role for glia in the regulation of post-embryonic neuroblast quiescence: glial cell secretion of the glycoprotein *anachronism* (Ana) is required for maintaining quiescence, while the fly Perlecan homologue, *terribly reduced optic lobes* (Trol), promotes proliferation downstream of *ana* [84–86]. Trol/Perlecan is expressed by glial cells in close proximity to larval central brain neuroblasts, but not by the neuroblasts themselves [80]. Insulin-like peptides 2/6 (dILP2/6) production from these surface glial cells is sufficient to induce neuroblast reactivation through Akt signaling [80]. However, the authors note that the extent of reactivation by ectopic dILP2/6 is lower than endogenous levels of proliferation—they suggest that another nutrient-dependent mitogen is required for full reactivation [80]. Recently Kanai et al. (2018) demonstrated that expression of *Dally-like* (Dlp; a heparin sulphate proteoglycan) by perineural glia is required for continued neuroblast proliferation in the larval brain lobes by promoting *glass bottom boat* (Gbb; a BMP homologue) signaling within neuroblasts [87]. To test the role of glial-derived secreted factors in promoting larval neuroblast proliferation, the authors used a temperature sensitive allele of *shibire* (*shi-ts*). *shibire* is required for exocytosis and its loss of function at the restrictive temperature prevents the release of secreted factors by glia. Kanai et al. (2018) determined that inhibition of glial exocytosis with the *shi-ts* allele reduced neuroblast BrdU incorporation, confirming that glial-derived secreted factors promote neuroblast proliferation. In this study, the *shi-ts* allele was inactivated at the restrictive temperature prior to neuroblast reactivation (24 h after egg laying). Some component of the reduced neuroblast proliferation in these larvae may therefore be due to a lack of cell cycle re-entry of the post-embryonic neuroblasts, in addition to a loss of factors that promote continued proliferation following reactivation.

While these studies demonstrate the role of dILPs and other secreted factors in larval neuroblast proliferation, the mechanism through which the glycoprotein Ana functions is less clear. *ana* transcript was first described to be expressed specifically in glia [84], but a recent study detected *ana* transcription in neuronal populations marked by the *elav-GAL4* driver [88]. The authors also determined that the *ana* mRNA is a target of *miR-124*, introducing an additional layer to the regulation of neuroblast proliferation [88]. It remains unclear whether Ana functions purely non-autonomously on neuroblasts, and which population of cells is the relevant source of secreted Ana protein.

As described briefly above, the glia-neuroblast signaling axis in the larval CNS lies downstream of a yet-to-be-identified "fat body-derived mitogen", which would connect incoming nutritional information from the digestive system of feeding larvae to dILP secretion by the insulin-producing cells within the central brain. Recent studies have identified fat body-secreted signals that induce insulin production, offering candidates for the fat body-derived mitogen (Figure 6C, reviewed in [89]). Delanoue et al. (2016) [90] determined that the secreted peptides SunA and SunB, gene products of *stunted*, are required for larval body growth and act through Methuselah, a GPCR found on insulin-secreting cells in the brain. Additionally, Koyama & Mirth (2016) [91] identified Growth Blocking Peptide (GBP) 1 and 2 as activators of insulin secretion. GBP1 was recently shown to physically interact with Methuselah-like receptor-10 (Mthl10; [92]), indicating a possible point of convergence of peptide signaling from the fat body to the Methuselah family of GPCRs in the brain. While these candidates are promising, their role in the reactivation of neuroblast proliferation has not yet been tested directly.

Neuroblast reactivation during larval development is ultimately dependent on the coupling of nutritional signals to the mitotic machinery to execute cell division. The recent observation that the majority of post-embryonic neuroblasts are arrested in G2 indicates that cell cycle re-entry of most larval neuroblasts occurs at the beginning of mitosis. Notably, *chromator* (Chro), a member of the spindle matrix complex, is essential for larval neuroblasts to respond to insulin/PI3K signaling in order to reactivate [93]. In addition to their role in the spindle matrix, both Chromator and its binding partner, East, associate with known transcriptional regulators and have been shown to modulate gene expression through their association with chromatin [94–98]. Notably, Chromator and East are required for expression of Grainyhead and repression of Prospero in larval central brain neuroblasts [93]. These observations could suggest a role for the spindle matrix complex to act as a signaling hub to integrate nutritional stimuli via insulin/PI3K and promote the execution of mitosis. Reactivation could potentially occur through direct interactions of the spindle matrix complex with other components of the kinetochore machinery, or through indirect transcriptional mechanisms.

6. Control of Stem Cell Apoptosis

Drosophila has been a hugely informative model for the study of programmed cell death (PCD) for decades, since the genetic control of apoptosis was initially described in the early nineties [2,5,99,100]. Neuroblast apoptosis is induced by expression of the cell death genes *reaper (rpr)*, *grim* and *sickle (skl)*, members of the Reaper, Hid and Grim (RHG) gene family of apoptosis regulators [101,102]. These genes are clustered within the genome and are required in various combinations for virtually all embryonic cell death [2,101,102]. In non-apoptotic conditions, caspases are constitutively kept inactive by the *Drosophila Inhibitor of Apoptosis Protein* (DIAP1, Figure 7A left; [103–105]). RHG proteins bind to DIAP1 and competitively displace the bound caspases, leading to their cleavage and activation, while also promoting increased turnover of DIAP1 (Figure 7A right, [105–111]). A benefit of *Drosophila* as a model system for studying PCD is the stereotyped pattern of cell death during development, allowing investigations into the upstream regulatory pathways responsible for dictating when and where cell death is initiated *in vivo*. The combined mechanisms of spatial and temporal signals that regulate neuroblast apoptosis control cell fate with exquisite precision. The regulation and execution of apoptosis in *Drosophila* neuroblasts has been the subject of recent in-depth reviews [112,113].

Most developmental apoptosis of neuroblasts occurs in the abdominal segments of the embryonic nervous system: 27 of 30 abdominal neuroblasts within each hemisegment will die by apoptosis late in embryonic development. A small number of thoracic neuroblasts die at the same time, but little is known about the regulation of their death. A second wave of apoptosis occurs during the third instar larval stage and eliminates the remaining abdominal neuroblasts. Thus, abdominal neuroblasts have proven a useful model for the study of upstream factors that regulate the induction of apoptosis. Our lab has identified an intergenic enhancer within the RHG gene locus called the Neuroblast Regulatory Region (NBRR) that is required for embryonic and larval neuroblast apoptosis by regulating *grim*, *rpr* and *skl* through long-range interactions (Figure 7B; [102,114]). The NBRR is required for full expression of these genes, and an intergenic deletion that removes the NBRR (*Df(3L)MM3*) rescues many, but not all, neuroblasts from apoptosis [102]. As some death still occurs in these mutants, there are likely additional unidentified enhancer regions within the RHG locus that contribute to the execution of apoptosis.

Perhaps unsurprisingly, the Hox gene *abdA*, specifically expressed in abdominal segments, has been shown to regulate neuroblast apoptosis at both embryonic and larval stages [3,20,114]. *abdA* is required for death of embryonic abdominal neuroblasts and is enough to induce apoptosis of thoracic neuroblasts. A reporter of NBRR activation is ectopically induced by *abdA*-misexpression, and this enhancer sequence was recently shown to be bound directly by AbdA in vitro [114,115]. Upstream of *abdA*, our lab showed that expression of the Notch ligand Delta by neural progeny is required for embryonic neuroblast death: progeny-induced Notch signaling within the neuroblast promotes AbdA expression and leads to activation of apoptosis through the NBRR [114]. The mechanism through which AbdA binding to the NBRR activates RHG gene expression remains unclear. We have recently shown that members of the cohesin complex are required for neuroblast death, consistent with a model of cohesin-mediated long-range interactions between the NBRR and the *rpr* and *grim* promoters (Figure 7C; [116]). In addition, we determined that the DNA-binding protein Cut is required for apoptosis, and loss of Cut protein is associated with an increase in the repressive heterochromatin mark H3K27me3 within the RHG locus, and globally in neuroblasts, during embryogenesis [116]. These findings support a role for chromatin architecture and long-range DNA interactions in the transcriptional regulation of the RHG genes.

While Hox genes determine the spatial dimension of apoptosis induction, the timing of embryonic abdominal neuroblast apoptosis is controlled by the temporal series. There is a fundamental link between the molecular clock of the temporal series and neuroblast fate: the transcription factors *Dichaete* (D) and *grainyhead* (Grh) both regulate apoptosis and are targets of the late embryonic temporal series factor Castor (see Figure 4A; [4,117,118]). Specifically, Grh is required for apoptosis and has recently been shown to bind to the NBRR [115], while Dichaete misexpression leads to ectopic neuroblast survival [4]. These results suggest that upon completion of the embryonic temporal series and expression of Cas, loss of D expression and gain of Grh expression would activate the NBRR and lead to RHG gene expression. However, Grh expression is constitutive in many embryonic neuroblasts that do not undergo apoptosis [118]. In addition, as the temporal series is shared across the thorax and abdomen, it is unclear how Grh expression leads to abdomen-specific induction of apoptosis. The most obvious explanation would involve combinatorial activity of Grh with Hox genes, leading to terminal division in the thorax and apoptosis in the abdomen. However, this model has not yet been directly tested, and the differences in Grh/Hox gene downstream targets across different regions of the VNC are unknown. In addition, the apoptosis of a small number of thoracic neuroblasts indicates additional levels of neuroblast-specific regulation of cell death, even within a given anatomical region.

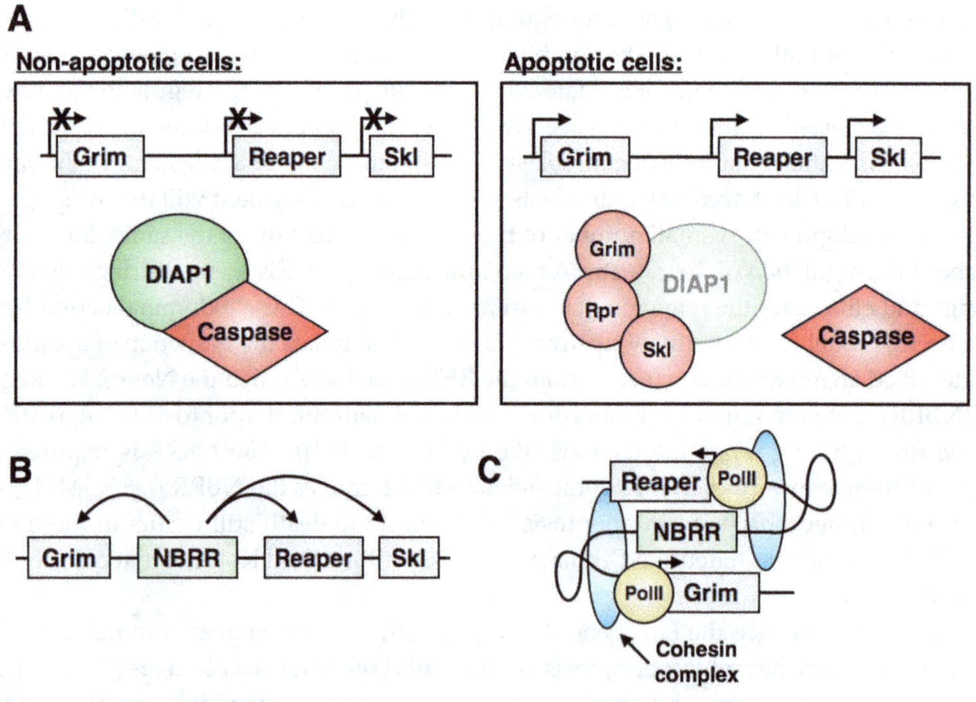

Figure 7. Regulation of neuroblast apoptosis. (**A**) The RHG genes *grim, reaper* and *sickle* are inactive in non-apoptotic cells, allowing DIAP1 to sequester and inhibit caspases. Apoptosis is induced following

transcriptional activation of the RHG genes. RHG binding to DIAP1 promotes its turnover and releases caspases to allow their activation. (**B**) The intergenic enhancer NBRR is required for neuroblast apoptosis and for expression of the surrounding RHG genes. (**C**) A proposed model for regulation of RHG genes by long-range DNA interactions between the NBRR and RHG promoters mediated by the cohesin complex.

7. Regulation of Terminal Division

While both embryonic and post-embryonic abdominal neuroblasts are eliminated by apoptosis, the post-embryonic thoracic and central brain neuroblasts undergo a terminal division in the larval CNS (see Figure 3A). The final symmetrical division of these neuroblasts is induced by nuclear accumulation of Prospero, which leads to down-regulation of self-renewal genes and promotes differentiation [4,77]. Similar to the apoptotic fate decision in the abdominal post-embryonic neuroblasts, terminal division of post-embryonic thoracic neuroblasts lies downstream of the temporal series. In *castor* and *seven-up* mutants, which do not complete the temporal series, larval neuroblasts in the central brain and thorax continue to proliferate during adulthood [4]. However, it was recently reported that loss of Castor within the NB5-6 lineage did not result in changes in Prospero expression or localization during mid-embryogenesis [119]. It remains to be tested whether this reflects lineage-specific responses to changes in the temporal series, or whether embryonic and post-embryonic neuroblasts differ in their sensitivity to the induction of nuclear Prospero.

Downstream of the temporal series, *hedgehog* (Hh) signaling is necessary and sufficient for cell cycle exit of larval central brain neuroblasts. Hh expression is promoted by Cas, and leads to nuclear Prospero accumulation and mitotic exit [120]. A more recent study implicated RNA-binding proteins *IGF-II mRNA-binding protein* (Imp) and *Syncrip* (Syp) in the regulation of Prospero localization in central brain neuroblasts (Figure 8A; [121]). The expression patterns of Imp and Syp vary through development: Imp is co-expressed with the early larval factors Castor, Seven-Up and Chinmo, while Syp is expressed with the late factors Broad and E93 (Figure 8B; [26,27,122]). *pros* mRNA was identified as a target of Syp by RNA-immunoprecipitation, suggesting a direct role in regulating Prospero protein levels [123]. While Imp is sufficient to inhibit terminal division in the larval CNS, Syp misexpression does not induce Prospero protein accumulation or early terminal division, indicating that it may act permissively rather than directly [121]. The molecular details that connect Imp/Syp and temporal signaling to the Hedgehog pathway remain unknown.

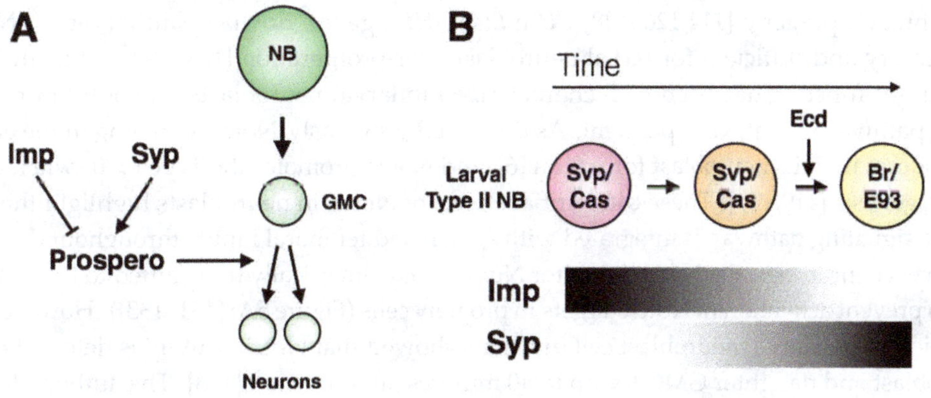

Figure 8. Regulation of neuroblast terminal division. (**A**) The RNA-binding proteins Imp and Syp have opposing effects on Prospero-mediated terminal division. Imp is sufficient to prevent cell cycle exit, but Syp may play an indirect permissive role for Prospero function. (**B**) Imp and Syp expression correlates to progression of the larval temporal series. Imp expression is down-regulated following the pulse of ecdysone, but Syp expression begins prior to this pulse.

Much focus has been placed on the transcription factor networks that regulate stem cell decisions, but, as seen with neuroblast reactivation, metabolic needs also influence proliferation of stem cells. The mushroom body neuroblasts, which continue to proliferate for 96 h in the pupal brain, are finally eliminated by apoptosis induced by insulin/PI3K signaling [124]. Similarly, the terminal division of larval central brain neuroblasts is driven by metabolic changes within the stem cell: release of the steroid hormone ecdysone promotes metamorphosis and pupariation, while inducing a switch in neuroblast growth mode [32]. Ecdysone signaling causes larval central brain neuroblasts to undergo "shrinkage", where each asymmetric division is accompanied by a reduction in neuroblast cell volume. In an unbiased screen, the authors identified the Mediator complex and oxidative phosphorylation (OxPhos) machinery as drivers of reductive neuroblast divisions. Loss of Mediator complex members, ecdysone signaling or OxPhos components all prevented shrinkage and increased neuroblast proliferation [32]. Similar to the central brain neuroblast shrinkage, reductive divisions are observed in embryonic neuroblasts [125]. However, these do not result in terminal division in the embryo, and it is not known whether embryonic shrinkage is ecdysone-dependent. In addition, it remains to be determined how the OxPhos switch is linked to Prospero localization.

Interestingly, terminal division and apoptosis seem to be mutually exclusive fates: apoptosis-deficient larval neuroblasts (homozygous for deletion of the RHG genes) display continuously proliferating abdominal neuroblasts [3]. This finding suggests that larval abdominal neuroblasts do not accumulate nuclear Prospero prior to their death. However, it remains unclear whether these ectopic abdominal neuroblasts continue to proliferate indefinitely, as terminal division of thoracic and central brain neuroblasts occurs more than 24 h later than larval abdominal neuroblast apoptosis, and this later time period was not examined in the study by Bello et al (2003) [3]. It is, therefore, unclear whether cell death-defective abdominal neuroblasts are competent to respond to systemic ecdysone signaling by undergoing terminal division, or whether this fate decision is truly restricted to larval neuroblasts of the thorax and brain.

8. Misregulation of Stem Cell and Progeny Identity

The previous sections have discussed endogenous differences in stem cell proliferation patterns across the *Drosophila* CNS, as well as the regulatory inputs that control quiescence and cell cycle re-entry. Throughout these dynamic changes, stem cell identity must be tightly regulated. Loss of stemness would inhibit tissue growth and could lead to hypoplasia, while gain of stemness in progeny cells could lead to uncontrolled proliferation and a tumor-like state. A potent inducer of the neuroblast identity is Notch: overexpression of the active Notch intracellular domain (N^{icd}) leads to massive tissue overgrowth and an increase in the number of neuroblasts at the expense of differentiated progeny [114,126–128]. The *E(spl)/HES* genes, downstream targets of Notch, are both necessary and sufficient for larval neuroblast over-proliferation [128,129]. While this effect of Notch gain-of-function has been well characterized in larval neuroblasts, the output of the Notch signaling pathway is context-dependent. As discussed previously, Notch signaling in the embryonic neuroectoderm inhibits neuroblast formation [6], and Notch promotes the Type I > 0 switch at the end of embryogenesis [119,130]. These differential effects of Notch in neuroblasts highlight the extent to which this signaling pathway is integrated with spatial and temporal inputs throughout development.

In larval neuroblasts, the Notch inhibitor Numb is asymmetrically transmitted to the GMC during mitosis to prevent activation of Notch targets in progeny cells (Figure 5A; [131–133]). However, a recent examination of live larval neuroblast cell divisions showed that Notch activity is detected equally in the neuroblast and daughter GMC for up to 40 minutes after division [128]. The authors determined that the super elongation complex (SEC) acts as a signal amplifier to strengthen Notch activity within the neuroblast: the SEC promotes expression of *E(spl)/HES* genes, which in turn promote expression of SEC members (Figure 9A). This model provides direct mechanistic evidence for how Notch activity is restricted to, and amplified within, neuroblasts.

Interestingly, the Notch targets *E(spl)* and *dpn* are required to maintain neuroblast identity through the post-embryonic quiescence period [129], but the molecular details are not clear. It was very recently reported that the v-ATPase complex, an intracellular transmembrane proton pump, and a v-ATPase-interacting protein Vap33, are required for Notch signaling and neuroblast regrowth following quiescence [134,135]. v-ATPase and Notch form a feedback regulatory loop [134], reminiscent of the SEC-Notch signaling connection. These examples highlight the central role of Notch signaling in neuroblast identity and offer a common model of regulatory positive feedback through which this identity can be maintained.

Loss of cell identity in differentiated cells can lead to a gain of stem-like behavior, associated with ectopic proliferation and loss of differentiation markers. A prominent example is the phenotype resulting from *prospero* loss of function, which leads to massive overgrowth of the nervous system due to lack of repression of cell cycle genes in GMCs and neurons [77]. This phenotype is consistent with an absence of differentiated cell identity, indicating that transmission of Prospero to daughter cells is necessary for the neuronal and/or glial cell identity program. In Type II neuroblast lineages, the INP identity must also be restricted with respect to the parental neuroblast, to prevent dedifferentiation back into a stem cell. The phenotype of *brat* mutants, which show unrestrained proliferation of daughter cells, is consistent with loss of INP identity and gain of stemness [136,137]. Brat promotes INP identity by antagonizing the transcription factor Klumpfuss, which is sufficient to convert INPs into neuroblasts [138]. Multiple additional regulators have been identified that are required for the restriction of INP identity: the transcriptional repressor Earmuff, the EGF signaling component PntP1, the transcription factor Buttonhead and multiple members of the SWI/SNF complex have all been implicated in preventing loss of the INP identity [59,139,140].

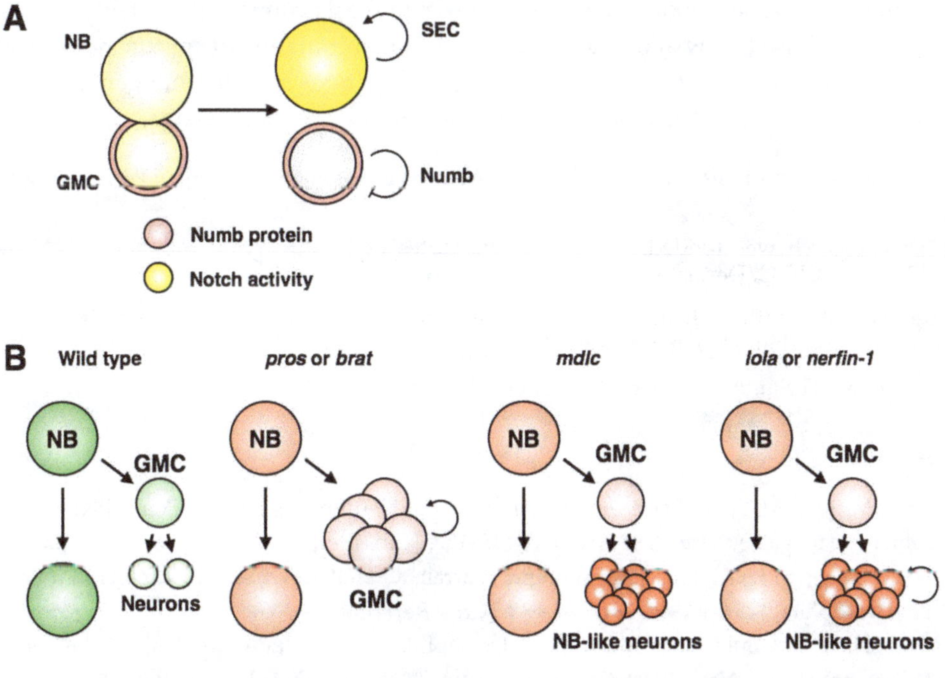

Figure 9. Misregulation of stem cell and differentiated cell identity. (**A**) Notch activity is detected equally in the stem cell and GMC following neuroblast division, but quickly becomes restricted to the neuroblast [128]. The super elongation complex (SEC) promotes a positive feedback loop with Notch signaling in the neuroblast, while Numb inhibits Notch specifically in the GMC. (**B**) Loss of differentiated cell identity can lead to unchecked proliferation in the nervous system. Two classes of dedifferentiation mutants have been identified: *pros* and *brat* mutants fail to restrict the GMC to a single terminal division, while *mdlc*, *lola* or *nerfin-1* mutants display ectopic expression of neuroblast markers in neurons. *lola* and *nerfin-1* mutants also result in ectopic proliferation of neuronal cells.

In contrast to these examples, another class of mutants has been identified in which *bona fide* dedifferentiation of neurons occurs (Figure 9B; reviewed in [141]). Neuronal dedifferentiation has been described following loss of function of each of the zinc finger-containing proteins *midlife crisis* (Mdlc), *longitudinals lacking* (Lola) and *nervous fingers 1* (Nerfin-1) [142–144]. While Mdlc was found to regulate Prospero expression through splicing [142], Nerfin-1 promotes neuronal differentiation in a Prospero-independent manner [144]. Lola shares many target genes with Prospero [143], suggesting that these genes may act redundantly. However, GMCs are able to differentiate properly in *lola* mutants, but not *pros* mutants, indicating that they are functionally distinct in the context of differentiation (Figure 8B; [143]). It remains unclear how these three examples cooperate to retain the differentiated state, or if there is cross-regulation between the pathways. Lola and Nerfin-1 proteins are excluded from neuroblasts, but Midlife Crisis is expressed broadly throughout the nervous system, indicating that neural stem cells use distinct mechanisms to repress the activity of these regulators of neuronal identity to prevent premature differentiation of the progenitor pool.

9. Concluding Thoughts

The concept of stem cell identity has in recent years shown increasing granularity: the advent of single-cell transcriptomics has unveiled the vast amount of heterogeneity within populations of cells previously thought to be homogenous. The *Drosophila* CNS offers a complex but experimentally amenable model for investigating the regulation of stem cell behavior. Populations of neural stem cells throughout the CNS have in common the ability to generate neural progeny, and they express many of the same molecular markers. Despite these shared features, their identities and cell-fate decisions are fundamentally distinct. The pathways and networks shared by *Drosophila* neural stem cells have been well characterized, providing a stepping-stone to begin to tease out the differences inherent to these populations. The genetic tools made available within the fly community, notably from the Janelia FlyLight Project [145–148], now allow identification and manipulation of specific neuroblasts *in vivo*. These tools offer an unprecedented opportunity for investigations into the endogenous heterogeneity of stem cell behavior and highlight the power of *Drosophila* as a model system.

Author Contributions: Writing–Original Draft, K.H. Writing–Review and Editing, K.W. Funding acquisition, K.W.

Funding: This research was funded by National Institutes of Health, grant numbers R03NS092063 and R01GM110477, and a CBRC/Dermatology seed grant to K.W.

Acknowledgments: The authors thank our colleagues and members of the White and Dyson labs for discussions and comments that contributed to this review.

Conflicts of Interest: The authors declare no conflict of interest.

References

1. Truman, J.W.; Bate, M. Spatial and Temporal Patterns of Neurogenesis in the Central Nervous System of Drosophila melanogaster. *Dev. Biol.* **1988**, *125*, 145–157. [CrossRef]
2. White, K.; Grether, M.E.; Abrams, J.M.; Young, L.; Farrell, K.; Steller, H. Genetic Control of Programmed Cell Death in Drosophila. *Science* **1994**, *264*, 677–683. [CrossRef] [PubMed]
3. Bello, B.C.; Hirth, F.; Gould, A.P. A Pulse of the Drosophila Hox Protein Abdominal-A Schedules the End of Neural Proliferation via Neuroblast Apoptosis. *Neuron* **2003**, *37*, 209–219. [CrossRef]
4. Maurange, C.; Cheng, L.; Gould, A.P. Temporal Transcription Factors and Their Targets Schedule the End of Neural Proliferation in Drosophila. *Cell* **2008**, *133*, 891–902. [CrossRef] [PubMed]
5. Abrams, J.M.; White, K.; Fessier, L.I.; Steller, H. Programmed cell death during Drosophila embryogenesis. *Development* **1993**, *117*, 29–43. [PubMed]
6. Campos-Ortega, J.A.; Jan, Y.N. Genetic and Molecular Bases of Neurogenesis in Drosophila melanogaster. *Ann. Rev. Neurosci.* **1991**, *14*, 399–420. [CrossRef] [PubMed]
7. Nambu, P.A.; Nambu, J.R. The Drosophila fish-hook gene encodes a HMG domain protein essential for segmentation and CNS development. *Development* **1996**, *122*, 3467–3475. [PubMed]

8. Overton, P.M.; Meadows, L.A.; Urban, J.; Russell, S. Evidence for differential and redundant function of the Sox genes Dichaete and SoxN during CNS development in Drosophila. *Development* **2002**, *129*, 4219–4228. [PubMed]
9. Zhao, G.; Skeath, J.B. The Sox-domain containing gene Dichaete/fish-hook acts in concert with vnd and ind to regulate cell fate in the Drosophila neuroectoderm. *Development* **2002**, *129*, 1165–1174. [PubMed]
10. Buescher, M.; Hing, F.S.; Chia, W. Formation of neuroblasts in the embryonic central nervous system of Drosophila melanogaster is controlled by SoxNeuro. *Development* **2002**, *129*, 4193–4203. [PubMed]
11. Aleksic, J.; Ferrero, E.; Fischer, B.; Shen, S.P.; Russell, S. The role of Dichaete in transcriptional regulation during Drosophila embryonic development. *BMC Genom.* **2013**, *14*, 861–878. [CrossRef] [PubMed]
12. Ashraf, S.I.; Ip, Y.T. The Snail protein family regulates neuroblast expression of inscuteable and string, genes involved in asymmetry and cell division in Drosophila. *Development* **2001**, *128*, 4757–4767. [PubMed]
13. Cai, Y.; Chia, W.; Yang, X. A family of snail-related zinc finger proteins regulates two distinct and parallel mechanisms that mediate Drosophila neuroblast asymmetric divisions. *EMBO J.* **2001**, *20*, 1704–1714. [CrossRef] [PubMed]
14. Bahrampour, S.; Gunnar, E.; Jonsson, C.; Ekman, H.; Thor, S. Neural Lineage Progression Controlled by a Temporal Proliferation Program. *Dev. Cell* **2017**, *43*, 332–348.e4. [CrossRef] [PubMed]
15. Doe, C.Q. Molecular markers for identified neuroblasts and ganglion mother cells in the Drosophila central nervous system. *Development* **1992**, *116*, 855–863. [PubMed]
16. Skeath, J.B. At the nexus between pattern formation and cell-type specification: The generation of individual neuroblast fates in the Drosophila embryonic central nervous system. *BioEssays* **1999**, *21*, 922–931. [CrossRef]
17. Becker, T.; Technau, G.M. Single cell transplantation reveals interspecific cell communication in Drosophila chimeras. *Development* **1990**, *109*, 821–832. [PubMed]
18. Prokop, A.; Technau, G.M. Early tagma-specific commitment of Drosophila CNS progenitor NB1-1. *Development* **1994**, *120*, 2557–2578.
19. Udolph, G.; Luer, K.; Bossing, T.; Technau, G. Commitment of CNS progenitors along the dorsoventral axis of Drosophila neuroectoderm. *Science* **1995**, *269*, 1278–1281. [CrossRef] [PubMed]
20. Prokop, A.; Bray, S.; Harrison, E.; Technau, G.M. Homeotic regulation of segment-specific differences in neuroblast numbers and proliferation in the Drosophila central nervous system. *Mech. Dev.* **1998**, *74*, 99–110. [CrossRef]
21. Furst, A.; Mahowald, A.P. Cell division cycle of cultured neural precursor cells from Drosophila. *Dev. Biol.* **1985**, *112*, 467–476. [CrossRef]
22. Birkholz, O.; Rickert, C.; Berger, C.; Urbach, R.; Technau, G.M. Neuroblast pattern and identity in the Drosophila tail region and role of doublesex in the survival of sex-specific precursors. *Development* **2013**, *140*, 1830–1842. [CrossRef] [PubMed]
23. Isshiki, T.; Pearson, B.; Holbrook, S.; Doe, C.Q. Drosophila Neuroblasts Sequentially Express Transcription Factors which Specify the Temporal Identity of Their Neuronal Progeny. *Cell* **2001**, *106*, 511–521. [CrossRef]
24. Grosskortenhaus, R.; Pearson, B.J.; Marusich, A.; Doe, C.Q. Regulation of Temporal Identity Transitions in Drosophila Neuroblasts. *Dev. Cell* **2005**, *8*, 193–202. [CrossRef] [PubMed]
25. Doe, C.Q. Temporal Patterning in the Drosophila CNS. *Annu. Rev. Cell Dev. Biol.* **2017**, *33*, 219–240. [CrossRef] [PubMed]
26. Syed, M.H.; Mark, B.; Doe, C.Q. Steroid hormone induction of temporal gene expression in Drosophila brain neuroblasts generates neuronal and glial diversity. *eLife* **2017**, *6*, e26287. [CrossRef] [PubMed]
27. Ren, Q.; Yang, C.P.; Liu, Z.; Sugino, K.; Mok, K.; He, Y.; Ito, M.; Nern, A.; Otsuna, H.; Lee, T. Stem Cell-Intrinsic, Seven-up-Triggered Temporal Factor Gradients Diversify Intermediate Neural Progenitors. *Curr. Biol.* **2017**, *27*, 1303–1313. [CrossRef] [PubMed]
28. Homem, C.C.F.; Repic, M.; Knoblich, J.A. Proliferation control in neural stem and progenitor cells. *Nat. Rev. Neurosci.* **2015**, *16*, 647. [CrossRef] [PubMed]
29. Bello, B.C.; Izergina, N.; Caussinus, E.; Reichert, H. Amplification of neural stem cell proliferation by intermediate progenitor cells in Drosophila brain development. *Neural Dev.* **2008**, *3*, 5. [CrossRef] [PubMed]
30. Bowman, S.K.; Rolland, V.; Betschinger, J.; Kinsey, K.A.; Emery, G.; Knoblich, J.A. The Tumor Suppressors Brat and Numb Regulate Transit-Amplifying Neuroblast Lineages in Drosophila. *Dev. Cell* **2008**, *14*, 535–546. [CrossRef] [PubMed]

31. Boone, J.Q.; Doe, C.Q. Identification of Drosophila type II neuroblast lineages containing transit amplifying ganglion mother cells. *Dev. Neurobiol.* **2008**, *68*, 1185–1195. [CrossRef] [PubMed]
32. Homem, C.C.F.; Steinmann, V.; Burkard, T.R.; Jais, A.; Esterbauer, H.; Knoblich, J.A. Ecdysone and Mediator Change Energy Metabolism to Terminate Proliferation in Drosophila Neural Stem Cells. *Cell* **2014**, *158*, 874–888. [CrossRef] [PubMed]
33. Baumgardt, M.; Karlsson, D.; Salmani, B.Y.; Bivik, C.; MacDonald, R.B.; Gunnar, E.; Thor, S. Global programmed switch in neural daughter cell proliferation mode triggered by a temporal gene cascade. *Dev. Cell* **2014**, *30*, 192–208. [CrossRef] [PubMed]
34. Baumgardt, M.; Karlsson, D.; Terriente, J.; Díaz-Benjumea, F.J.; Thor, S. Neuronal Subtype Specification within a Lineage by Opposing Temporal Feed-Forward Loops. *Cell* **2009**, *139*, 969–982. [CrossRef] [PubMed]
35. Karcavich, R.; Doe, C.Q. Drosophila neuroblast 7-3 cell lineage: A model system for studying programmed cell death, Notch/Numb signaling, and sequential specification of ganglion mother cell identity. *J. Comp. Neurol.* **2005**, *481*, 240–251. [CrossRef] [PubMed]
36. Sousa-Nunes, R.; Somers, W.G. Mechanisms of asymmetric progenitor divisions in the Drosophila central nervous system. *Adv. Exp. Med. Biol.* **2013**, *786*, 79–102. [CrossRef] [PubMed]
37. Yu, F.; Kuo, C.T.; Jan, Y.N. Drosophila Neuroblast Asymmetric Cell Division: Recent Advances and Implications for Stem Cell Biology. *Neuron* **2006**, *51*, 13–20. [CrossRef] [PubMed]
38. Knoblich, J.A.; Jan, Y.N. Asymmetric segregation of Numb and Prospero during cell division. *Nature* **1995**, *377*, 624–627. [CrossRef] [PubMed]
39. Hirata, J.; Nakagoshi, H.; Nabeshima, Y.; Matsuzaki, F. Asymmetric segregation of the homeodomain protein Prospero during Drosophila development. *Nature* **1995**, *377*, 627–630. [CrossRef] [PubMed]
40. Broadus, J.; Fuerstenberg, S.; Doe, C.Q. Staufen-dependent localization of *prospero* mRNA contributes to neuroblast daughter-cell fate. *Nature* **1998**, *391*, 792–795. [CrossRef] [PubMed]
41. Schuldt, A.J.; Adams, J.H.; Davidson, C.M.; Micklem, D.R.; Haseloff, J.; St Johnston, D.; Brand, A.H. Miranda mediates asymmetric protein and RNA localization in the developing nervous system. *Genes Dev.* **1998**, *12*, 1847–1857. [CrossRef] [PubMed]
42. Shen, C.P.; Knoblich, J.A.; Chan, Y.M.; Jiang, M.M.; Jan, L.Y.; Jan, Y.N. Miranda as a multidomain adapter linking apically localized Inscuteable and basally localized Staufen and Prospero during asymmetric cell division in Drosophila. *Genes Dev.* **1998**, *12*, 1837–1846. [CrossRef] [PubMed]
43. Ikeshima-Kataoka, H.; Skeath, J.B.; Nabeshima, Y.; Doe, C.Q.; Matsuzaki, F. Miranda directs Prospero to a daughter cell during Drosophila asymmetric divisions. *Nature* **1997**, *390*, 625–629. [CrossRef] [PubMed]
44. Matsuzaki, F.; Ohshiro, T.; Ikeshima-Kataoka, H.; Izumi, H. miranda localizes staufen and prospero asymmetrically in mitotic neuroblasts and epithelial cells in early Drosophila embryogenesis. *Development* **1998**, *125*, 4089–4098. [PubMed]
45. Wodarz, A.; Ramrath, A.; Kuchinke, U.; Knust, E. Bazooka provides an apical cue for Inscuteable localization in Drosophila neuroblasts. *Nature* **1999**, *402*, 544–547. [CrossRef] [PubMed]
46. Wodarz, A.; Ramrath, A.; Grimm, A.; Knust, E. Drosophila atypical protein kinase C associates with Bazooka and controls polarity of epithelia and neuroblasts. *J. Cell Biol.* **2000**, *150*, 1361–1374. [CrossRef] [PubMed]
47. Li, P.; Yang, X.; Wasser, M.; Cai, Y.; Chia, W. Inscuteable and Staufen mediate asymmetric localization and segregation of prospero RNA during Drosophila neuroblast cell divisions. *Cell* **1997**, *90*, 437–447. [CrossRef]
48. Schober, M.; Schaefer, M.; Knoblich, J.A. Bazooka recruits Inscuteable to orient asymmetric cell divisions in Drosophila neuroblasts. *Nature* **1999**, *402*, 548–551. [CrossRef] [PubMed]
49. Kraut, R.; Chia, W.; Jan, L.Y.; Jan, Y.N.; Knoblich, J.A. Role of inscuteable in orienting asymmetric cell divisions in Drosophila. *Nature* **1996**, *383*, 50–55. [CrossRef] [PubMed]
50. Kuchinke, U.; Grawe, F.; Knust, E. Control of spindle orientation in Drosophila by the Par-3-related PDZ-domain protein Bazooka. *Curr. Biol.* **1998**, *8*, 1357–1365. [CrossRef]
51. Mora, N.; Oliva, C.; Fiers, M.; Ejsmont, R.; Soldano, A.; Zhang, T.T.; Yan, J.; Claeys, A.; De Geest, N.; Hassan, B.A. A Temporal Transcriptional Switch Governs Stem Cell Division, Neuronal Numbers, and Maintenance of Differentiation. *Dev. Cell* **2018**, *45*, 53–66.e5. [CrossRef] [PubMed]
52. Skeath, J.B.; Carroll, S.B. Regulation of proneural gene expression and cell fate during neuroblast segregation in the Drosophila embryo. *Development* **1992**, *114*, 939–946. [PubMed]
53. Cabrera, C.V. Lateral inhibition and cell fate during neurogenesis in Drosophila: The interactions between scute, Notch and Delta. *Development* **1990**, *110*, 733–742. [PubMed]

54. Martin-Bermudo, M.D.; Martinez, C.; Rodriguez, A.; Jimenez, F. Distribution and function of the lethal of scute gene product during early neurogenesis in Drosophila. *Development* **1991**, *113*, 445–454. [PubMed]
55. Egger, B.; Boone, J.Q.; Stevens, N.R.; Brand, A.H.; Doe, C.Q. Regulation of spindle orientation and neural stem cell fate in the Drosophila optic lobe. *Neural Dev.* **2007**, *2*, 1. [CrossRef] [PubMed]
56. Hakes, A.E.; Otsuki, L.; Brand, A.H. A newly discovered neural stem cell population is generated by the optic lobe neuroepithelium during embryogenesis in Drosophila melanogaster. *Development* **2018**, 1–7. [CrossRef] [PubMed]
57. Walsh, K.T.; Doe, C.Q. Drosophila embryonic type II neuroblasts: Origin, temporal patterning, and contribution to the adult central complex. *Development* **2017**, *144*, 4552–4562. [CrossRef] [PubMed]
58. Álvarez, J.-A.; Díaz-Benjumea, F.J. Origin and specification of type II neuroblasts in the Drosophila embryo. *Development* **2018**, *145*, dev158394. [CrossRef] [PubMed]
59. Xie, Y.; Li, X.; Deng, X.; Hou, Y.; O'Hara, K.; Urso, A.; Peng, Y.; Chen, L.; Zhu, S. The Ets protein Pointed prevents both premature differentiation and dedifferentiation of Drosophila intermediate neural progenitors. *Development* **2016**, *143*, 3109–3118. [CrossRef] [PubMed]
60. Zhu, S.; Barshow, S.; Wildonger, J.; Jan, L.Y.; Jan, Y.N. Ets transcription factor Pointed promotes the generation of intermediate neural progenitors in Drosophila larval brains. *Proc. Natl. Acad. Sci. USA* **2011**, *108*, 20615–20620. [CrossRef] [PubMed]
61. An, H.; Ge, W.; Xi, Y.; Yang, X. Inscuteable maintains type I neuroblast lineage identity via Numb/Notch signaling in the Drosophila larval brain. *J. Genet. Genom.* **2017**, *44*, 151–162. [CrossRef] [PubMed]
62. Udolph, G.; Prokop, A.; Bossing, T.; Technau, G.M. A common precursor for glia and neurons in the embryonic CNS of Drosophila gives rise to segment-specific lineage variants. *Development* **1993**, *118*, 765–775. [PubMed]
63. Bossing, T.; Udolph, G.; Doe, C.Q.; Technau, G.M. The Embryonic Central Nervous System Lineages of Drosophila melanogaster: I. Neuroblast Lineages Derived from the Ventral Half of the Neuroectoderm. *Dev. Biol.* **1996**, *179*, 41–64. [CrossRef] [PubMed]
64. Schmid, A.; Chiba, A.; Doe, C.Q. Clonal analysis of Drosophila embryonic neuroblasts: Neural cell types, axon projections and muscle targets. *Development* **1999**, *126*, 4653–4689. [PubMed]
65. Cobeta, I.M.; Salmani, B.Y.; Thor, S. Anterior-Posterior Gradient in Neural Stem and Daughter Cell Proliferation Governed by Spatial and Temporal Hox Control. *Curr. Biol.* **2017**, *27*, 1161–1172. [CrossRef] [PubMed]
66. Salmani, B.Y.; Cobeta, I.M.; Rakar, J.; Bauer, S.; Curt, J.R.; Starkenberg, A.; Thor, S. Evolutionarily conserved anterior expansion of the central nervous system promoted by a common PcG-Hox program. *Development* **2018**, *145*, dev160747. [CrossRef] [PubMed]
67. Berger, C.; Pallavi, S.K.; Prasad, M.; Shashidhara, L.S.; Technau, G.M. A critical role for cyclin E in cell fate determination in the central nervous system of Drosophila melanogaster. *Nat. Cell Biol.* **2005**, *7*, 56–62. [CrossRef] [PubMed]
68. Berger, C.; Kannan, R.; Myneni, S.; Renner, S.; Shashidhara, L.S.; Technau, G.M. Cell cycle independent role of Cyclin E during neural cell fate specification in Drosophila is mediated by its regulation of Prospero function. *Dev. Biol.* **2010**, *337*, 415–424. [CrossRef] [PubMed]
69. Shen, Y.; Wang, L.; Hirose, S.; Zhou, Z.; Liu, Q. The transcriptional factor Apt regulates neuroblast differentiation through activating CycE expression. *Biochem. Biophys. Res. Commun.* **2018**, *499*, 889–894. [CrossRef] [PubMed]
70. Hartenstein, V.; Rudloff, E.; Ortega, J.A.C. The pattern of proliferation of the neuroblasts in the wild-type embryo of Drosophila melanogaster. *Rouxs Arch. Dev. Biol.* **1987**, *196*, 473–485. [CrossRef] [PubMed]
71. Ito, K.; Hotta, Y. Proliferation pattern of postembryonic neuroblasts in the brain of Drosophila melanogaster. *Dev. Biol.* **1992**, *149*, 134–148. [CrossRef]
72. Sousa-Nunes, R.; Yee, L.L.; Gould, A.P. Fat cells reactivate quiescent neuroblasts via TOR and glial insulin relays in Drosophila. *Nature* **2011**, *471*, 508–512. [CrossRef] [PubMed]
73. Otsuki, L.; Brand, A.H. Cell cycle heterogeneity directs the timing of neural stem cell activation from quiescence. *Science* **2018**, *360*, 99–102. [CrossRef] [PubMed]
74. Mata, J.; Curado, S.; Ephrussi, A.; Rørth, P. Tribbles coordinates mitosis and morphogenesis in Drosophila by regulating string/CDC25 proteolysis. *Cell* **2000**, *101*, 511–522. [CrossRef]

75. Grosshans, J.; Wieschaus, E. A genetic link between morphogenesis and cell division during formation of the ventral furrow in Drosophila. *Cell* **2000**, *101*, 523–531. [CrossRef]
76. Seher, T.C.; Leptin, M. Tribbles a cell-cycle brake that coordinates proliferation and morphogenesis during Drosophila gastrulation. *Curr. Biol.* **2000**, *10*, 623–629. [CrossRef]
77. Choksi, S.P.; Southall, T.D.; Bossing, T.; Edoff, K.; de Wit, E.; Fischer, B.E.; van Steensel, B.; Micklem, G.; Brand, A.H. Prospero Acts as a Binary Switch between Self-Renewal and Differentiation in Drosophila Neural Stem Cells. *Dev. Cell* **2006**, *11*, 775–789. [CrossRef] [PubMed]
78. Lai, S.-L.; Doe, C.Q. Transient nuclear Prospero induces neural progenitor quiescence. *eLife* **2014**, *3*, 379–12. [CrossRef] [PubMed]
79. Tsuji, T.; Hasegawa, E.; Isshiki, T. Neuroblast entry into quiescence is regulated intrinsically by the combined action of spatial Hox proteins and temporal identity factors. *Development* **2008**, *135*, 3859–3869. [CrossRef] [PubMed]
80. Chell, J.M.; Brand, A.H. Nutrition-Responsive Glia Control Exit of Neural Stem Cells from Quiescence. *Cell* **2010**, *143*, 1161–1173. [CrossRef] [PubMed]
81. Britton, J.S.; Edgar, B.A. Environmental control of the cell cycle in Drosophila: Nutrition activates mitotic and endoreplicative cells by distinct mechanisms. *Development* **1998**, *125*, 2149–2158. [PubMed]
82. Lin, S.; Marin, E.C.; Yang, C.P.; Kao, C.F.; Apenteng, B.A.; Huang, Y.; O'Connor, M.B.; Truman, J.W.; Lee, T. Extremes of lineage plasticity in the Drosophila brain. *Curr. Biol.* **2013**, *23*, 1908–1913. [CrossRef] [PubMed]
83. Sipe, C.W.; Siegrist, S.E. Eyeless uncouples mushroom body neuroblast proliferation from dietary amino acids in Drosophila. *eLife* **2017**, *6*, e26343. [CrossRef] [PubMed]
84. Ebens, A.J.; Garren, H.; Cheyette, B.N.R.; Zipursky, S.L. The Drosophila anachronism Locus: A Glycoprotein Secreted by Glia Inhibits Neuroblast Proliferation. *Cell* **1993**, *74*, 15–27. [CrossRef]
85. Datta, S. Control of proliferation activation in quiescent neuroblasts of the Drosophila central nervous system. *Development* **1995**, *121*, 1173–1182. [PubMed]
86. Voigt, A.; Pflanz, R.; Schäfer, U.; Jäckle, H. Perlecan participates in proliferation activation of quiescent Drosophila neuroblasts. *Dev. Dyn.* **2002**, *224*, 403–412. [CrossRef] [PubMed]
87. Kanai, M.I.; Kim, M.J.; Akiyama, T.; Takemura, M.; Wharton, K.; O'Connor, M.B.; Nakato, H. Regulation of neuroblast proliferation by surface glia in the Drosophila larval brain. *Sci. Rep.* **2018**, *8*, 3730. [CrossRef] [PubMed]
88. Weng, R.; Cohen, S.M. Drosophila miR-124 regulates neuroblast proliferation through its target anachronism. *Development* **2012**, *139*, 1427–1434. [CrossRef] [PubMed]
89. Droujinine, I.A.; Perrimon, N. Interorgan Communication Pathways in Physiology: Focus on Drosophila. *Annu. Rev. Genet.* **2016**, *50*, 539–570. [CrossRef] [PubMed]
90. Delanoue, R.; Meschi, E.; Agrawal, N.; Mauri, A.; Tsatskis, Y.; McNeill, H.; Léopold, P. Drosophila insulin release is triggered by adipose Stunted ligand to brain Methuselah receptor. *Science* **2016**, *353*, 1553–1556. [CrossRef] [PubMed]
91. Koyama, T.; Mirth, C.K. Growth-Blocking Peptides as Nutrition-Sensitive Signals for Insulin Secretion and Body Size Regulation. *PLoS Biol.* **2016**, *14*, e1002392. [CrossRef]
92. Sung, E.J.; Ryuda, M.; Matsumoto, H.; Uryu, O.; Ochiai, M.; Cook, M.E.; Yi, N.Y.; Wang, H.; Putney, J.W.; Bird, G.S.; et al. Cytokine signaling through Drosophila Mthl10 ties lifespan to environmental stress. *Proc. Natl. Acad. Sci. USA* **2017**, *114*, 13786–13791. [CrossRef] [PubMed]
93. Li, S.; Koe, C.T.; Tay, S.T.; Tan, A.L.; Zhang, S.; Zhang, Y.; Tan, P.; Sung, W.K.; Wang, H. An intrinsic mechanism controls reactivation of neural stem cells by spindle matrix proteins. *Nat. Commun.* **2017**, *8*, 122. [CrossRef] [PubMed]
94. Wasser, M.; Osman, Z.B.; Chia, W. EAST and Chromator control the destruction and remodeling of muscles during Drosophila metamorphosis. *Dev. Biol.* **2007**, *307*, 380–393. [CrossRef] [PubMed]
95. Guruharsha, K.G.; Rual, J.F.; Zhai, B.; Mintseris, J.; Vaidya, P.; Vaidya, N.; Beekman, C.; Wong, C.; Rhee, D.Y.; Cenaj, O.; et al. A Protein Complex Network of Drosophila melanogaster. *Cell* **2011**, *147*, 690–703. [CrossRef] [PubMed]
96. Bohla, D.; Herold, M.; Panzer, I.; Buxa, M.K.; Ali, T.; Demmers, J.; Krüger, M.; Scharfe, M.; Jarek, M.; Bartkuhn, M.; et al. A Functional Insulator Screen Identifies NURF and dREAM Components to Be Required for Enhancer-Blocking. *PLoS ONE* **2014**, *9*, e107765. [CrossRef] [PubMed]

97. Vogelmann, J.; Le Gall, A.; Dejardin, S.; Allemand, F.; Gamot, A.; Labesse, G.; Cuvier, O.; Nègre, N.; Cohen-Gonsaud, M.; Margeat, E.; et al. Chromatin Insulator Factors Involved in Long-Range DNA Interactions and Their Role in the Folding of the Drosophila Genome. *PLoS Genet.* **2014**, *10*, e1004544. [CrossRef] [PubMed]
98. Rhee, D.Y. Transcription Factor Networks in Drosophila melanogaster. *Cell Rep.* **2014**, *8*, 2031–2043. [CrossRef] [PubMed]
99. Chen, P.; Nordstrom, W.; Gish, B.; Abrams, J.M. grim, a novel cell death gene in Drosophila. *Genes Dev.* **1996**, *10*, 1773–1782. [CrossRef] [PubMed]
100. Grether, M.E.; Abrams, J.M.; Agapite, J.; White, K.; Steller, H. The head involution defective gene of Drosophila melanogaster functions in programmed cell death. *Genes Dev.* **1995**, *9*, 1694–1708. [CrossRef] [PubMed]
101. Peterson, C.; Carney, G.E.; Taylor, B.J.; White, K. reaper is required for neuroblast apoptosis during Drosophila development. *Development* **2002**, *129*, 1467–1476. [PubMed]
102. Tan, Y.; Yamada-Mabuchi, M.; Arya, R.; St Pierre, S.; Tang, W.; Tosa, M.; Brachmann, C.; White, K. Coordinated expression of cell death genes regulates neuroblast apoptosis. *Development* **2011**, *138*, 2197–2206. [CrossRef] [PubMed]
103. Yang, Y.; Fang, S.; Jensen, J.P.; Weissman, A.M.; Ashwell, J.D. Ubiquitin Protein Ligase Activity of IAPs and Their Degradation in Proteasomes in Response to Apoptotic Stimuli. *Science* **2000**, *288*, 874–877. [CrossRef] [PubMed]
104. Suzuki, Y.; Nakabayashi, Y.; Takahashi, R. Ubiquitin-protein ligase activity of X-linked inhibitor of apoptosis protein promotes proteasomal degradation of caspase-3 and enhances its anti-apoptotic effect in Fas-induced cell death. *Proc. Natl. Acad. Sci. USA* **2001**, *98*, 8662–8667. [CrossRef] [PubMed]
105. Chai, J.; Yan, N.; Huh, J.R.; Wu, J.W.; Li, W.; Hay, B.A.; Shi, Y. Molecular mechanism of Reaper-Grim-Hid-mediated suppression of DIAP1-dependent Dronc ubiquitination. *Nat. Struct. Biol.* **2003**, *10*, 892–898. [CrossRef] [PubMed]
106. Lisi, S.; Mazzon, I.; White, K. Diverse Domains of THREAD/DIAP1 Are Required to Inhibit Apoptosis Induced by REAPER and HID in Drosophila. *Genetics* **2000**, *154*, 669–678. [PubMed]
107. Wing, J.P.; Schwartz, L.M.; Nambu, J.R. The RHG motifs of Drosophila Reaper and Grim are important for their distinct cell death-inducing abilities. *Mech. Dev.* **2001**, *102*, 193–203. [CrossRef]
108. Zachariou, A.; Tenev, T.; Goyal, L.; Agapite, J.; Steller, H.; Meier, P. IAP-antagonists exhibit non-redundant modes of action through differential DIAP1 binding. *EMBO J.* **2003**, *22*, 6642–6652. [CrossRef] [PubMed]
109. Yokokura, T.; Dresnek, D.; Huseinovic, N.; Lisi, S.; Abdelwahid, E.; Bangs, P.; White, K. Dissection of DIAP1 Functional Domains via a Mutant Replacement Strategy. *J. Biol. Chem.* **2004**, *279*, 52603–52612. [CrossRef] [PubMed]
110. Ryoo, H.D.; Bergmann, A.; Gonen, H.; Ciechanover, A.; Steller, H. Regulation of Drosophila IAP1 degradation and apoptosis by reaper and ubcD1. *Nat. Cell Biol.* **2002**, *4*, 432–438. [CrossRef] [PubMed]
111. Yoo, S.J.; Huh, J.R.; Muro, I.; Yu, H.; Wang, L.; Wang, S.L.; Feldman, R.R.; Clem, R.J.; Müller, H.A.; Hay, B.A. Hid, Rpr and Grim negatively regulate DIAP1 levels through distinct mechanisms. *Nat. Cell Biol.* **2002**, *4*, 416–424. [CrossRef] [PubMed]
112. Arya, R.; White, K. Cell death in development: Signaling pathways and core mechanisms. *Semin. Cell Dev. Biol.* **2015**, *39*, 12–19. [CrossRef] [PubMed]
113. Pinto-Teixeira, F.; Konstantinides, N.; Desplan, C. Programmed cell death acts at different stages of Drosophila neurodevelopment to shape the central nervous system. *FEBS Lett.* **2016**, *590*, 2435–2453. [CrossRef] [PubMed]
114. Arya, R.; Sarkissian, T.; Tan, Y.; White, K. Neural stem cell progeny regulate stem cell death in a Notch and Hox dependent manner. *Cell Death Differ.* **2015**, *22*, 1378–1387. [CrossRef] [PubMed]
115. Khandelwal, R.; Sipani, R.; Rajan, S.G.; Kumar, R.; Joshi, R. Combinatorial action of Grainyhead, Extradenticle and Notch in regulating Hox mediated apoptosis in Drosophila larval CNS. *PLoS Genet.* **2017**, *13*, e1007043. [CrossRef] [PubMed]
116. Arya, R.; Gyonjyan, S.; Harding, K.; Sarkissian, T.; Li, Y.; Zhou, L.; White, K. A Cut/cohesin axis alters the chromatin landscape to facilitate neuroblast death. *bioRxiv* **2018**, 299164. [CrossRef]

117. Cenci, C.; Gould, A. Drosophila Grainyhead specifies late programmes of neural proliferation by regulating the mitotic activity and Hox-dependent apoptosis of neuroblasts. *Development* **2005**, *132*, 3835–3845. [CrossRef] [PubMed]
118. Almeida, M.S.; Bray, S.J. Regulation of post-embryonic neuroblasts by Drosophila Grainyhead. *Mech. Dev.* **2005**, *122*, 1282–1293. [CrossRef] [PubMed]
119. Ulvklo, C.; MacDonald, R.; Bivik, C.; Baumgardt, M.; Karlsson, D.; Thor, S. Control of neuronal cell fate and number by integration of distinct daughter cell proliferation modes with temporal progression. *Development* **2012**, *139*, 678–689. [CrossRef] [PubMed]
120. Chai, P.C.; Liu, Z.; Chia, W.; Cai, Y. Hedgehog Signaling Acts with the Temporal Cascade to Promote Neuroblast Cell Cycle Exit. *PLoS Biol.* **2013**, *11*, e1001494. [CrossRef] [PubMed]
121. Yang, C.P.; Samuels, T.J.; Huang, Y.; Yang, L.; Ish-Horowicz, D.; Davis, I.; Lee, T. Imp and Syp RNA-binding proteins govern decommissioning of Drosophila neural stem cells. *Development* **2017**, *144*, 3454–3464. [CrossRef] [PubMed]
122. Liu, Z.; Yang, C.P.; Sugino, K.; Fu, C.C.; Liu, L.Y.; Yao, X.; Lee, L.P.; Lee, T. Opposing intrinsic temporal gradients guide neural stem cell production of varied neuronal fates. *Science* **2015**, *350*, 317–320. [CrossRef] [PubMed]
123. McDermott, S.M.; Yang, L.; Halstead, J.M.; Hamilton, R.S.; Meignin, C.; Davis, I. Drosophila Syncrip modulates the expression of mRNAs encoding key synaptic proteins required for morphology at the neuromuscular junction. *RNA* **2014**, *20*, 1593–1606. [CrossRef] [PubMed]
124. Siegrist, S.E.; Haque, N.S.; Chen, C.H.; Hay, B.A.; Hariharan, I.K. Inactivation of Both foxo and reaper Promotes Long-Term Adult Neurogenesis in Drosophila. *Curr. Biol.* **2010**, *20*, 643–648. [CrossRef] [PubMed]
125. Fuse, N.; Hisata, K.; Katzen, A.L.; Matsuzaki, F. Heterotrimeric G proteins regulate daughter cell size asymmetry in Drosophila neuroblast divisions. *Curr. Biol.* **2003**, *13*, 947–954. [CrossRef]
126. Wang, H.; Somers, G.W.; Bashirullah, A.; Heberlein, U.; Yu, F.; Chia, W. Aurora-A acts as a tumor suppressor and regulates self-renewal of Drosophila neuroblasts. *Genes Dev.* **2006**, *20*, 3453–3463. [CrossRef] [PubMed]
127. Weng, M.; Golden, K.; Lee, C.-Y. dFezf/Earmuff Maintains the Restricted Developmental Potential of Intermediate Neural Progenitors in Drosophila. *Dev. Cell* **2010**, *18*, 126–135. [CrossRef] [PubMed]
128. Liu, K.; Shen, D.; Shen, J.; Gao, S.M.; Li, B.; Wong, C.; Feng, W.; Song, Y. The Super Elongation Complex Drives Neural Stem Cell Fate Commitment. *Dev. Cell* **2017**, *40*, 537–551.e6. [CrossRef] [PubMed]
129. Zacharioudaki, E.; Magadi, S.S.; Delidakis, C. bHLH-O proteins are crucial for Drosophila neuroblast self-renewal and mediate Notch-induced overproliferation. *Development* **2012**, *139*, 1258–1269. [CrossRef] [PubMed]
130. Bivik, C.; MacDonald, R.B.; Gunnar, E.; Mazouni, K.; Schweisguth, F.; Thor, S. Control of Neural Daughter Cell Proliferation by Multi-level Notch/Su(H)/E(spl)-HLH Signaling. *PLoS Genet.* **2016**, *12*, e1005984. [CrossRef] [PubMed]
131. Guo, M.; Jan, L.Y.; Jan, Y.N. Control of daughter cell fates during asymmetric division: Interaction of Numb and Notch. *Neuron* **1996**, *17*, 27–41. [CrossRef]
132. Spana, E.P.; Doe, C.Q. Numb antagonizes Notch signaling to specify sibling neuron cell fates. *Neuron* **1996**, *17*, 21–26. [CrossRef]
133. Rhyu, M.S.; Jan, L.Y.; Jan, Y.N. Asymmetric Distribution of Numb Protein during Division of the Sensory Organ Precursor Cell Confers Distinct Fates to Daughter Cells. *Cell* **1994**, *76*, 477–491. [CrossRef]
134. Wissel, S.; Harzer, H.; Bonnay, F.; Burkard, T.R.; Neumüller, R.A.; Knoblich, J.A. Time-resolved transcriptomics in neural stem cells identifies a v-ATPase/Notch regulatory loop. *J. Cell Biol.* **2018**, *217*, 3285–3300. [CrossRef] [PubMed]
135. Portela, M.; Yang, L.; Paul, S.; Li, X.; Veraksa, A.; Parsons, L.M.; Richardson, H.E. Lgl reduces endosomal vesicle acidification and Notch signaling by promoting the interaction between Vap33 and the V-ATPase complex. *Sci. Signal.* **2018**, *11*, eaar1976. [CrossRef] [PubMed]
136. Betschinger, J.; Mechtler, K.; Knoblich, J.A. Asymmetric Segregation of the Tumor Suppressor Brat Regulates Self-Renewal in Drosophila Neural Stem Cells. *Cell* **2006**, *124*, 1241–1253. [CrossRef] [PubMed]
137. Lee, C.-Y.; Wilkinson, B.D.; Siegrist, S.E.; Wharton, R.P.; Doe, C.Q. Brat Is a Miranda Cargo Protein that Promotes Neuronal Differentiation and Inhibits Neuroblast Self-Renewal. *Dev. Cell* **2006**, *10*, 441–449. [CrossRef] [PubMed]

138. Xiao, Q.; Komori, H.; Lee, C.-Y. klumpfuss distinguishes stem cells from progenitor cells during asymmetric neuroblast division. *Development* **2012**, *139*, 2670–2680. [CrossRef] [PubMed]
139. Eroglu, E.; Burkard, T.R.; Jiang, Y.; Saini, N.; Homem, C.C.; Reichert, H.; Knoblich, J.A. SWI/SNF Complex Prevents Lineage Reversion and Induces Temporal Patterning in Neural Stem Cells. *Cell* **2014**, *156*, 1259–1273. [CrossRef] [PubMed]
140. Koe, C.T.; Li, S.; Rossi, F.; Wong, J.J.; Wang, Y.; Zhang, Z.; Chen, K.; Aw, S.S.; Richardson, H.E.; Robson, P.; et al. The Brm-HDAC3-Erm repressor complex suppresses dedifferentiation in Drosophila type II neuroblast lineages. *eLife* **2014**, *3*, 2639. [CrossRef] [PubMed]
141. Froldi, F.; Cheng, L.Y. Understanding how differentiation is maintained: Lessons from the Drosophila brain. *Cell. Mol. Life Sci.* **2016**, *73*, 1641–1644. [CrossRef] [PubMed]
142. Carney, T.D.; Struck, A.J.; Doe, C.Q. midlife crisis encodes a conserved zinc-finger protein required to maintain neuronal differentiation in Drosophila. *Development* **2013**, *140*, 4155–4164. [CrossRef] [PubMed]
143. Southall, T.D.; Davidson, C.M.; Miller, C.; Carr, A.; Brand, A.H. Dedifferentiation of Neurons Precedes Tumor Formation in lola Mutants. *Dev. Cell* **2014**, *28*, 685–696. [CrossRef] [PubMed]
144. Froldi, F.; Szuperak, M.; Weng, C.-F.; Shi, W.; Papenfuss, A.T.; Cheng, L.Y. The transcription factor Nerfin-1 prevents reversion of neurons into neural stem cells. *Genes Dev.* **2015**, *29*, 129–143. [CrossRef] [PubMed]
145. Pfeiffer, B.D.; Ngo, T.T.; Hibbard, K.L.; Murphy, C.; Jenett, A.; Truman, J.W.; Rubin, G.M. Refinement of Tools for Targeted Gene Expression in Drosophila. *Genetics* **2010**, *186*, 735–755. [CrossRef] [PubMed]
146. Jenett, A.; Rubin, G.M.; Ngo, T.T.; Shepherd, D.; Murphy, C.; Dionne, H.; Pfeiffer, B.D.; Cavallaro, A.; Hall, D.; Jeter, J.; et al. A GAL4-driver line resource for Drosophila neurobiology. *Cell Rep.* **2012**, *2*, 991–1001. [CrossRef] [PubMed]
147. Birkholz, O.; Rickert, C.; Nowak, J.; Coban, I.C.; Technau, G.M. Bridging the gap between postembryonic cell lineages and identified embryonic neuroblasts in the ventral nerve cord of Drosophila melanogaster. *Biol. Open* **2015**, *4*, 420–434. [CrossRef] [PubMed]
148. Lacin, H.; Truman, J.W. Lineage mapping identifies molecular and architectural similarities between the larval and adult Drosophila central nervous system. *eLife* **2016**, *5*, e13399. [CrossRef] [PubMed]

Nutrient-Deprived Retinal Progenitors Proliferate in Response to Hypoxia: Interaction of the HIF-1 and mTOR Pathway

Helena Khaliullina *, Nicola K. Love and William A. Harris *

Department of Physiology, Development and Neuroscience, University of Cambridge, Cambridge CB2 3DY, UK; nicolakatelove@gmail.com
* Correspondence: hk402@cam.ac.uk (H.K.); wah20@cam.ac.uk (W.A.H.)

Academic Editors: Benjamin Yat-Ming Yung and Simon J. Conway

Abstract: At a cellular level, nutrients are sensed by the mechanistic Target of Rapamycin (mTOR). The response of cells to hypoxia is regulated via action of the oxygen sensor Hypoxia-Inducible Factor 1 (HIF-1). During development, injury and disease, tissues might face conditions of both low nutrient supply and low oxygen, yet it is not clear how cells adapt to both nutrient restriction and hypoxia, or how mTOR and HIF-1 interact in such conditions. Here we explore this question *in vivo* with respect to cell proliferation using the ciliary marginal zone (CMZ) of *Xenopus*. We found that both nutrient-deprivation and hypoxia cause retinal progenitors to decrease their proliferation, yet when nutrient-deprived progenitors are exposed to hypoxia there is an unexpected rise in cell proliferation. This increase, mediated by HIF-1 signalling, is dependent on glutaminolysis and reactivation of the mTOR pathway. We discuss how these findings in non-transformed tissue may also shed light on the ability of cancer cells in poorly vascularised solid tumours to proliferate.

Keywords: neural progenitors; hypoxia; nutrition; proliferation

1. Introduction

It is well known that normally cells depend on nutrients for growth. Indeed, nutrient deprivation (ND) blocks proliferation in many *in vitro* systems through a check-point in the cell cycle [1–4]. We have previously shown that, *in vivo*, cellular proliferation in the ciliary marginal zone (CMZ), the proliferating niche of the continuously growing retinas of lower vertebrates, is dependent on nutrition, and that mechanistic Target of Rapamycin (mTOR) acts as a restriction point regulator, such that in ND, the inactivation of mTOR blocks proliferation of early progenitors while allowing late progenitors to differentiate [5]. Interestingly, proliferating cells in the CMZ also show a Warburg metabolism, fuelling growth through glycolytic products and relying less on oxidative phosphorylation to meet their energy demands [6]. Cells that are exposed to hypoxia also show preference for glycolysis over oxidative phosphorylation. Many solid tumours lack vascularization, so they experience both ND and hypoxia. Intriguingly, hypoxia can induce a proliferative response in cancer cells [7,8], suggesting that these cells may be primed to respond in a different way to hypoxia than normal cells do. We therefore wondered how hypoxia affects the proliferation of nutrient deprived CMZ cells.

The adaptive responses to starvation and hypoxia depend largely on two major sensory proteins, mTOR and Hypoxia-Inducible Factor 1 (HIF-1). mTOR is activated by growth factor receptors, amino acids and other environmental stimuli that influence overall energy levels in tissues [9–14]. The activity of mTOR stimulates growth and proliferation. In the CMZ of *Xenopus*, nutrient deprivation inhibits the mTOR pathway leading to the cessation of cell proliferation that is normally continuously underway in the CMZ of the *Xenopus* retina [5]. HIF-1 activity is subject to tight regulation by oxygen concentration [15–19]. Hypoxia is also known to inhibit the cell cycle in non-transformed cells [20]. Some interactions between the HIF-1 pathway and the mTOR pathway have been described. For instance, the translation of the HIF-1 protein can be regulated by mTOR signalling [21–23] which itself can be activated [24] or inhibited by HIF-1 [25–27]. Yet it is unclear from these studies how proliferating cells would respond when faced with both ND and hypoxia. To approach this question, we used the CMZ of *Xenopus*, a stem cell niche where there is continuous proliferation throughout postembryonic growth. The advantage of using this system is that the effects of nutrient deprivation and hypoxia can be studied individually and *in vivo* in whole living animals. Nutrient deprivation can be achieved by dissection of yolk, which does not interfere with survival of the embryo (see Materials and Methods). Hypoxia can be induced by placing entire embryos in a hypoxic chamber, and be undertaken on either nutrient-deprived or normal-fed embryos to investigate different combinations of conditions. Under low nutrient conditions, progenitor cells in the CMZ are known to cease proliferating due to the inhibition of the mTOR pathway [5]. Here we show that this phenomenon can be reversed under hypoxic conditions, with ND cells in the CMZ upregulating mTOR signalling and increasing their proliferation in response to low oxygen. Furthermore, we demonstrate that this response is mediated by HIF-1 signalling and depends on both glutaminolysis and the reactivation of the mTOR pathway.

2. Materials and Methods

2.1. Animal Maintenance and Pharmacological Treatment

Xenopus laevis embryos, obtained by *in vitro* fertilization, were raised in 0.1× Modified Barth's solution (MBS) and staged according to [28]. For all experiments using nutrient-deprivation, except the glutamine re-feeding experiment using retinal explants, whole and alive embryos were used. Embryos were anaesthetized and nutrient deprived by yolk dissection at stage 35, and analysed at stage 38, after 24 h, as described previously [5]. Embryos were then maintained in 0.1× MBS throughout the experiment. In drug treatment conditions, 100 nM Echinomycin (Sigma-Aldrich, Gillingham, UK), 50 µM bis-2-(5-phenylacetamido-1,3,4-thiadiazol-2-yl)ethyl sulfide (BPTES) (Sigma) or 5 µM Rapamycin (Calbiochem/Merck-Millipore, Watford, UK) were bath-applied in 0.1× MBS.

2.2. Induction of Hypoxia

Whole and alive normal-fed and nutrient-deprived embryos at stage 38 were placed into a hypoxic bath chamber, which was maintained under a constant infusion of a mixture of 5% oxygen and 95% CO_2, for 5 h. These embryos were used for the Western blot and immunostaining. For 5-ethynyl-2′-deoxyuridine (EdU) incorporation, the embryos were incubated in the hypoxic bath chamber for 3 h followed by incubation in bath-applied EdU in the same hypoxic chamber for another 2 h. For the time-course of EdU incorporation, normal-fed and nutrient-deprived embryos were placed into the hypoxic bath chamber for different times and subsequently given a 1 h pulse of EdU bath-applied in the hypoxic chamber.

2.3. EdU Labelling

To mark cycling cells, 5 mM EdU was bath applied to embryos for 2 h prior to fixation. Embryos were fixed, sectioned and the EdU incorporation was detected on 14 µm sections using Click-iT chemistry kit performed in accordance with the manufacturer's instructions (Molecular Probes, Thermo Fisher, Paisley, UK). Fluorescent sections were visualized under the confocal microscope and EdU-positive cells were counted blind. Statistical analysis was determined by two-tailed Student's *t*-test (see Statistical Analysis).

2.4. *Ex Vivo Cultures*

An *ex vivo* culture method was adapted from that previously described by [29] and used only for the experiments with retinal explants. Embryos, nutrient-deprived by yolk dissection at stage 39, were grown for 24 h at 16 °C. Embryos were then washed in sterile 0.1× MBS with Penicillin/ Streptomycin/ Amphotericin (PSF), and anesthetized in MS222 solution. Retinas were removed under sterile conditions, washed and cultured at 20 °C on Parafilm to prevent adhesion, in 4 well or 35 mm culture dishes. For glutamine experiments, embryos were nutrient-deprived for 24 h at stage 36, grown to stage 41 and their retinas were then explanted and cultured overnight either in 60% L15 (Fisher Scientific, Loughborough UK) or 60% L15 without L-glutamine (Sigma) with/without substitution of 2.05 mM L-glutamine (Sigma).

2.5. *Immunostaining*

Immunostaining was performed on 14 μm sections using rabbit anti-Phospho Ribosomal protein S6 (Ser 235) (1:500, Cell Signaling) and Alexa Fluor 594 goat anti-rabbit (1:1000) (Invitrogen, Thermo Fisher, Paisley, UK) antibodies. Antigen retrieval was performed by steaming with 0.01 M Sodium Citrate, pH = 6, prior to staining with anti-pS6 antibody. Nuclei were labelled with 0.1 μg/mL 4′,6-diamidino-2-phenylindole (DAPI) (Invitrogen). Immunofluorescent sections were imaged with Laser Confocal Microscopy (Olympus Fluoview FV1000, Southend-on-Sea, UK). Sections examined, compared and depicted in all figures were all at the level of the optic nerve. All images were analysed using Volocity 6.3 (Perkin Elmer, Waltham, MA, USA).

2.6. *Statistical Analysis*

The number of independent experiments is indicated in each figure legend. For EdU quantification, a minimum of 8 cross-sections was used for analysis of each condition. Each examined section was at the level of optic nerve, therefore the number of examined sections corresponds to the number of examined retinas. Number of EdU-positive cells present in each section was counted blind. A two-tailed Student's t-test was used to calculate the P values between two conditions. The two conditions that were compared are always indicated by two hooked ends of horizontal line connecting the conditions. Above the line, * indicates a p value < 0.05, ** indicates a p value < 0.01, and *** indicates a p value < 0.001. Error bars indicate n independent experiments. In each experiment, between 8 and 17 retinas were examined.

2.7. *Western Blot*

Western blots were performed on retinal tissue lysates dissected from embryos treated as described, using at least 20 retinas per condition. Ten retinas for each condition were lysed on ice in RIPA buffer (Sigma) with 1:100 protease inhibitor (Sigma), homogenised by sonication and centrifuged. Supernatants were supplemented with SDS loading buffer and denatured at 80 °C for 5 min. Samples were run on precast 12% gels and transferred according to manufacturer's protocol (Biorad, Hertfordshire, UK). For detection, rabbit anti-HIF-1 (1:500, ab2185 Abcam), rabbit anti-pS6 Ser235/236 (1:1000, Cell Signalling), mouse anti-α-tubulin (1:5000, Abcam), goat anti-rabbit HRP-conjugated (1:1000 Abcam) and goat anti-mouse HRP-conjugated (1:1000; Abcam) antibodies were used. Western blots were visualized using the ECL system (Amersham, Buckinghamshire, UK) and quantified in FiJi where, the pixel intensities of each band were quantified and normalized to the pixel intensity of the background. The error bars represent the standard deviations between the same conditions in different experiments, the number of experiments is stated in the corresponding figure legend. Experimental antibodies were normalised to loading controls and then experimental conditions (e.g., ND) normalised to normal-fed untreated controls to determine the ratio to control that was depicted.

3. Results

3.1. *Hypoxia Regulates Proliferation in the CMZ Differently Dependent on Nutritional State*

To determine whether hypoxia affects proliferation in the CMZ, we incubated embryos in a hypoxic chamber maintained at 5% oxygen and monitored proliferation by incorporation of the nucleoside analogue of thymidine, EdU. Embryos were placed in a 2 mM solution of EdU for 2 h and proliferative cells were labelled by incorporation of EdU into DNA during active DNA synthesis. As expected, normoxic control retinas showed intense EdU staining in the CMZ, where proliferating progenitors reside (Figure 1A,B,F). After oxygen deprivation, a strong decrease in CMZ progenitor proliferation was observed (Figure 1C,F).

Having previously shown that CMZ cells are facultatively glycolytic, *i.e.*, they show Warburg metabolism in normoxia and sustained levels of ATP production when exposed to hypoxia [6], we wondered whether these cells might also be resistant to hypoxia with respect to proliferation. We also wondered whether the response to hypoxia on cell proliferation might depend on cellular context, particularly access to nutrients, similar to cancer cells that often proliferate in poorly vascularized niches with poor nutrient and oxygen availability [7,8,30]. To test how nutrient deprivation affects the proliferative response of CMZ cells to hypoxia, we repeated the EdU incorporation experiment using ND embryos in the hypoxic chamber. Strikingly, we found that in the CMZ of ND animals, proliferation was upregulated upon exposure to hypoxia (Figure 1B–F). This increase in proliferation was acute and greatest within 1 h, but was maintained above that found in ND-alone CMZs for many hours (Figure 1G). This result suggests that there are opposing effects of hypoxia on proliferation in nutrient replete *versus* starved condition.

3.2. *Activity of HIF-1 Signalling Regulates Proliferation of ND Retinal Progenitors*

One of the major regulators of cellular metabolism and proliferation under hypoxia is HIF-1, the α subunit of which is sensitive to oxygen concentration [31]. In proliferating CMZ cells of the *Xenopus* retina, HIF-1α is stabilized under hypoxia in both normal-fed and ND retinas (Figure 2A,B). To investigate, whether re-initiation of proliferation in hypoxic nutrient-deprived retinas depends on the function of HIF-1, we blocked HIF-1 transcriptional activity by treatment with the specific HIF-1-inhibitor Echinomycin [32] and monitored EdU incorporation into the CMZ. While treatment with Echinomycin did not rescue reduction of proliferation in normal-fed hypoxic samples (Figure 2C–E,I), it entirely prevented the increase in proliferation seen in the CMZs of ND retinas (Figure 2F–H,I), which was not due to increased apoptosis (Figure S1A,B,D,E,G). This result suggests that the increase of proliferation observed in ND retinas following hypoxia seems to rely on HIF-1 activity.

3.3. *Glutaminolysis is Critical for Proliferation in Response to Hypoxia in ND Retinal Progenitors*

One of the major effects of HIF-1 is the rewiring of cellular metabolism towards activation of glycolysis and glutaminolysis [33]. Glutaminolysis involves the deamination of glutamine and subsequent generation of α-ketoglutarate, which serves to replenish the TCA cycle [34,35]. Retinas from ND animals can be re-fed in an *ex vivo* explant culture system using L15 medium [5]. We re-fed ND retinal explants using L15 medium both with and without glutamine and found that proliferation was only rescued when glutamine was present in the culture medium (Figure 3), suggesting that glutamine is essential for proliferation of *Xenopus* CMZ. Glutaminolysis can be blocked by BPTES, a specific inhibitor of the key enzyme glutaminase. To investigate whether glutaminolysis is necessary for the proliferative response of ND CMZ cells exposed to low oxygen, we treated normal-fed and ND embryos under hypoxia with BPTES and monitored proliferation in the CMZ by EdU incorporation. Our results show that treatment with BPTES interferes with upregulation of proliferation in the CMZs of ND animals placed in hypoxia (Figure 4), which was not due to increased apoptosis (Figure S1A–C,D–F). Thus, glutaminolysis, presumably upregulated by HIF-1, is essential for the increase in proliferation induced by hypoxia in ND conditions.

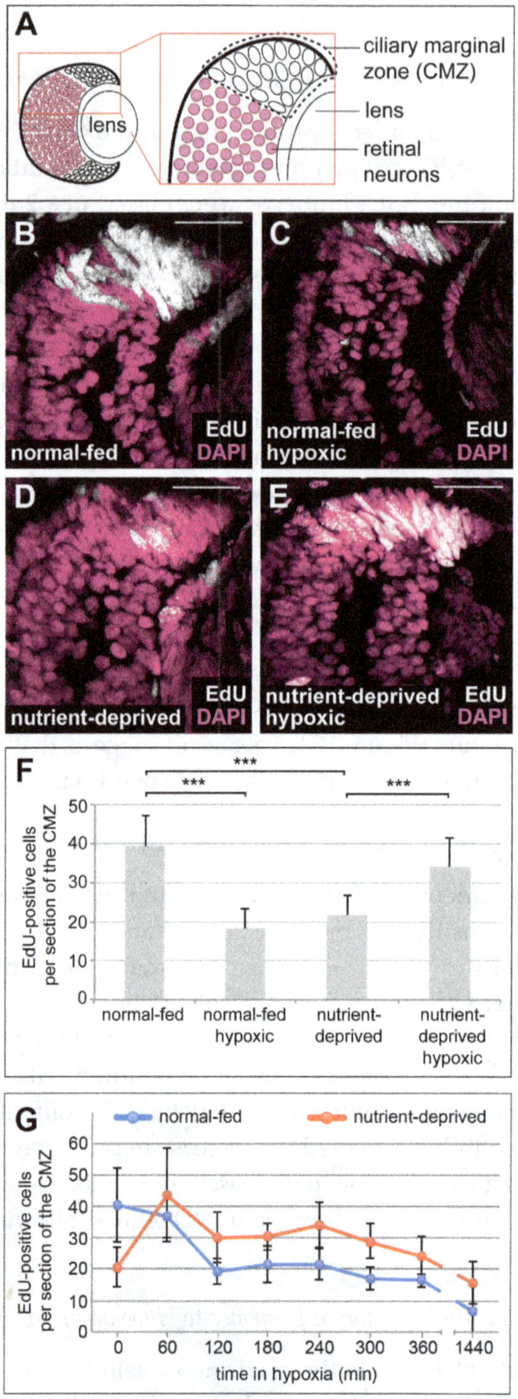

Figure 1. Hypoxia affects proliferation of progenitors in the retinal stem cell niche in the opposite manner, dependent on the nutritional status. (**A**) Schematic representation of a cross section through a 38-stage *Xenopus* retina, indicating the position of the ciliary marginal zone (CMZ). (**B–E**) EdU incorporation measured after a 2 h EdU pulse in DAPI-stained retinas from animals that have been normal-fed (**B**,**C**) or nutrient-deprived for 24 h (**D**,**E**) incubated under normoxia (**B**,**D**) or hypoxia (**C**,**E**). Nutrient deprivation or hypoxia reduced the number of proliferating cells in the CMZ, whereas hypoxic nutrient-deprived retinas show an increase of proliferation in the CMZ. Scale bars = 50 μm. For each condition, between 10 and 15 retinas were quantified. (**F**) Quantification of the experiment performed in (**B–E**). Error bars represent standard deviations. Two hooked ends of horizontal lines indicate the two conditions compared, *** p-value < 0.001; $n = 7$. (**G**) Quantification of EdU-positive cells after an 1 h EdU incorporation in retinas from normal-fed (blue line) or nutrient-deprived (red line) animals after various times in hypoxia (time in minutes indicated on the x-axis). Error bars represent standard deviations of two independent experiments ($n = 2$). For each condition and time point, a minimum of 10 retinas was quantified.

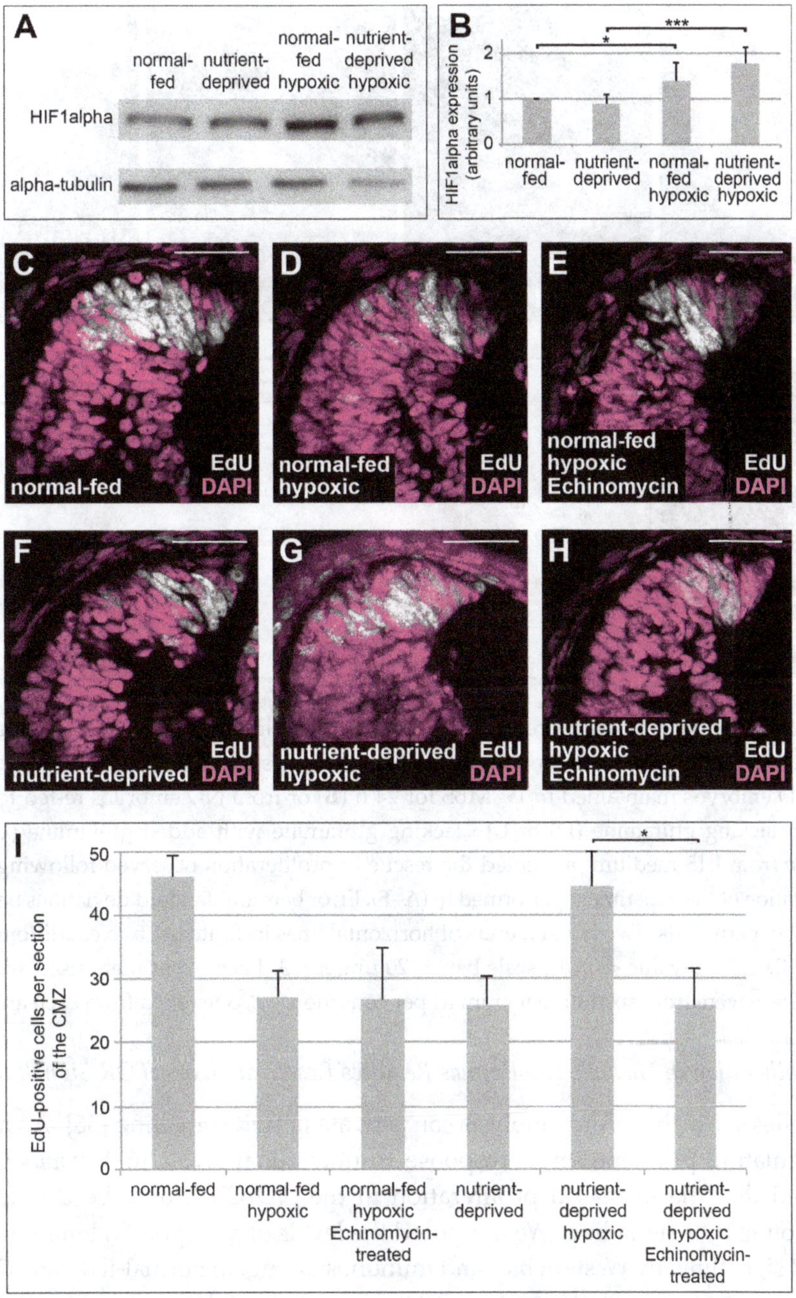

Figure 2. HIF-1 activity is required to induce re-initiation of proliferation in nutrient-deprived hypoxic retinas. (**A**) Western blot of retinas isolated from normal-fed or nutrient-deprived animals kept under normal oxygen concentrations or in hypoxia, probed for HIF-1α subunit and α-tubulin. HIF-1α is stabilized under hypoxia in both normal-fed and nutrient-deprived animals. At least 22 retinas were taken for each condition; all conditions in each experiment have the same number of retinas (n = 5). (**B**) Quantification of the Western blot performed in (**A**). Error bars represent standard deviations. Two hooked ends of horizontal lines indicate the two conditions compared, * p-value < 0.05, *** p-value < 0.001; n = 5. (**C**–**H**) EdU incorporation in DAPI-stained retinas from normal-fed animals (**C**–**E**) kept in normoxia (**C**), hypoxia (**D**) or incubated with Echinomycin under hypoxia (**E**) and in nutrient-deprived animals kept in normoxia (**F**), hypoxia (**G**) or incubated with Echinomycin under hypoxia (**H**). Nutrient-deprived retinas no longer resume their proliferation in hypoxia when function of HIF-1 is blocked by Echinomycin. Scale bars = 50 μm. In each condition, a minimum of 10 retinas was examined. (**I**) Quantification of the experiment performed in (**C**–**H**). Error bars represent standard deviations between three independent experiments. Two hooked ends of horizontal lines indicate the two conditions compared, * p-value < 0.05; n > 3. For each condition, between 10 and 15 retinas were quantified.

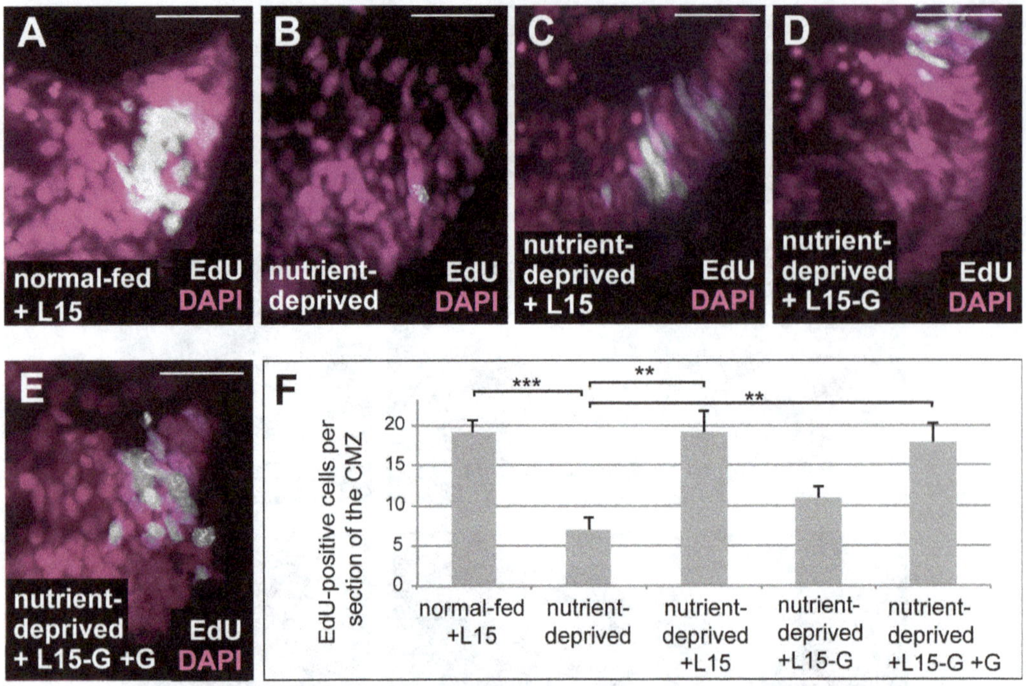

Figure 3. Glutamine is essential for retinal progenitor cell proliferation. (**A–E**) EdU incorporation in DAPI-stained *ex vivo* retinal explants from normal-fed embryos maintained in L15 medium for 24 h (**A**), from ND embryos maintained in 1× MBS for 24 h (**B**) or from ND embryos re-fed for 24 h with L15 (**C**), L15 lacking glutamine (**D**) or L15 lacking glutamine with added glutamine (**E**). Removal of glutamine from L15 medium prevented the rescue in proliferation observed following re-feeding. (**F**) Quantification of the experiment performed in (**A–F**). Error bars are standard deviations between three independent experiments. Two hooked ends of horizontal lines indicate the two conditions compared, ** p-values < 0.01, *** p-value < 0.001; scale bars = 20 μm; n = 3. Each experiment used six retinas per condition. These were from six different animals per condition (*i.e.*, one retina from each animal).

3.4. Increased Proliferation in Hypoxic ND Retinas Requires Reactivation of mTOR Signalling

Recent studies show that glutaminolysis can activate mTOR signalling [36]. As mTOR is closely involved in regulating proliferation in response to nutritional levels in *Xenopus* CMZ tissue [5], we hypothesized that the increased proliferation in the ND CMZ may be due to the effects of glutaminolysis on mTOR signalling. We measured the levels of phospho-S6 kinase (pS6), a target of activated mTOR signalling, by Western blot and immunostaining, in normal-fed and ND retinas under normoxia and hypoxia. As expected, normoxic ND retinas showed decreased pS6 staining in the CMZ, as well as lower levels of pS6 protein, compared to normal-fed retinas (Figure 5A,D,G,H and [5]). pS6 staining, however, remained high in the CMZ of normal-fed retinas upon induction of hypoxia (Figure 5B). In contrast, ND CMZs showed an increase of pS6 protein levels, as well as pS6 staining intensity in the CMZ when exposed to hypoxia (Figure 5E,G,H). Treatment with Echinomycin, a small molecule inhibitor of HIF-1 activity, interfered with this upregulation of pS6 staining in hypoxic ND CMZs, but not in normal-fed CMZs, implicating HIF-1 as critical in this process (Figure 5C,F).

Taken together, these results suggest that mTOR signalling may be required for reactivation of proliferation when starved CMZs are exposed to hypoxia. To further test this hypothesis, we monitored proliferation in ND CMZs exposed to hypoxia when the mTOR pathway was blocked by the addition of the mTOR inhibitor Rapamycin. Indeed, as expected, we found ND retinal progenitors could no longer reinitiate their proliferation in hypoxia when incubated in Rapamycin (Figure 5I). This suggests that under hypoxic conditions HIF-1 activates glutaminolysis, leading to the derepression of the mTOR pathway in nutrient-deprived proliferating cells and this accounts for the increase of their proliferation upon exposure to hypoxia.

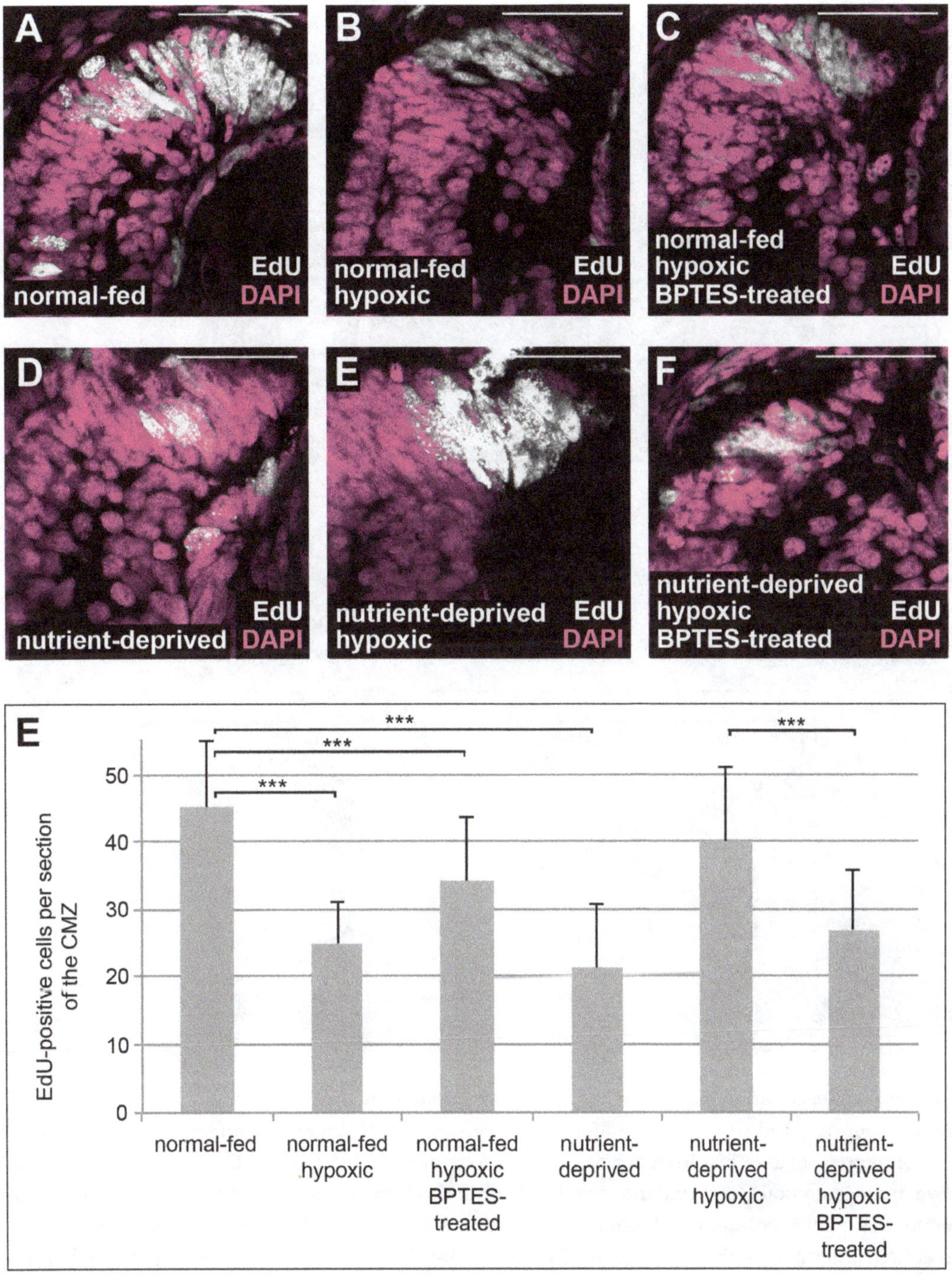

Figure 4. Block in glutaminolysis interferes with re-initiation of proliferation in nutrient-deprived hypoxic retinas. (**A–F**) EdU incorporation in DAPI-stained retinas from normal-fed animals (**A–C**) kept in normoxia (**A**), hypoxia (**B**) or incubated with BPTES under hypoxia (**C**) and in nutrient-deprived animals kept in normoxia (**D**), hypoxia (**E**) or incubated with BPTES under hypoxia (**F**). Nutrient-deprived retinas no longer resume their proliferation in hypoxia when glutaminolysis is blocked by BPTES. Scale bars = 50 μm. In each condition, a minimum of 8 retinas was observed. (**E**) Quantification of the experiment performed in (**A–F**). Error bars represent standard deviations between three independent experiments. Two hooked ends of horizontal lines indicate the two conditions compared, *** p-value < 0.001, n = 3. For each condition, between 8 and 17 retinas were quantified.

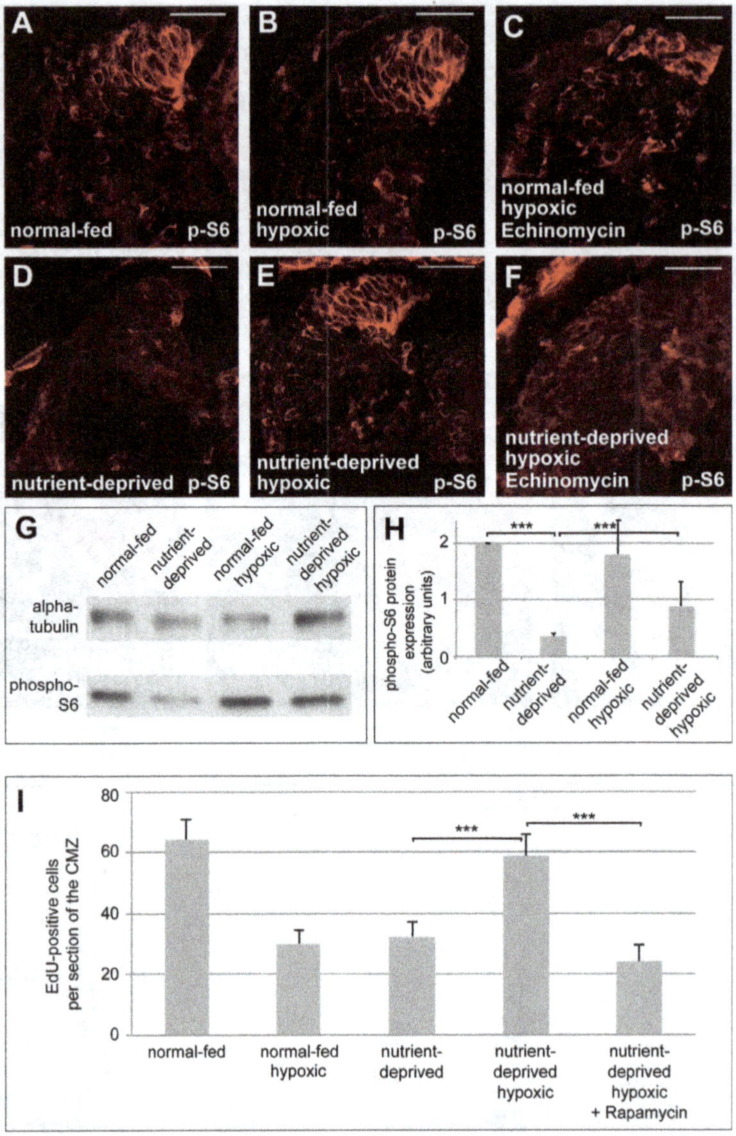

Figure 5. Reactivation of mTOR signalling, mediated by HIF-1 function, is required when nutrient-deprived retinas increase proliferation in hypoxia. (**A–D**) Immunostaining for the mTOR target phospho-S6 (p-S6) in retinas from normal-fed (**A–C**) or nutrient-deprived (**C–E**) animals kept in normal oxygen concentration (**A**,**D**), under hypoxia (**B**,**E**) or under hypoxia and treated with Echinomycin (**C**,**F**). Incubation of nutrient-deprived retinas in hypoxia restores p-S6 staining in the CMZ. Echinomycin treatment interferes with the increase of p-S6 staining in the CMZ of nutrient-deprived hypoxic retinas. Scale bars = 50 μm; n = 3. In each condition, a minimum of 8 retinas was examined. (**G**) Western blot of retinas dissected from normal-fed or nutrient-deprived normoxic or hypoxic animals, probed for phospho-S6 and α-tubulin. At least 22 retinas were taken for each condition, and all conditions in each experiment have the same number of retinas (n = 10). (**H**) Quantification of the Western blot shown in (**G**). Phospho-S6 levels, normalized to tubulin, are significantly reduced upon nutrient deprivation, unaffected by hypoxia alone, and increase to 40% of control levels upon nutrient deprivation in hypoxia. In Western blot quantification, error bars represent standard deviations between 10 independent experiments. Two hooked ends of horizontal lines indicate the two conditions compared, *** p-values < 0.001; n = 10. (**I**) Quantification of EdU incorporation in retinas from normal-fed or nutrient-deprived animals kept in normoxia, hypoxia or incubated with Rapamycin under hypoxia. Nutrient-deprived retinas no longer resume their proliferation in hypoxia when mTOR signalling is blocked by Rapamycin. Error bars represent standard deviations between two independent experiments. Two hooked ends of horizontal lines indicate the two conditions compared, *** p-values < 0.001; n = 2. For each condition, between 7 and 15 retinas were quantified.

4. Discussion

It seems sensible that starving cells should strive to decrease their metabolic activity, limit proliferation, and activate mechanisms essential for cell survival. We previously showed that the mTOR pathway is the main regulator of proliferation in response to starvation in the neurogenic CMZ of *Xenopus* [5]. Interestingly, we now show that the low rate proliferation in ND retinal progenitors *in vivo* is increased when these cells also become hypoxic. The finding is counterintuitive and suggests that ND progenitors have a different metabolic state than nourished ones. Unfortunately, the limitations of the *Xenopus* embryo system make it challenging to measure the specific metabolic profiles of CMZ cells. Our experiments suggest, however, that the effect involves HIF-1 signalling, which normally activates glycolysis and glutaminolysis. Both HIF-1 activity and glutaminolysis are essential for the increased proliferation in ND retinas exposed to hypoxia. Interestingly, cancer cells upregulate HIF-1 signalling and depend strongly on glutaminolysis for energy [16,37]. The response of normal proliferating cells to the conditions that cancer cells often experience suggests that it may be a natural adaptation for such cells to increase proliferative activity when confronted with both ND and hypoxia.

We show that hypoxia induces an upregulation of mTOR activity in ND CMZ cells. A similar mechanism has been recently proposed for umbilical cord blood-derived human mesenchymal stem cells (UCB-hMSCs), where hypoxia induces transient proliferation via upregulation of mTOR signalling [38]. It has also been shown that hypoxia-induced HIF-1 signalling activates fatty acid synthase to synthesize lipids [38] and that HIF-2 affects lipid handling in mouse retinal pigment epithelium cells [39]. Our findings in the CMZ are also in line with those of [24] who found that smooth muscle endothelial cells *in vitro* are induced to proliferate in hypoxia via a HIF-1-mediated increase in mTOR signalling.

Proliferating cells in the retina use a Warburg metabolism to fuel anabolic substrates for growth. This reduces the dependence on oxygen, but hypoxia often leads to the uncoupling of the electron transport chain, which can lead to the production of reactive oxygen species (ROS) [40,41]. ROS production regulates various signalling pathways related to proliferation and differentiation [42–47] and aids in tissue regeneration in *Xenopus* [48], yet ROS are known to be toxic [41]. Interestingly, a recent study shows that, in response to ND, lipid droplets that serve as ROS scavengers form in the glia cells surrounding the neuroblast niche, thereby protecting proliferating progenitors from ROS damage [49]. We have not investigated the role of ROS in the CMZ but it would be interesting to know what role ROS might play in the integration of the HIF-1 and mTOR pathways that affect proliferation. It is possible that HIF-1-mediated induction of glutaminolysis in hypoxia may also serve to limit ROS production through replenishment of glutathione, a major cellular ROS quencher [50,51]. Similar mechanisms may be at play in proliferative cancer cells that rely strongly on glutamine supply, but often display a high production of ROS to support proliferation [44,45,50,52,53].

Taken together, our study suggests that non-transformed cells *in vivo* during normal development may be naturally programmed to increase their proliferation when they are facing both nutrient and oxygen deprivation. In our experiments, a combination of ND and hypoxia has a beneficial effect on retinal proliferation, albeit for a limited period of time. This is an interesting finding, as ND hypoxic retinas show a reactivation of mTOR pathway, while mTOR upregulation is generally associated with increase in cellular senescence, and hypoxia suppresses geroconversion [54]. However, the relationship between aging, hypoxia and mTOR pathway is complex and additional studies are necessary to determine the specific interactions between HIF-1 and mTOR in non-transformed and cancer cells. While hypoxia may be a common feature of many solid tumours, cancer cells can react differently to chemotherapeutic drugs that influence different aspects of cellular metabolism, as a recent study by Strese and colleagues has shown [55]. Interestingly, their results also show that cultured cancer cells from solid tumours, such as ovarian and small lung cancer cells, were more sensitive to Rapamycin when treated under hypoxic conditions, suggesting that proliferation of transformed cells might be strongly supported by mTOR pathway in hypoxia. Whether this is adaptive response of normal proliferating cells or a maladaptive response that cancer cells in solid tumours take advantage of, remains an important but open question.

Supplementary Materials: The following are available online at www.mdpi.com/2221-3759/4/2/17/s1.

Acknowledgments: We would like to thank Andrew Murray, Michalis Agathocleous and Randall Johnson for their valuable insights and comments on the manuscript. This study was supported by Support from Wellcome Trust SIA Award 100329/Z/12/Z to W.A.H. and the DFG fellowship KH 376/1-1 awarded to H.K.

Author Contributions: H.K., N.K. and W.A.H. conceived and designed the experiments; H.K. performed the experiments; N.K. performed the experiments of Figure 3; and H.K. and W.A.H. analysed the data and wrote the paper.

Conflicts of Interest: The authors declare no conflict of interest.

Abbreviations

The following abbreviations are used in this manuscript:

ND	nutrient-deprived; nutrient deprivation
CMZ	ciliary marginal zone
mTOR	mechanistic Target of Rapamycin
HIF-1	Hypoxia-Inducible Factor 1
MBS	Modified Barth's solution
BPTES	bis-2-(5-phenylacetamido-1,3,4-thiadiazol-2-yl)ethyl sulfide
EdU	5-ethynyl-2′-deoxyuridine
PSF	Penicillin/ Streptomycin/ Amphotericin
L-15	Leibovitz medium
DAPI	4′,6-diamino-2-phenylindilone
RIPA buffer	radioimmunoprecipitation assay buffer
SDS	sodium dodecyl sulfate
HRP	horseradish peroxidase
HIF-1α	Hypoxia-Inducible Factor 1α
TCA	tricarboxylic acid
G	Glutamine
pS6	phospho-S6 kinase
ROS	reactive oxygen species
UCB-hMSCs	umbilical cord blood-derived human mesenchymal stem cells

References

1. Lanet, E.; Gould, A.P.; Maurange, C. Protection of neuronal diversity at the expense of neuronal numbers during nutrient restriction in the *Drosophila* visual system. *Cell Rep.* **2013**, *3*, 587–594. [CrossRef] [PubMed]
2. Fingar, D.C.; Blenis, J. Target of rapamycin (TOR): An integrator of nutrient and growth factor signals and coordinator of cell growth and cell cycle progression. *Oncogene* **2004**, *23*, 3151–3171. [CrossRef] [PubMed]
3. Britton, J.S.; Edgar, B.A. Environmental control of the cell cycle in *Drosophila*: Nutrition activates mitotic and endoreplicative cells by distinct mechanisms. *Development* **1998**, *125*, 2149–2158. [PubMed]
4. Pardee, A.B. A restriction point for control of normal animal cell proliferation. *Proc. Natl. Acad. Sci. USA* **1974**, *71*, 1286–1290. [CrossRef]
5. Love, N.K.; Keshavan, N.; Lewis, R.; Harris, W.A.; Agathocleous, M. A nutrient-sensitive restriction point is active during retinal progenitor cell differentiation. *Development* **2014**, *141*, 697–706. [CrossRef] [PubMed]
6. Agathocleous, M.; Love, N.K.; Randlett, O.; Harris, J.J.; Liu, J.; Murray, A.J.; Harris, W.A. Metabolic differentiation in the embryonic retina. *Nat. Cell. Biol.* **2012**, *14*, 859–864. [CrossRef] [PubMed]
7. Dang, C.V. Links between metabolism and cancer. *Genes Dev.* **2012**, *26*, 877–890. [CrossRef] [PubMed]
8. Vander Heiden, M.G. Targeting cancer metabolism: A therapeutic window opens. *Nat. Rev. Drug Discov.* **2011**, *10*, 671–684. [CrossRef] [PubMed]

9. Hall, M.N. mTOR-what does it do? *Transplant. Proc.* **2008**, *40* (Suppl. 10), S5–S8. [CrossRef] [PubMed]
10. Urban, J.; Soulard, A.; Huber, A.; Lippman, S.; Mukhopadhyay, D.; Deloche, O.; Wanke, V.; Anrather, D.; Ammerer, G.; Riezman, H.; *et al.* Sch9 is a major target of TORC1 in Saccharomyces cerevisiae. *Mol. Cell* **2007**, *26*, 663–674. [CrossRef] [PubMed]
11. Rohde, J.; Heitman, J.; Cardenas, M.E. The TOR kinases link nutrient sensing to cell growth. *J. Biol. Chem.* **2001**, *276*, 9583–9586. [CrossRef] [PubMed]
12. Desai, B.N.; Myers, B.R.; Schreiber, S.L. FKBP12-rapamycin-associated protein associates with mitochondria and senses osmotic stress via mitochondrial dysfunction. *Proc. Natl. Acad. Sci. USA* **2002**, *99*, 4319–4324. [CrossRef] [PubMed]
13. Crickmore, M.A.; Mann, R.S. The control of size in animals: Insights from selector genes. *Bioessays* **2008**, *30*, 843–853. [CrossRef] [PubMed]
14. Dechant, R.; Peter, M. Nutrient signals driving cell growth. *Curr. Opin. Cell Biol.* **2008**, *20*, 678–687. [CrossRef] [PubMed]
15. Gordan, J.D.; Simon, M.C. Hypoxia-inducible factors: Central regulators of the tumor phenotype. *Curr. Opin. Gene. Dev.* **2007**, *17*, 71–77. [CrossRef] [PubMed]
16. Semenza, G.L. Targeting HIF-1 for cancer therapy. *Nat. Rev. Cancer* **2003**, *3*, 721–732. [CrossRef] [PubMed]
17. Ratcliffe, P.J.; O'Rourke, J.F.; Maxwell, P.H.; Pugh, C.W. Oxygen sensing, hypoxia-inducible factor-1 and the regulation of mammalian gene expression. *J. Exp. Biol.* **1998**, *201 Pt 8*, 1153–1162. [PubMed]
18. Semenza, G.L. Hypoxia-inducible factors in physiology and medicine. *Cell* **2012**, *148*, 399–408. [CrossRef] [PubMed]
19. Jiang, B.H.; Semenza, G.L.; Bauer, C.; Marti, H.H. Hypoxia-inducible factor 1 levels vary exponentially over a physiologically relevant range of O_2 tension. *Am. J. Physiol.* **1996**, *271*, C1172–C1180. [PubMed]
20. Gardner, L.B.; Li, Q.; Park, M.S.; Flanagan, W.M.; Semenza, G.L.; Dang, C.V. Hypoxia inhibits G1/S transition through regulation of p27 expression. *J. Biol. Chem.* **2001**, *276*, 7919–7926. [CrossRef] [PubMed]
21. Jiang, B.H.; Jiang, G.; Zheng, J.Z.; Lu, Z.; Hunter, T.; Vogt, P.K. Phosphatidylinositol 3-kinase signalling controls levels of hypoxia-inducible factor 1. *Cell Growth Differ.* **2001**, *12*, 363–369. [PubMed]
22. Cramer, T.; Yamanishi, Y.; Clausen, B.E.; Forster, I.; Pawlinski, R.; Mackman, N.; Haase, V.H.; Jaenisch, R.; Corr, M.; Nizet, V.; *et al.* HIF-1alpha is essential for myeloid cell-mediated inflammation. *Cell* **2003**, *112*, 645–657. [CrossRef]
23. Majumder, P.K.; Febbo, P.G.; Bikoff, R.; Berger, R.; Xue, Q.; McMahon, L.M.; Manola, J.; Brugarolas, J.; McDonnell, T.J.; Golub, T.R.; *et al.* mTOR inhibition reverses Akt-dependent prostate intraepithelial neoplasia through regulation of apoptotic and HIF-1-dependent pathways. *Nat. Med.* **2004**, *10*, 594–601. [CrossRef] [PubMed]
24. Humar, R.; Kiefer, F.N.; Berns, H.; Resink, T.J.; Battegay, E.J. Hypoxia enhances vascular cell proliferation and angiogenesis *in vitro* via rapamycin (mTOR)-dependent signalling. *FASEB J.* **2002**, *16*, 771–780. [CrossRef] [PubMed]
25. Brugarolas, J.; Lei, K.; Hurley, R.L.; Manning, B.D.; Reiling, J.H.; Hafen, E.; Witters, L.A.; Ellisen, L.W.; Kaelin, W.G., Jr. Regulation of mTOR function in response to hypoxia by REDD1 and the TSC1/TSC2 tumor suppressor complex. *Genes Dev.* **2004**, *18*, 2893–2904. [CrossRef] [PubMed]
26. Reiling, J.H.; Hafen, E. The hypoxia-induced paralogs Scylla and Charybdis inhibit growth by down-regulating S6K activity upstream of TSC in *Drosophila*. *Genes Dev.* **2004**, *18*, 2879–2892. [CrossRef] [PubMed]
27. Vadysirisack, D.D.; Ellisen, L.W. mTOR activity under hypoxia. *Methods Mol. Biol.* **2012**, *821*, 45–58. [PubMed]
28. Nieuwkoop, P.D. The "organization centre". 3. Segregation and pattern formation in morphogenetic fields. *Acta Biotheor.* **1967**, *17*, 178–194. [CrossRef] [PubMed]
29. Harris, W.A.; Holt, C.E.; Smith, T.A.; Gallenson, N. Growth cones of developing retinal cells *in vivo*, on culture surfaces, and in collagen matrices. *J. Neurosci. Res.* **1985**, *13*, 101–122. [CrossRef] [PubMed]
30. Agathocleous, M.; Harris, W.A. Metabolism in physiological cell proliferation and differentiation. *Trends Cell Biol.* **2013**, *23*, 484–492. [CrossRef] [PubMed]

31. Huang, L.E.; Gu, J.; Schau, M.; Bunn, H.F. Regulation of hypoxia-inducible factor 1alpha is mediated by an O_2-dependent degradation domain via the ubiquitin-proteasome pathway. *Proc. Natl. Acad. Sci. USA* **1998**, *95*, 7987–7992. [CrossRef] [PubMed]
32. Kong, D.; Park, E.J.; Stephen, A.G.; Calvani, M.; Cardellina, J.H.; Monks, A.; Fisher, R.J.; Shoemaker, R.H.; Melillo, G. Echinomycin, a small-molecule inhibitor of hypoxia-inducible factor-1 DNA-binding activity. *Cancer Res.* **2005**, *65*, 9047–9055. [CrossRef] [PubMed]
33. Huang, D.; Li, C.; Zhang, H. Hypoxia and cancer cell metabolism. *Acta Biochim. Biophys. Sin.* **2014**, *46*, 214–219. [CrossRef] [PubMed]
34. Denko, N.C. Hypoxia, HIF-1 and glucose metabolism in the solid tumour. *Nat. Rev. Cancer* **2008**, *8*, 705–713. [CrossRef] [PubMed]
35. Smolkova, K.; Plecita-Hlavata, L.; Bellance, N.; Benard, G.; Rossignol, R.; Jezek, P. Waves of gene regulation suppress and then restore oxidative phosphorylation in cancer cells. *Int. J. Biochem. Cell Biol.* **2011**, *43*, 950–968. [CrossRef] [PubMed]
36. Duran, R.V.; Oppliger, W.; Robitaille, A.M.; Heiserich, L.; Skendaj, R.; Gottlieb, E.; Hall, M.N. Glutaminolysis activates Rag-mTORC1 signalling. *Mol. Cell* **2012**, *47*, 349–358. [CrossRef] [PubMed]
37. Kappler, M.; Taubert, H.; Schubert, J.; Vordermark, D.; Eckert, A.W. The real face of HIF-1alpha in the tumor process. *Cell Cycle* **2012**, *11*, 3932–3936. [CrossRef] [PubMed]
38. Lee, H.J.; Ryu, J.M.; Jung, Y.H.; Oh, S.Y.; Lee, S.J.; Han, H.J. Novel Pathway for Hypoxia-Induced Proliferation and Migration in Human Mesenchymal Stem Cells: Involvement of HIF-1alpha, FASN, and mTORC1. *Stem Cells* **2015**, *33*, 2182–2195. [CrossRef] [PubMed]
39. Kurihara, T.; Westenskow, P.D.; Gantner, M.L.; Usui, Y.; Schultz, A.; Bravo, S.; Aguilar, E.; Wittgrove, C.; Friedlander, M.S.; Paris, L.P.; *et al.* Hypoxia-induced metabolic stress in retinal pigment epithelial cells is sufficient to induce photoreceptor degeneration. *Elife* **2016**, *5*. [CrossRef] [PubMed]
40. Clanton, T. Yet another oxygen paradox. *J. Appl. Physiol.* **2005**, *99*, 1245–1246. [CrossRef] [PubMed]
41. Negre-Salvayre, A.; Coatrieux, C.; Ingueneau, C.; Salvayre, R. Advanced lipid peroxidation end products in oxidative damage to proteins. Potential role in diseases and therapeutic prospects for the inhibitors. *Br. J. Pharmacol.* **2008**, *153*, 6–20. [CrossRef] [PubMed]
42. Gorlach, A.; Dimova, E.Y.; Petry, A.; Martinez-Ruiz, A.; Hernansanz-Agustin, P.; Rolo, A.P.; Palmeira, C.M.; Kietzmann, T. Reactive oxygen species, nutrition, hypoxia and diseases: Problems solved? *Redox Biol.* **2015**, *6*, 372–385. [CrossRef] [PubMed]
43. Sing, A.; Tsatskis, Y.; Fabian, L.; Hester, I.; Rosenfeld, R.; Serricchio, M.; Yau, N.; Bietenhader, M.; Shanbhag, R.; Jurisicova, A.; *et al.* The atypical cadherin fat directly regulates mitochondrial function and metabolic state. *Cell* **2014**, *158*, 1293–1308. [CrossRef] [PubMed]
44. Le Belle, J.E.; Orozco, N.M.; Paucar, A.A.; Saxe, J.P.; Mottahedeh, J.; Pyle, A.D.; Wu, H.; Kornblum, H.I. Proliferative neural stem cells have high endogenous ROS levels that regulate self-renewal and neurogenesis in a PI3K/Akt-dependant manner. *Cell Stem Cell* **2011**, *8*, 59–71. [CrossRef] [PubMed]
45. Weinberg, F.; Hamanaka, R.; Wheaton, W.W.; Weinberg, S.; Joseph, J.; Lopez, M.; Kalyanaraman, B.; Mutlu, G.M.; Budinger, G.R.; Chandel, N.S. Mitochondrial metabolism and ROS generation are essential for Kras-mediated tumorigenicity. *Proc. Natl. Acad. Sci. USA* **2010**, *107*, 8788–8793. [CrossRef] [PubMed]
46. Fernandez-Fernandez, S.; Almeida, A.; Bolanos, J.P. Antioxidant and bioenergetic coupling between neurons and astrocytes. *Biochem. J.* **2012**, *443*, 3–11. [CrossRef] [PubMed]
47. Owusu-Ansah, E.; Banerjee, U. Reactive oxygen species prime *Drosophila* haematopoietic progenitors for differentiation. *Nature* **2009**, *461*, 537–541. [CrossRef] [PubMed]
48. Love, N.R.; Chen, Y.; Ishibashi, S.; Kritsiligkou, P.; Lea, R.; Koh, Y.; Gallop, J.L.; Dorey, K.; Amaya, E. Amputation-induced reactive oxygen species are required for successful Xenopus tadpole tail regeneration. *Nat. Cell Biol.* **2013**, *15*, 222–228. [CrossRef] [PubMed]
49. Bailey, A.P.; Koster, G.; Guillermier, C.; Hirst, E.M.; MacRae, J.I.; Lechene, C.P.; Postle, A.D.; Gould, A.P. Antioxidant Role for Lipid Droplets in a Stem Cell Niche of *Drosophila*. *Cell* **2015**, *163*, 340–353. [CrossRef] [PubMed]
50. Le, A.; Lane, A.N.; Hamaker, M.; Bose, S.; Gouw, A.; Barbi, J.; Tsukamoto, T.; Rojas, C.J.; Slusher, B.S.; Zhang, H.; *et al.* Glucose-independent glutamine metabolism via TCA cycling for proliferation and survival in B cells. *Cell Metab.* **2012**, *15*, 110–121. [CrossRef] [PubMed]

51. Jin, L.; Li, D.; Alesi, G.N.; Fan, J.; Kang, H.B.; Lu, Z.; Boggon, T.J.; Jin, P.; Yi, H.; Wright, E.R.; *et al.* Glutamate dehydrogenase 1 signals through antioxidant glutathione peroxidase 1 to regulate redox homeostasis and tumor growth. *Cancer Cell* **2015**, *27*, 257–270. [CrossRef] [PubMed]
52. DeBerardinis, R.J.; Lum, J.J.; Hatzivassiliou, G.; Thompson, C.B. The biology of cancer: Metabolic reprogramming fuels cell growth and proliferation. *Cell Metab.* **2008**, *7*, 11–20. [CrossRef] [PubMed]
53. Alberghina, L.; Gaglio, D. Redox control of glutamine utilization in cancer. *Cell Death Dis.* **2014**, *5*, e1561. [CrossRef] [PubMed]
54. Leontieva, O.V.; Blagosklonny, M.V. Hypoxia and gerosuppression: The mTOR saga continues. *Cell Cycle* **2012**, *11*, 3926–3931. [CrossRef] [PubMed]
55. Strese, S.; Fryknas, M.; Larsson, R.; Gullbo, J. Effects of hypoxia on human cancer cell line chemosensitivity. *BMC Cancer* **2013**, *13*, 331. [CrossRef] [PubMed]

13

Developmental Mechanism of Limb Field Specification along the Anterior–Posterior Axis during Vertebrate Evolution

Mikiko Tanaka

Graduate School of Bioscience and Biotechnology, Tokyo Institute of Technology, B-17, 4259 Nagatsuta-cho, Midori-ku, Yokohama 226-8501, Japan; mitanaka@bio.titech.ac.jp

Academic Editors: Vincenzo Zappavigna and Simon J. Conway

Abstract: In gnathostomes, limb buds arise from the lateral plate mesoderm at discrete positions along the body axis. Specification of these limb-forming fields can be subdivided into several steps. The lateral plate mesoderm is regionalized into the anterior lateral plate mesoderm (ALPM; cardiac mesoderm) and the posterior lateral plate mesoderm (PLPM). Subsequently, *Hox* genes appear in a nested fashion in the PLPM and provide positional information along the body axis. The lateral plate mesoderm then splits into the somatic and splanchnic layers. In the somatic layer of the PLPM, the expression of limb initiation genes appears in the limb-forming region, leading to limb bud initiation. Furthermore, past and current work in limbless amphioxus and lampreys suggests that evolutionary changes in developmental programs occurred during the acquisition of paired fins during vertebrate evolution. This review presents these recent advances and discusses the mechanisms of limb field specification during development and evolution, with a focus on the role of *Hox* genes in this process.

Keywords: *Hox*; retinoic acid; *Tbx4*; *Tbx5*; limb field specification; evolution

1. Introduction

Paleontological evidence suggests that acquisition of the first pairs of fin-like structures occurred in the lineage of ancestral agnathans (jawless vertebrates; Figure 1a) [1,2]. The cephalochordate amphioxus and the agnathan lampreys are thought to have diverged prior to the acquisition of paired fins (Figure 1a) [3], and thus these animals could be good models for developmental comparisons to gain insights into the evolutionary sequence that lead to the acquisition of the limb-forming fields in the lateral plate mesoderm.

In gnathostomes (jawed vertebrates), limb buds appear as small bulges protruding from the body trunk at a specific position along the body axis during the development of gnathostome embryos. Specification of limb fields, which is tightly correlated with the initiation of limb bud formation, involves multiple steps [4]. The lateral plate mesoderm is regionalized into the anterior lateral plate mesoderm (ALPM; cardiac mesoderm) and the posterior lateral plate mesoderm (PLPM), which includes the presumptive limb-forming fields [5,6]. Retinoic acid signaling seems to play pivotal roles in this process [5,6]. In the PLPM, *Hox* genes are expressed in a nested fashion along the anterior–posterior axis and appear to be involved in the regionalization of the PLPM into forelimb, interlimb flank and hindlimb fields [7,8]. Then, the lateral plate mesoderm thickens and splits into the somatic and splanchnic layers sequentially from the anterior to posterior region of the embryos (Figure 1b) [9]. Subsequently, expression of limb initiation genes appears in the presumptive limb-forming fields [10]. Initiation of limb buds needs to occur at a precise position along the body axis, although little has been known until recently about the molecular mechanism linking limb positioning and initiation. Analyses of chick and mouse embryos, however, revealed that the *Hox* proteins that define the axial position of the limb-forming fields directly activate the transcription of the forelimb initiation gene, *Tbx5* [11,12]. Furthermore, developmental analyses of limbless amphioxus and lampreys have provided evidence for the sequential events that occurred during the acquisition of limb-forming fields in the body of vertebrates during evolution [4]. This review highlights these recent advances and discusses the mechanisms of limb field specification during development and during vertebrate evolution, with a special focus on the function of *Hox* genes.

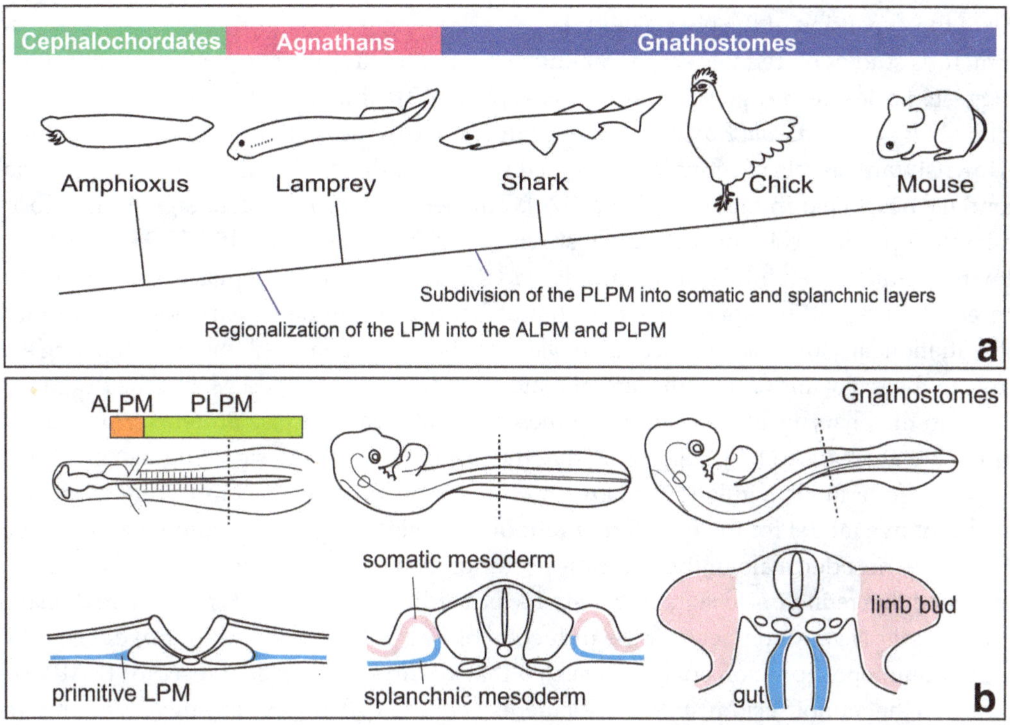

Figure 1. Model for the evolution of the lateral plate mesoderm (LPM); (**a**) a phylogenetic tree indicating the probable timing of the regionalization and subdivision of the LPM. See text for details. Modified after [4]. (**b**) development of the LPM in chick embryos. Bottom panels show schematic cross-sections at the wing level. The LPM is regionalized into the anterior lateral plate mesoderm (ALPM) and the posterior lateral plate mesoderm (PLPM). Then, the developing LPM splits into somatic and splanchnic mesoderm. Subsequently, somatic mesodermal cells proliferate and form limb buds, whereas splanchnic mesodermal cells contribute to gut formation. Modified after [9].

2. Regionalization of the Lateral Plate Mesoderm into the ALPM and the PLPM: Retinoic Acid and *Hox* Genes

During the development of gnathostome embryos, the lateral plate mesoderm is regionalized into the ALPM and the PLPM, including the presumptive limb-forming fields [5,6]. Recent developmental studies on zebrafish and mouse embryos suggested that such a regionalization of the lateral plate mesoderm is necessary to provide an environment permissive for forelimb-forming fields and that retinoic acid signaling has pivotal roles in this process [5,6]. In zebrafish, mutants for *retinaldehyde dehydrogenase 2* (*raldh2*) lack pectoral fin buds (forelimb buds) [13,14], and embryos treated with retinoic acid inhibitor show the posterior expansion of the heart field and fail to initiate formation of pectoral fin buds [5]. During this process, retinoic acid signaling induces the expression of *hoxb5b* within the forelimb fields [5]. As *hoxb5b* restricts the number of atrial cells, it also restricts the posterior extension of the heart field to determine the anterior boundary of forelimb-forming fields, although its function is dispensable for forelimb formation [5]. Chick and mouse *Hoxb5* orthologs have been implicated as direct targets of retinoic acid signaling [15–17]. Moreover, in mouse *Hoxb5* mutants, the shoulder girdles shift their position, although morphological abnormalities of the heart have not been reported [18]. Thus, it is suggested that *Hoxb5* mutants might have subtle cardiac defects, or be lethal, or other mouse *Hox* genes might have roles similar to those of zebrafish *hoxb5b* [5]. Multiple *Hox* genes are expressed in the heart fields of mouse and chick embryos [19,20]. However, heart phenotypes associated with the genetic inactivation of *Hox* genes have rarely been observed, which is probably due to the functional redundancy of paralogous *Hox* genes [21]. One notable exception is mouse *Hoxa3* null mutants, which exhibit heart defects, including hypertrophy of the atria and the enlargement of the chief veins [22]. Heart phenotypes seen in *Hoxa3* mutants, thus, have been implicated as a possible outcome of the expansion of the heart field [5].

Numerous studies on retinoic acid signaling in relation to the anterior-posterior patterning of the heart suggested roles for *Hox* genes in this process [21,23]. Treatment with retinoic acid enhances the expression of *Hoxa4, Hoxd3 and Hoxb5* in chicken cardiogenic tissue [20]. In addition, *Hoxb1, Hoxa1* and *Hoxa3*, downstream targets of retinoic acid signaling, define the distinct progenitor sub-domains of the secondary heart field in mouse embryos [24]. Furthermore, retinoic acid signaling is correlated with limb initiation and positioning in zebrafish, chick and mouse embryos [5,6,25,26]. Chick embryos treated with disulfiram, an inhibitor of retinoic acid synthesis, show hypoplasia, or a positional shift of the forelimb bud [25]. In mouse *Raldh2* mutants, heart-forming fields expanded posteriorly, and forelimb initiation fails [6], as seen in zebrafish *raldh2* mutants [5]. Thus, retinoic acid signaling seems to have a role in the regionalization of the lateral plate mesoderm into the ALPM and the PLPM [6,26,27]. In addition, in the pharyngeal endoderm and mesoderm of *Raldh2* mouse mutants, *Hoxa1* and *Hoxb1* transcript levels are reduced [27], suggesting that the role of retinoic acid signaling in regionalizing the lateral plate mesoderm via regulation of *Hox* genes is conserved in mouse embryos.

An alternative model for the function of retinoic acid signaling in the regionalization process of the lateral plate mesoderm along the anterior–posterior axis has been proposed by Duester's group. They suggested that retinoic acid signaling delimits the cardiac and epiblast *Fgf8*-positive domains, and provides a permissive environment for the induction of forelimb (but not hindlimb) development [6]. Chromatin immunoprecipitation analysis revealed that all three retinoic acid receptor (RAR) isoforms indeed bind to the retinoic acid responsive elements near the *Fgf8* promoter, suggesting that retinoic acid signaling directly represses *Fgf8* expression [28]. Interestingly, transgenic zebrafish embryos expressing ectopic Fibroblast growth factor (Fgf) signaling show expansion of the heart field and failure of initiation of pectoral fin development [29]. These results seem to support the view that FGF signaling is involved in the regionalization of the lateral plate mesoderm and provides a permissive environment for forelimb induction [6].

During embryonic development, regionalization of the lateral plate mesoderm into the ALPM and the PLPM provides a permissive environment for forelimb induction. Therefore, regionalization of the lateral plate mesoderm might have been a crucial step for acquisition of the limb-forming field during the evolution of vertebrates. In vertebrates, the heart is located at the anterior end of the body trunk, whereas the primitive heart of amphioxus—consisting of peristaltic blood vessels—is located at the posterior-ventral part of the body [30]. Recently, the distribution of the ALPM and the PLPM has been examined in amphioxus embryos by using molecular markers (Figure 2) [4]. In amphioxus, *AmphiHand*, an ortholog of *Hand1* and *Hand2*, which are expressed throughout the lateral plate mesoderm of zebrafish, chick and mouse embryos [31–33], is expressed throughout the ventral mesoderm (Figure 2) [4]. Interestingly, in amphioxus, both *AmphiNkx2-tin* and *AmphiTbx20*—orthologues of *Nkx2.5* and *Tbx20*, which are expressed in the cardiac mesoderm [32,34–36]—are also expressed throughout the ventral mesoderm. These results suggest that the amphioxus ventral mesoderm is not molecularly regionalized into cardiac *versus* posterior ventral mesoderm (Figure 2) [4]. In the limbless agnathan lamprey, *Lethenteron japonicum Tbx20* (*LjTbx20*) is expressed in the anterior part of the lateral plate mesoderm [37], whereas *LjMyb*, an orthologue of *c-myb* that is expressed in the PLPM, is expressed in mesodermal cells posterior to the *LjTbx20*-positive area (Figure 2) [4]. These results suggest that the lateral plate mesoderm of lamprey, a limbless agnathan, is regionalized into the ALPM and the PLPM as seen in gnathostomes (Figure 2) [4].

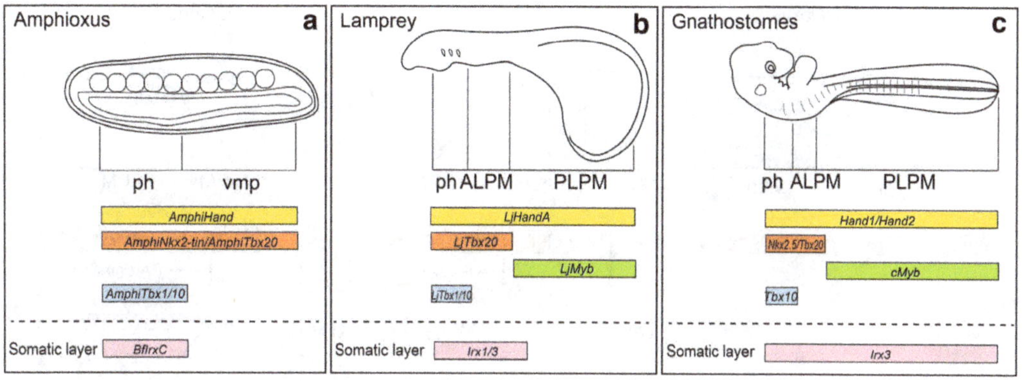

Figure 2. Schematic model for regionalization and differentiation of the ventral mesoderm and the lateral plate mesoderm (LPM) in (**a**) amphioxus, (**b**) lampreys and (**c**) gnathostomes. Yellow, orange, green and blue bars represent the distribution of molecular markers for the ventral mesoderm/LPM (*AmphiHand, LjHandA, Hand1, Hand2*), the pharyngeal mesoderm (ph) and ALPM (*AmphiNkx2-tin, AmphiTbx20, LjTbx20, Nkx2.5, Tbx20*), the PLPM (*LjMyb, cMyb*), and the pharyngeal mesoderm (*AmphiTbx1/10, LjTbx1/10, Tbx10*), respectively. Pink bars represent the distribution of *Irx3* orthologs (*BfIrxC, Irx1/3, Irx3*), which are expressed in the somatic layers of the LPM. vmp, ventral mesoderm posterior to the pharynx. See text for gene expression references. Modified after [4,38]

In lampreys, expression of *LjTbx1/10*, an ortholog of *Tbx1* and *Tbx10* that is a marker of pharyngeal mesoderm [39], is restricted to the pharyngeal mesoderm [40]. Importantly, expression of *AmphiTbx1/10* is also restricted to the pharyngeal ventral mesoderm [41], suggesting that the ventral mesoderm of amphioxus is indeed regionalized into the pharyngeal ventral mesoderm and ventral mesoderm caudal to the pharynx (Figure 2) [4]. Therefore, it seems that the system to regionalize the ventral mesoderm along the anterior–posterior axis was already established in the cephalochordate lineage, although the ventral mesoderm was not yet separated into the cardiac and posterior ventral mesoderm [4].

Previous studies suggest that retinoic acid signaling is involved in the patterning of various organs along the anterior–posterior axis both in amphioxus and lampreys [42–47]. RARs have been identified in amphioxus [42] and lampreys [43]. Furthermore, treatments with excess retinoic acid or retinoic acid inhibitors affect the anterior–posterior patterning of amphioxus and lampreys [42–47]. Interestingly, lamprey embryos treated with retinoic acid fail to form a heart [47], supporting the view that retinoic acid has retained its importance for the regionalization of the lateral plate mesoderm into the ALPM and the PLPM during evolution (Figure 3) [4]. Therefore, it is likely that retinoic acid signaling and its roles in the anterior–posterior patterning were present in the common ancestor of vertebrates and amphioxus. However, a recent study on amphioxus embryos revealed that the function of retinoic acid signaling in anterior–posterior patterning of somites in amphioxus might not be the same as in vertebrates [48]. Bertrand *et al.* showed that retinoic acid signaling has critical roles in somitogenesis in amphioxus; however, it does not antagonize Fgf signaling. Therefore, the anterior–posterior patterning system controlled by mutual inhibition between retinoic acid and Fgfs during somitogenesis might have been acquired after the divergence of amphioxus and ancestral vertebrates [48]. Zhao *et al.* proposed that, during mouse embryogenesis, antagonism between retinoic acid and *Fgf8* provides the permissive environment for the forelimb-forming field in the PLPM [6]. Thus, we should consider the possibility that alteration of the distribution of retinoic acid in the lateral plate mesoderm and/or acquisition of a novel antagonistic function of retinoic acids and Fgfs might have been critical for subdividing the lateral plate mesoderm into the ALPM and the PLPM during vertebrate evolution (Figure 3).

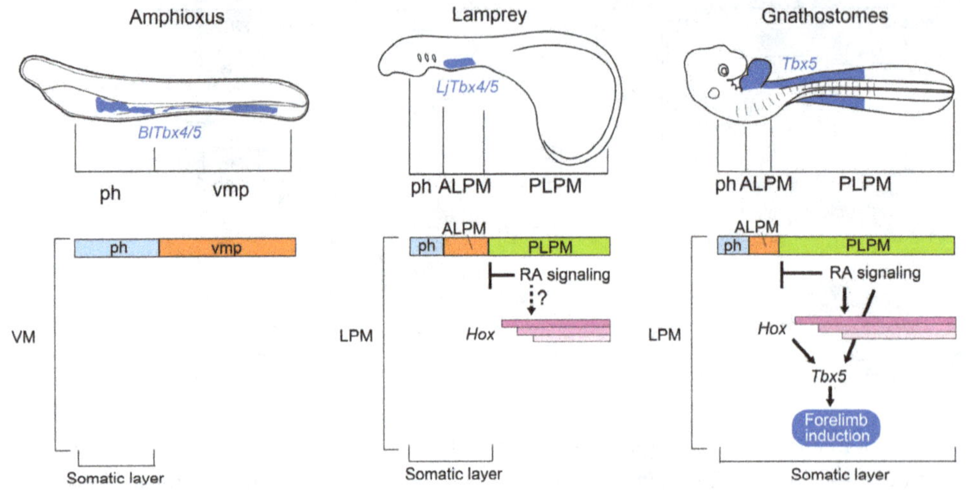

Figure 3. Schematic model for the acquisition of forelimb-forming fields during evolution. Light blue, orange and green bars represent the pharyngeal mesoderm (ph), ventral mesoderm posterior to the pharynx (vmp)/ALPM and PLPM, respectively (see Figure 2 for details). The ventral mesoderm (VM)/LPM is regionalized into the ALPM and PLPM in lampreys and gnathostomes, but not in amphioxus, and retinoic acid signaling seems to be involved in this process. Nested expression of *Hox* genes (pink bars) is observed in lampreys and gnathostomes, but not in amphioxus. Blue regions in the embryos are expression domains of *Tbx5* orthologs (*BlTbx4/5*, *LjTbx4/5*, *Tbx5*) in the VM/LPM. In amphioxus, *BlTbx4/5* (blue) appears in both ph and vmp, similar to cardiac markers (*AmphiNkx2.5-tin/AmphiTbx20*; see Figure 2). In lampreys, *LjTbx4/5* expression (blue) is expressed in the ALPM; it is not expressed in the unseparated PLPM. In gnathostomes, retinoic acid signaling, as well as Hox proteins, directly activates expression of *Tbx5* (blue) in the somatic layer of the PLPM to induce forelimb formation [11,49,50]. See text for additional references and details.

In gnathostomes, several lines of evidence indicate that retinoic acid signaling regulates the transcription of *Hox* genes and leads to the regionalization of the lateral plate mesoderm along the anterior–posterior axis [5,21,23,27]. However, transcripts of *Hox* genes have not yet been observed in the segmented mesoderm (the progenitor of the ventral mesoderm) of amphioxus (Figure 3) [51,52]. In amphioxus, retinoic acid signaling regulates the expression of *Hox* genes in anterior–posterior patterning of the endoderm, central nervous system and ectoderm at the neurula stage [45,53,54]. In addition, retinoic acid signaling directly regulates expression of *Hox* genes to pattern the anterior–posterior axis during gastrulation [55], suggesting that the regulatory mechanism of *Hox* gene transcription by retinoic acid signaling has already been established in the common ancestors of amphioxus and ancestral vertebrates. As the nested expression of *Hox* genes has been observed in the PLPM of both lamprey and gnathostome embryos [4,7], the acquisition of the novel expression domain for *Hox* genes in the lateral plate mesoderm—possibly via alteration of distribution of retinoic acid and/or the acquisition of a novel function of retinoic acid signaling—seems to be required for the regionalization of the lateral plate mesoderm into the ALPM and the PLPM (Figure 3).

In lamprey embryos, *Hox* genes are indeed expressed in a nested fashion in the PLPM [4], although limb formation is not initiated in this region (Figure 3). In gnathostomes, subsequent to the regionalization of the lateral plate mesoderm into the ALPM and the PLPM, the lateral plate mesodermal cells proliferate and are subdivided into the somatic and splanchnic layers beginning from the anterior toward the posterior part of the body (Figure 1b) [9]. Limb buds then appear as protrusions of the somatic layers in the PLPM. In contrast, in lampreys, the lateral plate mesoderm is subdivided into the somatic and splanchnic layers in the ALPM, but not in the PLPM (Figure 2) [4]. Thus, it is possible that lampreys cannot form limb buds because the underdeveloped PLPM does not split into the somatic and splanchnic layers [4]. Until recently, subdivision of the PLPM had not been observed during embryogenesis of any teleost fishes. However, a recent histological analysis of pufferfish (*Takifugu niphobles*) embryos at a very early stage (eight-somite stage) revealed that the lateral plate mesoderm is indeed subdivided into somatic and splanchnic layers both in the ALPM and the PLPM [56]. Importantly, cells from the PLPM of lampreys contribute to the formation of peritoneal epithelium and blood cells, but not the body wall [4,57]. Therefore, it is likely that the proliferation and differentiation of the PLPM, as well as formation of the coelom cavity, might have been a crucial step in acquiring paired appendages during the evolution of vertebrates [4,57]. In gnathostomes, *Forkhead box F* (*FoxF*) is expressed throughout the PLPM prior to the separation; subsequent to the subdivision, *FoxF* transcripts are restricted to the splanchnic mesoderm and *Iroquis homeobox 3* (*Irx3*) expression appears in the somatic mesoderm [9]. In lampreys, *LjIrx1/3* is expressed in the somatic layers of the ALPM, but not in the PLPM, although *LjFoxF* is expressed throughout the PLPM [4] (Figure 2). Furthermore, in amphioxus, *AmphiFoxF* is expressed from the pharynx to the posterior end of the body trunk [4], and *Branchiostoma floridae IrxC* (*BfIrxC*) is expressed in the ventral mesoderm at the level of the pharynx, but not in the ventral mesoderm posterior to the pharynx (Figure 2) [58]. Therefore, it is likely that the common ancestor of amphioxus and ancestral vertebrates had the *FoxF*-positive lateral plate mesoderm, and *Irx3*-positive somatic mesoderm seems to have appeared at the cardiac level after the divergence of ancestral chordates and vertebrates. The *FoxF*-positive unseparated PLPM seems to have been acquired prior to the lamprey lineages. The *Irx3*-positive somatic layers seem to have appeared in the PLPM after the divergence of lampreys and vertebrates [4], and provided permissive environment needed for emergence of fin-forming fields during evolution.

3. Specification of Limb-Forming Fields: *Tbx4*, *Tbx5* and *Hox* Genes

In gnathostomes, the lateral plate mesoderm is divided into the ALPM and the PLPM, and *Hox* genes seem to be involved in this process [5]. Although regionalization of the PLPM along the anterior–posterior axis by *Hox* genes has not been studied as extensively as in the nervous system or in somites, it has been proposed that expression patterns of *Hox* genes in the lateral plate mesoderm are somehow related to the specification of limb-forming fields [7]. Furthermore, the positioning of the forelimb buds at approximately the anterior border of the expression domains of *Hox5* and *Hox6* in the lateral plate mesoderm is conserved among zebrafish, chick and mouse [5,59–61]. Recently, it was shown that Hox proteins are involved in the regulation of the transcription of *T-box transcription factor 5* (*Tbx5*) (Figure 3) [11,12,19]. *Tbx5* is expressed in the anterior paired appendages of zebrafish, chick and mouse embryos [62–64] and initiates forelimb development [12,65–69]. In humans, mutations in *TBX5* cause Holt–Oram syndrome (OMIM 142900), an autosomal-dominant disorder characterized by abnormalities of the upper limbs and cardiac defects [70,71]. Zebrafish with mutations in *pbx4*, which encodes a Hox protein co-factor, fail to induce *tbx5* expression or initiate formation of pectoral fin buds [72]. In addition, developmental and molecular analyses of chick and mouse embryos revealed that rostral *Hox* genes expressed in the lateral plate mesoderm directly regulate the expression of *Tbx5* and thereby control the axial position of forelimb-forming fields (Figure 3) [11].

Recent analyses of mouse mutants showed that *Hox5* and *Hox9* genes are involved in establishing the anterior and posterior domains of the forelimb, respectively [73,74]. Loss of function of all three *Hox5* genes leads to anterior forelimb defects resulting from derepression of *Shh* expression [73]. In the anterior forelimbs, *Hox5* proteins interact with promyelocytic leukemia zinc finger (Plzf) and cooperatively mediate repression of *Shh* expression [73]. In contrast, an analysis of a *Hox9* quadruple knockout revealed that axial *Hox9* paralogs establish the posterior field of forelimbs by triggering the posteriorly restricted expression of *Hand2*, which directly activates *Shh* at the posterior margin [74].

In chick and zebrafish, Wnt/β-catenin signaling has key roles in the initiation of limb formation [67,75]. In chick embryos, *Wnt2b* is expressed in the intermediate mesoderm and the lateral plate mesoderm in the presumptive forelimb-forming region, whereas *Wnt8c* is expressed in the presumptive hindlimb-forming region [75], and the implantation of *Wnt2b*- or *Wnt8c*-expressing cells into the interlimb flank region leads to the formation of an additional limb via induction of ectopic *Fgf10* expression [75]. Consistent with this result, zebrafish embryos injected with *wnt2b* morpholinos show downregulation of *tbx5* expression and fail to form pectoral fin buds [67]. In mouse embryos, however, Wnt ligands that act upstream of *Fgf10* in the limb-forming field have not yet been found. Furthermore, double knockout mice for both *Lef1* and *Tcf1*, which mediate Wnt signaling by forming a complex with β-catenin, form normal limb buds [76]. However, a recent study in mice showed that conditional removal of *β-catenin* from forelimb-forming regions leads to the reduction of *Tbx5* expression and incomplete forelimbs, suggesting that the function of β-catenin to regulate *Tbx5* transcription is conserved among zebrafish, chick and mouse [50]. The T-cell factor/lymphoid enhancer factor (TCF/LEF) binding sequence is required for the enhancer activity of the forelimb-specific *Tbx5* regulatory element, suggesting that β-catenin directly regulates *Tbx5* expression [50]. Furthermore, retinoic acid signaling directly activates transcription of *Tbx5* in the forelimb-forming fields (Figure 3) [50]. Taken together, retinoic acid signaling, β-catenin/LEF/TCF and anteriorly nested *Hox* proteins seem to cooperatively activate *Tbx5* transcription for inducing forelimb bud formation [50].

It has also been shown that caudally expressed *Hox* proteins, such as *Hoxc8*, *Hoxc9* and *Hoxc10*, can form a repressive complex and bind to the forelimb-specific regulatory sequences of *Tbx5* and repress its expression [49]. Interestingly, in chick embryos, transplantation of an Fgf-soaked bead into the interlimb flank region turns off *Hoxc9* expression when an ectopic wing bud is induced, whereas it turns on *Hoxc9* expression when an ectopic leg bud is induced [7]. Thus, these previous findings seem to support the view that *Hoxc9* acts as a repressor of *Tbx5* transcription [49]. In addition, rostrally expanded *Hoxc8* distribution is present in the body trunk of a limbless vertebrate, the python [77], supporting the view that *Hoxc8* represses *Tbx5* expression [49].

Molecular mechanisms that specify the hindlimb-forming fields are not as well characterized as those of the forelimb-forming fields. *Tbx4*, which is closely related to *Tbx5*, is specifically expressed in the presumptive hindlimb forming fields of chick, mouse, teleosts and catshark (*Scyliorhinus canicula*) [62–64,78,79], and its expression is suggested to be regulated by *Hox* genes [12] and *Pitx1* [12,80]. In addition, studies on transgenic mice, in which *Tbx5* is replaced by *Tbx4*, revealed that *Tbx4* is able to substitute for *Tbx5* function in the initiation of the forelimb bud [12], suggesting that Tbx4 has a similar role in the initiation of hindlimb formation. However, unlike *Tbx5* null mutants, mice deficient for *Tbx4* are able to initiate *Fgf10* expression in hindlimb fields and form truncated hindlimbs [81]. Additional studies in mice deficient for *Tbx4* revealed that *Tbx4* is crucial for initiating hindlimb outgrowth, but an additional factor(s) other than *Tbx4* is likely to be involved in this process [82]. *Islet1*, a LIM-homeodomain transcription factor that is specifically expressed in the hindlimb progenitors, was recently shown to be required for hindlimb initiation [83,84]. Early inactivation of *Islet1* as well as β-catenin in the lateral plate mesoderm leads to a failure to initiate hindlimb development [84]. Furthermore, *Islet1* is required for the nuclear accumulation of β-catenin and thereby for activation of the β-catenin pathway, and that the β-catenin pathway is required for the maintenance of *Islet1* expression [84]. *Islet1* seems to be a missing piece in the regulation of hindlimb-specific initiation [84].

Pitx1, which encodes a member of the paired-like family of homeodomain transcription factors, is expressed in the presumptive hindlimb-forming fields in chick, mouse, stickleback (*Gasterosteus aculeatus*) and catshark embryos [79,85–87]. In mice, *Pitx1* is required for normal initiation of hindlimb development by regulating transcription levels of *Tbx4* [85,88,89]. It has recently been suggested that in Liebenberg syndrome (OMIM 186550) [90], an autosomal-dominant malformation of the upper limbs, the upper limbs have undergone a partial homeotic hindlimb transformation, and that this phenotype is associated with genomic rearrangements at the *PITX1* locus [91].

In teleosts, the roles of *tbx4* and *pitx1* in pelvic fin formation seem to be more critical than their roles in the hindlimb formation of amniotes. In *tbx4* mutant zebrafish, expression of *fgf10* is not induced, and thus pelvic fin buds fail to initiate their formation at a very early stage of fin initiation [92]. In the spine-deficient morph of the three-spined sticklebacks, expressions of neither *pitx1* nor *tbx4* are detected in the lateral plate mesoderm at the level of the pelvic fin, and pelvic fin buds fail to initiate their formation [87]. Subsequently, the responsible regulatory sequence for pelvic spine deficiency was found within the pelvic fin-specific enhancer of *pitx1* in three-spined stickleback [93,94]. In teleosts, pelvic fin formation is significantly delayed as it is triggered by thyroid hormone at the time of the larva-juvenile transition [95] (for instance, three weeks post–fertilization in zebrafish). Thus, the levels of molecules that affect the limb initiation process in the pelvic fin-forming region of teleosts at the time of pelvic fin initiation are not identical to those in the presumptive hindlimb-forming region of early chick and mouse embryos. Although we cannot exclude the possibility that the functions of Pitx1 and/or *Tbx4* in the initiation of pelvic fin/hindlimb development in teleosts are different from those in mice, it is possible that the Pitx1-*Tbx4* pathway is sufficient to activate the initiation of pelvic fin outgrowth in teleost fishes.

A shift in limb position is also observed in chick and mouse embryos in which expression of *Gdf11* has been manipulated [96,97]. Mutant mice carrying a targeted deletion of *Gdf11* exhibit posterior displacement of the hindlimbs, and patterns of expression of *Hoxc6*, *Hoxc10* and *Hoxc11* in spinal cord and paraxial mesoderm also shift posteriorly [96]. In chick embryos, overexpression of *Gdf11* in the spinal cord leads to the rostral shift of both fore- and hindlimb [97]. In the spinal cord of chicks that overexpress *Gdf11*, expression of *Hoxc6*, *Hoxc8*, *Hoxc9* and *Hoxc10* shift rostrally and lead to the rostral shift of the lateral motor columns [97], which are defined by Hox6 and Hox10 [98–100]. Thus, the shift in limb buds in chick following the introduction of *Gdf11*-expressing constructs into the spinal cord was interpreted to be caused by a shift in the lateral motor columns [97]. *Gdf11* expression in the caudal part of the body is conserved in chick, mouse, zebrafish and Nile tilapia embryos [101–103]. Furthermore, in zebrafish injected with *gdf11* morpholinos, the pelvic fin shifts caudally, along with a posterior shift in *Hoxc10* expression in the spinal cord [103], suggesting that the Gdf11-controlled mechanism that positions the hindlimb along the body axis is likely to be conserved in gnathostomes [103,104]. Gdf11 signaling is a central modulator of the trunk-to-tail transition during mouse development [105]. In *Gdf11* mutants, the specification of the cloaca and the induction of the hindlimbs are significantly delayed, whereas precocious activation of Gdf11 signaling in the epiblast anteriorizes these structures with a concomitant reduction in trunk length [105]. In contrast, precocious activation of genes of the *Hox9*, *Hox10* and *Hox11* groups in the epiblast do not alter the phenotype, with one exception: transgenic mice expressing *Hoxb9* show only a slight anterior displacement of the hindlimbs by one somite [105]. Thus, Jurberg and his colleagues proposed that *Hox* genes do not play a major role in specifying the hindlimb position along the anterior–posterior axis during the trunk-to-tail transition [105]. In early stages of hindlimb induction, *Islet1* has a major role in establishing the posterior hindlimb field by regulating the Hand2-Shh pathway [84,106]. During the trunk-to-tail transition, Gdf11 directly activates *Islet1* to promote formation of the most posterior progenitors of the lateral plate mesoderm, such as the hindlimbs and cloacal tissue [105].

It has been suggested that the nested *Hox* gene expression in the lateral plate mesoderm is necessary for acquiring the first pair of appendages during the evolution of vertebrates [107]. In limbless amphioxus species *Branchiostoma floridae* and *Branchiostoma lanceolatum*, *Tbx4/5* is expressed throughout the ventral mesoderm (Figure 3) [108,109]. Interestingly, amphioxus *Tbx4/5* has the ability to initiate forelimb formation in mouse transgenic lines [110]. This apparent contradiction can be explained by the regionalization of the ventral mesoderm and expression of *Hox* genes mentioned above. In amphioxus, the ventral mesoderm is not molecularly regionalized into the cardiac and posterior ventral mesoderm [4], and *Hox* genes are not expressed in the ventral mesoderm progenitors [51,52], suggesting that *Tbx4/5* in the ventral mesoderm may contribute only to heart formation in amphioxus [4]. In lampreys, the lateral plate mesoderm is regionalized into the ALPM and the PLPM, and *LjHox5i* and *LjHox6w* are expressed in the PLPM in a nested fashion [4]. Nevertheless, lamprey *LjTbx4/5* transcripts are not present in the PLPM but are in the ALPM [4,37]. Although we do not know whether lamprey *Hox* genes have the conserved function of inducing *Tbx4/5* in the PLPM, one of the main reasons for limblessness seems to be the lack of separation of the PLPM into the somatic and splanchnic layers (Figure 3), which results in the contribution of cells from the PLPM to the peritoneal epithelium but not to the body wall [4,57].

4. Conclusions

In gnathostomes, the lateral plate mesoderm is regionalized into the ALPM and PLPM, and *Hox* genes are expressed in a nested fashion along the anterior–posterior axis in the PLPM. Then, the lateral plate mesoderm splits into the somatic and splanchnic layers, and the expression of limb initiation genes appears in the limb-forming fields. Figure 4 summarizes the genes implicated in the specification of limb-forming fields in the PLPM during mouse development. *Tbx5* and *Tbx4* are expressed in the forelimb- and hindlimb-forming fields, respectively [63,64], and play crucial roles in the initiation of limb outgrowth [12,65–69]. In the forelimb-forming field, retinoic acid signaling and β-catenin signaling act cooperatively with rostrally expressed *Hox* genes to directly regulate *Tbx5* expression [11,50], whereas caudally expressed *Hox* genes repress *Tbx5* to restrict its expression to this field [49]. In the hindlimb-forming field, *Tbx4* expression, which is crucial for the induction of hindlimb development, is suggested to be activated by *Hox* genes [12] and *Pitx1* [12,85]. *Islet1*-mediated activation of β-catenin signaling is also required for hindlimb induction [84]. In the early forelimb field, *Hox9* paralogous genes are required for the expression of *Hand2* to establish the posterior domain [74], whereas *Hox5* genes and *Plzf* cooperatively mediate repression of *Shh* expression in the anterior domain [73]. In contrast, the hindlimb posterior domain is established by *Islet1* to regulate *Hand2* expression in this region [106].

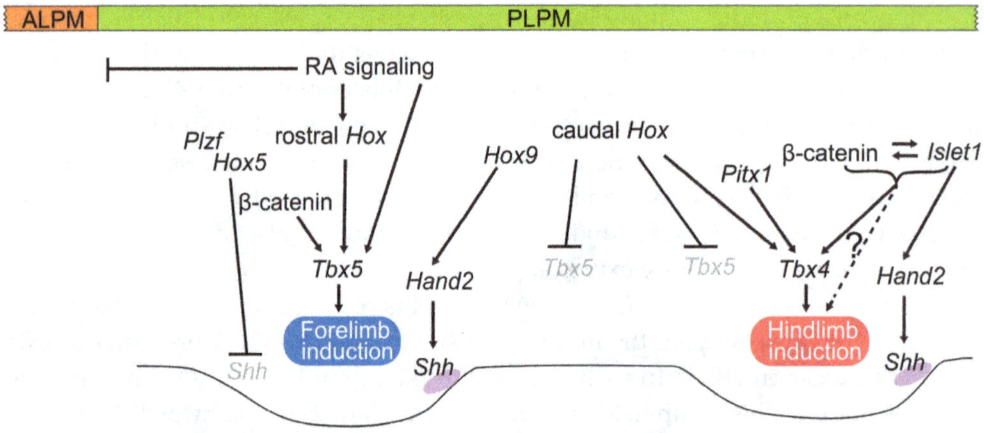

Figure 4. Genes implicated in the specification of the limb field during development of mouse embryos. Retinoic acid signaling is involved in the regionalization of the LPM into the ALPM and PLPM. In the PLPM, *Tbx5* and *Tbx4*, which are expressed in forelimb- and hindlimb-forming regions, respectively, play pivotal roles in limb initiation. Nested expression of *Hox* genes seems to be involved in the regionalization of the PLPM into forelimb, interlimb flank and hindlimb fields. In the forelimb level, rostrally expressed *Hox* proteins, retinoic acid signaling and β-catenin signaling cooperatively activate *Tbx5* transcription. Whereas caudally expressed *Hox* genes repress *Tbx5* to restrict its expression to the forelimb level, *Hox* genes and *Pitx1* have been suggested to activate *Tbx4* expression at the hindlimb level. *Islet1* is also required for hindlimb initiation. *Islet1* is required for the nuclear accumulation of β-catenin, and β-catenin is required for the maintenance of *Islet1* expression. In the forelimb posterior domain, *Hox9* genes are required to trigger *Hand2* expression in this region. In the forelimb anterior domain, *Hox5* genes and *Plzf* cooperatively mediate repression of *Shh* expression. In the hindlimb posterior domain, *Islet1* activates *Hand2* expression. It should be noted that the roles of *Hox* genes, RA signaling and β-catenin signaling are disputed. See text for references and details.

Developmental analyses of limbless amphioxus and lampreys have provided clues to understand the evolutionary scenarios that led to the acquisition of limb-forming fields in the vertebrate body [4].

In amphioxus, the ventral mesoderm is not molecularly regionalized into cardiac versus posterior ventral mesoderm. In lamprey, the lateral plate mesoderm is regionalized into the ALPM and the PLPM, and *Hox* genes are expressed in a nested fashion in the PLPM [4], but the PLPM is not separated into the somatic and splanchnic layers [4]. These past and current findings suggest that, to acquire the limb-forming fields, novel expression domains of limb initiation genes in the lateral plate mesoderm, regionalization along its anterior–posterior axis, and its subdivision into somatic and splanchnic layers were likely to be required during vertebrate evolution.

Acknowledgments: The author thanks past and current members of the Tanaka laboratory and many collaborators for their contribution to the studies cited here. Portions of the studies conducted in the Tanaka laboratory were supported by a Grant-in-Aid for Scientific Research (B) 25291086, the Takeda Foundation, the Naito Foundation and the Inamori Foundation.

Conflicts of Interest: The author declares no conflict of interest. The funding sponsors had no role in the design of the study; in the collection, analyses, or interpretation of data; in the writing of the manuscript; and in the decision to publish the results.

References

1. Coates, M.I. The origin of vertebrate limbs. *Dev. Suppl.* **1994**, 169–180.
2. Janvier, P. *Early Vertebrates*; Clarendon Press: Oxford, UK, 1996.
3. Donoghue, P.C.; Forey, P.L.; Aldridge, R.J. Conodont affinity and chordate phylogeny. *Biol. Rev. Camb. Philos. Soc.* **2000**, *75*, 191–251. [CrossRef] [PubMed]
4. Onimaru, K.; Shoguchi, E.; Kuratani, S.; Tanaka, M. Development and evolution of the lateral plate mesoderm: Comparative analysis of amphioxus and lamprey with implications for the acquisition of paired fins. *Dev. Biol.* **2011**, *359*, 124–136. [CrossRef] [PubMed]
5. Waxman, J.S.; Keegan, B.R.; Roberts, R.W.; Poss, K.D.; Yelon, D. *Hoxb5b* acts downstream of retinoic acid signaling in the forelimb field to restrict heart field potential in zebrafish. *Dev. Cell* **2008**, *15*, 923–934. [CrossRef] [PubMed]
6. Zhao, X.; Sirbu, I.O.; Mic, F.A.; Molotkova, N.; Molotkov, A.; Kumar, S.; Duester, G. Retinoic acid promotes limb induction through effects on body axis extension but is unnecessary for limb patterning. *Curr. Biol.* **2009**, *19*, 1050–1057. [CrossRef] [PubMed]
7. Cohn, M.J.; Patel, K.; Krumlauf, R.; Wilkinson, D.G.; Clarke, J.D.; Tickle, C. *Hox9* genes and vertebrate limb specification. *Nature* **1997**, *387*, 97–101. [CrossRef] [PubMed]
8. Nowicki, J.L.; Burke, A.C. *Hox* genes and morphological identity: Axial versus lateral patterning in the vertebrate mesoderm. *Development* **2000**, *127*, 4265–4275. [PubMed]
9. Funayama, N.; Sato, Y.; Matsumoto, K.; Ogura, T.; Takahashi, Y. Coelom formation: Binary decision of the lateral plate mesoderm is controlled by the ectoderm. *Development* **1999**, *126*, 4129–4138. [PubMed]
10. Logan, M. Finger or toe: The molecular basis of limb identity. *Development* **2003**, *130*, 6401–6410. [CrossRef] [PubMed]
11. Minguillon, C.; Nishimoto, S.; Wood, S.; Vendrell, E.; Gibson-Brown, J.J.; Logan, M.P. Hox genes regulate the onset of *Tbx5* expression in the forelimb. *Development* **2012**, *139*, 3180–3188. [CrossRef] [PubMed]
12. Minguillon, C.; del Buono, J.; Logan, M.P. *Tbx5* and *Tbx4* are not sufficient to determine limb-specific morphologies but have common roles in initiating limb outgrowth. *Dev. Cell* **2005**, *8*, 75–84. [CrossRef] [PubMed]
13. Grandel, H.; Lun, K.; Rauch, G.J.; Rhinn, M.; Piotrowski, T.; Houart, C.; Sordino, P.; Kuchler, A.M.; Schulte-Merker, S.; Geisler, R.; *et al.* Retinoic acid signalling in the zebrafish embryo is necessary during pre-segmentation stages to pattern the anterior-posterior axis of the CNS and to induce a pectoral fin bud. *Development* **2002**, *129*, 2851–2865. [PubMed]
14. Gibert, Y.; Gajewski, A.; Meyer, A.; Begemann, G. Induction and prepatterning of the zebrafish pectoral fin bud requires axial retinoic acid signaling. *Development* **2006**, *133*, 2649–2659. [CrossRef] [PubMed]
15. Bruce, A.E.; Oates, A.C.; Prince, V.E.; Ho, R.K. Additional *hox* clusters in the zebrafish: Divergent expression patterns belie equivalent activities of duplicate *hoxB5* genes. *Evol. Dev.* **2001**, *3*, 127–144. [CrossRef] [PubMed]

16. Oosterveen, T.; Niederreither, K.; Dolle, P.; Chambon, P.; Meijlink, F.; Deschamps, J. Retinoids regulate the anterior expression boundaries of 5′ *Hoxb* genes in posterior hindbrain. *EMBO J.* **2003**, *22*, 262–269. [CrossRef] [PubMed]
17. Sharpe, J.; Nonchev, S.; Gould, A.; Whiting, J.; Krumlauf, R. Selectivity, sharing and competitive interactions in the regulation of hoxb genes. *EMBO J.* **1998**, *17*, 1788–1798. [CrossRef] [PubMed]
18. Rancourt, D.E.; Tsuzuki, T.; Capecchi, M.R. Genetic interaction between *hoxb-5* and *hoxb-6* is revealed by nonallelic noncomplementation. *Genes Dev.* **1995**, *9*, 108–122. [CrossRef] [PubMed]
19. Ryckebusch, L.; Wang, Z.; Bertrand, N.; Lin, S.C.; Chi, X.; Schwartz, R.; Zaffran, S.; Niederreither, K. Retinoic acid deficiency alters second heart field formation. *Proc. Natl. Acad. Sci. USA* **2008**, *105*, 2913–2918. [CrossRef] [PubMed]
20. Searcy, R.D.; Yutzey, K.E. Analysis of *Hox* gene expression during early avian heart development. *Dev. Dyn.* **1998**, *213*, 82–91. [CrossRef]
21. Lo, P.C.; Frasch, M. Establishing A-P polarity in the embryonic heart tube: A conserved function of hox genes in *Drosophila* and vertebrates? *Trends Cardiovasc. Med.* **2003**, *13*, 182–187. [CrossRef]
22. Chisaka, O.; Capecchi, M.R. Regionally restricted developmental defects resulting from targeted disruption of the mouse homeobox gene *hox-1.5*. *Nature* **1991**, *350*, 473–479. [CrossRef] [PubMed]
23. Rosenthal, N.; Xavier-Neto, J. From the bottom of the heart: Anteroposterior decisions in cardiac muscle differentiation. *Curr. Opin. Cell Biol.* **2000**, *12*, 742–746. [CrossRef]
24. Bertrand, N.; Roux, M.; Ryckebusch, L.; Niederreither, K.; Dolle, P.; Moon, A.; Capecchi, M.; Zaffran, S. *Hox* genes define distinct progenitor sub-domains within the second heart field. *Dev. Biol.* **2011**, *353*, 266–274. [CrossRef] [PubMed]
25. Stratford, T.; Horton, C.; Maden, M. Retinoic acid is required for the initiation of outgrowth in the chick limb bud. *Curr. Biol.* **1996**, *6*, 1124–1133. [CrossRef]
26. Mic, F.A.; Haselbeck, R.J.; Cuenca, A.E.; Duester, G. Novel retinoic acid generating activities in the neural tube and heart identified by conditional rescue of Raldh2 null mutant mice. *Development* **2002**, *129*, 2271–2282. [PubMed]
27. Niederreither, K.; Subbarayan, V.; Dolle, P.; Chambon, P. Embryonic retinoic acid synthesis is essential for early mouse post-implantation development. *Nat. Genet.* **1999**, *21*, 444–448. [CrossRef] [PubMed]
28. Cunningham, T.J.; Zhao, X.; Sandell, L.L.; Evans, S.M.; Trainor, P.A.; Duester, G. Antagonism between retinoic acid and fibroblast growth factor signaling during limb development. *Cell Rep.* **2013**, *3*, 1503–1511. [CrossRef] [PubMed]
29. Marques, S.R.; Lee, Y.; Poss, K.D.; Yelon, D. Reiterative roles for fgf signaling in the establishment of size and proportion of the zebrafish heart. *Dev. Biol.* **2008**, *321*, 397–406. [CrossRef] [PubMed]
30. Holland, N.D.; Venkatesh, T.V.; Holland, L.Z.; Jacobs, D.K.; Bodmer, R. *Amphink2-tin*, an amphioxus homeobox gene expressed in myocardial progenitors: Insights into evolution of the vertebrate heart. *Dev. Biol.* **2003**, *255*, 128–137. [CrossRef]
31. Charite, J.; McFadden, D.G.; Olson, E.N. The bHLH transcription factor dHand controls *Sonic hedgehog* expression and establishment of the zone of polarizing activity during limb development. *Development* **2000**, *127*, 2461–2470. [PubMed]
32. Deimling, S.J.; Drysdale, T.A. Retinoic acid regulates anterior-posterior patterning within the lateral plate mesoderm of Xenopus. *Mech. Dev.* **2009**, *126*, 913–923. [CrossRef] [PubMed]
33. Yelon, D.; Ticho, B.; Halpern, M.E.; Ruvinsky, I.; Ho, R.K.; Silver, L.M.; Stainier, D.Y. The bHLH transcription factor hand2 plays parallel roles in zebrafish heart and pectoral fin development. *Development* **2000**, *127*, 2573–2582. [PubMed]
34. Buchberger, A.; Pabst, O.; Brand, T.; Seidl, K.; Arnold, H.H. Chick NKx-2.3 represents a novel family member of vertebrate homologues to the Drosophila homeobox gene tinman: Differential expression of cNKx-2.3 and cNKkx-2.5 during heart and gut development. *Mech. Dev.* **1996**, *56*, 151–163. [CrossRef]
35. Kraus, F.; Haenig, B.; Kispert, A. Cloning and expression analysis of the mouse T-box gene *tbx20*. *Mech. Dev.* **2001**, *100*, 87–91. [CrossRef]
36. Yamagishi, T.; Nakajima, Y.; Nishimatsu, S.; Nohno, T.; Ando, K.; Nakamura, H. Expression of *tbx20* rna during chick heart development. *Dev. Dyn.* **2004**, *230*, 576–580. [CrossRef] [PubMed]

37. Kokubo, N.; Matsuura, M.; Onimaru, K.; Tiecke, E.; Kuraku, S.; Kuratani, S.; Tanaka, M. Mechanisms of heart development in the japanese lamprey, lethenteron japonicum. *Evol. Dev.* **2010**, *12*, 34–44. [CrossRef] [PubMed]
38. Tanaka, M. Molecular and evolutionary basis of limb field specification and limb initiation. *Dev. Growth Differ.* **2013**, *55*, 149–163. [CrossRef] [PubMed]
39. Garg, V.; Yamagishi, C.; Hu, T.; Kathiriya, I.S.; Yamagishi, H.; Srivastava, D. *Tbx1*, a digeorge syndrome candidate gene, is regulated by sonic hedgehog during pharyngeal arch development. *Dev. Biol.* **2001**, *235*, 62–73. [CrossRef] [PubMed]
40. Tiecke, E.; Matsuura, M.; Kokubo, N.; Kuraku, S.; Kusakabe, R.; Kuratani, S.; Tanaka, M. Identification and developmental expression of two *Tbx1/10*-related genes in the agnathan lethenteron japonicum. *Dev. Genes Evol.* **2007**, *217*, 691–697. [CrossRef] [PubMed]
41. Meulemans, D.; Bronner-Fraser, M. Insights from amphioxus into the evolution of vertebrate cartilage. *PLoS ONE* **2007**, *2*, e787. [CrossRef] [PubMed]
42. Escriva, H.; Bertrand, S.; Germain, P.; Robinson-Rechavi, M.; Umbhauer, M.; Cartry, J.; Duffraisse, M.; Holland, L.; Gronemeyer, H.; Laudet, V. Neofunctionalization in vertebrates: The example of retinoic acid receptors. *PLoS Genet.* **2006**, *2*, e102. [CrossRef] [PubMed]
43. Castillo, H.A.; Cravo, R.M.; Azambuja, A.P.; Simoes-Costa, M.S.; Sura-Trueba, S.; Gonzalez, J.; Slonimsky, E.; Almeida, K.; Abreu, J.G.; de Almeida, M.A.; *et al.* Insights into the organization of dorsal spinal cord pathways from an evolutionarily conserved raldh2 intronic enhancer. *Development* **2010**, *137*, 507–518. [CrossRef] [PubMed]
44. Koop, D.; Chen, J.; Theodosiou, M.; Carvalho, J.E.; Alvarez, S.; de Lera, A.R.; Holland, L.Z.; Schubert, M. Roles of retinoic acid and *Tbx1/10* in pharyngeal segmentation: Amphioxus and the ancestral chordate condition. *EvoDevo* **2014**, *5*, 36. [CrossRef] [PubMed]
45. Schubert, M.; Holland, N.D.; Laudet, V.; Holland, L.Z. A retinoic acid-*Hox* hierarchy controls both anterior/posterior patterning and neuronal specification in the developing central nervous system of the cephalochordate amphioxus. *Dev. Biol.* **2006**, *296*, 190–202. [CrossRef] [PubMed]
46. Murakami, Y.; Pasqualetti, M.; Takio, Y.; Hirano, S.; Rijli, F.M.; Kuratani, S. Segmental development of reticulospinal and branchiomotor neurons in lamprey: Insights into the evolution of the vertebrate hindbrain. *Development* **2004**, *131*, 983–995. [CrossRef] [PubMed]
47. Kuratani, S.; Ueki, T.; Hirano, S.; Aizawa, S. Rostral truncation of a cyclostome, lampetra japonica, induced by all-trans retinoic acid defines the head/trunk interface of the vertebrate body. *Dev. Dyn.* **1998**, *211*, 35–51. [CrossRef]
48. Bertrand, S.; Aldea, D.; Oulion, S.; Subirana, L.; de Lera, A.R.; Somorjai, I.; Escriva, H. Evolution of the role of RA and FGF signals in the control of somitogenesis in chordates. *PLoS ONE* **2015**, *10*, e0136587. [CrossRef] [PubMed]
49. Nishimoto, S.; Minguillon, C.; Wood, S.; Logan, M.P. A combination of activation and repression by a colinear Hox code controls forelimb-restricted expression of *Tbx5* and reveals hox protein specificity. *PLoS Genet.* **2014**, *10*, e1004245. [CrossRef] [PubMed]
50. Nishimoto, S.; Wilde, S.M.; Wood, S.; Logan, M.P. RA acts in a coherent feed-forward mechanism with Tbx5 to control limb bud induction and initiation. *Cell Rep.* **2015**, *12*, 879–891. [CrossRef] [PubMed]
51. Holland, P.W.; Holland, L.Z.; Williams, N.A.; Holland, N.D. An amphioxus homeobox gene: Sequence conservation, spatial expression during development and insights into vertebrate evolution. *Development* **1992**, *116*, 653–661. [PubMed]
52. Wada, H.; Garcia-Fernandez, J.; Holland, P.W. Colinear and segmental expression of amphioxus *Hox* genes. *Dev. Biol.* **1999**, *213*, 131–141. [CrossRef] [PubMed]
53. Schubert, M.; Yu, J.K.; Holland, N.D.; Escriva, H.; Laudet, V.; Holland, L.Z. Retinoic acid signaling acts via *Hox1* to establish the posterior limit of the pharynx in the chordate amphioxus. *Development* **2005**, *132*, 61–73. [CrossRef] [PubMed]
54. Schubert, M.; Holland, N.D.; Escriva, H.; Holland, L.Z.; Laudet, V. Retinoic acid influences anteroposterior positioning of epidermal sensory neurons and their gene expression in a developing chordate (amphioxus). *Proc. Nat. Acad. Sci. USA* **2004**, *101*, 10320–10325. [CrossRef] [PubMed]

55. Koop, D.; Holland, N.D.; Semon, M.; Alvarez, S.; de Lera, A.R.; Laudet, V.; Holland, L.Z.; Schubert, M. Retinoic acid signaling targets *Hox* genes during the amphioxus gastrula stage: Insights into early anterior-posterior patterning of the chordate body plan. *Dev. Biol.* **2010**, *338*, 98–106. [CrossRef] [PubMed]
56. Tanaka, M.; Yu, R.; Kurokawa, D. Anterior migration of lateral plate mesodermal cells during embryogenesis of the pufferfish Takifugu niphobles: Insight into the rostral positioning of pelvic fins. *J. Anat.* **2015**, *227*, 81–88. [CrossRef] [PubMed]
57. Tulenko, F.J.; McCauley, D.W.; Mackenzie, E.L.; Mazan, S.; Kuratani, S.; Sugahara, F.; Kusakabe, R.; Burke, A.C. Body wall development in lamprey and a new perspective on the origin of vertebrate paired fins. *Proc. Nat. Acad. Sci. USA* **2013**, *110*, 11899–11904. [CrossRef] [PubMed]
58. Kaltenbach, S.L.; Holland, L.Z.; Holland, N.D.; Koop, D. Developmental expression of the three iroquois genes of amphioxus (BfIrxa, BfIrxb, and BfIrxc) with special attention to the gastrula organizer and anteroposterior boundaries in the central nervous system. *Gene Expr. Patterns* **2009**, *9*, 329–334. [CrossRef] [PubMed]
59. Oliver, G.; Wright, C.V.; Hardwicke, J.; de Robertis, E.M. A gradient of homeodomain protein in developing forelimbs of Xenopus and mouse embryos. *Cell* **1988**, *55*, 1017–1024. [CrossRef]
60. Becker, D.; Eid, R.; Schughart, K. The limb/lpm enhancer of the murine *Hoxb6* gene: Reporter gene analysis in transgenic embryos and studies of DNA-protein interactions. *Pharm. Acta Helv.* **1996**, *71*, 29–35. [CrossRef]
61. Nelson, C.E.; Morgan, B.A.; Burke, A.C.; Laufer, E.; DiMambro, E.; Murtaugh, L.C.; Gonzales, E.; Tessarollo, L.; Parada, L.F.; Tabin, C. Analysis of *Hox* gene expression in the chick limb bud. *Development* **1996**, *122*, 1449–1466. [PubMed]
62. Tamura, K.; Yonei-Tamura, S.; Belmonte, J.C. Differential expression of *Tbx4* and *Tbx5* in zebrafish fin buds. *Mech. Dev.* **1999**, *87*, 181–184. [CrossRef]
63. Gibson-Brown, J.J.; Agulnik, S.I.; Chapman, D.L.; Alexiou, M.; Garvey, N.; Silver, L.M.; Papaioannou, V.E. Evidence of a role for T-box genes in the evolution of limb morphogenesis and the specification of forelimb/hindlimb identity. *Mech. Dev.* **1996**, *56*, 93–101. [CrossRef]
64. Isaac, A.; Rodriguez-Esteban, C.; Ryan, A.; Altabef, M.; Tsukui, T.; Patel, K.; Tickle, C.; Izpisua-Belmonte, J.C. Tbx genes and limb identity in chick embryo development. *Development* **1998**, *125*, 1867–1875. [PubMed]
65. Ahn, D.G.; Kourakis, M.J.; Rohde, L.A.; Silver, L.M.; Ho, R.K. T-box gene tbx5 is essential for formation of the pectoral limb bud. *Nature* **2002**, *417*, 754–758. [CrossRef] [PubMed]
66. Garrity, D.M.; Childs, S.; Fishman, M.C. The heartstrings mutation in zebrafish causes heart/fin Tbx5 deficiency syndrome. *Development* **2002**, *129*, 4635–4645. [PubMed]
67. Ng, J.K.; Kawakami, Y.; Buscher, D.; Raya, A.; Itoh, T.; Koth, C.M.; Rodriguez Esteban, C.; Rodriguez-Leon, J.; Garrity, D.M.; Fishman, M.C.; *et al.* The limb identity gene Tbx5 promotes limb initiation by interacting with Wnt2b and Fgf10. *Development* **2002**, *129*, 5161–5170. [PubMed]
68. Agarwal, P.; Wylie, J.N.; Galceran, J.; Arkhitko, O.; Li, C.; Deng, C.; Grosschedl, R.; Bruneau, B.G. Tbx5 is essential for forelimb bud initiation following patterning of the limb field in the mouse embryo. *Development* **2003**, *130*, 623–633. [CrossRef] [PubMed]
69. Rallis, C.; Bruneau, B.G.; del Buono, J.; Seidman, C.E.; Seidman, J.G.; Nissim, S.; Tabin, C.J.; Logan, M.P. Tbx5 is required for forelimb bud formation and continued outgrowth. *Development* **2003**, *130*, 2741–2751. [CrossRef] [PubMed]
70. Basson, C.T.; Bachinsky, D.R.; Lin, R.C.; Levi, T.; Elkins, J.A.; Soults, J.; Grayzel, D.; Kroumpouzou, E.; Traill, T.A.; Leblanc-Straceski, J.; *et al.* Mutations in human TBX5 [corrected] cause limb and cardiac malformation in holt-oram syndrome. *Nat. Genet.* **1997**, *15*, 30–35. [CrossRef] [PubMed]
71. Li, Q.Y.; Newbury-Ecob, R.A.; Terrett, J.A.; Wilson, D.I.; Curtis, A.R.; Yi, C.H.; Gebuhr, T.; Bullen, P.J.; Robson, S.C.; Strachan, T.; *et al.* Holt-oram syndrome is caused by mutations in TBX5, a member of the Brachyury (T) gene family. *Nat. Genet.* **1997**, *15*, 21–29. [CrossRef] [PubMed]
72. Popperl, H.; Rikhof, H.; Chang, H.; Haffter, P.; Kimmel, C.B.; Moens, C.B. Lazarus is a novel pbx gene that globally mediates *hox* gene function in zebrafish. *Mol. Cell* **2000**, *6*, 255–267. [CrossRef]
73. Xu, B.; Hrycaj, S.M.; McIntyre, D.C.; Baker, N.C.; Takeuchi, J.K.; Jeannotte, L.; Gaber, Z.B.; Novitch, B.G.; Wellik, D.M. *Hox5* interacts with Plzf to restrict *Shh* expression in the developing forelimb. *Proc. Nat. Acad. Sci. USA* **2013**, *110*, 19438–19443. [CrossRef] [PubMed]
74. Xu, B.; Wellik, D.M. Axial *Hox9* activity establishes the posterior field in the developing forelimb. *Proc. Nat. Acad. Sci. USA* **2011**, *108*, 4888–4891. [CrossRef] [PubMed]

75. Kawakami, Y.; Capdevila, J.; Buscher, D.; Itoh, T.; Rodriguez Esteban, C.; Izpisua Belmonte, J.C. Wnt signals control FGF-dependent limb initiation and aer induction in the chick embryo. *Cell* **2001**, *104*, 891–900. [CrossRef]
76. Galceran, J.; Hsu, S.C.; Grosschedl, R. Rescue of a wnt mutation by an activated form of LEF-1: Regulation of maintenance but not initiation of brachyury expression. *Proc. Nat. Acad. Sci. USA* **2001**, *98*, 8668–8673. [CrossRef] [PubMed]
77. Cohn, M.J.; Tickle, C. Developmental basis of limblessness and axial patterning in snakes. *Nature* **1999**, *399*, 474–479. [CrossRef] [PubMed]
78. Tanaka, M.; Munsterberg, A.; Anderson, W.G.; Prescott, A.R.; Hazon, N.; Tickle, C. Fin development in a cartilaginous fish and the origin of vertebrate limbs. *Nature* **2002**, *416*, 527–531. [CrossRef] [PubMed]
79. Tanaka, M.; Hale, L.A.; Amores, A.; Yan, Y.L.; Cresko, W.A.; Suzuki, T.; Postlethwait, J.H. Developmental genetic basis for the evolution of pelvic fin loss in the pufferfish takifugu rubripes. *Dev. Biol.* **2005**, *281*, 227–239. [CrossRef] [PubMed]
80. Marcil, A.; Dumontier, E.; Chamberland, M.; Camper, S.A.; Drouin, J. Pitx1 and Pitx2 are required for development of hindlimb buds. *Development* **2003**, *130*, 45–55. [CrossRef] [PubMed]
81. Naiche, L.A.; Papaioannou, V.E. Loss of Tbx4 blocks hindlimb development and affects vascularization and fusion of the allantois. *Development* **2003**, *130*, 2681–2693. [CrossRef] [PubMed]
82. Naiche, L.A.; Papaioannou, V.E. Tbx4 is not required for hindlimb identity or post-bud hindlimb outgrowth. *Development* **2007**, *134*, 93–103. [CrossRef] [PubMed]
83. Yang, L.; Cai, C.L.; Lin, L.; Qyang, Y.; Chung, C.; Monteiro, R.M.; Mummery, C.L.; Fishman, G.I.; Cogen, A.; Evans, S. Isl1cre reveals a common bmp pathway in heart and limb development. *Development* **2006**, *133*, 1575–1585. [CrossRef] [PubMed]
84. Kawakami, Y.; Marti, M.; Kawakami, H.; Itou, J.; Quach, T.; Johnson, A.; Sahara, S.; O'Leary, D.D.; Nakagawa, Y.; Lewandoski, M.; *et al.* Islet1-mediated activation of the beta-catenin pathway is necessary for hindlimb initiation in mice. *Development* **2011**, *138*, 4465–4473. [CrossRef] [PubMed]
85. Lanctot, C.; Moreau, A.; Chamberland, M.; Tremblay, M.L.; Drouin, J. Hindlimb patterning and mandible development require the Ptx1 gene. *Development* **1999**, *126*, 1805–1810. [PubMed]
86. Logan, M.; Tabin, C.J. Role of Pitx1 upstream of tbx4 in specification of hindlimb identity. *Science* **1999**, *283*, 1736–1739. [CrossRef] [PubMed]
87. Cole, N.J.; Tanaka, M.; Prescott, A.; Tickle, C. Expression of limb initiation genes and clues to the morphological diversification of threespine stickleback. *Curr. Biol.* **2003**, *13*, R951–R952. [CrossRef] [PubMed]
88. Szeto, D.P.; Rodriguez-Esteban, C.; Ryan, A.K.; O'Connell, S.M.; Liu, F.; Kioussi, C.; Gleiberman, A.S.; Izpisua-Belmonte, J.C.; Rosenfeld, M.G. Role of the bicoid-related homeodomain factor Pitx1 in specifying hindlimb morphogenesis and pituitary development. *Genes Dev.* **1999**, *13*, 484–494. [CrossRef] [PubMed]
89. Duboc, V.; Logan, M.P. Pitx1 is necessary for normal initiation of hindlimb outgrowth through regulation of *Tbx4* expression and shapes hindlimb morphologies via targeted growth control. *Development* **2011**, *138*, 5301–5309. [CrossRef] [PubMed]
90. Liebenberg, F. A pedigree with unusual anomalies of the elbows, wrists and hands in five generations. *S. Afr. Med. J.* **1973**, *47*, 745–748. [PubMed]
91. Spielmann, M.; Brancati, F.; Krawitz, P.M.; Robinson, P.N.; Ibrahim, D.M.; Franke, M.; Hecht, J.; Lohan, S.; Dathe, K.; Nardone, A.M.; *et al.* Homeotic arm-to-leg transformation associated with genomic rearrangements at the pitx1 locus. *Am. J. Hum. Genet.* **2012**, *91*, 629–635. [CrossRef] [PubMed]
92. Don, E.K.; de Jong-Curtain, T.A.; Doggett, K.; Hall, T.E.; Heng, B.; Badrock, A.P.; Winnick, C.; Nicholson, G.A.; Guillemin, G.J.; Currie, P.D.; *et al.* Genetic basis of hindlimb loss in a naturally occurring vertebrate model. *Biol. Open* **2016**, *5*, 359–366. [CrossRef] [PubMed]
93. Shapiro, M.D.; Marks, M.E.; Peichel, C.L.; Blackman, B.K.; Nereng, K.S.; Jonsson, B.; Schluter, D.; Kingsley, D.M. Genetic and developmental basis of evolutionary pelvic reduction in threespine sticklebacks. *Nature* **2004**, *428*, 717–723. [CrossRef] [PubMed]
94. Chan, Y.F.; Marks, M.E.; Jones, F.C.; Villarreal, G., Jr.; Shapiro, M.D.; Brady, S.D.; Southwick, A.M.; Absher, D.M.; Grimwood, J.; Schmutz, J.; *et al.* Adaptive evolution of pelvic reduction in sticklebacks by recurrent deletion of a Pitx1 enhancer. *Science* **2010**, *327*, 302–305. [CrossRef] [PubMed]

95. Tanaka, M. Revealing the mechanisms of the rostral shift of pelvic fins among teleost fishes. *Evol. Dev.* **2011**, *13*, 382–390. [CrossRef] [PubMed]
96. McPherron, A.C.; Lawler, A.M.; Lee, S.J. Regulation of anterior/posterior patterning of the axial skeleton by growth/differentiation factor 11. *Nat. Genet.* **1999**, *22*, 260–264. [PubMed]
97. Liu, J.P. The function of growth/differentiation factor 11 (gdf11) in rostrocaudal patterning of the developing spinal cord. *Development* **2006**, *133*, 2865–2874. [CrossRef] [PubMed]
98. Dasen, J.S.; Liu, J.P.; Jessell, T.M. Motor neuron columnar fate imposed by sequential phases of *Hox-c* activity. *Nature* **2003**, *425*, 926–933. [CrossRef] [PubMed]
99. Shah, V.; Drill, E.; Lance-Jones, C. Ectopic expression of *Hoxd10* in thoracic spinal segments induces motoneurons with a lumbosacral molecular profile and axon projections to the limb. *Dev. Dyn.* **2004**, *231*, 43–56. [CrossRef] [PubMed]
100. Wu, Y.; Wang, G.; Scott, S.A.; Capecchi, M.R. Hoxc10 and Hoxd10 regulate mouse columnar, divisional and motor pool identity of lumbar motoneurons. *Development* **2008**, *135*, 171–182. [CrossRef] [PubMed]
101. Nakashima, M.; Toyono, T.; Akamine, A.; Joyner, A. Expression of growth/differentiation factor 11, a new member of the BMP/TGFbeta superfamily during mouse embryogenesis. *Mech. Dev.* **1999**, *80*, 185–189. [CrossRef]
102. Liu, J.P.; Laufer, E.; Jessell, T.M. Assigning the positional identity of spinal motor neurons: Rostrocaudal patterning of *Hox-c* expression by fgfs, gdf11, and retinoids. *Neuron* **2001**, *32*, 997–1012. [CrossRef]
103. Murata, Y.; Tamura, M.; Aita, Y.; Fujimura, K.; Murakami, Y.; Okabe, M.; Okada, N.; Tanaka, M. Allometric growth of the trunk leads to the rostral shift of the pelvic fin in teleost fishes. *Dev. Biol.* **2010**, *347*, 236–245. [CrossRef] [PubMed]
104. Murakami, Y.; Tanaka, M. Evolution of motor innervation to vertebrate fins and limbs. *Dev. Biol.* **2011**, *355*, 164–172. [CrossRef] [PubMed]
105. Jurberg, A.D.; Aires, R.; Varela-Lasheras, I.; Novoa, A.; Mallo, M. Switching axial progenitors from producing trunk to tail tissues in vertebrate embryos. *Dev. Cell* **2013**, *25*, 451–462. [CrossRef] [PubMed]
106. Itou, J.; Kawakami, H.; Quach, T.; Osterwalder, M.; Evans, S.M.; Zeller, R.; Kawakami, Y. Islet1 regulates establishment of the posterior hindlimb field upstream of the Hand2-Shh morphoregulatory gene network in mouse embryos. *Development* **2012**, *139*, 1620–1629. [CrossRef] [PubMed]
107. Coates, M.I.; Cohn, M. Vertebrate axial and appendicular patterning: The early development of paired appendages. *Amer. Zool.* **1999**, *39*, 676–685. [CrossRef]
108. Horton, A.C.; Mahadevan, N.R.; Minguillon, C.; Osoegawa, K.; Rokhsar, D.S.; Ruvinsky, I.; de Jong, P.J.; Logan, M.P.; Gibson-Brown, J.J. Conservation of linkage and evolution of developmental function within the Tbx2/3/4/5 subfamily of T-box genes: Implications for the origin of vertebrate limbs. *Dev. Genes Evol.* **2008**, *218*, 613–628. [CrossRef] [PubMed]
109. Pascual-Anaya, J.; Albuixech-Crespo, B.; Somorjai, I.M.; Carmona, R.; Oisi, Y.; Alvarez, S.; Kuratani, S.; Munoz-Chapuli, R.; Garcia-Fernandez, J. The evolutionary origins of chordate hematopoiesis and vertebrate endothelia. *Dev. Biol.* **2013**, *375*, 182–192. [CrossRef] [PubMed]
110. Minguillon, C.; Gibson-Brown, J.J.; Logan, M.P. Tbx4/5 gene duplication and the origin of vertebrate paired appendages. *Proc. Nat. Acad. Sci. USA* **2009**, *106*, 21726–21730. [CrossRef] [PubMed]

14

Regulation of Silk Genes by Hox and Homeodomain Proteins in the Terminal Differentiated Silk Gland of the Silkworm *Bombyx mori*

Shigeharu Takiya [1,2,*], Takuya Tsubota [3] and Mai Kimoto [2]

[1] Shigeharu Takiya, Division of Biological Sciences and Center for Genome Dynamics, Faculty of Science, Hokkaido University, North 10, West 8, Kita-ku, Sapporo, Hokkaido 060-0810, Japan

[2] Graduate School of Life Science, Hokkaido University, North 10, West 8, Kita-ku, Sapporo 060-0810, Japan; kimotomai870310@yahoo.co.jp

[3] Transgenic Silkworm Research Unit, Institute of Agrobiological Sciences, National Agriculture and Food Research Organization, 1-2 Owashi, Tsukuba, Ibaraki 305-8634, Japan; tsubota@affrc.go.jp

* Correspondence: takiya@sci.hokudai.ac.jp

Academic Editors: Vincenzo Zappavigna and Simon J. Conway

Abstract: The silk gland of the silkworm *Bombyx mori* is a long tubular organ that is divided into several subparts along its anteroposterior (AP) axis. As a trait of terminal differentiation of the silk gland, several silk protein genes are expressed with unique regional specificities. Most of the *Hox* and some of the homeobox genes are also expressed in the differentiated silk gland with regional specificities. The expression patterns of *Hox* genes in the silk gland roughly correspond to those in embryogenesis showing "colinearity". The central Hox class protein Antennapedia (Antp) directly regulates the expression of several middle silk gland–specific silk genes, whereas the Lin-1/Isl-1/Mec3 (LIM)-homeodomain transcriptional factor Arrowhead (Awh) regulates the expression of posterior silk gland–specific genes for silk fiber proteins. We summarize our results and discuss the usefulness of the silk gland of *Bombyx mori* for analyzing the function of *Hox* genes. Further analyses of the regulatory mechanisms underlying the region-specific expression of silk genes will provide novel insights into the molecular bases for target-gene selection and regulation by Hox and homeodomain proteins.

Keywords: Antp; *Bombyx mori*; fibroin; homeodomain; *Hox* genes; promoter element; sericin; silk gland; silkworm

---

1. Introduction

The pioneering genetic studies on homeotic mutations in the silkworm *Bombyx mori*, which change the identity of the larval thoracic segments and abdominal segments, were mostly carried out in the first half of the 20th century [1–3] and therein. More than 30 homeotic mutations in Nc and E loci have been described in these studies. *Hox* genes responsible for some of these homeotic mutations, and those corresponding to the genes in the Antennapedia (Antp) complex and bithorax complex of *Drosophila*, were subsequently mapped in the *Bombyx* genome structure [4–11].

Hox genes are well-known master regulators for determining regional identities along the anteroposterior (AP) axis in most multicellular animals [12–15]. *Hox* genes encode transcriptional factors, are expressed in different but often overlapping regions along the AP axis during embryogenesis, and regulate different sets of target genes, leading to a different identity of each region in animal bodies. However, the 60-amino-acid-long DNA-binding domains near the C-terminus of Hox proteins, termed homeodomains, are highly conserved and commonly bind to similar adenine (A) and thymine (T)-rich DNA stretches containing the ATTA core, at least *in vitro* [16–20].

This discrepancy is referred to as the "Hox paradox" [21–23]. The coordination of Hox proteins with another class of homeodomain proteins, Extradenticle (Exd)/Pbx and Homothorax (Hth)/Meis, increases their DNA-binding specificities [24–30]. However, these complexes still retain the ability to bind to divergent sequence elements [21,31]. Although genome-wide approaches using microarray or chromatin immunoprecipitation (ChIP) technologies have identified large numbers of *Hox* downstream gene candidates [23,32–35], the molecular mechanisms by which Hox proteins select and regulate these candidates have not been analyzed in detail. Furthermore, the mechanisms of those leading to functional differences in a Hox protein on target genes also have not been elucidated. Hox proteins can sometimes activate a set of target genes and otherwise repress another set of target genes. Various mechanisms including cofactor-mediated and cofactor-independent DNA-binding steps in addition to DNA-binding-independent steps appear to be involved in the functional regulation of Hox proteins [28,36–45]. The identification of more target genes and their regulatory elements that are actually recognized by Hox proteins will be useful for understanding more clearly the molecular bases underlying the *in vivo* functions of Hox proteins. We previously reported that the regional specific expression of silk protein genes in the fully differentiated silk gland of the silkworm *Bombyx mori* was directly regulated by several homeodomain-containing transcriptional factors including the Hox protein Antp, and identified their regulatory elements recognized by these homeodomain proteins [46–51].

Larvae of the silkworm *Bombyx mori* possess a pair of silk glands. The silk gland is a long tubular organ consisting of a single cell layer of approximately 1000 substantially large, polyploidy cells. Silk gland placodes are induced in the ectoderm of the labial segment by *Sex comb reduced* (*Scr*) during embryogenesis, and the expression of *Scr* disappears from the invaginating silk gland with the *fork head* (*fkh*) and *POU-M1* genes being expressed there instead [52–55]. The expression of the *fibroin-heavy-chain* gene (*fibH*) is detectable immediately after the completion of silk gland development at around stage 25 of embryogenesis [56]. Silk glands in the silkworm extend from a spinneret on the head to the seventh abdominal segment, and are divided into several subparts along the AP axis (Figure 1) [57–62]. Silk fiber proteins are produced in the posterior silk gland (PSG), secreted into the lumen, and then transferred to the middle silk gland (MSG). Several glue proteins are produced in the MSG, secreted into the lumen, and then coat silk fiber proteins. Silk proteins are transferred forward through the cuticle-lined anterior part of the silk gland (ASG), and are spun as silk threads from the spinneret. The MSG is further divided into three subparts: the anterior (MSG-A), middle (MSG-M) and posterior (MSG-P) portions, based on natural turns and their functions [46,47,51,55,57,59,62,63].

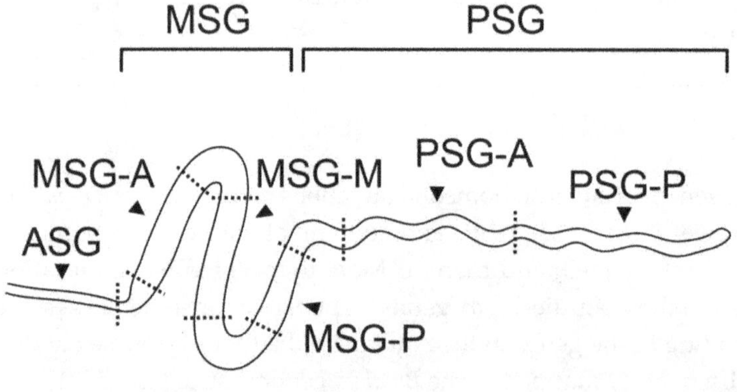

Figure 1. Schematic of the silk gland of *Bombyx mori.* ASG; anterior silk gland, MSG; middle silk gland, MSG-A; anterior portion of the MSG, MSG-M; middle portion of the MSG, MSG-P; posterior portion of the MSG, PSG; posterior silk gland, PSG-A; anterior half of the PSG; PSG-P; posterior half of the PSG. In order to avoid contamination of the neighbor regions, each portion was prepared as shown in this figure. The structural border is not observed in the PSG, and so the long tubular tissue was divided simply into the anterior and posterior halves.

Although the silkworms produce a large amount of silk proteins at the end of the last instar to make cocoons, a certain amount of silk proteins is produced constantly from the first instar, except during the molting stages, possibly to keep the larvae stable on mulberry leaves and other substrata. Therefore, the expression of silk protein genes is repeatedly turned on and off in a manner depending on the larval molting cycles [56,58,64–67]. The mechanisms of this stage-specific expression involve temporal-specific regulation of the activity of Hox and homeodomain-containing transcriptional factors by the insect hormones ecdysone and juvenile hormone, regardless of whether directly or indirectly. We herein describe and discuss the roles of Hox and other homeodomain-containing transcriptional factors in regulating the region-specific expression of silk genes in the terminally differentiated silk gland at the last instar.

2. Region-Specific Expression of Silk Genes in the Silk Gland

The *fibroin-heavy-chain* (*fibH*), *fibroin-light-chain* (*fibL*) and *fibrohexamerin* (*fhx*) genes for silk fiber proteins are expressed specifically in PSG cells (Table 1) [64,67–70]. Silk fiber proteins are assembled into a large elementary unit consisting of Fib-H, Fib-L and Fhx in a 6:6:1 molecular ratio [71], and are then secreted into the lumen. In contrast, the sericin-1 (*ser1*), sericin-2 (*ser2*) and sericin-3 (*ser3*) genes for glue proteins are expressed in MSG cells with sublocal specificities (Table 1) [46,47,51,55,57,59,62,63]. *ser1* is expressed in the MSG-M and MSG-P at the last instar, *ser2* is mainly expressed in the MSG-A, and *ser3* is expressed in the MSG-A and MSG-M. *ser1* mRNAs are differentially spliced depending on the larval stages and the subparts of the MSG, and several isotypes of Ser1 proteins are produced [47,57,72]. *ser2* mRNAs are also differentially spliced [57,62]. In addition to these silk genes, Tsubota *et al.* [51] recently found that other silk genes, *fibrohexamerin-like 4* and *5* (*fhxh4* and *5*), are specifically expressed in the MSG-M and MSG-P as *ser1*. Fhxh4 and Fhxh5 were previously identified as abundant proteins in cocoons [73]. Thus, the expression of silk protein genes is regulated strictly in region-specific manners in the silk gland of *Bombyx mori*.

3. Region-Specific Expression of *Hox* and Several Homeobox Genes in the Fully Differentiated Silk Gland

Similar to the silk protein genes, *Hox* and several homeobox genes are known to be expressed with sublocal specificities along the AP axis of the silk gland (Table 1) [6,46,47,55,74–76]. The anterior *Hox* genes *labial* (*lab*) and *Scr* are mainly expressed in the ASG, whereas the expression of central *Hox* genes such as *Antp*, *Ultrabithorax* (*Ubx*) and *abdominal-A* (*abd-A*) is absent in the ASG [47]. *Antp* was previously found to be expressed in the MSG, and *abd-A* was expressed in the MSG and PSG. The expression level of *Ubx* was very low, at least at the last instar, but the RT-PCR products were barely detected in the MSG-P and PSG. The posterior *Hox* gene *Abdominal-B* (*Abd-B*) was shown to be expressed in the ASG and posterior half of the PSG [47]. Although we have not yet detected the structural border within the PSG, it may be divided into subparts with different functions. The segment polarity gene *engrailed* (*en*) and its paralogue *invected* (*in*) are strongly expressed in the ASG and posterior portions of the MSG [47,74,76]. The Pit/Oct/Unc-86 (POU)-homeodomain protein gene *POU-M1* is also strongly expressed in the ASG and MSG-A [46,54]. The *Arrowhead* gene (*Awh*), which codes a transcriptional factor belonging to the Lin-1/Isl-1/Mec3 (LIM)-homeodomain family, was shown to be expressed specifically in the PSG [48,49]. Expression of the maternal homeobox gene in *Drosophila* embryogenesis, caudal (cad), was detected in ASG and MSG. These *Hox* and homeobox genes have been expected to serve in the regional specificity of the silk gland and the region-specific expression of silk protein genes.

Table 1. Region-specific expression of the genes for silk proteins, and Hox and some homeodomain-containing transcriptional factors in the silk gland.

Silk Gene	ASG (1)	MSG			PSG		Reference
		Anterior (1)	Middle (1)	Posterior (1)	Anterior (1),(2)	Posterior (1),(2)	
Silk Fiber Protein							
fibH	–	–	–	–	+++ (3)	+++	[48]
fibL	–	–	–	–	+++	+++	[48]
fhx	–	–	–	–	+++	+++	[48]
Glue Protein							
ser1	–	–	+++	+++	–	–	[46,47,51]
ser2	–	+++	++	–	–	–	[51]
ser3	–	+++	+++	–	–	–	[51]
fhxh4	–	–	++	+++	–	–	[51]
fhxh5	–	–	++	+++	–	–	[51]
Hox Gene							
lab	++	+/–	+	+	+	+	[47]
Scr	++	+/–	–	–	–	–	[47]
Antp	–	++ (4)	++	++	–	–	[47]
Ubx	–	–	–	+/–	+/–	+/–	[47]
abd-A	–	+	+	+	+	+	[47]
Abd-B	+	+/–	–	–	–	+	[47]
Homeobox Gene (5)							
Awh	–	–	–	–	++	++	[48]
En	+	+/–	+	+	–	–	[47]
in	++	–	+	++	–	–	[47]
Pou-m1	++	++	+	–	–	–	[46,47]
cad	+	+/–	+/–	+/–	–	–	[47]
exd	++	++	++	++	++	++	[47]
hth	+	+/–	++	++	+	+	[47]

(1) Total RNA prepared from each region of the silk gland at two days or three days (in the reference [51]) of the fifth instar was used for RT-PCR; (2) The structral border is not observed within the PSG. The long tissue was divided simply into the anterior and posterior halves.; (3) Results of RT-PCR are summarized. (+++) detectable with less than 25 cycles, (++) detectable with 30 cycles, (+) detectable with 35 cycles, (+/–) barely detectable with more than 35 cycles, and (–) undetectable. Expression levels were not be compared between different genes.; (4) The *Antp* gene is efficiently transcribed, whereas the Antp protein was not detected in the MSG-A with Western blotting.; (5) Genes encoding homeodomain-containing proteins except for the *Hox* genes.

4. Antp Regulates the MSG-Specific Expression of *ser1*

The expression of *ser1* is restricted to the MSG-M and MSG-P in the last instar [46,47,55]. We attempted to identify the promoter elements responsible for the region-specific expression of *ser1*. By using a gene gun, reporter constructs containing the *ser1* promoter were introduced into silk glands dissected out from the last instar larvae, and silk glands were then transplanted into other silkworms at the same stage. After being cultivated for one or two days, the transplanted silk glands were removed from the recipient larvae, and the promoter activity of each construct was assessed. We found that the −70 region of the *ser1* promoter (the transcription start site is denoted +1) was essential for its MSG-specific expression [50]. A protein factor or factor complex binding to the −70 element was detected in extracts prepared from the intermolt MSG-P using the electrophoretic mobility sift assay (EMSA), but not in extracts from the PSG or MSG-P at the molting stage, which is consistent with the spatio-temporal specificity of *ser1* expression. This factor was designated as the MSG-intermolt-specific complex (MIC). The MIC also bound regions further upstream (−1350, −320 and −180) of the *ser1* promoter, and these MIC-binding elements were found to commonly contain the ATTA-core AT-rich sequences recognized by homeodomain proteins (Table 2) [50].

Table 2. Nucleotide sequence around the regulatory elements of the silk gene promoter.

Gene	Position	Strand	Factor	Sequence	Reference
Silk Fiber Protein					
fibH	−1620	−	SGF-2	CTTG CAATTA AGCACTTATTC	[77]
	−200	+	SGF-2	GAT CAATTA AAT CATAATTA ATC (3)	[48,49,78]
	−110	−	SGF-2	GATA CAATTA CATAG AAATTA ATC (3)	[48,49,78]
	−270	−	Fkh (1)	GTAA TATTTAAAGA ACTTA	
	−130	−	Fkh, FMBP-1	ATCT TTTTATTTAACAT AACAA (4)	[79,80]
	−70	+	Fkh	TAGA TGTTTATTCT ATCG	[78]
	−60	−	Fkh (2)	GACG TATTTACTTT CGAT	
	−130	+	POU-M1	TGTT ATGTTAAA TAAA	[78]
	−110	+	POU-M1	TTCT ATGTAATT GTATC	[78]
fibL	−230	+	SGF-2 (1)	GGAT CAATTA GATCGCTTTG	
	−50	−	SGF-2	AAGA CAATTA AAA TAAATA TC (3)	[48]
	−50	+	Fkh (1), FMBP-1 (1)	TTGA TATTTATTTT AATTG (4)	
	−30	−	Fkh (1), FMBP-1 (1)	CCAC TATTTATATA TAAAA (5)	
fhx	−30	+	SGF-2	GGAA CAAT ACTTG TATAATTA ATGTTG (6)	[48,81]
	+100	+	SGF-2 (1)	GGT CAATTA TAACTAC	
	−70	+	Fkh, FMBP-1 (1)	ACGC TATTTATTTA ACGT (4)	[81–83]
Glue Protein					
ser1	−1350	+	MIC	TAATGC AATTAATATC GTATC	[50]
	−310	+	MIC	AATTCC AATTAATTAT AGTCG	[50]
	−180	+	MIC	GAAATC AATTAATAAC ATAAA	[50]
	−70	+	MIC	GCGAA AATTTATTAC TCTCT	[50]
	−160	−	Fkh (1)	ATTT TGTTTGCCTA TTTTA	
	−120	+	Fkh (1)	AGAA CGTTTGTTGA ACAA	
	−90	+	Fkh	ACAT TGTTTGCACA AATGTT	[84–86]
	−200	+	POU-M1	AGCC ATGAATAA ATTAG	[85,86]
	−140	−	POU-M1	CTCT ATGTAAAT GGTTT	[85,86]
ser3	−90	+	MIC, Fkh (2), FMBP-1 (1)	AAAT AATTAATTATTTATTT ATTG (4)	[51]
	−100	−	Fkh (2)	TAAT TATTTGTTTA ATACAC	
	−30	−	Fkh (1)	CGGC TATTTATACT AATTT (7)	
	−70	−	POU-M1 (1)	CTTT ATGAATAA ACAG	
fhxh4	−1660	−	MIC	AAATT GATTTATGAC AGAG	[51]
	−180	−	Fkh (1)	TTTT TGTTTAATTA TTAT	
	−160	−	Fkh (1)	TTTT TGTTTAATTT TTTT	
	−60	+	Fkh (1)	TAAA TGTTTATTTT CTTAT	
	−40	−	Fkh (1)	ACTG TGTTTAAATT ATGTT	
	−30	+	Fkh (1)	TGCT TATTTATATG TAAG (8)	
fhxh5	−300	−	MIC	AATGA GATTTATAAT ATTGAT	[51]
	−200	+	Fkh (1)	CTTG TATTTAGATT ATTTA	
	−130	+	Fkh (1)	ATATAA TATTTAATGT AAACG	
	−60	−	Fkh (1)	GTGTAA TATTTGCTGG ATATTA	
	−30	−	Fkh (1)	AAACT AGTTTGTATA ATTCC	

(1) Presumption from consensus sequences; (2) Takiya, unpublished results; (3) Paired SGF-2-binding element; (4) The FMBP-1-binding element "ATNTWTNTA" or its variations overlapped; (5) The Fkh-binding element and FMBP-1-binding element overlapped with the TATA box; (6) The incomplete paired SGF-2-binding element overlapped with the TATA box; (7) The Fkh-binding element overlapped with the TATA box; (8) The TATA box of *fhxh4* might be a weak Fkh-binding element. Affinity of Fkh to various sequence elements was estimated experimentally in reference [87].

Since *Antp*, *en* and *in* are specifically expressed in the MSG [47,74,76], we prepared recombinant proteins of Antp and En, and determined whether these proteins bind to the −70 element. Although both proteins bound to the −70 element of the *ser1* promoter, Antp but not En showed a similar DNA-binding specificity to that of MIC [47]. The apparent molecular size of the Antp-probe DNA complex detected with EMSA was markedly smaller than that of the MIC-probe complex. Hox proteins are known to make a complex with the TALE class homeodomain proteins extradenticle (Exd) and homothorax (Hth) via the conserved hexapeptide motif that resides on the N-terminal side of the homeodomain of Hox proteins [28,29]. The addition of recombinant Exd and Hth proteins together with Antp into EMSA reactions led to the formation of a MIC-like complex with a similar size to that of MIC. Using antibodies specific to Antp, Exd or Hth in EMSA, we confirmed that the MIC in MSG-P extracts included these Hox and homeodomain proteins [47], though the −70 element of *ser1* does not contain apparent binding sequences for Exd or Hth, except the ATTA core commonly recognized by homeodomain proteins. Although a hexapeptide-independent interaction between Exd and several Hox proteins has been reported [88,89], the YPWM motif of Antp appeared to be necessary for the formation of MIC with the cofactors Exd and Hth on the −70 element of the *ser1* promoter.

When the recombinant Antp protein was added to extracts from the PSG, the formation of a MIC-like complex was observed via EMSA. In addition, the genes for Exd and Hth were expressed in the PSG as in the MSG. We investigated whether the misexpression of Antp in the PSG induces ectopic expression of *ser1*. The expression of *ser1* was induced in the PSG of transgenic silkworms expressing *Antp*. When the binding of MIC to the −70 element was disturbed by two-nucleotide mutations, the promoter activity of *ser1* was mostly lost [50]. Thus, the central Hox protein Antp is an essential activator of *ser1* expressed specifically in the middle portions (MSG-M and MSG-P) of the fully differentiated silk gland of normal silkworms [47], and the activation function of Antp is achieved through binding to the −70 element of *ser1* promoter.

5. Antp Activates Other MSG-Specific Silk Genes

Tsubota *et al.* [51] recently found the novel Antp-induced silk genes *fhxh4* and *fhxh5*. *fhxh4* and *fhxh5* are specifically expressed in the MSG-M and MSG-P as *ser1* in wild-type silkworms, whereas the ectopic expression of these genes was induced in the PSG when *Antp* was misexpressed in transgenic silkworm lines. The ectopic expression of *ser3* in the PSG induced by Antp was also observed. Consensus MIC-binding sequences have been identified in the upstream regions of these genes (Table 2), and the binding of MIC to these elements was confirmed with EMSA, suggesting that these MSG-specific silk genes are direct targets of Antp, similar to *ser1* [51]. *Antp* appears to function as a master regulator for MSG-specific gene expression. The expression of *ser3* is restricted to the MSG-A and MSG-M in wild-type silkworms. mRNAs for *Antp* are detected in the MSG-A, but the Antp protein was not detected with Western blotting, and even with the addition of recombinant Antp into MSG-A extracts, the MIC-like complex was not formed in EMSA [47]. Therefore, the expression of *ser3* in the MSG-A appears to be regulated by an Antp-independent mechanism.

6. The LIM-Homeodomain Protein, Arrowhead, Regulates the PSG-Specific Expression of Silk Fiber Protein Genes

Of the PSG-specific silk fiber protein genes, the promoter of *fibH* has been analyzed in detail using cell-free transcription systems prepared from PSG and other tissues of the silkworm [79,90–96]. The activity of natural DNA of the *fibH* promoter obtained from the PSG without cloning technology has also been assessed in cell-free systems [97]. Regulatory elements bound by various transcription factors such as Fork head/silk gland factor-1 (Fkh/SGF-1), silk gland factor-2 (SGF-2), POU-M1/SGF-3 and fibroin-modulator-binding protein-1 (FMBP-1) were detected (Table 2) [78–80,84,92,98]. In these studies, the SGF-2-binding elements present from −238 to −73 were found to be essential for efficient transcription of *fibH* in PSG extracts [78,91,94,95]. Furthermore, Shimizu *et al.* [77] demonstrated that the combination of two far-upstream enhancer elements with promoter elements is indispensable for the full activation of *fibH in vivo* using a gene gun system, and an SGF-2-binding sequence at approximately −1620 was a key element in the far-upstream enhancer.

The LIM-homeodomain protein Arrowhead (Awh) was found to be an essential factor regulating PSG-specific expression of *fibH*, *fibL* and *fhx*. SGF-2 was purified from PSG extracts by Ohno *et al.* [49], and it was shown to be composed of Awh, LIM-domain binding protein (Ldb) and a member of a single-stranded DNA-binding protein family (LIM-homeodomain and Ldb complex associated factor; Lcaf). In transgenic *Drosophila*, a reporter gene having *Bombyx fibH* or the *fhx* promoter are expressed in a restricted region around the imaginal ring of the salivary gland, which resembles the expression region of the *Drosophila Awh* gene [99–101]. The salivary gland of *Drosophila* develops from the labial segment under the regulation of *Scr*, as does the *Bombyx* silk gland; however, the *Bombyx* salivary gland is derived from the mandibular segment. In *Bombyx mori*, *Awh* is expressed in a strictly specific manner in the PSG, but not in the salivary gland in the last instar, whereas *Ldb* and *Lcaf* are expressed in most tissues [48,49]. The misexpression of *Awh* in transgenic silkworms has been shown to induce the ectopic expression of *fibH*, *fibL*, and *fhx* in the MSG, and SGF-2-binding elements have also been detected around the promoter of *fibL* and *fhx* (Table 2). We previously reported that the core promoter of *fhx* itself is an SGF-2-binding element, while the core promoter of *fibL* has the ability to function as a binding element for both Fkh and FMBP-1. The core promoter of *fhx* and *fibL* may act as a tissue-specific regulatory element.

Ldb and *Lcaf* are expressed in most tissues in last-instar larvae, and the misexpression of *Awh* was induced in the whole body from a heat-shock promoter in transgenic silkworms. However, the ectopic expression of *fibH* was induced only in the MSG. Under the same conditions, ectopic expression of *fibL* and *fhx* was induced in the ASG in addition to the MSG [48]. The findings suggest that SGF-2 directly regulates the expression of these PSG-specific silk genes, and that coordination with other factors such as Bmsage, which is a silk gland–specific basic helix-loop-helix transcription factor interacting with Fkh, may be necessary [102]. The coordination of these transcription factors will be interpreted by the context (*i.e.*, structure) of each promoter. The expressions of *fibH* and *ser1* were not induced in the ASG under the misexpression of *Awh* and *Antp*, respectively, and the expression level of *fibH* in the MSG-P induced by the misexpression of *Awh* was very low [47,48]. *POU-M1*, *en* and *in*, which are efficiently expressed in these portions, may exert a suppressive function on these genes [46,55,74,76,98,103]. A typical POU-M1-binding element was detected in the promoter of *fibH* and *ser-1* (Table 2), but not around the promoter of *fibL* and *fhx*. In *Drosophila*, En has the ability to recruit the corepressor Groucho and repress target genes [104–106]. For example, En suppresses the expression of *fkh* under coordination with Abd-A in *Drosophila*.

7. Perspective

We have demonstrated that the homeodomain proteins Antp and Awh are involved in regulating the region-specific expression of silk genes in the silk gland of *Bombyx mori* by direct binding to the regulatory elements of the silk genes [47–51]. The region-specific regulatory functions of Antp and Awh on the silk genes are achieved through region-specific expression of *Antp* and *Awh*. Therefore, it is important to analyze the upstream regulatory mechanisms responsible for the region-specific expression of *Antp* and *Awh*. In the silk gland, other *Hox* and homeobox genes are also expressed with unique regional specificities along the AP axis, and the homeodomain-containing transcriptional factors encoded by these genes may be involved in the cascades regulating the regional specific expression of silk genes. A large number of ATTA core AT-rich elements consist of the consensus sequences recognized by homeodomain proteins in and around the silk genes' promoters [50,51,78,107,108]. Several AT-rich stretches have also been found around the promoter of *Awh* (Takiya, unpublished results). Therefore, the silk gland of *Bombyx mori* is an appropriate system for analyzing how Hox and other homeodomain proteins select and regulate their specific target genes *in vivo*, in spite of the relative lack of DNA-binding specificity *in vitro*. This will provide an insight into the mechanisms by which *Hox* genes determine regional identities along the AP axis of animal bodies.

The expression of *Antp* mRNA was detectable in the entire MSG, whereas the Antp protein was only found in the MSG-M and MSG-P and was scarcely detectable in the MSG-A with Western blotting [47]. These findings suggest that post-transcriptional regulatory mechanisms also restrict the function of *Antp* in the silk gland. *Antp* mRNA of *Drosophila* has an internal ribosomal entry site (IRES), and the activity of *Antp* is regulated not only at the step of transcription, but also at the step of translation. [109]. The translational regulation of some *Hox* mRNAs is necessary for the normal development of vertebrates [110,111]. Therefore, it is possible that other *Hox* genes expressed in the silk gland are regulated at translational and/or post-translational steps, as is *Antp*.

The expression pattern of *Hox* genes along the AP axis during embryogenesis corresponds to the tandem array on the chromosome and is called "colinearity" [12,112]. Although colinearity is conserved in many animals, the biological significance of maintaining the colinearity during animal evolution remains unclear. Expression levels of *Hox* genes appear to be affected by the distance from one enhancer, and the distance from the remote enhancer determines the chromatin structure affecting the expression level of each *Hox* gene [112–114]. When *Hox* genes are expressed in overlapped regions, the more posterior *Hox* genes tend to govern specificity of the regions. The phenomenon is termed "posterior prevalence". It is not clear at present whether the mechanisms underlying posterior prevalence play a role in the silk gland. However, *Hox* genes are expressed in regionally specific, but partly overlapping patterns that roughly correspond to those during embryogenesis. The mechanisms of posterior prevalence may be involved in the differentiation of each part of the silk gland and also in the region-specific expression of silk protein genes.

The silk gland of *Bombyx mori* is a large and single-cell layered organ, and it is easy to obtain many materials for the preparation of proteins, RNAs and other cell components produced in a region-specific manner. We have the ability to analyze the promoter activities of various constructs transiently in the silk gland and other tissues using a gene gun system [50,77,81,101,115,116]. Technologies to prepare transgenic silkworms and manipulate the genome have been established and are being improved [117–121]. Therefore, the silk gland is an appropriate system to analyze the issues described above both biochemically and genetically. The silkworm *Bombyx mori* in which Mendel's law was first confirmed in the animal kingdom [122] is promising to provide novel insight into the molecular basis underlying the regulation of animal body plans by *Hox* genes.

Acknowledgments: We are grateful to the members of our laboratories for their support of the research, which led to the main body of information in the present review. Our studies referred in this review were supported by the Ministry of Education, Culture, Sports, Science and Technology (MEXT), and the Ministry of Agriculture, Forestry and Fisheries (MAFF) of Japan.

Author Contributions: ST wrote the manuscript. TT and MK read and approved the final manuscript.

Conflicts of Interest: The authors declare no conflict of interest.

References

1. Tazima, Y. *The Genetics of the Silkworm*; Logos Press: London, UK; Academic Press: London, UK, 1964.
2. Suzuki, Y. Genes that are involved in *Bombyx* body plan and silk gene regulation. *Int. J. Dev. Biol.* **1994**, *38*, 231–235. [PubMed]
3. Ueno, K.; Nagata, T.; Suzuki, Y. Roles of homeotic genes in the *Bombyx* body plan. In *Molecular Model Systems in the Lepidoptera*; Goldsmith, M.R., Wilkins, A.S., Eds.; Cambridge University Press: New York, NY, USA, 1995; pp. 165–180.
4. Chai, C.-L.; Zhang, Z.; Huang, F.-F.; Wang, X.-Y.; Yu, Q.-Y.; Liu, B.-B.; Tian, T.; Xia, Q.-Y.; Lu, C.; Xiang, Z.-H. A genomewide survey of homeobox genes and identification of novel structure of the Hox cluster in the silkworm *Bombyx mori*. *Insect Biochem. Mol. Biol.* **2008**, *38*, 1111–1120. [CrossRef] [PubMed]
5. Masumoto, M.; Yaginuma, T.; Niimi, T. Functional analysis of *Ultrabithorax* in the silkworm, *Bombyx mori* using RNAi. *Dev. Genes Evol.* **2009**, *219*, 437–444. [CrossRef] [PubMed]
6. Nagata, T.; Suzuki, Y.; Ueno, K.; Kokubo, H.; Xu, X.; Hui, C.-C.; Hara, W.; Fukuta, M. Developmental expression of the *Bombyx Antennapedia* homologue and homeotic changes in the Nc mutant. *Genes Cells* **1996**, *1*, 555–568. [CrossRef] [PubMed]
7. Pan, M.-H.; Wang, X.-Y.; Chai, C.-L.; Zhang, C.-D.; Lu, C.; Xiang, Z.-H. Identification and function of *abdominal-A* in the silkworm, *Bombyx mori*. *Insect Mol. Biol.* **2009**, *18*, 155–160. [CrossRef] [PubMed]
8. The International Silkworm Genome Consortium. The genome of a lepidopteran model insect, the silkworm *Bombyx mori*. *Insect Biochem. Mol. Biol.* **2008**, *38*, 1036–1045.
9. Tomita, S.; Kikuchi, A. *Abd-B* suppresses lepidopteran proleg development in posterior abdomen. *Dev. Biol.* **2009**, *328*, 403–409. [CrossRef] [PubMed]
10. Ueno, K.; Hui, C.-C.; Fukuta, M.; Suzuki, Y. Molecular analysis of the deletion mutants in the E homeotic complex of the silkworm *Bombyx mori*. *Development* **1992**, *114*, 555–563. [PubMed]
11. Yasukochi, Y.; Ashakumary, L.A.; Wu, C.; Yoshido, A.; Nohata, J.; Mita, K.; Sahara, K. Organization of the Hox gene cluster of the silkworm, *Bombyx mori*: A split of the Hox cluster in a non-*Drosophila* insect. *Dev. Genes Evol.* **2004**, *214*, 606–614. [CrossRef] [PubMed]

12. Lewis, E.B. A gene complex controlling segmentation in *Drosophila*. *Nature* **1978**, *276*, 565–570. [CrossRef] [PubMed]
13. Mann, R.S.; Carroll, S.B. Moleculae mechanisms of selector gene function and evolution. *Curr. Opin. Genet. Dev.* **2002**, *12*, 592–600. [CrossRef]
14. Mann, R.S.; Morata, G. The developmental and molecular biology of genes that subdivide the body of *Drosophila*. *Annu. Rev. Cell Dev. Biol.* **2000**, *16*, 243–271. [CrossRef] [PubMed]
15. McGinnis, W.; Krumlauf, R. Homeobox genes and axial patterning. *Cell* **1992**, *68*, 283–302. [CrossRef]
16. Berger, M.F.; Badis, G.; Gehrke, A.R.; Talukder, S.; Philippakis, A.A.; Penta-Castillo, L.; Alleyene, T.M.; Mnaimneh, S.; Botvinnik, O.B.; Chan, E.T.; *et al.* Variation in homeodomain DNA binding revealed by high-resolution analysis of sequence preferences. *Cell* **2008**, *133*, 1266–1276. [CrossRef] [PubMed]
17. Ekker, S.C.; Jackson, D.G.; Kessler, D.P.; Sun, B.I.; Young, K.E.; Beachy, P.A. The degree of variation in DNA sequence recognition among four *Drosophila* homeotic proteins. *EMBO J.* **1994**, *13*, 3551–3560. [PubMed]
18. Gehring, W.J.; Qian, Y.Q.; Billeter, M.; Furukubo-Tokunaga, K.; Schier, A.F.; Resendez-Perez, D.; Affolter, M.; Otting, G.; Wuthrich, K. Homeodomain-DNA recognition. *Cell* **1994**, *78*, 211–223. [CrossRef]
19. Mann, R.S. The specificity of homeotic gene function. *BioEssays* **1995**, *17*, 855–863. [CrossRef] [PubMed]
20. Noyes, M.B.; Christensen, R.G.; Wakabayashi, A.; Stormo, G.D.; Brodsky, M.H.; Wolfe, S.A. Analysis of homeodomain specificities allows the family-wide prediction of preferred recognition sites. *Cell* **2008**, *133*, 1277–1289. [CrossRef] [PubMed]
21. Ebner, A.; Cabernard, C.; Affolter, M.; Merabet, S. Recognition of distinct target sites by a unique Labial/Extradenticle/Homothorax complex. *Development* **2005**, *132*, 1591–1600. [CrossRef] [PubMed]
22. Hayashi, S.; Scott, M.P. What determines the specificity of action of *Drosophila* homeodomain proteins? *Cell* **1990**, *63*, 883–894. [CrossRef]
23. Huber, S.D.; Lohmann, I. Shaping segments: *Hox* gene function in the genomic age. *BioEssays* **2008**, *30*, 965–979. [CrossRef] [PubMed]
24. Chang, C.-P.; Brocchieri, L.; Shen, W.-F.; Largman, C.; Cleary, M.L. Pbx modulation of Hox homeodomain amino-terminal arms establishes different DNA-binding specificities across the *Hox* locus. *Mol. Cell Biol.* **1996**, *16*, 1734–1745. [CrossRef] [PubMed]
25. Chan, S.-K.; Jaffe, L.; Capovilla, M.; Botas, J.; Mann, R.S. The DNA binding specificity of Ultrabithorax is modulated by cooperative interactions with extradenticle, another homeoprotein. *Cell* **1994**, *78*, 603–615. [CrossRef]
26. Chan, S.-K.; Mann, R.S. A structural model for a homeotic protein-extradenticle-DNA complex accounts for the choice of HOX protein in the heterodimer. *Proc. Natl. Acad. Sci. USA* **1996**, *93*, 5223–5228. [CrossRef] [PubMed]
27. Joshi, R.; Passner, J.M.; Rohs, R.; Jain, R.; Sosinsky, A.; Crickmore, M.A.; Jacob, V.; Aggarwal, A.K.; Honig, B.; Mann, R. Functional specificity of a Hox protein Mediated by the recognition of minor groove structure. *Cell* **2007**, *131*, 530–543. [CrossRef] [PubMed]
28. Mann, R.S.; Lelli, K.M.; Joshi, R. Hox specificity: Unique roles for cofactors and collaborators. *Curr. Top. Dev. Biol.* **2009**, *88*, 63–101. [PubMed]
29. Moens, C.B.; Selleri, L. Hox cofactors in vertebrate development. *Dev. Biol.* **2006**, *291*, 193–206. [CrossRef] [PubMed]
30. Slattery, M.; Riley, T.; Liu, P.; Abe, N.; Gomez-Alcala, P.; Dror, I.; Zhou, T.; Rohs, R.; Honig, B.; Bussemaker, H.J.; *et al.* Cofactor binding evokes latent differences in DNA binding specificity between Hox proteins. *Cell* **2011**, *147*, 1270–1282. [CrossRef] [PubMed]
31. Uhl, J.D.; Cook, T.A.; Gebelein, B. Comparing anterior and posterior Hox complex formation reveals guidelines for predicting *cis*-regulatory elements. *Dev. Biol.* **2010**, *343*, 154–166. [CrossRef] [PubMed]
32. Agrawal, P.; Habib, F.; Yelagandula, R.; Shashidhara, L.S. Genome-level identification of targets of Hox protein Ultrabitholax in *Drosophila*: Novel mechanisms for target selection. *Sci. Rep.* **2011**. [CrossRef] [PubMed]
33. Hersh, B.M.; Nelson, C.E.; Stoll, S.J.; Norton, J.E.; Albert, T.J.; Caroll, S.B. The UBX-regulated network in the haltere imaginal disc of *D. melanogaster*. *Dev. Biol.* **2007**, *302*, 717–727. [CrossRef] [PubMed]
34. Hueber, S.D.; Bezdan, D.; Henz, S.R.; Blank, M.; Wu, H.; Lohmann, I. Comparative analysis of Hox downstream genes in *Drosophila*. *Development* **2007**, *134*, 381–392. [CrossRef] [PubMed]

35. Slattery, M.; Ma, L.; Negre, N.; White, K.P.; Mann, R.S. Genome-wide tissue-specific occupancy of the Hox protein Ultrabithorax and Hox cofactor Homothorax in Drosophila. *PLoS ONE* **2011**. [CrossRef] [PubMed]
36. Galant, R.; Caroll, S.B. Evolution of a transcriptional repression domain in an insect Hox protein. *Nature* **2002**, *415*, 910–913. [CrossRef] [PubMed]
37. Galant, R.; Walsh, C.M.; Caroll, S.B. Hox repression of a target gene: extradenticle-independent, additive action through multiple monomer binding sites. *Development* **2002**, *129*, 3115–3126. [PubMed]
38. Krasnow, M.A.; Saffman, E.E.; Kornfeld, K.; Hogness, D.S. Transcriptional activation and repression by *Ultrabithorax* proteins in cultured *Drosohila* cells. *Cell* **1989**, *57*, 1031–1043. [CrossRef]
39. Li, X.; Murre, C.; McGinnis, W. Activity regulation of a Hox protein and a role for the homeodomain in inhibiting transcriptional activation. *EMBO J.* **1999**, *18*, 198–211. [CrossRef] [PubMed]
40. Merabet, S.; Kambris, Z.; Capovilla, M.; Berenger, H.; Pradel, J.; Graba, Y. The hexapeptide and linker regions of the AbdA Hox protein regulate its activating and repressive functions. *Dev. Cell* **2003**, *4*, 761–768. [CrossRef]
41. Papadopoulos, D.K.; Resendez-Perez, D.; Cardenas-Chavez, D.L.; Villanueva-Segura, K.; Canales-del-Castillo, R.; Felix, D.A.; Funfschilling, R.; Ghering, W.J. Functional synthetic *Antennapedia* genes and the dual roles of YPWM motif and linker size in transcriptional activation and repression. *Proc. Natl. Acad. Sci. USA* **2011**, *108*, 11959–11964. [CrossRef] [PubMed]
42. Prince, F.; Katuyama, T.; Oshima, Y.; Plaza, S.; Resendez-Perez, D.; Berry, M.; Kurata, S.; Ghering, W.J. The YPWM motif links Antennapedia to the basal transcriptional machinery. *Development* **2008**, *135*, 1669–1679. [CrossRef] [PubMed]
43. Ronshaugen, M.; McGinnis, N.; McGinnis, W. Hox protein mutation and macroevolution of the insect body plan. *Nature* **2002**, *415*, 914–917. [CrossRef] [PubMed]
44. Saadaoi, M.; Merabet, S.; Litim-Mecheri, I.; Arbeille, E.; Sambrani, N.; Damen, W.; Brena, C.; Pradel, J.; Graba, Y. Selection of distinct Hox-Extradenticle interaction modes fine-tunes Hox protein activity. *Proc. Natl. Acad. Sci. USA* **2011**, *108*, 2276–2281. [CrossRef] [PubMed]
45. Winslow, G.M.; Hayashi, S.; Krasnow, M.; Hogness, D.S.; Scott, M.P. Transcriptional activation by the *Antennapedia* and *fushi tarazu* proteins in cultured *Drosophila* cells. *Cell* **1989**, *57*, 1017–1030. [CrossRef]
46. Kimoto, M.; Kitagawa, T.; Kobayashi, I.; Nakata, T.; Kuroiwa, A.; Takiya, S. Inhibition of the binding of MSG-intermolt-specific complex, MIC, to the *sericin-1* gene promoter and *sericin-1* gene expression by POU-M1/SGF3. *Dev. Genes Evol.* **2012**, *222*, 351–359. [CrossRef] [PubMed]
47. Kimoto, M.; Tsubota, T.; Uchino, K.; Sezutsu, H.; Takiya, S. Hox transcription factor Antp regulates *sericin-1* gene expression in the terminal differentiated silk gland of *Bombyx mori*. *Dev. Biol.* **2014**, *386*, 64–71. [CrossRef] [PubMed]
48. Kimoto, M.; Tsubota, T.; Uchino, K.; Sezutsu, H.; Takiya, S. LIM-homeodomain transcription factor Awh is a key component activating all three fibroin genes, *fibH*, *fibL* and *fhx*, in the silk gland of the silkworm, *Bombyx mori*. *Insect Biochem. Mol. Biol.* **2015**, *56*, 29–35. [CrossRef] [PubMed]
49. Ohno, K.; Sawada, J.; Takiya, S.; Kimoto, M.; Matsumoto, A.; Tsubota, T.; Uchino, K.; Hui, C.-C.; Sezutsu, H.; Handa, H.; *et al.* Silk gland factor-2, involved in fibroin gene transcription, consists of LIM homeodomain, LIM-interactiong, and single-stranded DNA-binding proteins. *J. Biol. Chem.* **2013**, *288*, 31581–31591. [CrossRef] [PubMed]
50. Takiya, S.; Inoue, H.; Kimoto, M. Novel enhancer and promoter elements indispensable for the tissue-specific expression of the *sericin-1* gene of the silkworm *Bombyx mori*. *Insect Biochem. Mol. Biol.* **2011**, *41*, 592–601. [CrossRef] [PubMed]
51. Tsubota, T.; Tomita, S.; Uchino, K.; Kimoto, M.; Takiya, S.; Kajiwara, H.; Yamazaki, T.; Sezutsu, H. A *Hox* gene, *Antennapedia*, regulates expression of multiple major silk protein genes in the silkworm *Bombyx mori*. *J. Biol. Chem.* **2016**, *291*, 7087–7096. [CrossRef] [PubMed]
52. Kokubo, H.; Takiya, S.; Mach, V.; Suzuki, Y. Spatial and temporal expression pattern of *Bombyx fork head*/SGF-1 gene in embryogenesis. *Dev. Genes Evol.* **1996**, *206*, 80–85. [CrossRef] [PubMed]
53. Kokubo, H.; Ueno, K.; Amanai, K.; Suzuki, Y. Involvement of the *Bombyx Scr* gene in development of the embryonic silk gland. *Dev. Biol.* **1997**, *186*, 46–57. [CrossRef] [PubMed]
54. Kokubo, H.; Xu, P.-X.; Xu, X.; Matsunami, K.; Suzuki, Y. Spatial and temporal expression pattern of POU-M1/SGF-3 in *Bombyx mori* embryogenesis. *Dev. Genes Evol.* **1997**, *206*, 494–502. [CrossRef]

55. Matsunami, K.; Kokubo, H.; Ohno, K.; Suzuki, Y. Expression pattern analysis of SGF-3/POU-M1 in relation to sericin-1 gene expression in the silk gland. *Dev. Growth Differ.* **1998**, *40*, 591–597. [CrossRef] [PubMed]
56. Ohta, S.; Suzuki, Y.; Hara, W.; Takiya, S.; Suzuki, T. Fibroin gene transcription in the embryonic stages of the silkworm, *Bombyx mori*. *Dev. Growth Differ.* **1988**, *30*, 293–299. [CrossRef]
57. Couble, P.; Michaille, J.-J.; Garel, A.; Couble, M.-L.; Prudohmme, J.-C. Developmental switches of sericin mRNA splicing in individual cells of *Bombyx mori* silk gland. *Dev. Biol.* **1987**, *124*, 431–440. [CrossRef]
58. Ishikawa, E.; Suzuki, Y. Tissue-and stage-specific expression of sericin genes in the middle silk gland of *Bombyx mori*. *Dev. Growth Differ.* **1985**, *27*, 73–82. [CrossRef]
59. Michaille, J.-J.; Garel, A.; Prudhomme, J.-C. Cloning and characterization of the highly polymorphic *Ser2* gene of *Bombyx mori*. *Gene* **1990**, *86*, 177–184. [CrossRef]
60. Obara, T.; Suzuki, Y. Temporal and spatial control of silk gene transcription analyzed by nuclear run-on assay. *Dev. Biol.* **1988**, *127*, 384–391. [CrossRef]
61. Suzuki, Y.; Takiya, S.; Suzuki, T.; Hui, C.-C.; Matsuno, K.; Fukuta, M.; Nagata, T.; Ueno, K. Developmental regulation of silk gene expression in *Bombyx mori*. In *Molecular Insect Science*; Hagedorn, H.H., Hildebrand, J.G., Kidwell, M.G., Law, J.H., Eds.; Plenum Press: New York, NY, USA, 1990.
62. Takasu, Y.; Yamada, H.; Tamura, T.; Sezutsu, H.; Mita, K.; Tsubouchi, K. Identification and characterization of a novel sericin gene expressed in the anterior middle silk gland of the silkworm *Bombyx mori*. *Insect Biochem. Mol. Biol.* **2007**, *37*, 1234–1240. [CrossRef] [PubMed]
63. Takasu, Y.; Hata, T.; Uchino, K.; Zhan, Q. Identification of Ser2 proteins as major sericin components in the non-cocoon silk of *Bombyx mori*. *Insect Biochem. Mol. Biol.* **2010**, *40*, 339–344. [CrossRef] [PubMed]
64. Kimura, K.; Oyama, F.; Ueda, H.; Mizuno, S.; Shimura, K. Molecular cloning of the fibroin light chain complementary DNA and its use in the study of the expression of the light chain gene in the posterior silk gland of *Bombyx mori*. *Experientia* **1985**, *41*, 1167–1171. [CrossRef] [PubMed]
65. Maekawa, H.; Suzuki, Y. Repeated turn-off and turn-on of fibroin gene transcription during silk gland development of *Bombyx mori*. *Dev. Biol.* **1980**, *78*, 394–406. [CrossRef]
66. Suzuki, Y.; Giza, P.E. Accentuated expression of silk fibroin genes *in vivo* and *in vitro*. *J. Mol. Biol.* **1976**, *107*, 183–206. [CrossRef]
67. Suzuki, Y.; Suzuki, E. Quantitative measurements of fibroin messenger RNA synthesis in the posterior silk gland of normal and mutant *Bombyx mori*. *J. Mol. Biol.* **1974**, *88*, 393–407. [CrossRef]
68. Bello, B.; Horard, B.; Couble, P. The selective expression of silk-protein encoding genes in *Bombyx mori* silk gland. *Bull. Inst. Pasteur.* **1994**, *92*, 81–100.
69. Couble, P.; Moine, A.; Garel, A.; Prudhomme, J.-C. Developmental variations of a nonfibroin mRNA of *Bombyx mori* silkgland, encoding for a low-molecular-weight silk protein. *Dev. Biol.* **1983**, *97*, 398–407. [CrossRef]
70. Couble, P.; Chevillard, M.; Moine, A.; Revel-Chapuis, P.; Prudhomme, J.-C. Structural organization of the P25 gene of *Bombyx mori* and comparative analysis of its 5′ flanking DNA with that of the fibroin gene. *Nucleic Acids Res.* **1985**, *13*, 1801–1814. [CrossRef] [PubMed]
71. Inoue, S.; Tanaka, K.; Arisaka, F.; Kimura, S.; Ohtomo, K.; Mizuno, S. Silk fibroin of *Bombyx mori* is secreted, assembling a high molecular elementary unit consisting of H-chain, L-chain, and P25, with a 6:6:1 molar ratio. *J. Biol. Chem.* **2000**, *275*, 40517–40528. [CrossRef] [PubMed]
72. Garel, A.; Deleage, G.; Prudhomme, J.-C. Structure and organization of the *Bombyx mori* sericin 1 gene and of the sericins1 deduced from the sequence of the ser 1B cDNA. *Insect Biochem. Mol. Biol.* **1997**, *27*, 469–477. [CrossRef]
73. Zhang, Y.; Zhao, P.; Dong, Z.; Wang, D.; Guo, P.; Guo, X.; Song, Q.; Zhang, W.; Xia, Q. Comparative proteome analysis of multi-layer cocoon of the silkworm, *Bombyx mori*. *PLoS ONE* **2015**, *10*, e0123403. [CrossRef] [PubMed]
74. Dhawan, S.; Gopinathan, K.P. Expression pattern of *Cubitus interruptus* from the mulberry silkworm *Bombyx mori* in late developmental stages. *Dev. Genes Evol.* **2003**, *213*, 166–177. [PubMed]
75. Dhawan, S.; Gopinathan, K.P. Expression profiling of homeobox genes in silk gland development in the mulberry silkworm *Bombyx mori*. *Dev. Genes Evol.* **2003**, *213*, 523–533. [CrossRef] [PubMed]
76. Hui, C.-C.; Matsuno, K.; Ueno, K.; Suzuki, Y. Molecular characterization and silkgland expression pattern of *Bombyx* engrailed and invected genes. *Proc. Natl. Acad. Sci. USA* **1992**, *89*, 167–171. [CrossRef] [PubMed]

77. Shimizu, K.; Ogawa, S.; Hino, R.; Adachi, T.; Tomita, M.; Yoshizato, K. Structure and function of 5′-flanking regions of a novel transcription enhancing element with a homeodomain protein-binding motif. *Insect Biochem. Mol. Biol.* **2007**, *37*, 713–725. [CrossRef] [PubMed]
78. Hui, C.-C.; Matsuno, K.; Suzuki, Y. Fibroin gene promoter contains a cluster of homeodomain binding sites that interact with three silk gland factors. *J. Mol. Biol.* **1990**, *213*, 651–670. [CrossRef]
79. Takiya, S.; Kokubo, H.; Suzuki, Y. Transcriptional regulatory elements upstream and intron of the fibroin gene bind three specific factors POU-M1, Bm Fkh and FMBP-1. *Biochem. J.* **1997**, *321*, 645–653. [CrossRef] [PubMed]
80. Takiya, S.; Ishikawa, T.; Ohtsuka, K.; Nishita, Y.; Suzuki, Y. Fibroin-modulator-binding protein-1 (FMBP-1) contains a novel DNA-binding domain, repeats of the score and three amino acid peptode (STP), conserved from *Caenorhabditis elegans* to humans. *Nucleic Acids Res.* **2005**, *33*, 786–795. [CrossRef] [PubMed]
81. Horard, B.; Julien, E.; Nony, P.; Garel, A.; Couble, P. Differential binding of the Bombyx silk gland-specific factor SGFB to its target DNA sequence drives posterior-cell-restricted expression. *Mol. Cell. Biol.* **1997**, *17*, 1572–1579. [CrossRef] [PubMed]
82. Durand, B.; Drevet, J.; Couble, P. P25 gene regulation in *Bombyx mori* silk gland: Two promoter-binding factors have distinct tissue and developmental specificities. *Mol. Cell. Biol.* **1992**, *12*, 5768–5777. [CrossRef] [PubMed]
83. Julien, E.; Bordeaux, M.-C.; Garel, A.; Couble, P. Fork head alternative binding drives stage-specific gene expression in the silk gland of *Bombyx mori*. *Insect Biochem. Mol. Biol.* **2002**, *32*, 377–387. [CrossRef]
84. Mack, V.; Takiya, S.; Ohno, K.; Handa, H.; Imai, T.; Suzuki, Y. Silk gland factor-1 involved in the regulation of *Bombyx* sericin-1 gene contain fork head motif. *J. Biol. Chem.* **1995**, *270*, 9340–9346.
85. Matsuno, K.; Hui, C.-C.; Takiya, S.; Suzuki, T.; Ueno, K.; Suzuki, Y. Transcriptional signals and protein binding sites for sericin gene transcription *in vitro*. *J. Biol. Chem.* **1989**, *264*, 18707–18713. [PubMed]
86. Matsuno, K.; Takiya, S.; Hui, C.-C.; Fukuta, M.; Ueno, K.; Suzuki, Y. Transcriptional stimulation via SC site of *Bombyx* sericin-1 gene through an interaction with a DNA binding protein SGF-3. *Nucleic Acids Res.* **1990**, *18*, 1853–1858. [CrossRef] [PubMed]
87. Takiya, S.; Gazi, M.; Mach, V. The DNA binding of insect Fork head factor is strongly influenced by the negative cooperation of neighbouring bases. *Insect Biochem. Mol. Biol.* **2003**, *33*, 1145–1154. [CrossRef]
88. Foos, N.; Maurel-Zaffran, C.; Mate, M.J.; Vincentelli, R.; Hainaut, M.; Berenger, H.; Pradel, J.; Saurin, A.J.; Ortiz-Lombardia, M.; Graba, Y. A flexible extension of the *Drosophila* Ultrabithorax homeodomain defines a novel Hox/PBC interaction mode. *Structure* **2015**, *23*, 270–279. [CrossRef] [PubMed]
89. Hudry, B.; Remacle, S.; Delfini, M.-C.; Rezsohazy, R.; Graba, Y.; Merabet, S. Hox proteins display a common and ancestral ability to diversify their interaction mode with PBC class cofactors. *PLoS Biol.* **2012**, e1001351. [CrossRef] [PubMed]
90. Hirose, S.; Suzuki, Y. *In vitro* transcription of eukaryotic genes is affected differently by the degree of DNA supercoiling. *Proc. Natl. Acad. Sci. USA* **1988**, *85*, 718–722. [CrossRef] [PubMed]
91. Suzuki, Y.; Tsuda, M.; Takiya, S.; Hirose, S.; Suzuki, E.; Kameda, M.; Ninaki, O. Tissue-specific transcription enhancement of the fibroin gene characterized by cell-free systems. *Proc. Natl. Acad. Sci. USA* **1986**, *83*, 9522–9526. [CrossRef] [PubMed]
92. Takiya, S.; Hui, C.-C.; Suzuki, Y. A contribution of the core-promoter and its surrounding regions to the preferential transcription of the fibroin gene in posterior silk gland extracts. *EMBO J.* **1990**, *9*, 489–496. [PubMed]
93. Tsuda, M.; Suzuki, Y. Faithful transcription initiation of fibroin gene in a homologous cell-free system reveals an enhancing effect of 5′ flanking sequence far upstream. *Cell* **1981**, *27*, 175–182. [CrossRef]
94. Tsuda, M.; Suzuki, Y. Transcription modulation *in vitro* of the fibroin gene exerted by a 200-base-pair region upstream from the "TATA" box. *Proc. Natl. Acad. Sci. USA* **1983**, *80*, 7442–7446. [CrossRef] [PubMed]
95. Tsuda, M.; Hirose, S.; Suzuki, Y. Participation of the upstream region of the fibroin gene in the formation of transcription complex *in vitro*. *Mol. Cell. Biol.* **1986**, *6*, 3928–3933. [CrossRef] [PubMed]
96. Tsujimoto, Y.; Suzuki, Y. The DNA sequence of Bombyx mori fibroin gene including the 5′ flanking, mRNA coding, entire intervening and fibroin protein coding regions. *Cell* **1979**, *18*, 591–600. [CrossRef]
97. Tsujimoto, Y.; Suzuki, Y. Natural fibroin genes purified without using cloning procedures from fibroin-producing and nonproducing tissues reveal indistinguishable structure and function. *Proc. Natl. Acad. Sci. USA* **1984**, *81*, 1644–1648. [CrossRef] [PubMed]

98. Fukuta, M.; Matsuno, K.; Hui, C.-C.; Nagata, T.; Takiya, S.; Xu, P.-X.; Ueno, K.; Suzuki, Y. Molecular cloning of a POU domain-containing factor involved in the regulation of the Bombyx sericin-1 gene. *J. Biol. Chem.* **1993**, *268*, 19471–19475. [PubMed]
99. Bello, B.; Couble, P. Specific expression of a silk-encoding gene of *Bombyx* in the anterior salivary gland of *Drosophila*. *Nature* **1990**, *346*, 480–482. [CrossRef] [PubMed]
100. Curtiss, J.; Heilig, J.S. Establishment of *Drosophila* imaginal precursor cells controlled by *Arrowhead* gene. *Development* **1995**, *121*, 3819–3828. [PubMed]
101. Nony, P.; Prudhomme, J.-C.; Couble, P. Regulation of the P25 gene transcription in the silk gland of *Bombyx*. *Biol. Cell.* **1995**, *84*, 43–52. [CrossRef]
102. Zhao, X.-M.; Liu, C.; Li, Q.-Y.; Hu, W.-B.; Zhou, M.-T.; Zhang, Y.-X.; Peng, Z.-C.; Zhao, P.; Xia, Q.-Y. Basic helix-loop-helix transcription factor Bmsage is involved in regulation of *fibroin-H-chain* gene via interaction with SGF1 in *Bombyx mori*. *PLoS ONE* **2014**. [CrossRef] [PubMed]
103. Xu, P.-X.; Fukuta, M.; Takiya, S.; Matsuno, K.; Xu, X.; Suzuki, Y. Promoter of the POU-M1/SGF-3 gene involved in the expression of *Bombyx* silk genes. *J. Biol. Chem.* **1994**, *269*, 2733–2742. [PubMed]
104. Tolkunova, E.N.; Fujioka, M.; Kobayashi, M.; Deka, D.; Jaynes, J.B. Two distinct types of repression domain in engrailed: one interacts with the groucho corepressor and is preferentially active on integrated target genes. *Mol. Cell. Biol.* **1998**, *18*, 2804–2814. [CrossRef] [PubMed]
105. Gebelein, B.; McKay, D.J.; Mann, R.S. Direct integration of *Hox* and segmentation gene inputs during *Drosophila* development. *Nature* **2004**, *431*, 653–659. [CrossRef] [PubMed]
106. Merabet, S.; Pradel, J.; Graba, Y. Getting a molecular grasp on Hox contextual activity. *Trends Genet.* **2005**, *21*, 477–480. [CrossRef] [PubMed]
107. Hui, C.-C.; Suzuki, Y.; Kikuchi, Y.; Mizuno, S. Homeodomain binding sites in the 5′ flanking region of the *Bombyx mori* silk fibroin light-chain gene. *J. Mol. Biol.* **1990**, *213*, 395–398. [CrossRef]
108. Hui, C.-C.; Suzuki, Y. Homeodomain binding sites in the promoter region of silk protein genes. *Dev. Growth Differ.* **1990**, *32*, 263–273. [CrossRef]
109. Oh, S.-K.; Scott, M.P.; Sarnow, P. Homeotic gene *Antennapedia* mRNA contains 5′-noncoding sequences that confer translational initiation by internal ribosome binding. *Genes Dev.* **1992**, *6*, 1643–1653. [CrossRef] [PubMed]
110. Kondrashov, N.; Pusic, A.; Stumpf, C.R.; Shimizu, K.; Hsieh, A.C.; Xue, S.; Ishijima, J.; Shiroishi, T.; Barna, M. Ribosome-mediated specificity in Hox mRNA translation and vertebrate tissue patterning. *Cell* **2011**, *145*, 383–397. [CrossRef] [PubMed]
111. Xue, S.; Tian, S.; Fujii, K.; Kladwang, W.; Das, R.; Barna, M. RNA regulons in *Hox* 5′ UTRs confer ribosome specificity to gene regulation. *Nature* **2015**, *517*, 33–38. [CrossRef] [PubMed]
112. Kmita, M.; Duboule, D. Organizing axes in time and space; 25 years of colinear tinkering. *Science* **2003**, *301*, 331–333. [CrossRef] [PubMed]
113. Noordermeer, D.; Leleu, M.; Splinter, E.; Rougemont, J.; de Laat, W.; Duboule, D. The dynamic architecture of *Hox* gene clusters. *Science* **2011**, *334*, 222–225. [CrossRef] [PubMed]
114. Zakany, J.; Kmita, M.; Duboule, D. A dual role for *Hox* genes in limb anterior-posterior asymmetry. *Science* **2004**, *304*, 1669–1672. [CrossRef] [PubMed]
115. Horard, B.; Mange, A.; Pelisser, B.; Couble, P. Bombyx gene promoter analysis in transplanted silk gland transformed by particle delivery system. *Insect Mol. Biol.* **1994**, *3*, 261–265.
116. Wang, H.-B.; Nita, M.; Iwanaga, M.; Kawasaki, H. βFTZ-F1 and Broad-complex positively regulate the transcription of the wing cuticle protein gene, *BMWCP5*, in wing discs of *Bombyx mori*. *Insect Biochem. Mol. Biol.* **2009**, *39*, 624–633. [CrossRef] [PubMed]
117. Tamura, T.; Kuwabara, N.; Uchino, K.; Kobayashi, I.; Kanda, T. An improved DNA injection method for silkworm eggs drastically increases the efficiency of producing transgenic silkworms. *J. Insect Biotech. Sericol.* **2007**, *76*, 155–159.
118. Imamura, M.; Nakai, J.; Inoue, S.; Quan, G.X.; Kanda, T.; Tamura, T. Targeted gene expression using the *GAL4/UAS* system in the silkworm *Bombyx mori*. *Genetics* **2003**, *165*, 1329–1340. [PubMed]
119. Tamura, T.; Thibert, C.; Royer, C.; Kanda, T.; Eppen, A.; Kamba, M.; Komoto, N.; Thomas, J.L.; Mauchamp, B.; Chavancy, G.; *et al.* Germline transformation of the silkworm *Bombyx mori L.* using a *piggyBac* transposon-derived vector. *Nat. Biotechnol.* **2000**, *18*, 81–84. [PubMed]

120. Takasu, Y.; Sajwan, S.; Damion, T.; Osanai-Futanashi, M.; Uchino, K.; Sezutsu, H.; Tamura, T.; Zurovec, M. Efficient TALEN construction for *Bombyx mori* gene targeting. *PLoS ONE* **2013**, *8*, e73458. [CrossRef]
121. Nakade, S.; Tsubota, T.; Sakano, Y.; Kume, S.; Sakamoto, N.; Obara, M.; Daimon, T.; Sezutsu, H.; Yamamoto, T.; Sakuma, T.; *et al.* Microhomology-mediated end-joining-dependent integration of donor DNA in cells and animals using TALENs and CRISPR/Cas9. *Nat. Commun.* **2014**, *5*, 5560. [CrossRef] [PubMed]
122. Toyama, K. Mendel's low of heredity as applied to the silkworm crosses. *Biol. Zbl.* **1906**, *26*, 321–334.

15

Solving Classification Problems for Large Sets of Protein Sequences with the Example of Hox and ParaHox Proteins

Stefanie D. Hueber * and Tancred Frickey

Department of Biology, University of Konstanz, Konstanz 78464, Germany; tancred.frickey@uni-konstanz.de
* Correspondence: stefanie.hueber@uni-konstanz.de

Academic Editors: Vincenzo Zappavigna and Andy Wessels

Abstract: Phylogenetic methods are key to providing models for how a given protein family evolved. However, these methods run into difficulties when sequence divergence is either too low or too high. Here, we provide a case study of Hox and ParaHox proteins so that additional insights can be gained using a new computational approach to help solve old classification problems. For two (Gsx and Cdx) out of three ParaHox proteins the assignments differ between the currently most established view and four alternative scenarios. We use a non-phylogenetic, pairwise-sequence-similarity-based method to assess which of the previous predictions, if any, are best supported by the sequence-similarity relationships between Hox and ParaHox proteins. The overall sequence-similarities show Gsx to be most similar to Hox2–3, and Cdx to be most similar to Hox4–8. The results indicate that a purely pairwise-sequence-similarity-based approach can provide additional information not only when phylogenetic inference methods have insufficient information to provide reliable classifications (as was shown previously for central Hox proteins), but also when the sequence variation is so high that the resulting phylogenetic reconstructions are likely plagued by long-branch-attraction artifacts.

Keywords: Hox; ParaHox; Cdx; caudal; Pdx; Xlox; Gsx; Gsh; intermediate neuroblasts defective; molecular function; sequence-to-function

1. Introduction

One key feature of classifying protein sequences is that the classification provides the best surrogate measure for predicting likely functional properties of novel protein sequences by comparing and transferring information from better described proteins in closely related groups or clades. It is important to stress that, when focusing solely on the protein coding sequence of a gene, the resulting functional inferences for the corresponding proteins cannot take into account the expression pattern (location and concentration) and can therefore only provide information about the likely "molecular" or "biochemical" function a given protein sequence will exhibit. The underlying assumption here is that more sequence-similar proteins are also more likely to have similar biochemical properties. All functional predictions, be they based on phylogenetic or sequence-similarity analyses, only ever provide a best-guess and need to be tested experimentally.

Depending on the precise question to be resolved, different approaches to classification and prediction of protein function are used. For example, a very rudimentary prediction of protein function can be based on comparisons of domain compositions (e.g., [1]); all homeodomain-containing proteins are presumed to, at least, have the ability to bind to DNA. For those proteins with highly similar domain compositions, additional sequence features can be subsequently employed to refine the classification and prediction of protein function (e.g., see [2]). Such additional features might be the complete amino-acid sequence of a protein or, if additional structural knowledge is available, a focused analysis of the active site residues or residues known to be involved in substrate or interaction partner binding motifs.

Most protein classifications are based on phylogenetic inference methods. These are well-established standard methods aimed at answering questions by placing the data in an explicit evolutionary context (generation of an evolutionary tree of the protein family). For phylogenetic inference methods to work well, a number of conditions have to be met: The protein sequences to be classified must have evolved in accordance with the underlying model of evolution the phylogenetic inference method assumes, and the multiple sequence alignment from which the tree is inferred must be correct (or as correct as possible). Because of the latter, multiple sequence alignments are often truncated to include only unambiguously alignable regions prior to phylogenetic inference. Generating such unambiguous alignments is only possible for long stretches of highly conserved sequence regions, which can cause problems for many protein families. In the case of Hox and ParaHox proteins, the only unambiguously alignable sequence region is the homeodomain and short conserved adjacent regions. While this approach results in a robust, reproducible classification for most Hox proteins, there are limitations. The central Hox proteins (Hox4–8), for example, are so similar in their extended homeodomain that multiple sequence alignments do not provide sufficient sequence variation across these sequences for phylogenetic inference methods to resolve their relationships (reviewed in Hueber *et al.* 2010 [3]). An alternative approach was developed that is based on Profile-Hidden-Markov-Models (HMM) [4]. The advantage of HMM-based approaches is that it is possible to generate different profiles that are specific to subsets of a family. This makes it possible to include some regions that are conserved in a specific sub-group, but that do not need to be present or conserved across all groups. The HMM profile approach, while very fast and useful for verifying to which Hox protein group a specific sequence may belong in general (checking for mis-annotations), did not help resolve the relationships between vertebrate and arthropod central Hox proteins. This is not surprising as, even with this approach, not all sequence regions can be included. A multiple sequence alignment provides the basis for the derived HMM's and, ideally, only unambiguously alignable regions should be employed. The quality of any multiple alignment-based classification will also greatly depend on the skill of the researcher generating the multiple alignment, as human perception and experience can increase alignment quality and human, computational or sequence biases can decrease it.

Advances in experimental and computational resources provide us with access to a multitude of tools and data, as well as the computational power to make better sense of the biological data we generate. So far, the only method that successfully classified the central Hox proteins based on sequence information employed a fairly recently developed method called CLANS [5]. CLANS is based on all-against-all pairwise sequence similarities and thus circumvents the need for an multiple alignment entirely. The CLANS-based approach uncovered that Hox7 and Antp protein sequence-similarity groups are conserved throughout all bilaterian clades, and that the Hox6 and Hox8 protein groups are specific to the vertebrate lineage and are markedly different from the most similar relatives present in the chordate amphioxus. It also detected two new groups of central Hox proteins, termed Echi/Hemi 7 and Echi/Hemi 8 [6].

In contrast to having the problem of insufficient sequence variation across the homeodomain, the Hox/ParaHox field has likely encountered the opposite problem. So far, a number of different predictions are available for how the respective Hox and ParaHox genes are thought to have evolved from an ancestral Ur-*Hox/ParaHox* gene or gene-cluster. The most commonly depicted view is that Gsx and Hox1/2, Pdx/Xlox and Hox3 as well as Cdx and the abdominal Hox proteins (Hox9–13) are likely to have been derived from the same genes in the ancestral *Hox/ParaHox* gene-cluster (see Figure 1). While the assignment of Pdx/Xlox sequences as sequences most closely related to Hox3 seem to be consistent, no matter which method is used, the assignment of other ParaHox proteins to prospective Hox counterparts varies. For Gsx the assignments vary: the traditional view of Hox1–2 [7–12], a further undefined Hox1–3 group [13], a clear assignment of Hox2 based on the depicted tree [14] or a specific assignment of Hox3-like sequences as close relatives [4]. The assignments of Cdx in the literature, in contrast, differs from the commonly depicted view (see Figure 1) only in such a way that Cdx either cannot be assigned as closely related to any specific group, or that authors with similar data interpret Cdx as derived from Pdx/Xlox [13,15]. The only analysis providing a noticeably different result is the HMM-based analysis carried out by Thomas-Chollier *et al.* [4]. In this case, Cdx was assigned as most likely to be a Hox4–8 derivative. What can cause such a drastic difference in assignment between HMM-based and phyogenetic-based methods? Considering that the posterior Hox proteins are known to be highly divergent in their sequence composition [3,16], it is possible that the ParaHox *vs.* Hox classification field is suffering from a phenomenon called long-branch attraction (LBA). In LBA, systematic classification artifacts are generated due to too much variation in a subset of the aligned sequences. To make the scenario for LBA likely would, however, require Cdx as well as abdominal Hox proteins to be highly divergent in their sequences when compared to other Hox and ParaHox proteins.

To gain a different, non-multiple-alignment-based view of the relationships between these sequences, we used the same approach that resolved the relationships of central Hox proteins across bilaterians. Here, we show that this approach can improve classifications not only when multiple sequence alignment-based methods have insufficient information to provide reliable classifications, but also when the sequence variation is so high that it induces systematic artifacts in the resulting phylogentic reconstructions (see variations in assignment of Cdx in Figure 1), in this example, there is a putative long branch attraction (LBA).

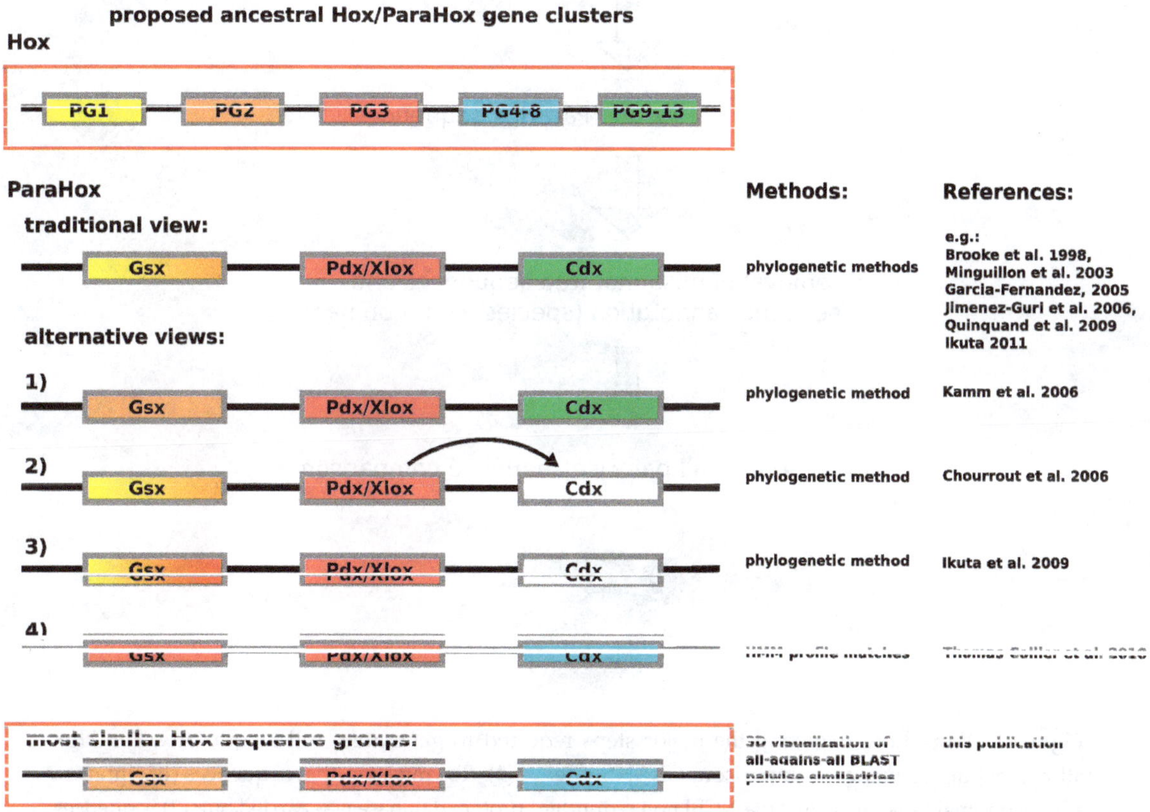

Figure 1. Depiction of the currently proposed relationships between *ParaHox* and *Hox* genes. Similar colors between *Hox* and *ParaHox* reflect the different assignments. The traditional view proposes that the ancestral *ParaHox* cluster contained genes whose proteins corresponded to at least one anterior (Hox1/2), one Hox3 and one posterior (Hox9+/Abd-B) type Hox protein [7–12]. Alternative views propose slightly different scenarios, except for the classification based on hidden Markov models in which Gsx is assigned to the Hox3 group and Cdx to the central Hox protein group (Hox4–8) [4,13–15].

2. Experimental Section

A flow chart of the approach is depicted in Figure 2, and a detailed step-by-step guide of the approach is available in the Supplementary Materials as "Approach.pdf".

Similar to the approach described in Hueber *et al.* 2010 and Hueber *et al.* 2013 [3,6], we searched the NCBI non-redundant protein database (non redundant = "nr" downloaded 16 September 2015) for all sequences containing a Hox/ParaHox-like homeodomain. The alignment we based our search strategy on is available in the Supplementary Materials as "HD62.aln," the original unaligned sequences as "seedsequences.txt." The alignment was generated using MUSCLE (version 3.7-r1) and manually curated in AlnEdit. A Profile-Hidden-Markov-Model was derived from this alignment using HMMER version 3.0 [17]. This HMM was used to search the non-redundant protein database for sequences resembling our set of Hox/ParaHox proteins. For this set of sequences, we then generated a pairwise-sequence-similarity map using CLANS, a pairwise-sequence-similarity visualization tool (which allows visualization in both a 2D and 3D space—the 3D graphs provide more visual information to the viewer, but are harder to incorporate into a printed manuscript).

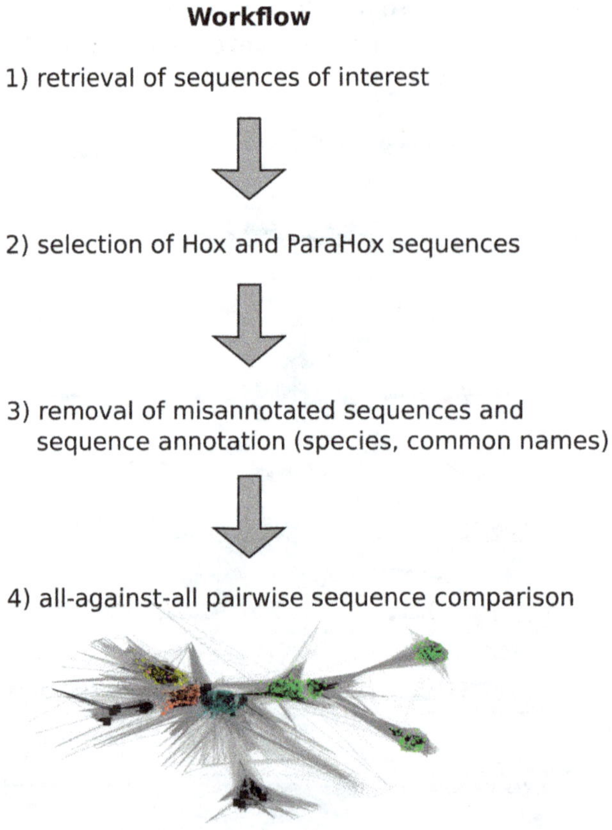

Figure 2. Work flow depicting the major steps required to generate a full-length Hox/ParaHox all-against-all sequence comparison CLANS file. (**1**) To retrieve the sequences of interest, a homeodomain search against the NCBI non redundant protein database was carried out; (**2**) Sequences belonging to the groups of Hox and ParaHox proteins were identified and extracted based on a first sequence-similarity map; (**3**) As some "Hox" sequences were apparent mis-annotations, we removed those sequences from the dataset (e.g., sequences that consisted of a concatenation of multiple different Hox proteins), and the remaining sequences were further annotated by appending common names and the species the sequences originated from; (**4**) The final step was the visualization and analysis of the all-against-all pairwise sequence similarities for the group of Hox and ParaHox sequences in 2D and 3D space.

To generate the all-against-all pairwise sequence-similarities, Blast version 2.2.30 was used (downloaded on 8 October 2014) [18] as part of the CLANS program [5]. The precise version used to perform the all-against-all pairwise sequence comparison is provided in the Supplementary Materials (clans.core.jar). Based on the resulting sequence-similarity-maps generated by CLANS, sequence clusters containing Hox and ParaHox sequences were selected. As mis-annotated Hox/ParaHox sequences containing multiple homeodomains were detected, we identified those sequences containing more than a single 60 amino acid homeodomain using HMMER version 2.3.2 and excluded them from the analysis. In addition to the already present unique sequence identifier (gi number) and the sequence itself, the remaining sequences were further annotated with common names, e.g., Hox1–13 or Lab-AbdB, as well as the species the sequences originated from.

The Supplementary Materials as "Approach.pdf" provides specific details and step-by-step instructions to reproduce the above analysis.

As the nomenclature for the proteins we are working with can vary between organisms (due to tradition or due to mis-annotations), we would like to explicitly state that for deuterostomes we employ the vertebrate nomenclature (using *Mus musculus* as a standard), and for protostomes we employ the nomenclature based on *Drosophila melanogaster* (which only has two ParaHox protein types) and *Capitella teleta* (which has all three ParaHox protein types) as a standard for annotating the resulting cluster groups.

3. Results and Discussion

In situations where different variants of an approach yield varying results, the use of an independent method may provide additional insights. In this case we opted for a comparison of all-against-all pairwise sequence similarities of full-length Hox/ParaHox sequences to shed light on the classification of ParaHox proteins in relation to Hox-protein groups. The aim of this method is to avoid the introduction of user-based biases induced by generating and analyzing user-defined multiple sequence alignments (we use the alignment and HMM-searches only to retrieve all sequences of interest and then carry out the analysis with unaligned sequences). In addition, as there is not much information available on how precisely ParaHox proteins carry out their molecular functions and what sequence elements precisely are relevant to their function, we decided to exclude potential user-induced (Hox-research focused scientists) biases and analyze the full-length protein sequences rather than focusing on specific domains or subsections of these proteins.

Our basic approach to identifying Hox/ParaHox proteins is likely to have ensured a full inclusion of all Hox and ParaHox protein sequences as we retrieved plant TALE homeodomain sequences as well as Pax sequences, which are well beyond the stated scope of this study. As a first step, we generously selected all sequence-similarity clusters containing one or more of our seed sequences (Figure 3a, red dots). For the second step, we extracted these sequences to a separate file, re-inflated the dataset by reversing the CD-hit filtering (as described in the Supplementary Materials methods section) and removed known non-Hox/ParaHox sequences (e.g., NK) in order to provide us with all sequences available for our groups of interest. For the resulting set of sequences, the similarities of each ParaHox group to the most similar Hox protein group were analyzed both in 2D and 3D space (the figures here can only depict 2D).

The default map, *i.e.* the map that provided us with the best resolution for the ParaHox/Hox analysis, was calculated at an E-value cutoff of 1e-33 using sequence similarities calculated over full-length sequences.

The results are as follows (see Figure 3b,c).

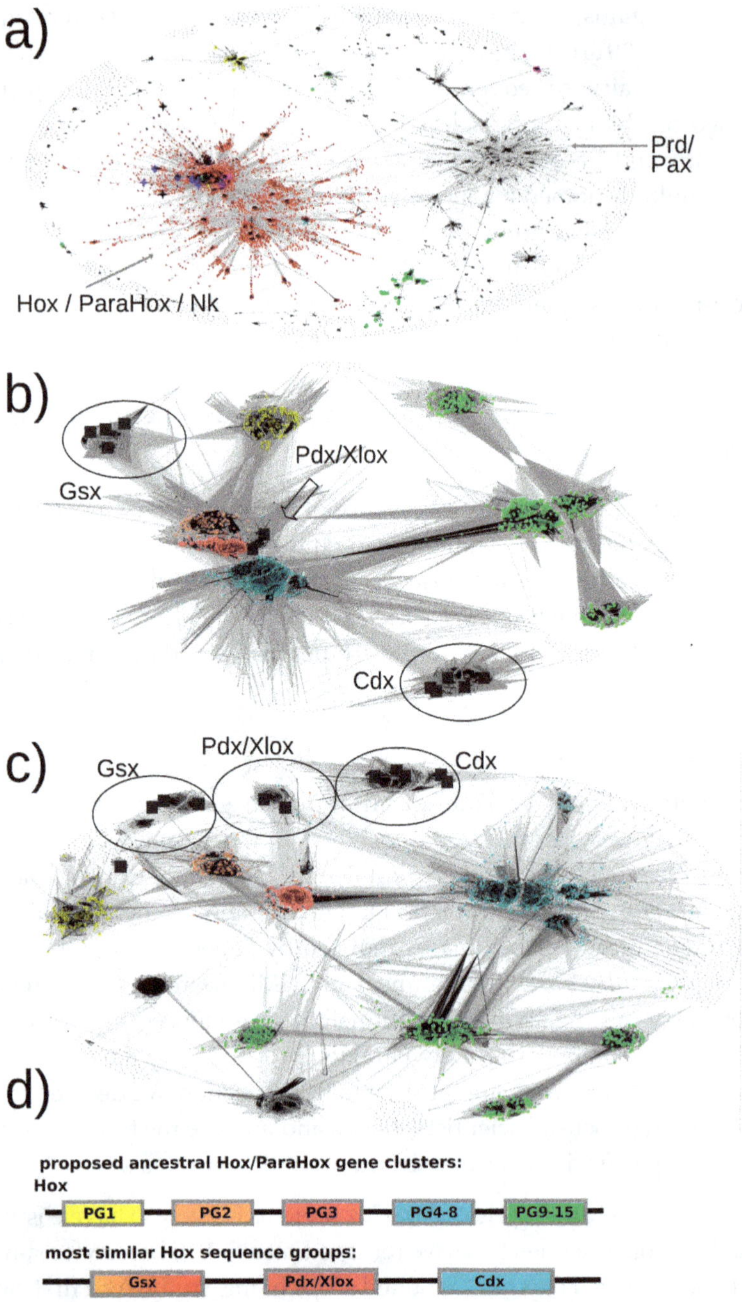

Figure 3. All-against-all pairwise-sequence-similarity maps of full-length protein sequences mapped in 2D. Each sequence is represented by a dot in the 2D space. The lines connecting the dots represent pairwise sequence similarities. The darker the lines, the more significant (lower) the E-value of the reciprocal blast hits. (**a**) The sequence-similarity map of all sequences of potential interest to us (see methods, step 7) with the red dots representing the selected Hox/ParaHox/Nk cluster. (**b** + **c**) All sequences of the Hox/ParaHox cluster, with (**b**) depicting the map at an E-value cutoff of 1e-33 and (**c**) depicting the map at an E-value cutoff of 1e-40. Yellow dots represent Hox1/Lab sequences, orange dots Hox2/Pb, red dots Hox3, blue dots central Hox proteins (Hox4–8/Antp/Abd-A/Ubx) and green dots represent posterior Hox proteins (Hox9+/Abd-B). The grey squares represent the ParaHox proteins from *Mus musculus*, *Drosophila melanogaster* and *Capitella teleta*. (**d**) Summary of which ParaHox proteins are most similar to which Hox groups. Please note that the coloring only represent the observed highest similarity in sequence across the respective groups and is not meant to provide a prediction of functional similarity (see text for more details).

3.1. *Pdx Is the ParaHox Protein Most Similar to Any of the Hox Protein Groups and, Across Hox Proteins, Most Similar to Hox3*

Based on the sequence-similarity map in Figure 3b, Pdx sequences appear very similar to Hox2/3 sequences and cluster in close proximity. To assess whether there is a sequence-similarity signal present that might allow us to further resolve the placement of Pdx, a more stringent E-value cutoff was selected and the dataset re-clustered. At an E-value cutoff of 1e-40, a greater number of high-significance sequence similarities could be observed between Pdx and the Hox3 paralog group than to any other group (see Figure 3c). These results are consistent with all previous analyses we are aware of (reviewed in [12,19]).

3.2. *Gsx Is Similar to Hox2 and 3*

As visible in Figure 3b, the sequence-similarity data indicates a clustering of Gsx group proteins with proteins from the Hox paralogy groups 2 and 3. There are also some individual proteins showing noticeable similarity to the central group Hox proteins (defined as PG4–8) (mainly to the Dfd/Hox4 group) as well as to the Hox1 group. However, the most consistent signal (number of connections and darkest lines) indicates Gsx proteins, as a group, being most similar to Hox2 and Hox3 proteins.

Gsx has previously either been assigned as similar to Hox1 or Hox2 (with some exceptions assigning it to Hox2 alone, Hox3 alone or Hox1-3-like) (Figure 1). Our data does not support the notion of a higher similarity being present between Gsx and Hox1 proteins, but rather support an assignment to Hox2 and Hox3 (both Hox2 and Hox3 are also more similar to each other than to Hox1). Using more stringent E-value cutoffs up to 1e-40 (at which point most connections are lost), we also examined whether there is a signal in the sequence-similarity data preferentially assigning Gsx to either Hox2 or Hox3. However, the connections seem roughly equal in number and quality, which indicates that, at least for full-length sequences, the sequence assignment of Gsx to being either more Hox2 or more Hox3-like cannot be resolved using this approach.

3.3. *Cdx Is Most Similar to the "Central-Group" Hox Proteins*

Previous methods either assigned Cdx to the posterior Hox protein group (Hox9–13/Abd-B) or did not provide a clear assignment due to insufficient resolution or conflicting data. In contrast, our sequence-similarity map clearly indicates a much higher similarity of Cdx group proteins to the central-group Hox proteins than to any other group of Hox proteins. In addition, it should be noted that the Cdx- and Hox-posterior-group proteins are located the most distant from the center of the map, indicating a greater than average dissimilarity in sequence than the rest of the Hox/ParaHox sequences. The distance from the center and absence of connections between the Cdx-group and any of the Hox-posterior-group sequences further supports the idea that these two groups of proteins have accumulated more changes in sequence than the other Hox/ParaHox proteins and indicates that they have done so independently of each other.

Assuming a higher than average rate and independence of accumulated mutations for these groups, helps explain the discrepancy between our results and the claim that Cdx proteins are similar to the posterior-type Hox proteins. Fortunately, the literature also indicates that this claim is not universally accepted. One publication depicts the sequence deviation of Cdx from other Hox and ParaHox proteins in an unrooted tree (and, perhaps because of this, even suggested that the *Cdx* gene might have been independently duplicated in the *ParaHox* gene-cluster), another publication used a different method and actually stated that there is a similarity of Cdx to central Hox proteins [4,15]. The latter result, however, seems to have gone unnoticed by recent reviews [12,19]. One explanation for the difficulty in achieving a consistent classification of Cdx and posterior-group Hox proteins in relation to the other Hox/ParaHox proteins using phylogenetic approaches lies in the notorious long-branch-attraction (LBA) artifact. The LBA-problem is especially prevalent when groups of highly similar proteins are being compared to much more dissimilar proteins, as classification methods have a tendency to group the similar sequences together, thereby leading to an artificial, exclusion-based, grouping of the more dissimilar sequences. If this effect is not considered and corrected for, it can lead to serious artifacts in phylogenetic reconstructions. Based on the sequence-similarity data available to us, we would postulate that LBA-effects are likely to have contributed to the differences observed between previous attempts at classifying these protein families

4. Conclusions

A number of hypotheses, with varying degrees of support, are present in the current literature to explain how the *Hox* and *ParaHox* genes arose from an ancestral UR-*Hox/ParaHox* gene or gene-cluster (Figure 1). An UR-*Hox/ParaHox* gene must have duplicated or triplicated to give rise to the proto-*Hox/ParaHox* gene-cluster. The number of suggested proto *Hox/ParaHox* genes, however, varies between 2 and 4 [19]. The most commonly depicted view is that there must have been at least three genes, one anterior *Hox*1/2/*Gsx* like, one *Hox*3/*Xlox*/*Pdx* like and one *Cdx*/*Hox*9+ like (see Figure 1) [12]. This gene-cluster then duplicated giving rise to the *ParaHox* and *Hox* clusters, the latter of which further duplicated some of its constituent genes.

Using a method analyzing pairwise sequence similarities across all available Hox/ParaHox full-length protein sequences provides us with more information.

While our results support the hypothesis that Pdx/Xlox and Hox PG3 proteins are encoded by genes that arose from the same ancestral gene in the proto *Hox/ParaHox* cluster, the assignments of Gsx to Hox1–2 and Cdx to abdominal class Hox proteins (PG9+/Abd-B), however, cannot be supported and may have to be reevaluated. Previous analyses suggested Gsx being similar to Hox1–3, in varying combinations (see Figure 1). While Gsx does share similarity with Hox1, it does so to the same degree as to central Hox proteins. By far the strongest sequence-similarity signal is to the group of Hox2/3 proteins. This result allows us to postulate that the corresponding *Hox*2/3 and *Gsx* genes are most likely to have diverged from a common gene in the ancestral *Hox/ParaHox* gene-cluster (and that other pairing are less likely), a result similar to, yet slightly different from, previous analyses. As for Cdx, a consistent higher sequence similarity to central group Hox proteins (here defined as PG4–8) than to any other group of Hox proteins (including PG9+/Abd-B) indicates that the central Hox and Cdx protein coding genes arose from the same ancestral gene in the proto *Hox/ParaHox* cluster. A simple explanation as to why some previous analyses assigned Cdx to posterior Hox lies in their increased difference in sequence when compared to all other Hox/ParaHox sequences. This increased sequence dissimilarity can cause some methods to suffer from the long-branch-attraction artifact, causing the least similar sequences to artificially group together in a tree.

In terms of the number of proto ParaHox/Hox proteins that must have been encoded in the gene-cluster before the spit into *Hox* and *ParaHox* gene-clusters, we can see evidence of at least two different protein types: Hox2/3/Gsx/Xlox/Pdx and central Hox/Cdx. Whether Gsx/Hox2–3 and Xlox/Pdx/Hox3 may have been originally encoded by two different genes (meaning the proto *ParaHox/Hox* cluster contained three genes in total) or by a single gene with *Xlox/Pdx* representing a gene duplication from *Gsx* (or vice versa), we cannot say given the current data.

In terms of how far we can expect Hox and ParaHox proteins to show a similarity in their molecular function (e.g., DNA and protein binding abilities and preferences), it seems plausible that Pdx/Xlox will exhibit similar molecular functions to Hox3 proteins in roughly the same way that Hox2 proteins behave similarly to Hox3 proteins (we cannot yet quantify how similar that is, as we do not have any quantitative comparative experimental data available). Considering the high sequence divergence of Cdx, we would presume Cdx to behave significantly differently from all other Hox and ParaHox proteins. Somewhere in between these two extremes is Gsx. Since Hox2 and 3 are more similar to each other than they are to Gsx, it is possible that Gsx differs significantly in a number of molecular functions. We suggest that, in cases where Gsx does act similarly to Hox2/3, whether Gsx will act more like a Hox2 or more like a Hox3 protein will depend on the precise sequence region involved in a given molecular function.

No other set of transcription factors provides us with a set of sequences that are as similar across regions known to be relevant to protein function (e.g., the DNA binding homeodomain), and yet whose members induce such wildly different effects, even when expressed in the same concentration, location and space. If we ever want to fully understand precisely how homeodomain transcription factors (and perhaps other transcription factors) achieve their specific function, the set of Hox and ParaHox proteins is probably the best set of proteins to study. One first step is to identify linear motifs [20] and figure out what each putative motif contributes to the protein function. Another way in which the Hox/ParaHox field should excel is an expansion of the existing "top-down" approaches of generating rescue experiments, as well as comparative and quantitative data of phenotypes or expression data for Hox and ParaHox proteins (e.g., [21–26]). This expansion of our traditional "top down" approaches could be done by including comparable analyses of the molecular functions of the proteins, e.g., qualitative and quantitative protein-protein and protein-DNA interaction studies, in order to establish a thorough overview of which sequence regions of the proteins are involved in achieving which molecular functions.

Supplementary Materials: The following are available online at www.mdpi.com/2221-3759/4/1/8/s1, Files: Approach.pdf, seedsequence.txt, HD62.aln, clans.core.jar, ParaHox_CLANS_sequences.txt, perl (folder), reference.txt. The Supplementary Materials above as well as the Clans file used for the analysis are also available for download: http://bioinformatics.uni-konstanz.de/HueberHox/ParaHox/.

Acknowledgments: We would like to thank the University Konstanz YSF (grant 83943711 and 83990512) and AFF (grant 15902014).

Author Contributions: Stefanie D. Hueber designed the experiment, and, with Tancred Frickey, designed and wrote perl scripts, carried out the analysis and prepared the manuscript.

Conflicts of Interest: The authors declare no conflict of interest.

References

1. Bürglin, T.R.; Affolter, M. Homeodomain proteins: An update. *Chromosoma* **2015**, 1–25. [CrossRef] [PubMed]
2. Merabet, S.; Galliot, B. The TALE face of Hox proteins in animal evolution. *Front. Genet.* **2015**, *6*, 267. [CrossRef] [PubMed]
3. Hueber, S.D.; Weiller, G.F.; Djordjevic, M.A.; Frickey, T. Improving Hox protein classification across the major model organisms. *PLoS ONE* **2010**, *5*, e10820. [CrossRef] [PubMed]
4. Thomas-Chollier, M.; Ledent, V.; Leyns, L.; Vervoort, M. A non-tree-based comprehensive study of metazoan Hox and ParaHox genes prompts new insights into their origin and evolution. *BMC Evol. Biol.* **2010**, *10*, 73. [CrossRef] [PubMed]
5. Frickey, T.; Lupas, A. CLANS: A Java application for visualizing protein families based on pairwise similarity. *Bioinformatics* **2004**, *20*, 3702–3704. [CrossRef] [PubMed]
6. Hueber, S.D.; Rauch, J.; Djordjevic, M.A.; Gunter, H.; Weiller, G.F.; Frickey, T. Analysis of central Hox protein types across bilaterian clades: On the diversification of central Hox proteins from an Antennapedia/Hox7-like protein. *Dev. Biol.* **2013**, *383*, 175–185. [CrossRef] [PubMed]
7. Brooke, N.M.; Garcia-Fernàndez, J.; Holland, P.W. The ParaHox gene cluster is an evolutionary sister of the Hox gene cluster. *Nature* **1998**, *392*, 920–922. [PubMed]
8. Minguillón, C.; Garcia-Fernàndez, J. Genesis and evolution of the Evx and Mox genes and the extended Hox and ParaHox gene clusters. *Genome Biol.* **2003**, *4*, R12. [CrossRef] [PubMed]
9. Garcia-Fernandez, J. Hox, ParaHox, ProtoHox: Facts and guesses. *Heredity* **2005**, *94*, 145–152. [CrossRef] [PubMed]
10. Jiménez-Guri, E.; Paps, J.; García-Fernández, J.; Saló, E. Hox and ParaHox genes in Nemertodermatida, a basal bilaterian clade. *Int. J. Dev. Biol.* **2006**, *50*, 675–679. [CrossRef] [PubMed]
11. Quiquand, M.; Yanze, N.; Schmich, J.; Schmid, V.; Galliot, B.; Piraino, S. More constraint on ParaHox than Hox gene families in early metazoan evolution. *Dev. Biol.* **2009**, *328*, 173–187. [CrossRef] [PubMed]
12. Ikuta, T. Evolution of invertebrate deuterostomes and Hox/ParaHox genes. *Genom. Proteom. Bioinform.* **2011**, *9*, 77–96. [CrossRef]

13. Ikuta, T.; Miyamoto, N.; Saito, Y.; Wada, H.; Satoh, N.; Saiga, H. Ambulacrarian prototypical Hox and ParaHox gene complements of the indirect-developing hemichordate Balanoglossus simodensis. *Dev. Genes Evol.* **2009**, *219*, 383–389. [CrossRef] [PubMed]
14. Kamm, K.; Schierwater, B.; Jakob, W.; Dellaporta, S.L.; Miller, D.J. Axial patterning and diversification in the cnidaria predate the Hox system. *Curr. Biol.* **2006**, *16*, 920–926. [CrossRef] [PubMed]
15. Chourrout, D.; Delsuc, F.; Chourrout, P.; Edvardsen, R.B.; Rentzsch, F.; Renfer, E.; Jensen, M.F.; Zhu, B.; de Jong, P.; Steele, R.E. Minimal ProtoHox cluster inferred from bilaterian and cnidarian Hox complements. *Nature* **2006**, *442*, 684–687. [CrossRef] [PubMed]
16. Feiner, N.; Ericsson, R.; Meyer, A.; Kuraku, S. Revisiting the origin of the vertebrate Hox14 by including its relict sarcopterygian members. *J. Exp. Zoolog. B Mol. Dev. Evol.* **2011**, *316*, 515–525. [CrossRef] [PubMed]
17. *HMMER*, version 3.0; A new generation of sequence homology search software. Available online: http://hmmer.org (accessed on 3 February 2016).
18. Altschul, S.F.; Madden, T.L.; Schäffer, A.A.; Zhang, J.; Zhang, Z.; Miller, W.; Lipman, D.J. Gapped BLAST and PSI-BLAST: A new generation of protein database search programs. *Nucleic Acids Res.* **1997**, *25*, 3389–3402. [CrossRef] [PubMed]
19. Biscotti, M.A.; Canapa, A.; Forconi, M.; Barucca, M. Hox and ParaHox genes: A review on molluscs. *Genesis* **2014**, *52*, 935–945. [CrossRef] [PubMed]
20. Ogishima, S.; Tanaka, H. Missing link in the evolution of Hox clusters. *Gene* **2007**, *387*, 21–30. [CrossRef] [PubMed]
21. McGinnis, N.; Kuziora, M.A.; McGinnis, W. Human Hox-4.2 and Drosophila deformed encode similar regulatory specificities in Drosophila embryos and larvae. *Cell* **1990**, *63*, 969–976. [CrossRef]
22. Zhao, J.J.; Lazzarini, R.A.; Pick, L. The mouse Hox-1.3 gene is functionally equivalent to the Drosophila Sex combs reduced gene. *Genes Dev.* **1993**, *7*, 343–354. [CrossRef] [PubMed]
23. Lutz, B.; Lu, H.C.; Eichele, G.; Miller, D.; Kaufman, T.C. Rescue of Drosophila labial null mutant by the chicken ortholog Hoxb-1 demonstrates that the function of Hox genes is phylogenetically conserved. *Genes Dev.* **1996**, *10*, 176–184. [CrossRef] [PubMed]
24. Hirth, F.; Loop, T.; Egger, B.; Miller, D.F.B.; Kaufman, T.C.; Reichert, H. Functional equivalence of Hox gene products in the specification of the tritocerebrum during embryonic brain development of Drosophila. *Development* **2001**, *128*, 4781–4788. [PubMed]
25. Hueber, S.D.; Bezdan, D.; Henz, S.R.; Blank, M.; Wu, H.; Lohmann, I. Comparative analysis of Hox downstream genes in Drosophila. *Development* **2007**, *134*, 381–392. [CrossRef] [PubMed]
26. Hudry, B.; Thomas-Chollier, M.; Volovik, Y.; Duffraisse, M.; Dard, A.; Frank, D.; Technau, U.; Merabet, S. Molecular insights into the origin of the Hox-TALE patterning system. *Elife* **2014**, *3*, e01939. [CrossRef] [PubMed]

Permissions

All chapters in this book were first published in JDB, by MDPI; hereby published with permission under the Creative Commons Attribution License or equivalent. Every chapter published in this book has been scrutinized by our experts. Their significance has been extensively debated. The topics covered herein carry significant findings which will fuel the growth of the discipline. They may even be implemented as practical applications or may be referred to as a beginning point for another development.

The contributors of this book come from diverse backgrounds, making this book a truly international effort. This book will bring forth new frontiers with its revolutionizing research information and detailed analysis of the nascent developments around the world.

We would like to thank all the contributing authors for lending their expertise to make the book truly unique. They have played a crucial role in the development of this book. Without their invaluable contributions this book wouldn't have been possible. They have made vital efforts to compile up to date information on the varied aspects of this subject to make this book a valuable addition to the collection of many professionals and students.

This book was conceptualized with the vision of imparting up-to-date information and advanced data in this field. To ensure the same, a matchless editorial board was set up. Every individual on the board went through rigorous rounds of assessment to prove their worth. After which they invested a large part of their time researching and compiling the most relevant data for our readers.

The editorial board has been involved in producing this book since its inception. They have spent rigorous hours researching and exploring the diverse topics which have resulted in the successful publishing of this book. They have passed on their knowledge of decades through this book. To expedite this challenging task, the publisher supported the team at every step. A small team of assistant editors was also appointed to further simplify the editing procedure and attain best results for the readers.

Apart from the editorial board, the designing team has also invested a significant amount of their time in understanding the subject and creating the most relevant covers. They scrutinized every image to scout for the most suitable representation of the subject and create an appropriate cover for the book.

The publishing team has been an ardent support to the editorial, designing and production team. Their endless efforts to recruit the best for this project, has resulted in the accomplishment of this book. They are a veteran in the field of academics and their pool of knowledge is as vast as their experience in printing. Their expertise and guidance has proved useful at every step. Their uncompromising quality standards have made this book an exceptional effort. Their encouragement from time to time has been an inspiration for everyone.

The publisher and the editorial board hope that this book will prove to be a valuable piece of knowledge for researchers, students, practitioners and scholars across the globe.

List of Contributors

Kyle Jewhurst and Kelly A. McLaughlin
Department of Biology, Tufts University, Medford, MA 02155, USA

Shinsuke Ohba
Department of Bioengineering, Graduate School of Engineering, The University of Tokyo, 7-3-1 Hongo, Bunkyo-ku, Tokyo 113-0033, Japan

Christoph Gerhardt, Antonia Wiegering, Tristan Leu and Ulrich Rüther
Institute for Animal Developmental and Molecular Biology, Heinrich-Heine University Düsseldorf, 40225 Düsseldorf, Germany

Hugo Fernandes-Silva
Life and Health Sciences Research Institute (ICVS), School of Medicine, University of Minho, 4710-057 Braga, Portugal
ICVS/3B's-PT Government Associate Laboratory, 4710-057 Braga/Guimarães, Portugal

Jorge Correia-Pinto
Life and Health Sciences Research Institute (ICVS), School of Medicine, University of Minho, 4710-057 Braga, Portugal
ICVS/3B's-PT Government Associate Laboratory, 4710-057 Braga/Guimarães, Portugal
Department of Pediatric Surgery, Hospital de Braga, 4710-243 Braga, Portugal

Rute Silva Moura
Life and Health Sciences Research Institute (ICVS), School of Medicine, University of Minho, 4710-057 Braga, Portugal
ICVS/3B's-PT Government Associate Laboratory, 4710-057 Braga/Guimarães, Portugal
Biology Department, School of Sciences, University of Minho, 4710-057 Braga, Portugal

Tanushree Pandit and Stacey K. Ogden
Department of Cell and Molecular Biology, St. Jude Children's Research Hospital, Memphis, TN 38105, USA

Margery G. Smelkinson
Biological Imaging Section, Research Technologies Branch, National Institute of Allergy and Infectious Diseases, National Institutes of Health, Bethesda, MD 20892, USA

Chih-Ning Chang and Chrissa Kioussi
Department of Pharmaceutical Sciences, College of Pharmacy, Oregon State University, Corvallis, OR 97331, USA
Molecular Cell Biology Graduate Program, Oregon State University, Corvallis, OR 97331, USA

Jennifer M. Feenstra, Jonathan M. Van Arsdale, Karndeep S. Rai-Bhatti, Diana J. H. Kim, Ashley S. Coggins, Gennaya L. Mattison, Stephen Yoo, Eric D. Steinman, Charmaine U. Pira and Kerby C. Oberg
1 Department of Pathology and Human Anatomy, School of Medicine, Loma Linda University, Loma Linda, CA 92354, USA

Billy A. Watson
Department of Pathology and Human Anatomy, School of Medicine, Loma Linda University, Loma Linda, CA 92354, USA
Division of Microbiology and Molecular Genetics, Department of Basic Sciences, School of Medicine, Loma Linda University, Loma Linda, CA 92354, USA

Brendan R. Gongol
Department of Cardiopulmonary Sciences, School of Allied Health Professions, Loma Linda University, Loma Linda, CA 92354, USA

Joana F. Monteiro, Sandra Martins, Matheus Farias, Telma Costa and Ana Catarina Certal
Fish Platform, Champalimaud Center for the Unknown, 1400-038 Lisbon, Portugal

Cynthia D. Cooper
School of Molecular Biosciences, Washington State University Vancouver, Vancouver, WA 98686, USA
College of Arts and Sciences, Washington State University Vancouver, Vancouver, WA 98686, USA

Steve D. Erickson, Scott Yin, Trevor Moravec and Brian Peh
College of Arts and Sciences, Washington State University Vancouver, Vancouver, WA 98686, USA

Kevin Curran
Department of Biology, University of San Diego, San Diego, CA 92110, USA

Katherine Harding and Kristin White
Massachusetts General Hospital Cutaneous Biology Research Center, Harvard Medical School, Boston, MA 02129, USA

Helena Khaliullina, Nicola K. Love and William A. Harris
Department of Physiology, Development and Neuroscience, University of Cambridge, Cambridge CB2 3DY, UK

Mikiko Tanaka
Graduate School of Bioscience and Biotechnology, Tokyo Institute of Technology, B-17, 4259 Nagatsuta-cho, Midori-ku, Yokohama 226-8501, Japan

Shigeharu Takiya
Shigeharu Takiya, Division of Biological Sciences and Center for Genome Dynamics, Faculty of Science, Hokkaido University, North 10, West 8, Kita-ku, Sapporo, Hokkaido 060-0810, Japan
Graduate School of Life Science, Hokkaido University, North 10, West 8, Kita-ku, Sapporo 060-0810, Japan

Mai Kimoto
Graduate School of Life Science, Hokkaido University, North 10, West 8, Kita-ku, Sapporo 060-0810, Japan

Takuya Tsubota
Transgenic Silkworm Research Unit, Institute of Agrobiological Sciences, National Agriculture and Food Research Organization, 1-2 Owashi, Tsukuba, Ibaraki 305-8634, Japan

Stefanie D. Hueber and Tancred Frickey
Department of Biology, University of Konstanz, Konstanz 78464, Germany

Tristan Aguero, Yi Zhou and Mary Lou King
Department of Cell Biology, University of Miami School of Medicine, Miami, FL 33136, USA

Malgorzata Kloc
Houston Methodist Research Institute, The Houston Methodist Hospital, Bertner Avenue, Houston, TX 77030, USA

Patrick Chang and Evelyn Houliston
Laboratoire de Biologie du Développement de Villefranche-sur-mer (LBDV), Sorbonne Universités, UPMC Univ Paris 06, CNRS, Villefranche-sur-mer 06230, France

Antonia Wiegering, Ulrich Rüther and Christoph Gerhardt
Institute for Animal Developmental and Molecular Biology, Heinrich-Heine University Düsseldorf, 40225 Düsseldorf, Germany

Index

CPSIA information can be obtained
at www.ICGtesting.com
Printed in the USA
BVHW011101280822
645597BV00026B/371

9 781639 892587